W9-BZE-727

Praise for *Engineering Methods for Robust Product Design*

"If you want to learn from application based on theory, from practice with a blend of engineering, statistics, and business disciplines, and from on-the-job bottom-line responsibility integrated with classroom experience, then *Engineering Methods for Robust Product Design* is the text for you. It is about practicing robust design in a total quality environment."

Albert J. Simone, Ph.D.
President
Rochester Institute of Technology

"Having helped Kodak to start making the modern improvements in their equipment commercialization process, it is very gratifying to me to see this book by Bill Fowlkes and Skip Creveling emerge from that activity. The successful practice of Robust Design is critical to better products. This book starts with the basics, and thus can be used by engineers without prior experience in Robust Design or related topics to become proficient practitioners. By including software and many examples, this book provides a complete package for successful practice."

Don Clausing, Ph.D.
MIT

"For the novice to Robust Design or an experienced practitioner, this text is a fountain of knowledge on Dr. Taguchi's methods. Authors Bill Fowlkes and Skip Creveling are among the minority of writers on Taguchi Methods/Robust Design that truly understand Dr. Taguchi's design paradigm. This text is testimony to their appreciation of the finer points of Robust Design and their strong belief in its corporate value."

Bill Bellows, Ph.D.
Rocketdyne Division
Rockwell Aerospace

"Fowlkes and Creveling present the principles of Robust Design in a practical and understandable manner that is based on their substantial experience with on-the-job engineering applications at Kodak. Robust Design methods are powerful. They extract technical information from engineering studies efficiently and effectively and provide the analytic framework for making product and process decisions based on total cost to society. Industrial organizations that apply the principles outlined in this book will increase engineering productivity and improve customer satisfaction."

John J. King
Manager, Engineering Education
Ford Design Institute
Ford Motor Company

# ENGINEERING METHODS FOR ROBUST PRODUCT DESIGN

# ENGINEERING PROCESS IMPROVEMENT SERIES

*John W. Wesner, Ph.D., P.E., Consulting Editor*

Lou Cohen, *Quality Function Deployment: How to Make QFD Work for You*

William Y. Fowlkes/Clyde M. Creveling, *Engineering Methods for Robust Product Design: Using Taguchi Methods® in Technology and Product Development*

Maureen S. Heaphy/Gregory F. Gruska, *The Malcolm Baldrige National Quality Award: A Yardstick for Quality Growth*

Jeanenne LaMarsh, *Changing the Way We Change: Gaining Control of Major Operational Change*

William J. Latzko/David M. Saunders, *Four Days with Dr. Deming: A Strategy for Modern Methods of Management*

Mary G. Leitnaker/Richard D. Sanders/Cheryl Hild, *The Power of Statistical Thinking: An Integral Building Block for Improving Industrial Processes*

Rohit Ramaswamy, *Design and Management of Service Processes*

Richard C. Randall, *Randall's Practical Guide to ISO 9000: Implementation, Registration, and Beyond*

John W. Wesner/Jeffrey M. Hiatt/David C. Trimble, *Winning with Quality: Applying Quality Principles in Product Development*

# ENGINEERING METHODS FOR ROBUST PRODUCT DESIGN

*Using Taguchi Methods® in Technology and Product Development*

*William Y. Fowlkes*

*Clyde M. Creveling*

**with**
**WinRobust Lite Software®**
**written by**
**John Derimiggio**

**Addison-Wesley Publishing Company**
Reading, Massachusetts   Menlo Park, California   New York
Don Mills, Ontario   Wokingham, England   Amsterdam   Bonn
Sydney   Singapore   Tokyo   Madrid   San Juan
Paris   Seoul   Milan   Mexico City   Taipei

The publisher offers discounts on this book when ordered in quantity for special sales.

For more information, please contact:

Corporate & Professional Publishing Group
Addison-Wesley Publishing Company
One Jacob Way
Reading, Massachusetts 01867

Library of Congress Cataloging-in-Publication Data
Fowlkes, William Y., 1954–
Engineering methods for robust product design : using Taguchi methods in technology and product development / William Y. Fowlkes, Clyde M. Creveling ; with WinRobust software written by John Derimiggio.
p. cm. — (Engineering process improvement series)
Includes bibliographical references and index.
ISBN 0-201-63367-1 (alk. paper)
1. Taguchi methods (Quality control) 2. Design, Industrial. 3. New products. 4. WinRobust. I. Creveling, Clyde M., 1956– . II. Title. III. Series.
TS156.F68 1995
658.5′75—dc20 95-15970
CIP



ISBN 0-201-63367-1

1 2 3 4 5 6 7 8 9-CRW-98979695

First printing August 1995

*This book is dedicated to our families:*

***Susan, Hannah and Helen, William and Carolyn***
*and*
***Kathy and Neil***

*Their love, encouragement and patience*
*were greater than the number of words in this book express.*

# ENGINEERING PROCESS IMPROVEMENT SERIES

*Consulting Editor, John Wesner, Ph.D., P.E.*

Global competitiveness is of paramount concern to the engineering community worldwide. As customers demand ever-higher levels of quality in their products and services, engineers must keep pace by continually improving their processes. For decades, American business and industry have focused their quality efforts on their end products rather than on the processes used in the day-to-day operations that create these products and services. Experts across the country now agree that focusing on continuous improvements of the core business and engineering processes within an organization will lead to the most meaningful, long-term improvements and production of the highest-quality products.

Whether your title is researcher, designer, developer, manufacturer, quality or business manager, process engineer, student, or coach, you are responsible for finding innovative, practical ways to improve your processes and products in order to be successful and remain world-class competitive. The **Engineering Process Improvement Series** takes you beyond the ideas and theories, focusing in on the practical information you can apply to your job for both short-term and long-term results. These publications offer current tools and methods and useful how-to advice. This advice comes from the top names in the field; each book is both written and reviewed by the leaders themselves, and each book has earned the stamp of approval of the series consulting editor, John W. Wesner.

Key innovations by industry leaders in process improvement include work in benchmarking, concurrent engineering, robust design, customer-to-customer cycles, process management, and engineering design. Books in this series will discuss these vital issues in ways that help engineers of all levels of experience become more productive and increase quality significantly.

All of the books in the series share a unique graphic cover design. Viewing the graphic blocks descending, you see random pieces coming together to build a solid structure, signifying the ongoing effort to improve processes and produce quality products most satisfying to the customer. If you view the graphic blocks moving upward, you see them breaking through barriers—just as engineers and companies today must break through traditional, defining roles to operate more effectively with concurrent systems. Our mission for this series is to provide the tools, methods, and practical examples to help you hurdle the obstacles, so that you can perform simultaneous engineering and be successful at process and product improvement.

The series is divided into three categories:

**Process Management and Improvement** This includes books that take larger views of the field, including major processes and the end-to-end process for new product development.

**Improving Functional Processes** These are the specific functional processes that are combined to form the more inclusive processes covered in the first category.

**Special Process Topics and Tools** These are methods and techniques that are used in support of improving the various processes covered in the first two categories.

# CONTENTS

# FOREWORD

*George M. C. Fisher*
*Chairman, President, and CEO*
*Eastman Kodak Company*

In the battle for global market share, the company that creates customer confidence with on-time delivery of defect-free, reliable products and services is the company that will succeed. Customer satisfaction cannot be a frill or a fluke; it is an imperative for companies wishing to earn or maintain world-class stature.

Quality is the gateway to American business success in the global free-for-all for customer satisfaction and loyalty. Quality is no longer a nicety; it's a necessity. Productivity may have been boss in the 20th century, but as J. M. Juran has predicted, the 21st century will be the century of quality. If we are to seize the phenomenal growth opportunities on the horizon, we must clarify our quality goals and quantify the means to attain them.

Robust product design offers a reliable, quantifiable and efficient means to attain six sigma quality in all stages of product commercialization. By defining on-target performance and variability, robust design methods provide engineers with the means to make products and processes insensitive to sources of variability. Robust design helps us quantify the cost of quality. From an engineering perspective, it provides us with hard numbers for measuring where we are, where we want to be, and how to get there.

Though time and chance—or variability, as it is called in the engineering world—encumber all of us, robust design gives us the tools to minimize the effects of chance and to maximize control of our processes, and thus our success. The benefits are far-reaching: Cycle time is reduced, rework is eliminated, and products are optimized to satisfy customer needs.

As Albert Einstein said, significant problems "cannot be solved at the same level of thinking we were at when we created them." Robust design offers a new way of thinking that opens the way to the highest levels of quality and broadens our opportunities.

*Engineering Methods for Robust Product Design* offers clear, comprehensive, and up-to-date guidelines for practical application of robust design processes. The book covers basic principles and practices of robust design through advanced techniques, using case studies to facilitate understanding and application.

The contents of this book reflect the authors' considerable experience in product design methodologies and processes. As engineers at Kodak, they put theory into practice in the day-to-day reality of the production environment. As instructors in robust design, they must synthesize their knowledge and communicate it in clear, understandable ways to engineering colleagues and university students alike. This experience, plus solid grounding in the fundamentals of quality leadership, has resulted in a book that offers definite value to students as well as seasoned engineering practitioners.

An exceptional feature of *Engineering Methods* is the integration of a unique, custom software program with the text—the first book solely on robust design to do so. The software automates analysis of data, freeing the reader to concentrate on the engineering issues rather than computation. The fully functional, partially featured software was authored by John Derimiggio, a development engineer in Office Imaging Research and Technology Development at Kodak.

Bill Fowlkes and Skip Creveling have successfully applied the robust design process to mechanical and electrical systems, electrophotographic process optimization, and chemical process optimization. Fowlkes is Laboratory Head of Thermal Print Equipment for Thermal Imaging Products Research and Advanced Development. He earned a Six Sigma Black Belt Certification from the Six Sigma Research Institute in 1992, and won a Promotion Award in 1992 and Best Paper Award in 1993 from the American Supplier Institute. He was a panel speaker on robust engineering for the November 1994 meeting of the Industry Advisory Board of the American Society of Mechanical Engineers, and is also a founding member of the North American Quality Engineering Forum.

Creveling is Staff Engineer and Team Leader of the Electrophotographic Process Hardware Group at Kodak. He has co-authored courses on the Taguchi approach to robust design and product optimization, and to rational tolerancing and tolerance design. He won a Promotion Award in 1992 from the American Supplier Institute, and the Dan Totsu Award at the 1993 Quality Through Experimental Design (QED) conference sponsored by the Center of Quality and Applied Statistics at Rochester Institute of Technology (RIT). Creveling teaches engineering mechanics and machine design at RIT.

Fowlkes and Creveling have also team-taught a course on Taguchi methods of robust design at RIT for three years. Both were instrumental in writing the custom training materials which are used to train Kodak staff in the principles and methods of robust design. *Engineering Methods for Robust Product Design* is the culmination of a three-year effort to develop state-of-the-art training materials in robust design as it exists in industrial practice. Applied assiduously, it can be an integral driving force in the push to make American industry competitive and successful well into the next century.

*—Rochester, New York, February 1995*

# PREFACE

Quality in products and product related processes is now, more than ever, a critical requirement for success in manufacturing. In fact, for many successful companies, such as Motorola, Toyota, Ford, Bausch & Lomb, Xerox, and Kodak, it is fair to say that quality is a corporate priority. These companies have realized that to obtain customer loyalty, their products have to be perceived as nearly flawless. In addition, to be competitive, their product development process must minimize waste, cycle time, and rework.

The practices adopted by companies that are succeeding in the quality competition vary, but two common elements can be found. Careful attention to the customer is absolutely paramount. Products must satisfy a diverse customer base, with features accurately targeted to customer requirements. Technology must serve customer needs and wants, or the latest and greatest widget will languish on the shelf. Also, continuous improvement, applied to both products and business processes, is ubiquitous.

At the Eastman Kodak Company, the authors have been participants in the ongoing effort to improve the equipment development process. The result has been a world-class process for developing products. This process features *Quality Function Deployment* (QFD) for capturing the voice of the customer, *Robust Design* (Quality Engineering) to deliver the level of quality demanded by the customer, and a disciplined engineering process for managing the business of product commercialization. Much of the Kodak Equipment Commercialization Process is described in Professor Don Clausing's book *Total Quality Development* (ASME Press, 1994).

Physics and engineering principles are the basis for beginning a good product design or fixing problems with a design that is already in existence. Any graduate from engineering school knows these fundamental subjects well. They have been used effectively by many generations of engineers. However, they alone are no longer enough. The current competitive situation requires a disciplined engineering *process* that ties together the multitude of engineering tools currently being taught and practiced.

The need to further define the process for linking the principles of engineering and physics to product commercialization inspired the writing of this book. The authors' experiences in applying Robust Design to mechanical and electrical systems, electrophotographic process optimization, and chemical process optimization at Kodak have demonstrated convincingly that Dr. Taguchi's design optimization techniques are extremely effective in reducing cycle time and rework. Every company that employs Robust Design does so in the context of their own internal culture. Only the books written by Dr. Taguchi follow his views in totality. This book is a reflection of how we have internalized Dr. Taguchi's insights and teachings into our culture at Kodak. In this industrial environment, we have found broad acceptance and a strong willingness to employ Taguchi methods when practiced in an engineering context. This, of course, is exactly how Dr. Taguchi and those who have listened to him over the years approach the topic—as an *engineering process.* The successes experienced at Kodak and at many other companies we have encountered are derived from Dr. Taguchi's advice [T6]: *"Spend about 80% of your time in engineering analysis and planning and about 20% actually running experiments and evaluating the results."*

Recently, engineering process improvement has been introduced into the academic arena. Courses on Robust Design, QFD, Six Sigma, and other quality processes can now be found at an increasing number of schools. Some of the leaders in this new trend include the Massachusetts Institute of Technology, Stanford University, Georgia Institute of Technology, and Michigan Technological Institute. Rochester Institute of Technology (RIT), where we teach and serve on the Industrial Advisory Board, recently adopted elements of a quality engineering curriculum as mechanical engineering electives. This book is largely based on our experience in teaching the Robust Design course at RIT in an *engineering* department. This is unique, because much of the academic attention given to Dr. Taguchi's methods has come from the statistics community as a result of Dr. Taguchi's use of empirical statistical techniques, particularly design of experiments. This has led to a misunderstand-ing of robust design as being statistical in nature. This book takes an entirely fresh look at robust design as an engineering process, where the emphasis is on using engineering analysis to improve product performance.

This book offers simple, yet effective, guidelines on how to practice robust design in the context of a total quality development effort. In these pages, the fundamental metrics of quality engineering are fully developed, and the rationale behind them is explained. Designing low-cost solutions is a given requirement. We discuss the impact of robust design on the cost of a design, as well as how cost and quality can be co-optimized using Dr. Taguchi's *Quality Loss Function.* The fundamental statistical tools (e.g., design of experiments, analysis of variance, and analysis of interactions) are explained in what we hope will be an intuitive yet mathematically precise way. A healthy balance exists between the statistical sciences and the engineering sciences. In this book, we try to introduce practical insight into the statistical side of Robust Design, while maintaining the highest priority in basing the experimental approach on sound engineering principles. The most important element of engineering success is clear thinking, planning, analysis, and communication. For this reason, we offer this book primarily as a guide on how to invest your time efficiently in the 80% up-front engineering effort required, particularly as it pertains to technology development and product commercialization.

## The Structure of This Book

Chapter 1 is organized to provide a broad introduction to Quality Engineering and to establish the fundamental concepts needed to build the reader's understanding for work presented in later chapters.

The rest of the book is presented in three major parts. The first is an introduction to *Quality Engineering Metrics.* It consists of Chapters 2 through 6. Robust Design is a data driven process. Chapter 2 goes through *Introductory Data Analysis for Robust Design* and is presented to establish a context for how data will be treated throughout the rest of the book. Chapter 3 presents the theory and derivation of the various forms of the *Quality Loss Function.* An application of the quality loss function to tolerance design is also included. Chapter 4 presents the fundamental knowledge behind the *Signal-to-Noise Ratio.* The static and dynamic signal-to-noise ratios are fully discussed with numerous examples in Chapters 5 and 6, respectively.

The second part delves into the details of the parameter design process with a special emphasis on achieving *additivity.*[1] Additivity is a property of a design that reduces harmful *interactions,*[1] thus simplifying the optimization process. Chapter 7 is a practical *Introduction to Designed Experiments.* Without the use of designed experiments, the process of optimizing a product becomes a time consuming endeavor laced with rework and unwanted surprises due to interactions. Chapter 8 is focused on a thorough discussion concerning the *Selection of the Quality Characteristics.* Few choices in the process of quality engineering are as critical as the selection of the physical responses to be measured during the designed experiments. Chapter 9 provides a sound basis for the *Selection and Testing of Noise Factors* to stress the design during the development of robustness. Constructing viable noise factor experiments is an indispensable step in preparing for credible and realistic optimization experiments. Chapter 10 completes the discussion on the selection of experimental parameters by giving strategies for the *Selection of Control Factors.* Chapter 11 shows how to lay out the *Parameter Optimization Experiment* and is followed in Chapter 12 with the *Analysis and Verification of the Parameter Optimization Experiment.* Quantifying the individual control factor effects on the overall design performance is highly prized information to the engineering team. In summary, Chapters 7–12 are designed to take you through a comprehensive process of planning, experimenting, and verifying optimized parameter performance.

This book is intended to be useful for teaching and learning the principles and practices of robust design. Chapter 13 demonstrates the parameter design process by covering, in detail, three examples that are actually used as workshop problems by the authors during courses in robust design at the Rochester Institute of Technology and at Eastman Kodak Company. These simple examples are good illustrations of the techniques and can be performed by the reader to practice the method. Chapter 14 demonstrates the parameter design process by presenting three actual Kodak case studies, previously unpublished. Real design problems always take on additional complexity that is intentionally avoided in heuristic examples. These case studies show how parameter design is effective at real-life problem solving. The performance improvements are significant and lasting.

The third and final part of the book is geared toward the engineering practitioner who is interested in more advanced techniques of Robust Design. Chapter 15 provides the necessary information to allow the engineering team to modify arrays to aid in the optimization of unique cases of parameter design. *Working with Interactions* (Chapter 16) is probably the most controversial topic among the methods of Robust Design. We have produced a balanced approach to maintaining statistical validity of experimentation while promoting the use of engineering knowledge and experience during the construction and analysis of designed experiments that may contain interactive control factors. Chapter 17 teaches the method of the *Analysis of Variance,* an advanced tool for analyzing designed experiments. Chapter 18 completes the book with a discussion of three special topics within the field of Robust Design. They are the relationship of Robust Design to (1) Quality Function Deployment, (2) Classical Design of Experiments, and (3) Six Sigma.

---

1. The terms *additivity* and *interaction* are defined in the glossary.

Because the empirical methods of Robust Design require statistical analysis of large amounts of data, WinRobust Lite software is included with this book. Numerous examples are provided to introduce the reader to the many helpful features contained in this PC-based Windows software package. This is the first book of its kind to integrate a custom software package with the text. This union with computer-aided Robust Design techniques will provide you with a comprehensive set of tools that will simplify the tedious process of computation, thus freeing your efforts to focus on the essential engineering issues behind the functional performance you seek to optimize.

## Acknowledgments

The authors would like to acknowledge, with great respect, our *sensei* and teachers of Robust Design, Dr. Genichi Taguchi and his son Shin Taguchi; Dr. Madhav Phadke; Prof. Don Clausing (Massachusetts Institute of Technology); and Prof. Tom Barker (Rochester Institute of Technology). Each of these individuals has been very generous in sharing his unique insights and knowledge of quality engineering. We would like to thank the management at Kodak who supported our efforts to learn, teach, and advise others within Kodak in the proficient use of these methods. In particular we would like to recognize Tom Plutchak and Martin Berwick, who supported us in bringing Dr. Madhav Phadke to Kodak to get us started on our journey, and then encouraged us to *get busy* and make something valuable happen with our new-found knowledge.

We have been very fortunate to have had the opportunity to create and teach one of the first undergraduate-level courses in the United States specifically on the topic of Robust Design. This book is, in large part, a product of our course notes. We greatly appreciate the support and encouragement of Bob Merrill, the current chairman, and Ron Amberger, the past chairman, of the Department of Mechanical Engineering Technology within the College of Applied Science and Technology at the Rochester Institute of Technology. Additional support came from the members of the Industrial Advisory Board of the Department of Mechanical Engineering Technology, and in particular from the advisory board chairman, John Shannon of the Bausch & Lomb Corporation. Thanks to each of them.

Special recognition is due to George Walgrove and Tom Foster, engineers who through their frequent and imaginative use of parameter design have helped make Kodak world-class in the Robust Design process. We would also like to thank the individuals who contributed case studies to this book: Chuck Bennett, Marc Bermel, Atsushi Hatakeyama, Shigeomi Koshimizu, Mike Parsons, Allen Rushing, Steve Russel, Markus Weber, Reinhold Weltz, and Mark Zaretsky. We wish we could list all the other contributors who have added their experiences to ours to help make this book possible, but it is impossible to be comprehensive in such a list, so we simply thank them collectively.

Finally, we would like to thank Susan Baruch for her help in preparing the manuscript, our editor Jennifer Joss, our technical reviewers, and the entire staff at Addison-Wesley for their support.

# CHAPTER 1

# Introduction to Quality Engineering

## 1.1 An Overview

Quality engineering is viewed differently in Japan and the West. In the West, quality engineering is considered to be a task separate from development, design, and manufacturing. Quality engineering is often the responsibility of staff engineers who are not on the product development team. These engineers are responsible for the independent and objective testing and evaluation of product prototypes and early production machines. The task can include statistical analysis of data or evaluation and maintenance of test equipment, and can even include advising management on how best to improve the quality emphasis in an organization [J1]. The premier professional society of Western quality experts is the ASQC, the American Society of Quality Control.

In Japan, quality is considered the responsibility of all engineers and management. This is typical of the Eastern holistic approach. Therefore, there is no separate Japanese Society of Quality Control. Rather, the Japanese Union of Scientists and Engineers (JUSE) includes quality in its domain. In this context, Dr. Genichi Taguchi has developed an engineering method of quality improvement referred to as *Quality Engineering* in Japan and as *Robust Design* in the West. Regardless of the name, Robust Design is a disciplined engineering process that seeks to find the best expression of a product design. "Best" is carefully defined to mean that the design is the lowest-cost solution to the product design specification, which is based on the customer needs. Cost includes more than just manufacturing cost. Also included are life-cycle costs and losses to society (e.g., cost incurred by the manufacturer, the customer, and all others affected by any deficiencies of the product). Dr. Taguchi has introduced a holistic approach to the traditional engineering tasks of minimizing cost and maximizing quality by including the quality of the product as one more dimension of cost. This is done by measuring the quality of the product by its life-cycle costs and losses to society. High-quality products minimize these costs by performing consistently.

The ability to view delivering the features, cost, and quality of a product as a systemic problem is becoming more and more necessary in today's product commercialization environment. Companies that view these as separate issues risk losing business to well-integrated competitors who have better time-to-market, fewer start-up problems, and lower operating costs (including warranty and replacement losses). Innovative technologies and manufacturing processes are used only after they have been demonstrated to be insensitive against known sources of variation. Using the process discussed in this book, parameter optimization, these product concepts are optimized to demonstrate stable, on-target performance even under extreme stress conditions. A design is completed only after thorough cost-based tolerancing has been done to assure that the proper balance between quality and cost has been achieved. Production tolerance limits, statistical process control, and corrective action costs are tuned to optimize quality and cost. It is impossible, over the long run, to compete against such an engineering approach. For that reason, Dr. Taguchi's methods have received considerable attention in the United States and Europe over the last 10 to 15 years.

Robust product design focuses on the product concept selection and parameter optimization. These are done by reducing the measured variation of key quality characteristics and ensuring that those characteristics can be easily adjusted onto the nominal value or target. Minimizing variation or making the system less sensitive to variation makes it possible to decrease cost, as expensive means for *controlling* quality are no longer necessary. Providing functionality to put the system on target facilitates process control. Design reusability and flexible manufacturing processes are enabled by tuning or adjustment capability. To analyze a system's robustness and adjustability, unique metrics that are central to Robust Design, referred to as *signal-to-noise ratios,* have been created. These simple metrics make it possible for practicing engineers to use powerful experimental analysis methods to analyze and optimize product concepts.

## 1.2 The Concept of Noise in Robust Design

Once the customer starts using a product, the quality of the product (and therefore its performance) can vary for many reasons. The causes of variability are called *noise factors.* Noise factors are defined as anything that causes a functional characteristic or response to deviate from its target value:

- The effect of high-power transmission lines on an AM radio receiver
- Water in automobile gasoline
- Wear of a punch-press die

Another way to consider noise has to do with the conversion of energy. Whether it be kinetic, chemical, thermal, potential, or any other form, engineers are very interested in the *efficiency* of the transformation of energy from one form to another. Most products either convert energy in order to deliver some desired function to the customer or are themselves the result of a transformation of energy. Engineers employing Robust Design account for and make detailed use of the transformations of energy.

Engineers seek to employ the laws of nature to make things that people are willing to pay for. Variability is simply an outcome of something obeying a natural physical law. When things change (a loose definition of variability), they do so because the physics active in the design promote the change.

In order to develop a process of designing products that provides on-target performance and maintains it in the face of variability, we must first have a clear understanding of the nature of noise. There are three types of noise factors:

1. External Noise Factors
2. Unit-to-Unit Noise Factors
3. Deterioration Noise Factors

*External noise factors* are defined as sources of variability that come from outside the product. Examples of external sources of variability include the following:

- The temperature and relative humidity in which a product is used
- The load to which a product is subjected
- Any unintended input of energy (heat, vibration, radiation) into a design to which the system is sensitive
- Human error including misuse (ignorance) and abuse (intentional)
- Dust in the environment
- Input voltage variation
- Electromagnetic interference
- Ultraviolet light

*Unit-to-unit noise* is a result of never being able to make any two items exactly alike. Manufacturing processes and materials are major sources of unit-to-unit variability in product components. Process nonuniformity and process drift are common sources of variability in this context. Some typical examples of unit-to-unit variability include the following:

- Camshaft weights
- Amount of torque applied to bolts
- Resistance of electrical resistors
- Dimensions from any metal-forming or removal process
- Batch-to-batch concentrations of chemicals
- Thickness variations in coated products

*Deterioration noise* is often referred to as an internal noise factor. This is so because something changes internally within the product or process. It is common for certain products to "age" during use or storage such that their performance deteriorates. The following are examples of internal deterioration noise factors:

- Compression set or plastic creep of a washer in a faucet
- Loss of mass of a light-bulb filament
- Loss of plasticizer in an auto dashboard
- Total current passed through a car battery
- Weathering of paint on a house
- Mileage on a car

## 1.3 Product Reliability and Quality Engineering

The result of noise is usually characterized as some form of reliability problem. Product failure can be characterized in relation to time. Three categories are usually identified: early-life failures, failures during normal design life, and end-of-life failures. Classical reliability methods account for these failures in a graph called a "bathtub curve," which is shown in Figure 1.1.

This curve states that the probability of failure is higher early or late in life, while a majority of the product enjoys a fairly low failure rate during the normal period of design life. Early-life failures are attributed to latent failures—that is, defects due to variation in the manufacturing process that are not apparent during inspection. This unit-to-unit noise causes failure in the field when the latent failure is subjected to external noise. The unit-to-unit noise represents manufacturing variation that results in

**Figure 1.1** The reliability bathtub curve

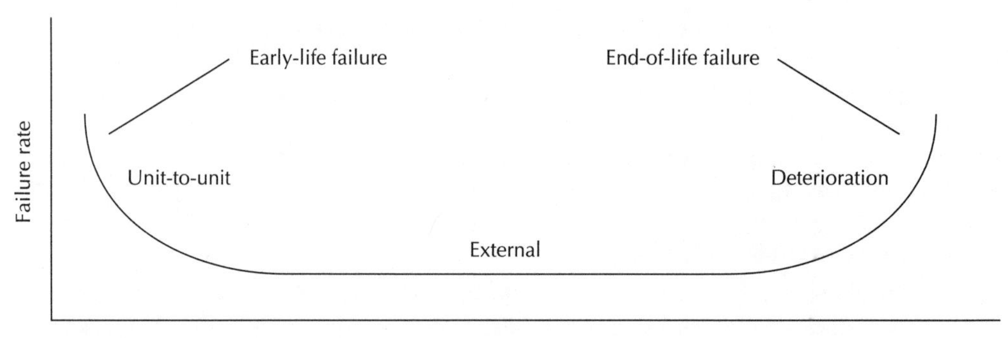


some product being near to failure—a little additional external noise, and the unit with a latent failure can go into failure prematurely. As the off-target units fail and are eliminated from the population, the failure rate declines.

The most common type of noise encountered is external noise. External noises are present, with varying intensities, at all times. Compensation for them, in the form of feedback control, is expensive and adds complexity to the design. Eliminating external noise requires that restrictions be placed on how the customer can use the product—for example, by specifying the operating environment or duty cycle. The random failure rate that characterizes the bottom of the bathtub curve is attributed to external noise.

Eventually the cumulative effects of deterioration noises and external noises cause end-of-life failure. As a result, the failure rate goes up after a long time in service.

There is a lot more to quality than preventing failure. As practitioners of Robust Design, engineers seek to identify the noises that promote not only failure, but also the less obvious off-target performance during these stages of product life. The engineering team must be thorough in its consideration of all types of noises and how they interact with the physics of the design as it progresses through its intended life. If the team can't experimentally evaluate a known set of noises, they probably have an inherently poor design that cannot benefit from the robustness process. Paraphrasing Lord Kelvin's well-known maxim: "If you can't measure it, then you can't optimize it."

It is important to build a strong multidisciplinary team of broadly experienced engineers from many functional areas. Considering the subtle sources of variability can be complex. Having experienced people from various technical and business backgrounds is recommended [P1]. The engineering team needs to make their design insensitive to noises that are active throughout the product's life as well as noises that may only exhibit their behavior as nonrandom events early or late in the product's life.

## 1.4 What Is Robustness?

The user's perception of the quality of a design is very closely related to the sensitivity of the design to noise. The engineering team needs to minimize the effect noise has on the performance of the design. There are two ways to minimize variability:

1. Eliminate the actual source of noise.
2. Eliminate the product's sensitivity to the source of noise.

It can be very costly and time-consuming to eliminate the noise factors themselves, because some noise factors just cannot be controlled and some noise factors are too expensive or difficult to control.

This leads to the fundamental definition of Robust Design:

> *A product or process is said to be robust when it is insensitive to the effects of sources of variability, even though the sources themselves have not been eliminated.*

The objective of the engineering team is to develop a product or process that functions as intended under a wide range of conditions for the duration of its design life. Robust Design is a process to obtain product performance that is minimally affected by noise.

## 1.5 What Is Quality?

Robust Design requires a quantitative definition of what *quality* is for products and processes. This is necessary to allow the team to compare different concepts and to find the optimum configuration of a design. But how is quality to be defined? Who defines quality? Engineering? Management? Customers? Quality as defined by the customer is very difficult to describe. Very often quality is perceived as a *feeling* of satisfaction. You'll know what satisfies you when you're feeling satisfied. Of course, this is not a very useful approach to Quality Engineering.

Concepts such as low cost, high reliability, and consistent performance are obviously important elements in determining quality. These are the aspects of quality that the engineering team improves quantitatively through the application of Robust Design. More recently, the TQM (Total Quality Management) movement has convinced industry to give the customers a major voice in determining what they want in products and services and at what quality level. It is clear that the ultimate answer to the question of who defines quality must be the customer. Customers are a very diverse group. Giving them the features that they want has led to product specialization, short product lifetimes, and the need for flexible manufacturing.

The quality guru J. M. Juran [J1] defines two fundamental components of quality, which include two major elements of customer satisfaction and Robust Design:

1. A product's *features*
2. A product's *conformance* to those features

A product's features are the performance characteristics that make the product attractive to the customer. Conformance to those features means the ability of the product to consistently deliver *on-target performance* each time it's used, under all intended operating conditions, throughout its intended life, with no harmful side effects.

Conformance to customer-defined performance levels means quality improves when variation of performance is minimized. Consider two designs, A and B, having the distributions shown in Figure 1.2, describing the variation in a performance characteristic (e.g., color saturation of an image, water flow from a hose, acceleration of a car). Which has better quality? Design B, of course, would be perceived by the customer as having higher quality *regardless of tolerance limits.*

Robust Design is the process discussed in this book for changing Design A to achieve the performance shown for Design B, i.e., achieving more consistent performance of product features.

**Figure 1.2** Quality comparison by distribution width

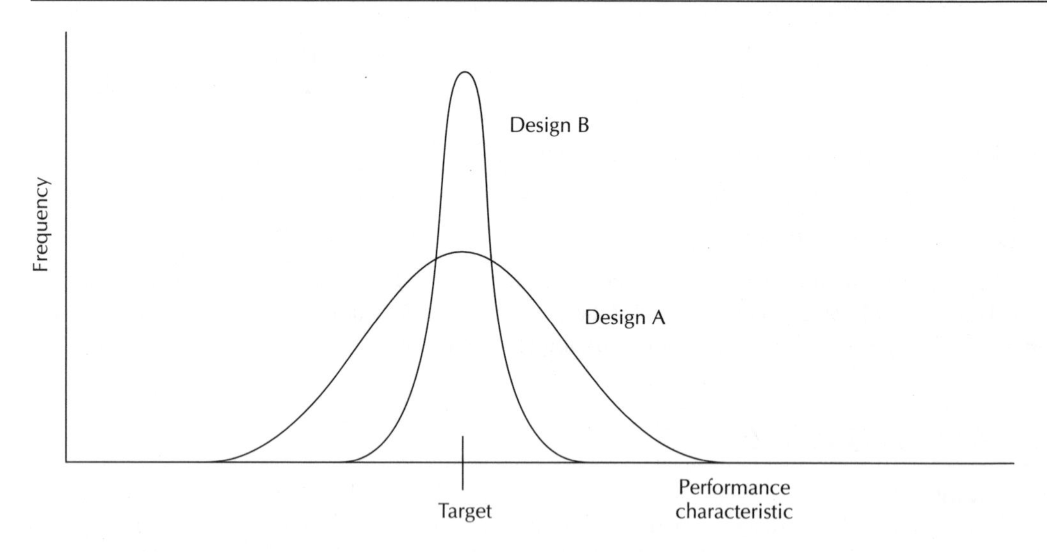


Selecting the features to offer in a product is the responsibility of marketing and management. It is the engineer's task, using *engineering methods*, to deliver a robust design that consistently performs on target [T3].

## 1.6 On-Target Engineering

Conformance to features leaves open the question of how far from the target a product can stray before the customer perceives that it is no longer of high quality. In the Western approach to quality engineering, much attention is given to finding out what the customer's tolerance is to deviations from targeted performance. This approach defines quality in terms of the customer's tolerance. Thus, we hear engineering terms such as conformance to tolerance specifications, yield, fraction defective, and capability index.

The approach represented by these terms compares the actual variation to the "allowable" variation. In contrast, Robust Design considers *all* deviation from the target to represent inferior quality—regardless of how wide the allocation for tolerance. The approach of developing products and processes that are insensitive to noise and are readily tunable to a specific mean performance is referred to as *On-Target Engineering*.

## 1.7 How Is Quality Measured?

Dr. Taguchi believes that loss of quality results in an ultimate *cost* to society. In very measurable ways, both customers *and* manufacturers lose time and money when products perform off-target. This is the comprehensive cost of quality.

**Figure 1.3** Quality loss increases with deviation from target

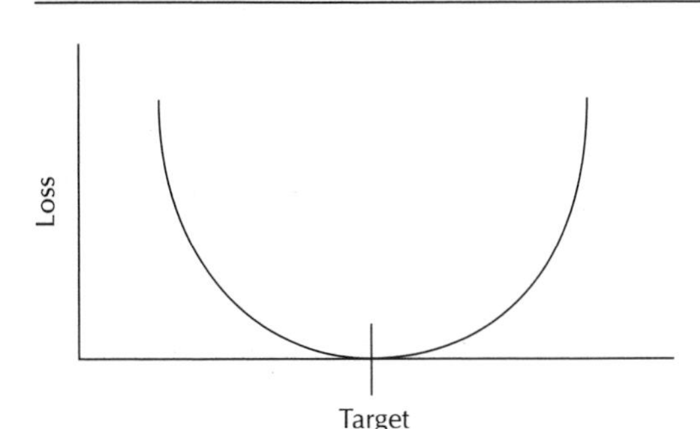


Today's engineering teams must learn to account for cost in more dimensions than their predecessors. They must account for the monetary losses that customers incur when a product's performance is degraded in the presence of noise. Accounting for the effects of noise in the factory is no longer enough.

A quantitative description of quality, consistent with on-target engineering, can be developed based upon two principles. The first principle is that deviation from on-target performance results in a loss to the customer, a loss to the manufacturer and ultimately a loss to society, as shown in Figure 1.3. How do we visualize this? Customers can perceive variation even though the product performs to requirements. Products that perform to nominal requirements under low-noise conditions may fail under higher levels of noise. The total percentage of units that fail to meet specifications (the farthest points out on the distribution of data) may be relatively small, but in a high-volume product may result in substantial repair and warranty costs.

The second principle is that monetary losses due to deviations from on-target performance form the basis for a quantitative description of quality. We can use *cost* to measure and quantify quality:

- *High quality is freedom from costs associated with poor quality.*

The costs may be hard to quantify, but they are real. Part of an engineer's job is to minimize the total cost of a design—unit manufacturing cost, life cycle cost, and quality loss. This last cost may be harder to quantify than the others, but it must not be overlooked. The three dimensions of cost can be visualized as a balanced cost structure, as shown in Figure 1.4.

*Unit Manufacturing Cost* is the most inward-looking measure of cost a company can use. This simply accounts for the manufacturing cost content of a product. Items such as material, labor, and burden costs associated with manufacturing the product make up the UMC. Making business decisions solely on the basis of UMC results in poor quality, poor reliability, high life-cycle costs, and a poor reputation.

*Life-Cycle Costs* include measurable costs that the manufacturer and customer incur as a result of the operation of the product. For example, consumables and energy required to operate a product must be purchased. Repair and warranty costs give a broader picture of the cost of ownership over the life of the product. It is an important step forward for engineering teams to take these costs into account when developing a product.

*Quality Loss Costs* take into account less tangible costs incurred by the manufacturer, the customer, and society. This form of cost "accounting" directly relates customer costs to deviations of func-

**Figure 1.4** The total cost of quality triangle

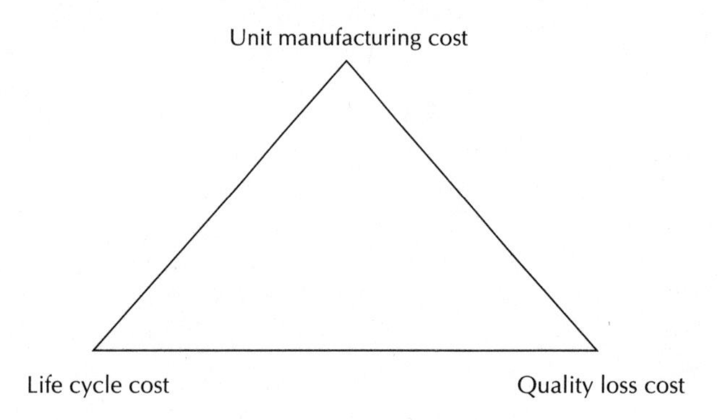


tional performance. A product's quality loss costs are based on the economic consequences of deviation from on-target performance.

For example, losses to the customer include the following:

- Time and effort taken to work around minor failures
- Lost profits due to a nonfunctioning machine
- Rental cost to replace a machine being repaired
- Nonwarranty service costs and service contract costs

Losses to the manufacturer include the following:

- Inspection, scrap, and rework
- Warranty costs
- Returns and recall costs
- Lost sales and customers
- Lawsuits

Losses to society include the following:

- Pollution and other waste that must be disposed of
- Personal injury or loss of life
- Disruption of communications and transportation

In Chapter 3 we use these concepts to develop the Quality Loss Function (Equation 3.1), which describes the monetary losses as a function of the deviation from the target performance.

## 1.8 The Phases of Quality Engineering in Product Commercialization

How and when is Robust Design applied during product commercialization? There are always sources of variability influencing the performance of products and processes even as they are being developed. Engineering processes designed to improve quality need to be used at every step in product development, design, and manufacturing. These processes vary in the tools they employ. Their common goal is

**Figure 1.5** Robust design processes aligned with product commercialization processes

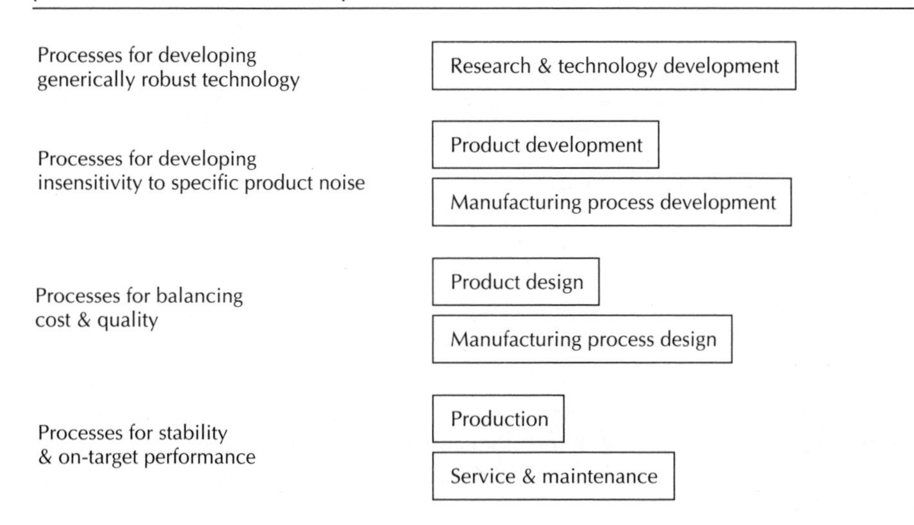


to ensure on-target performance in spite of the many sources of variability active in each stage of product commercialization.

A chronology of robustness opportunities can be established by considering the various stages of product commercialization, and then aligning the processes employed in Robust Design with each stage, as shown in Figure 1.5.

Broadly speaking, quality engineering can be divided into two phases. *Off-Line Quality Engineering* refers to the activities that take place during the product development and design processes. *On-Line Quality Engineering* refers to processes that take place during production. The engineering processes for reducing variability and thus improving quality differ markedly in these two phases. The following section establishes the relationship between the previous chronology of commercialization stages and these two phases of quality engineering.

## 1.9 Off-Line Quality Engineering

Off-line quality engineering consists of three phases of product development and design, as shown in Figure 1.6. Each of these phases has unique processes that enable the engineering team to quantify the results of their planning and analysis.

- *Concept Design* is the phase in which product development teams define a system that functions under an initial set of nominal conditions. The system should only use technology that has been shown to be robust.
- *Parameter Design* is the phase in which product development teams optimize the concept design by identifying specific control factor set points (levels) that make the system least sensitive to noise, thereby enhancing the system's robustness.
- *Tolerance Design* is the phase in which product design teams specify the allowable deviations in parameter values, loosening tolerances where possible and tightening where neces-

**Figure 1.6** The three stages of off-line engineering

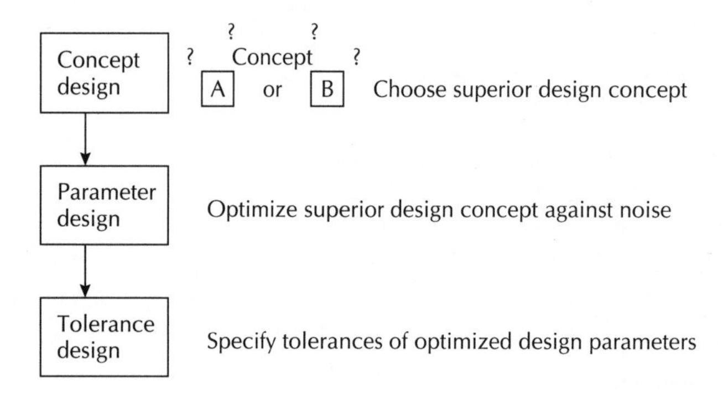


sary. This is done to balance the cost of the product with the quality of the product—within the context of satisfying customer demands. With this information, the design team can complete the detailed design.

### 1.9.1 The Concept Design Phase

During this phase, various alternative solutions to the engineering problem (product concept) are studied. The ideal solution is a flexible product built upon inherently robust technology. The concept may be a new invention that surprises both the customer and the competition, a direct response to a known customer need, or an incremental competitive offering. It is normal in this stage of design to conduct detailed research, invention, and technology development. The noises used in this phase are generic in nature, because the specific product is still evolving.

These are the Tools of Concept Design:

- *Quality Function Deployment:* Translating the customer wants into engineering terms and targets [A2, C2].
- *Dynamic Signal-to-Noise Optimization:* This Robust Design technique optimizes the engineering function, resulting in robust, tunable technology.
- *Theory of Inventive Problem Solving:* A collection of tools derived from analysis of the patent literature that are useful for generating innovative solutions to technical problems [A3].
- *Design of Experiments:* Full factorial and partial factorial experiments exploring the effects of several parameters simultaneously.
- *Competitive Technology Assessment:* Benchmarking (comparing) the inherent robustness of internally and externally developed technologies.
- *Pugh Concept Selection Process:* Gathering and displaying information from a team of experts, comparing the features and quality of a variety of concepts, *and* evolving those concepts until a clearly superior winner emerges [P3].

### 1.9.2 The Parameter Design Phase

During this phase, the specific controllable factors in the engineering concept that affect the system's robustness are identified. Their optimum set points are then determined and verified.

**Figure 1.7** The system P-diagram

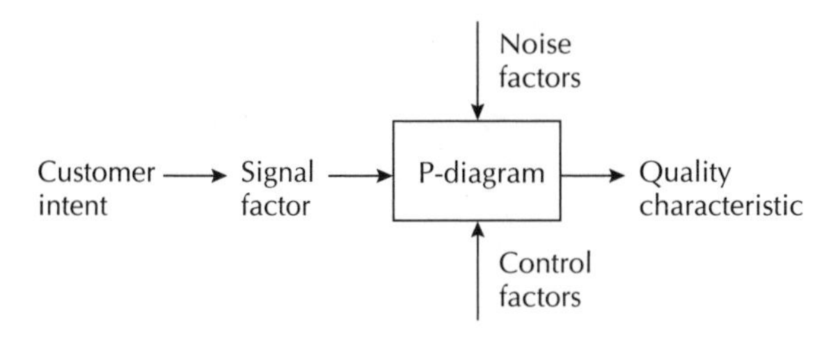


These are the Tools of Parameter Design:

- *Engineering Analysis:* Using training, experience, and experimentation to discover the sources of variability and effective countermeasures.
- *The System P-Diagram [P2]:* A powerful tool for representing and characterizing the various parameters that influence the system's output (shown in Figure 1.7).
- *Dynamic and Static Signal-to-Noise Optimization:* Optimizing the design parameters to reduce variability as measured using a signal-to-noise ratio.
- *Crossed Array Experiments*: A particular experimental design that exploits the interactions between control factors and noise factors in order to make the system more robust.

### 1.9.3 The Tolerance Design Phase

During this phase the design specifications are completed by determining the level of allowable control factor variability and, if necessary and possible, the limits on external noise levels required to achieve the product's quality requirements. The goal here is to co-optimize unit manufacturing cost (UMC), life-cycle cost (consumables, service, pollution), and quality loss (deviations from target).

These are the Tools of Tolerance Design:

- *The Quality Loss Function:* The equation relating the cost of product performance variation to the level of deviation from the target.
- *Analysis of Variance (ANOVA):* A statistical technique that quantitatively determines the contribution to the total variation made by each noise and control factor that is studied in a designed experiment.
- *Design of Experiments:* Full factorial and partial factorial experiments exploring the effects of several parameters simultaneously.

## 1.10 On-Line Quality Engineering

On-line engineering is the means of maintaining consistency within the production and assembly processes so as to minimize unit-to-unit variation. It is important to keep costs low and quality high simultaneously. Rapid problem-solving (fire-fighting) is critical during production.

These are the Tools of On-Line Engineering:

- *Statistical Process Control:* Monitoring a process's output and testing its variations to identify random variation versus variation assignable to special causes. Only the latter type of variation results in adjustment to the process.

- *Static Signal-to-Noise Optimization:* Further reduction of variation by the application of Robust Design to problematic production parts and processes.
- *Compensation:* Various control schemes for keeping a process on aim, such as feed-forward and feedback loops.
- *Loss Function–Based Process Control:* The reduction of all costs including unit part costs, inspection and adjustment costs needed for process control, and the quality loss attributable to the remaining variation in the process output.

## 1.11 The Link between Sir Ronald Fisher and Dr. Genichi Taguchi

Traditionally, people have tended to think of Robust Design as a statistical method. This is a misconception. It is an *engineering* method that uses formal and informal methods adapted from the discipline of statistical design of experiments. Dr. Taguchi blends methods of experimental design based in part on Sir Ronald Fisher's unique approach of applying matrices to aid in the discipline, precision, and productivity of evaluating and optimizing agricultural performance.

Fisher did his work in the early 1920s in the agricultural industry of England. Fisher and a mathematician named Frank Yates developed much of what we know today as Statistical Design of Experiments (DOE). Dr. Taguchi is a student of the methods that evolved from the Fisher/Yates designed experiment approach and has brought them into the world of engineering. He has applied the discipline and productivity of designed experiments based on the application of orthogonal (balanced) arrays to product and process development. His contribution is a unique engineering process that unifies the optimization of cost, quality, and product development cycle time. He has done this on a level that practicing engineers can readily understand and implement in their daily activities. With this methodology, the engineering team can rapidly commercialize products that have been exposed to the rigors of empirical experimentation representative of the actual manufacturing and customer use environments—including the variability associated with those environments. Not only is it important to have an engineering process that routinely produces completed designed experiments that optimize performance, but also it is important that the experimental results have a direct bearing on minimizing customer costs throughout the life of the product. The methods of Robust Design account for both of these goals.

Robust Design seeks to enable a strategy that is focused on *rapid* product development, design, and manufacturing. In product commercialization, time-to-market is an important constraint that limits one's ability to do research and technology development. The priority for Robust Design is to optimize *known* technology for specific commercial applications. It is not a substitute for understanding a technology. Robust Design requires the evaluation of product or process control factors (parameters engineers can control) in the noisy environment from which the classical DOE method seeks isolation. Robust Design intentionally induces large amounts of nonrandom and random variability during the experimental process. The outcome is uniquely useful experimental data capable of providing the kind of information the engineer can use to optimize the product's performance in light of the variation that is active during the experimental process. The product engineering environment requires a process suited for producing optimized product performance against sources of variability, while obtaining minimized costs for delivering economical, on-target product performance. It is this set of requirements that the methods of Robust Design are designed to fulfill.

Statistical DOE is traditionally more at home in research and technology development environments. In research and development it is essential to find the time and resources necessary to rigorously evaluate fundamental phenomena and create mathematical models. These can then be used by product design engineers to practice Robust Design as part of their product development process.

Dr. Taguchi's approach is to simplify the complex world of DOE and put the power of practical experimental design into the hands of the engineer. In the Robust Design method, the engineers plan, design, analyze, and verify their own experiments within their own teams. Thus, the ability of the engineer to personally gain efficiency in the product development process by obtaining the "required information with the least expenditure of resources" [B3] is available in the methods of Robust Design. One of the strengths of Robust Design is having the engineering team control the designed experiment by making use of their combined education and experience. This makes the experiment highly efficient, because it is not necessary to evaluate things already known to the engineering team.

## 1.12 A Brief History—The Taguchi Method of Quality Engineering

In 1949, the Electrical Communications Laboratory of Japan's Nippon Telephone & Telegraph Company undertook a project to upgrade the nation's telecommunications system. Known for his engineering skills and his evolving methods for applying designed experiments in a uniquely productive way, Dr. Genichi Taguchi was given the responsibility of promoting productivity within the Research & Development section of the Electrical Communications Laboratory. It was here that the foundations of what Dr. Taguchi calls Quality Engineering began to take form. The stage was set for a 40-year period of engineering process development that has earned the respect of the world's industrial and academic communities.

As a result of his contributions, Dr. Taguchi has received the Deming award four times. The Emperor of Japan has recognized Dr. Taguchi for his contributions to Japan's industrial productivity. In the U.S., Dr. Taguchi was recently inducted into the prestigious Engineering Hall of Fame alongside other renowned names such as Chester Carlson—the inventor of xerography.

A number of industrial leaders, technical historians, and quality consultants regard Dr. Taguchi as one of the most significant contributors to the advances of late 20th-century engineering and product quality improvement processes.

Dr. Taguchi came to the U.S. in 1980 to assist the American telecommunications industry in applying his methods as a gesture of gratitude for America's help in rebuilding Japan after the war. His first association was with AT&T Bell Laboratories in New Jersey. Dr. Taguchi's methods were adopted then, in the early 1980s, by a few companies—most notably by Ford Motor Company and the Xerox Corporation.

Dr. Taguchi has profoundly influenced a few key individuals to carry the methods of Quality Engineering forward. Dr. Madhav Phadke, formerly with AT&T, is the author of *Quality Engineering Using Robust Design* (Prentice-Hall, 1989). This work is the first in-depth presentation of Dr. Taguchi's concepts from an American author. Dr. Don Clausing, formerly an engineering manager at Xerox, and currently Bernard M. Gordon Adjunct Professor of Engineering Innovation at MIT, is the author of a comprehensive work on the total engineering management process entitled *Total Quality Development* (ASME Press, 1994). Dr. Clausing is an early student of robustness methods and currently is a leading authority on the methods. He has been instrumental in assisting Dr. Taguchi in publishing his work in English. He is the editor of Dr. Taguchi's classic two-volume work—*System of Experimental Design* (ASI Press, 1987). Dr. Clausing played a major role in institutionalizing Taguchi methods at Xerox and much later at Eastman Kodak Company.

Through his relationship with Ford Motor Company, Dr. Taguchi joined with Larry Sullivan to form the American Supplier Institute (ASI)—a nonprofit organization, located in Detroit, Michigan, designed to aid American industry in the methods of Quality Engineering. Dr. Taguchi's son, Shin Taguchi, is an ASI vice president and consultant traveling the globe teaching and advising on advanced applications of the methods. Another key figure in the successful development of ASI is Dr. Taguchi's longtime associate and interpreter, Yuin Wu. ASI is known throughout the world as the premier source of Taguchi methods for applications in industrial settings. For a complete history of Dr. Taguchi and his work, we recommend Lance Ealey's book *Quality by Design: Taguchi Methods and U.S. Industry* (ASI Press, 1988).

Today Dr. Taguchi serves as the Executive Director of ASI, frequently lectures, consults in industry, and writes books and articles that promote the evolving methods of Quality Engineering. The American Supplier Institute sponsors an annual Symposium on Taguchi Methods, including the presentation of industrial case studies and tutorials that feature the state of the art in Quality Engineering. The latest advances in the concepts and applications of Quality Engineering are reviewed and summarized at these meetings. The case studies and tutorials presented each year are available in a series of symposium proceedings published by ASI. We can think of no better source of outstanding examples of actual case studies. The best account of the world's working history of Quality Engineering can be found in these symposium proceedings.

## 1.13 Concluding Remarks

In this book we concentrate primarily on the engineering tools that are applied to off-line engineering processes. In particular, the focus is on the Parameter Design process. This is where we believe industry can make the greatest progress in improving product quality and reducing costs. Manufacturing is, of course, critical, and many of the tools we teach in this book are directly applicable to manufacturing process development. Remember that tremendous improvements in cost and quality have been achieved by manufacturing firms in Japan and elsewhere by concentrating their creativity and resources on manufacturing. Through a dedicated focus on on-target engineering, any company can make substantial improvements in their product commercialization performance.

## Exercises for Chapter 1

1. List several reasons why the methods of Robust Design can aid in the successful commercialization of products and processes.
2. List several reasons why practicing analytical and empirical engineering methods jointly is advisable. What can happen if one or the other is neglected?
3. List several examples of each of the three kinds of noise, and explain how they might affect the performance of a product or process.
4. Select a common consumer product and describe how each of the three kinds of noise could affect its performance.
5. Select a common manufacturing process and

describe how each of the three kinds of noise could affect its performance.

6. For an internal combustion engine, define how energy is transformed from gasoline to vehicle acceleration. Be very specific by using engineering terms and units to phrase your definition.
7. For question 6, what energy flow paths can occur that would promote inefficient performance in the engine?
8. Why is it desirable to choose an “inherently robust” concept (a criterion for superiority) at the beginning of the technology development process?
9. What can happen to a product’s performance if immature (nonrobust) technology is included in the product’s architecture?
10. What tools are used to reduce product variability during the production process?
11. What is the definition of a robust design?
12. Why is it better for engineering teams to focus on implementing the methods of robust design than to spend time predicting the amount of reliability needed to launch a product commercialization project?
13. What is the purpose of the classical Design of Experiments process in comparison to the Taguchi robustness process? What are the specific outcomes from each process?

# PART I

# Quality Engineering Metrics

The first part of this book discusses the quality metrics used for Robust Design—the mean and variance, the loss function, and the signal-to-noise ratio. The maxim that you can only improve what you can measure certainly applies to Robust Design. Good engineering performance requires that key responses, the quality characteristics, have low variability and be on target. The variance and mean are the fundamental measures used to determine the amount of variation and the average value of a quantitative response. These basic statistical measures are introduced in Chapter 2. Choosing the appropriate quality characteristic to measure and obtaining good metrology for that response is also critical to the robustness effort. We will defer the discussion of that issue to Part II, Parameter Design.

The variance and the deviation of the mean from the target allow one to determine the quality of a design. Quality is measurable with a clear relationship to both engineering units and monetary units. The relationship between the cost of poor quality—cost that is a result of waste and customer dissatisfaction—is described by the quality loss function. This equation uses a parabolic form to relate the variation of a quality characteristic to the expected losses due to off-target performance. The conversion of the effect of variation to dollars is done using the customer's tolerance limit. The cost of corrective action that is incurred when the quality characteristic exceeds the tolerance limit is used to determine an economic proportionality constant.

The quality loss function is the quantitative description of the value system that drives the entire Taguchi process of quality improvement. On-line economic quality control uses the loss function to optimize the testing and corrective intervals for statistical process control and for field maintenance. Tolerance design is done using the loss function to determine rational safety factors, based on the relationship between the cost of failure in the field and the corrective action cost in the factory. These topics are beyond the scope of this book and will only be touched on lightly in Chapter 3. In Chapter 4, the connection between the loss function and the signal-to-noise ratio is shown.

The signal-to-noise ratio is the fundamental metric to be used for design optimization. The loss function or the signal-to-noise ratio can be used to determine quality or to compare performance. However, design optimization must be done free of on-target constraints to allow variance reduction to take precedence over putting the system on target. This is the fundamental property of the *S/N* ratio. Static signal-to-noise ratios are discussed in Chapter 5, and dynamic signal-to-noise ratios in Chapter 6. For the remainder of the book, we will focus on how to combine design of experiments and the signal-to-noise ratio to improve product and process designs.

CHAPTER 2

# Introductory Data Analysis for Robust Design

## 2.1 The Nature of Data

The practice of applying experimental methods to optimize the robustness of a product or process always requires the collection of data. The ease of working with data to optimize some form of functional performance depends on the type of data. In general, it is difficult to make the necessary determinations to improve a product's performance by looking at the raw data collected during an experiment. The data must be organized into a form that is useful for building knowledge and ultimately drawing conclusions that lead to improved product performance.

Data can consist of numbers that represent *discrete* variables such as whole integers, e.g., 1, 2, 3, 4, or *continuous* variables such as rational numbers, e.g., 1.245, 2.657, 3.912, 4.003. While discrete values such as counting events, gradation and classification, digital values (zeros and ones) and pass/fail can be used in Robust Design activities, data having continuous values such as measured deviations from a target, engineering units (scalars and vectors) and direct measures of physical performance are preferred. In statistical terms, data may be classified by two features: A data set either is a sample from an unknown population, or is made up of all possible values from an entire population. Examples of a sample of data and a population of data are displayed in Figure 2.1.

Data that represent a sample from a population are capable of being mathematically processed into descriptive values called *sample statistics*. Data that comprise an entire population can be mathematically processed into descriptive values called *population statistics*. Some population statistics can never be quantified: They are so large that they are uncountable, or they are ongoing, as in a manufacturing process, so that it is not possible to gather the entire data set.

**Figure 2.1** Illustration of data—sample vs. population

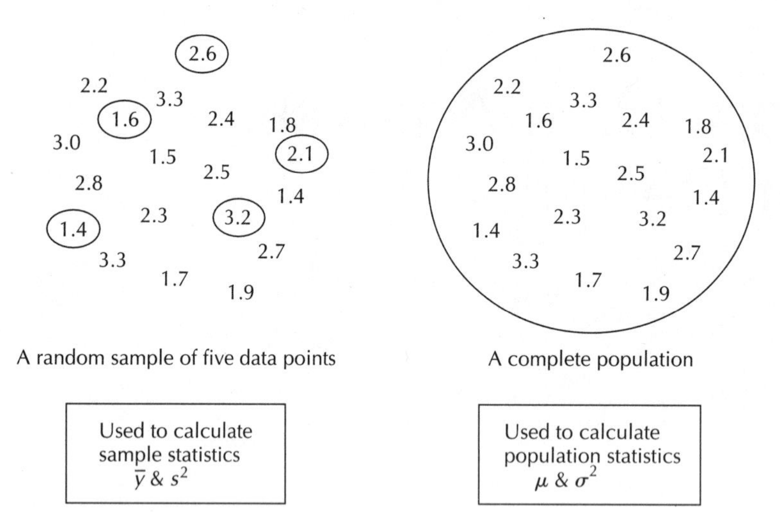


## 2.2 Graphical Methods of Data Analysis

### 2.2.1 The Frequency Distribution Table [F4, I1]

Whenever a number of measurements are taken as data, it is almost certain that some amount of variation will exist among the values. It is also typical to find the experimental measurements tending to fall around a value that suggests a "central tendency" in the data. Another way of describing the nature of the data is to graphically display the shape or distribution of the data. The distribution displays both the variability in the data and the propensity of the data to be spread around a "central" value. A set of data can be organized and displayed by the *frequency* of occurrence of specific values. From such a display three visual observations can be made:

1. The variation that exists between the data points
2. The tendency of the data to fall around a specific central value
3. The shape of the data as it is distributed about the point of central tendency

To illustrate these three characteristics, let's look at a sample of 100 data points (diameter measurements) taken from a batch of several hundred one-inch tubes that were extruded through an aging extrusion die. The data are shown in Table 2.1.

In order to visualize the data, it is necessary to process it from its raw form into a summary form called a *frequency distribution table.* A frequency distribution table is used to generate a plot of the frequency of sample or population data values by placing them within *ranges* of numbers in a graphical format. The distribution is established by counting how many data points fall into each range. In Table 2.2, the 100 data points are counted in 11 equally sized ranges of 0.005 inch starting at 0.975 inch and ending at 1.030 inches.

**Table 2.1** Table of data for one-inch nominal diameter tubes

| | | | | | | | | | |
|---|---|---|---|---|---|---|---|---|---|
| 1.013 | 1.010 | 1.014 | 1.009 | 0.996 | 1.003 | 0.997 | 1.000 | 1.007 | 0.996 |
| 1.007 | 0.984 | 1.012 | 0.996 | 0.991 | 0.995 | 1.006 | 0.988 | 1.005 | 0.992 |
| 0.993 | 1.002 | 1.005 | 1.008 | 0.982 | 1.015 | 1.010 | 1.004 | 0.987 | 1.014 |
| 1.000 | 0.985 | 1.006 | 0.994 | 1.001 | 1.012 | 1.006 | 0.983 | 1.004 | 0.994 |
| 0.998 | 1.008 | 0.985 | 0.995 | 1.009 | 0.992 | 1.002 | 0.986 | 0.995 | 1.030 |
| 1.019 | 1.001 | 1.021 | 0.993 | 1.015 | 0.997 | 0.993 | 0.994 | 1.008 | 0.990 |
| 0.994 | 1.007 | 0.998 | 0.994 | 0.996 | 1.005 | 0.986 | 1.018 | 1.003 | 1.013 |
| 1.009 | 0.990 | 0.990 | 0.993 | 0.995 | 1.017 | 1.000 | 1.009 | 1.006 | 1.005 |
| 1.020 | 1.005 | 1.003 | 1.005 | 0.998 | 0.999 | 1.000 | 0.997 | 1.000 | 0.995 |
| 1.007 | 1.005 | 1.015 | 0.985 | 0.989 | 1.015 | 1.005 | 1.011 | 0.992 | 0.984 |

## 2.2.2 The Histogram

The frequency distribution table can be shown using a graphical display called a *histogram.* Histograms display the shape, dispersion, and central tendency of the distribution of a population or sample of data. The extrusion histogram shown in Figure 2.2 has the same 11 ranges that are 0.005 inch wide, begin at 0.975 inch, and continue up to 1.030 inches. The ranges have been numbered from 1 to 11 to act as place-holders for the data points that fall within each range. The vertical axis represents the frequency of data occurrence. The histogram shows a clear picture of the shape, dispersion, and central tendency of the sample data.

**Table 2.2** The frequency distribution table

| *Range no.* | *Range boundaries* | *Frequency* |
|---|---|---|
| 1 | $0.975 \leq y < 0.980$ | 0 |
| 2 | $0.980 \leq y < 0.985$ | 4 |
| 3 | $0.985 \leq y < 0.990$ | 8 |
| 4 | $0.990 \leq y < 0.995$ | 17 |
| 5 | $0.995 \leq y < 1.000$ | 15 |
| 6 | $1.000 \leq y < 1.005$ | 16 |
| 7 | $1.005 \leq y < 1.010$ | 21 |
| 8 | $1.010 \leq y < 1.015$ | 9 |
| 9 | $1.015 \leq y < 1.020$ | 7 |
| 10 | $1.020 \leq y < 1.025$ | 2 |
| 11 | $1.025 \leq y < 1.030$ | 1 |

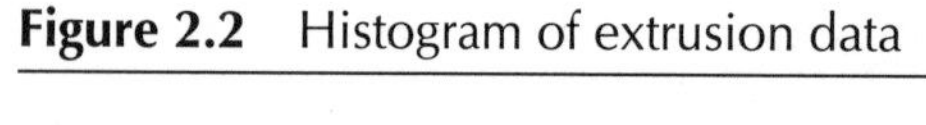

**Figure 2.2** Histogram of extrusion data

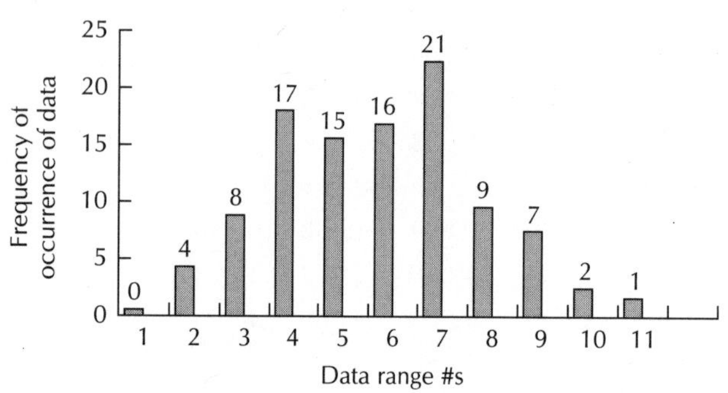


## 2.3 Quantitative Methods of Data Analysis

The graphical display of the central tendency and distribution width of a set of data is a useful analysis tool. But there is much more to be gained by quantifying the nature of the data—specifically, its variability and central tendency. A quantitative description of the data using just a few summary statistics allows us to develop the key quality measures, the quality loss and the signal-to-noise ratio, that are extremely powerful in design optimization. Let us leave the graphical methods of analyzing a data set for now and focus on applying mathematics to quantify the data in terms of sample statistics.

### 2.3.1 The Mean

The central tendency of the sample data is given by the *sample mean,* $\bar{y}$, shown in Equation 2.1.

$$\bar{y} = \frac{1}{n} \sum_{i=1}^{n} y_i \qquad \textbf{(2.1)}$$

This expression states that a measure of the sample mean or average of all the sample data can be calculated by adding up ($\Sigma$) all the individual data points ($y_i$) and then dividing by the total number of individual data points ($n$).

### 2.3.2 The Variance

The variability of sample data is given by the *sample variance,* $S^2$, shown in Equation 2.2:

$$S^2 = \frac{1}{n-1} \sum_{i=1}^{n} (y_i - \bar{y})^2 \qquad \textbf{(2.2)}$$

For the five random samples shown in Figure 2.1:

$\bar{y} = \frac{1}{5}(1.6 + 2.6 + 1.4 + 2.1 + 3.2) = 2.18$ . . . *the sample mean.*

For the five samples from Figure 2.1:

$$S^2 = \frac{1}{5-1}\sum_{i=1}^{5}\left[(1.6-2.18)^2 + (2.6-2.18)^2 + (1.4-2.18)^2 + (2.1-2.18)^2 + (3.2-2.18)^2\right]$$

$S^2 = 0.542$ . . . *the sample variance.*

This expression states that a measure of the dispersion or scatter of the data is based on the average of the squared deviations of the individual data points from the sample mean. A deviation, also referred to as a residual, $(y_i - \bar{y})$, is given by the difference between the individual data point, $y_i$, and the sample mean, $\bar{y}$. The value $n - 1$ has been shown to give a more accurate value of the sample variance (i.e., closer to the actual value of the variance for the complete population). If this is not done, the variance tends to understate what would actually be the case if many samples (>30) were taken[1] [F5 p. 73, B2].

The definition of the population variance is (where $\mu$ is the population mean):

$$\sigma^2 = \frac{1}{n}\sum_{i=1}^{n}(y_i - \mu)^2 \qquad \textbf{(2.3)}$$

## 2.3.3 The Standard Deviation

Another measure of variability in the data is commonly used. The square root of the variance is a statistic called the *sample standard deviation, S:*

$$S = \sqrt{S^2} \qquad \textbf{(2.4)}$$

The Standard Deviation is the *root* mean square deviation of the data from the sample mean. For the five samples from Figure 2.1, $S = 0.736$.

## 2.3.4 A Note on the Statistical Symbols Used in This Text

It is rare that complete population data are collected during a robustness experiment. Consequently, this text uses the sample mean $(\bar{y})$ and the sample variance $(S^2)$ as statistics for the calculation of the Loss Functions (Chapter 3) and the signal-to-noise ratios (Chapters 4–6). In some texts on Taguchi methods the symbols for population parameters ($\mu$ and $\sigma$) are used for sample statistics (cf. Section 5.9, A Note on Notation). In the literature, particularly from Japan, including Dr. Taguchi's writings, other symbols such as $V_e$ are used for the sample variance. The authors have chosen to use the standard

1. Note that it makes little difference from an engineering point of view whether you use $n - 1$ in the variance calculation (sample statistics) or $n$ in the variance calculation (population statistics), as long as you choose one convention and use it for all calculations. The engineering goal in robust design is to minimize the variation; the exact value obtained is not critical to any decision-making process discussed in this book.

statistical notation for sample statistic symbols ($\bar{y}$ and $S^2$) exclusively in this text so that there is no confusion as to the nature of the data (sample vs. population) being considered. When using a handheld calculator or spreadsheet program, find out the convention employed for the variance. The authors have found that the standard notation is often not used. WinRobust Lite, the software package included with this book, uses sample statistics in all the calculations.

### 2.3.5 The Coefficient of Variation

When the nature of the design requires a response that can have variability on either side of a target value, the mean and the standard deviation can have a meaningful relation to one another. This relationship can be expressed numerically as a ratio between the standard deviation and the mean. This ratio is called the coefficient of variation (COV):

$$\mathrm{COV} = \frac{S}{\bar{y}} \tag{2.5}$$

The COV expresses the variation in the data as a fraction of the mean value. This permits one to say, for example, that the variation may be 1% or 10% or 50% of the mean response being measured. This highlights the relative size of the variation with respect to the mean. Knowing this relationship enhances the engineer's understanding of the nature of the variability taking place in the experiments.

Which of the two measures of variation, $S^2$ and COV, should be used depends upon the nature of the system being optimized. Which one to use is an engineering decision. It is helpful to examine the nature of the relationship between variability and the mean.

If the COV is *constant* as the mean rises and falls in value for a particular factor within the experimental design being studied, then the standard deviation is proportional to the mean. This means that optimizing the system based on only the variance may be in conflict with putting the mean value onto the desired target for a particular factor. This can greatly complicate the analysis and can be avoided by optimizing the COV, which is independent for certain factors that can be used for adjustment of the mean.

If the COV is *changing* as the mean increases or decreases in value as the experimental design space is studied, then another quality metric, such as $S^2$, should be used to evaluate the system's quality. The point here is that the type of data to be measured and how that data should be analyzed is entirely an engineering decision. Much more will be said later concerning these issues.

### 2.3.6 The Use of Distributions in Robust Design

It is beyond the scope of this book to develop the details concerning the statistical analysis of data. Nor is it necessary to go much further with basic statistics for Robust Design. Just the few specific statistical quantities given in Equations 2.1–2.5 are needed to define the metrics used for analyzing the robustness of a product or process in later chapters. Understanding these fundamental statistical measures is a requirement for properly applying the methods Dr. Taguchi has developed.

How to commercialize products in spite of the presence of variability is considered at length in this book. Most things have a definite "pattern" to their variability. The specific shape of the distribution of a sample of data dictates how the data can and should be analyzed. Many products are made up of components that vary randomly. Their distribution forms what is called the *normal or Gaussian distribution,* shown in Figure 2.3.

This pattern can represent how manufacturing variability affects part dimensions. It can also represent how external sources of customer use and external environmental conditions can affect a prod-

**Figure 2.3** The normal distribution

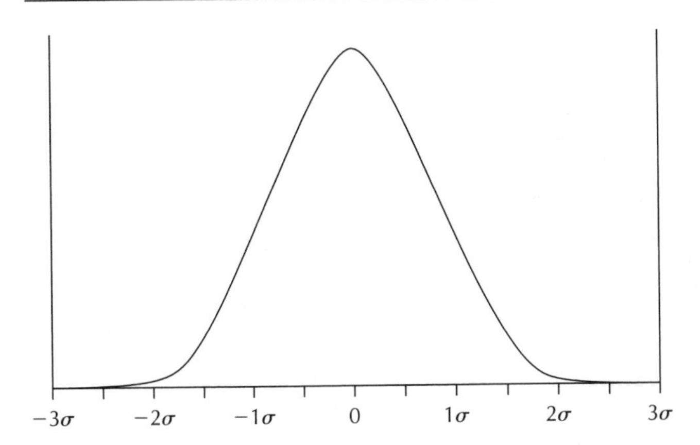


uct's performance. There are numerous other types of distributions found in nature. It is important for you to know that they exist, even though their description is beyond the scope of this book [E3].

The two basic sample statistics ($\overline{y}$ and $S^2$) used to describe the distribution's central location and width are fundamental to quantifying variability. In Robust Design, it is not necessary to assume that the data are coming from a normal distribution, or even from a distribution that is a reasonable approximation of a normal distribution. The simple statistics discussed here are particularly useful because they are good measures regardless of the exact distribution shape. In fact, through the application of noise factors to purposefully induce variation, it is common to apply these statistics to non-Gaussian distributions.

## 2.4 An Introduction to the Two-Step Optimization Process

In Robust Design, emphasis is placed on how a product is performing relative to its *targeted* response. Often, nominal control-factor set points for a design are determined through engineering analysis. A prototype is constructed and tested to see if the mean response is close to the targeted response. Usually the mean response is quite close to the target, but displays relatively large amounts of variability when the prototype is exposed to certain noises. If the sample mean response is not on the target, action will need to be taken. It is difficult and costly to reduce variability *after* the design parameters have been used to put the mean response on the target. In order to keep the mean close to the target in the actual product, the variation in the response must first be reduced. Once the optimal control factor set points are identified for variation reduction, the engineering focus can shift to adjusting the mean response onto the target the customer expects for the product. This is called the Two-Step Optimization Process, Figure 2.4.

Figure 2.4 illustrates how the distributions, shown here graphically, of a key performance measure are used in the two-step optimization process. The initial distribution, labeled A in the figure, describes the performance of the system at its initial set points. Rather than shifting this distribution onto the target and then working to reduce variation, experimentation is done to find the narrowest distribution possible, *Step 1*. This is illustrated by the distribution labeled B in the figure. The experimentation also reveals which factors can be used to adjust the mean with the least effect on the width of

**Figure 2.4** Shifting the narrow distribution onto the target

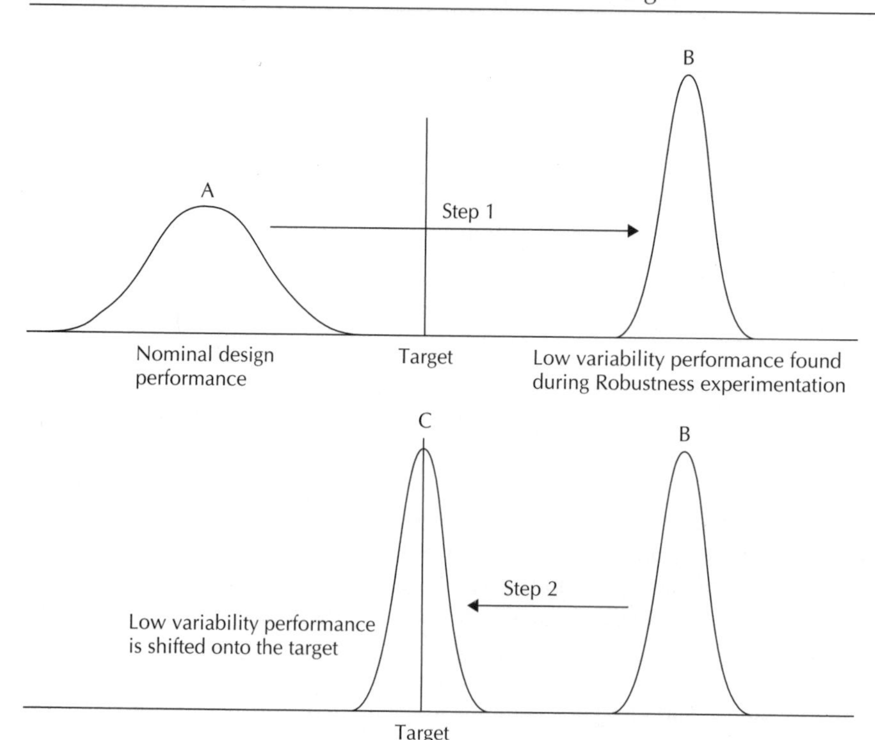


the distribution. Such a factor is used in *Step 2,* to put the distribution onto the target, labeled C in the figure.

Thus, the two-step optimization process breaks the complex problem of achieving on-target performance into two simpler problems. The first objective of robustness experimentation is to produce data that will permit the engineer to find control-factor set points that produce the narrowest response distribution in the presence of noise. Frequently this combination of set points is some distance from the desired target. The second objective is to identify one factor that is good at allowing the mean performance to be adjusted onto the target while marginally affecting the width of the distribution being adjusted. The fact that the data can be used to tell the engineer when certain control factors have a constant COV, and other control factors do not possess a constant COV, helps simplify the two-step optimization process. This is a major theme in Chapters 4 and 5.

## 2.5 Summary

Robust Design procedures use experimentation to produce data that define the distribution width of key performance characteristics while noise is active during the experiment. The engineer's job is to discover ways to narrow these distributions and to put them on target in the design or process being optimized. Thus, the statistical techniques for describing data distributions are key concepts for Robust Design. These include graphical methods such as the histogram, but much more important are the quantitative descriptions, e.g., the mean, the variance, the standard deviation, and the coefficient of

**Table 2.3** Formulas for sample summary statistics

| *Name* | *Symbol* | *Formula* |
|---|---|---|
| Mean | $\bar{y}$ | $\bar{y} = \frac{1}{n}\sum_{i=1}^{n} y_i$ |
| Variance | $S^2$ | $S^2 = \frac{1}{n-1}\sum_{i=1}^{n} (y_i - \bar{y})^2$ |
| Standard Deviation | $S$ | $S = \sqrt{S^2}$ |
| Coefficient of Variation | COV | $\text{COV} = \frac{S}{\bar{y}}$ |

variation. These are summarized in Table 2.3. The quantitative statistics, which are calculated from sample data, are necessary for the application of design of experiments to the optimization process. The responses that will be analyzed, especially the signal-to-noise ratio, are based upon these fundamental statistical quantities.

In concluding this chapter on data and how the engineer can begin to process it, a few words are in order on the general strategy robustness practitioners employ in producing useful data. When conducting experiments to gather data, how variability is *allowed* to occur is crucial to effective optimization (see Chapter 9). In order to optimize a design or process, the same sources of variation that occur during component manufacturing, product assembly, and customer use must be present in the experiment. Taking random samples of data without attempting to stimulate the various types of noises that can occur is an inadequate practice. In this text, random samples are not a priority. The focus is on producing and analyzing strategically biased samples of data where the engineer is intentionally forcing variability to occur by controlling many different kinds of noise during the experimentation.

## Exercises for Chapter 2

1. What is meant by the term "continuous variable"?
2. What is meant by the term "discrete variable"?
3. Define what is meant by sample data and suggest an example of sample data.
4. Construct a frequency distribution from the following data:

| | | | | | | | |
|---|---|---|---|---|---|---|---|
| 2 | 4 | 1 | 3 | 3 | 6 | 5 | 4 |
| 1 | 5 | 2 | 4 | 2 | 1 | 3 | 3 |
| 7 | 5 | 4 | 2 | 1 | 3 | 7 | 6 |
| 3 | 6 | 4 | 2 | 4 | 3 | 2 | 3 |
| 3 | 1 | 2 | 2 | 3 | 8 | 1 | 2 |
| 4 | 1 | 2 | 5 | 3 | 6 | 4 | 3 |

5. Calculate the mean and standard deviation for the data set provided in exercise 4.
6. What is the coefficient of variation for the data set provided in exercise 4?
7. Explain what is meant by the two-step optimization process.
8. What is the difference between $\bar{y}$ and $\mu$, $S$ and $\sigma$, and $S^2$ and $\sigma^2$?
9. What three characteristics do all distributions possess?
10. What is the difference between a sample statistic and a population statistic?
11. What is a histogram and what value does it add to the data analysis process?

12. Why is $n - 1$ used in the computation of the sample variance and sample standard deviation?
13. Why is it not desirable to force a prototype design to hit its targeted performance first and then worry about reducing the variability in the production design later?
14. Explain the logic behind conducting experiments where variability is intentionally induced.
15. In Figure 2.4, why is the distribution obtained some distance away from the target preferred?

CHAPTER 3

# The Quality Loss Function

## 3.1 The Nature of Quality

In Chapter 1 we discussed in qualitative terms the concept of a product's quality in relation to the costs to customers, manufacturers, and society due to off-target product performance (quality loss). Since the same product is used by many different types of customers, in many different ways, and under many different environmental conditions, it can be difficult to measure a product's quality loss. However, Chapter 2 illustrates that it is possible to measure a product's performance variation using data generated under controlled noise conditions. How do such data relate to the quality loss? Can the quality loss be estimated as performance variation is estimated?

It is important to be able to place a dollar value on quality to decide whether alternative product or process designs would be more cost-effective. Ideally, there should be a rapid and inexpensive engineering process capable of determining the cost of quality (loss). The process should yield reasonably accurate results close to the target where most of the action is occurring.

## 3.2 Relating Performance Distributions to Quality

Consider an example of three different part suppliers. Imagine that you are playing the role of a product manufacturer whose product quality and reputation are dependent on the quality of the subassemblies you buy from your suppliers. There are three suppliers who want to do business with you—which, by the way, means they are doing business with your customers. They are producing a subassembly that must perform on target in order that the rest of the product function properly. You are sent in as part of a team of quality auditors to measure how capable each supplier actually is. Your task is to construct representative distributions of performance based on 30 samples. The distributions found are shown in Figure 3.1.

**Figure 3.1** Performance distributions for three suppliers

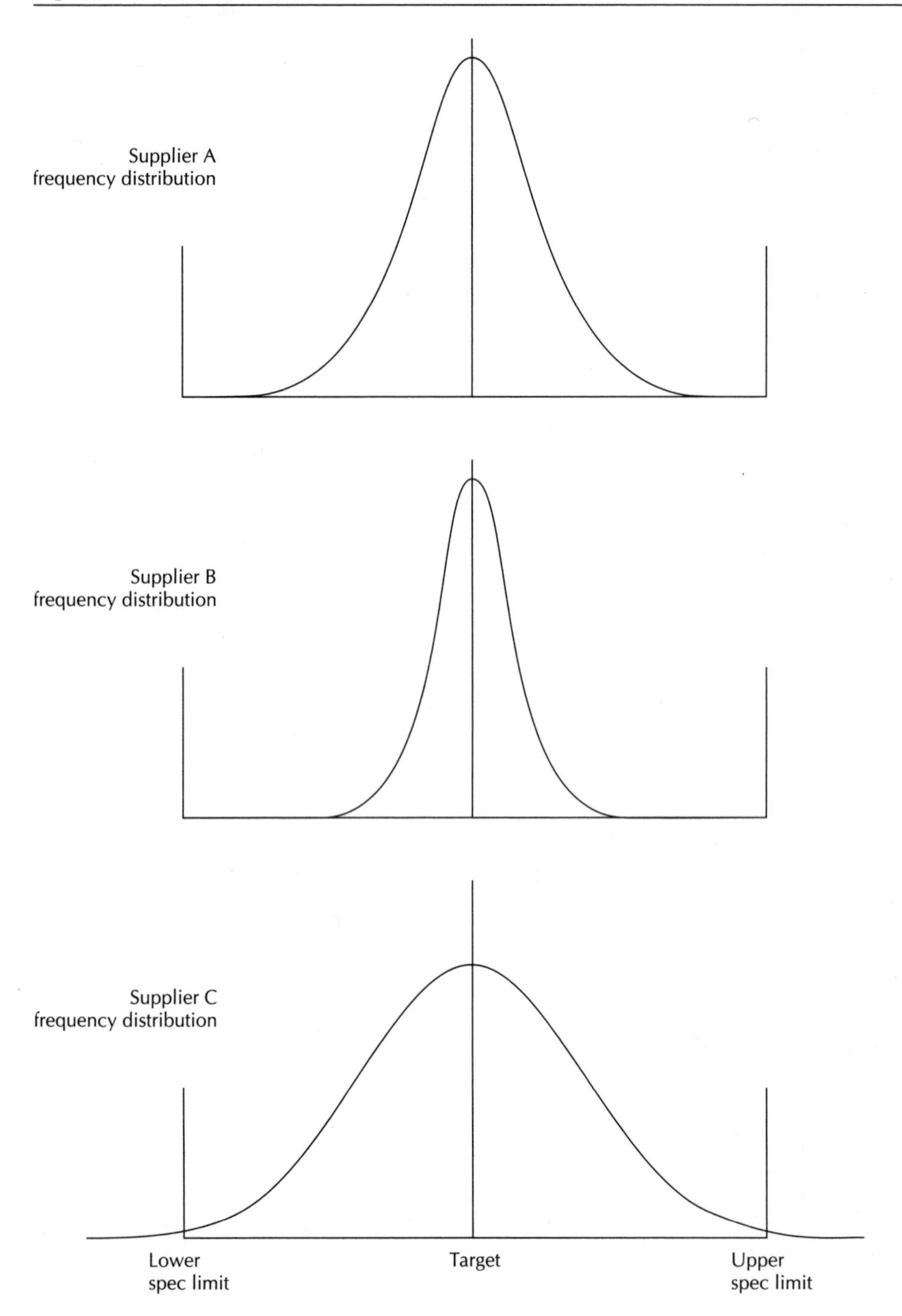

As you can see, all three suppliers are able to produce most of the subassemblies within the specification limits. All the suppliers have "methods" to assure that no out-of-specification subassemblies are delivered to our plant. In fact, the worst performer, Supplier C, has more than 99% of the subassemblies within specifications. Which supplier can best help to satisfy your customers? Most people would focus on the percentage of subassemblies that fall outside of the specification limits. It is quite common in most industries to consider only the costs associated with units that are outside the specification limits and focus on what can be done to minimize that number. Instead, consider *all* subassemblies, those within specification as well as those outside of the specification limits. This point of view is not just one of getting all units within the specification limits—but of getting all units *on-target*. This is the first of many key insights you need to internalize in order to become an effective practitioner of the methods of Robust Design.

Measuring only the number of parts that are within specification is a misleading and incomplete method of quality measurement. It implies that all products that meet specifications are equally good. It also implies that all products that are found to be out of specification are equally bad.

A major problem with "within specification" type measures of quality is that, from a customer's perspective, a product that is just within specification is not as good as a product that is perfectly on target. Let's think this through by considering how our educational system uses a pass/fail criterion for examinations. Is there really much difference between a student who passes a test with a 65% grade and a student who fails the test with a 64% grade? In a similar fashion, is there much difference between a product that is barely in specification and one that is barely out of specification? On the other hand, consider the difference between a student who scores a 100% grade on a test and the student who also passes, but with a 65% grade. There is a substantial difference in performance between these two students, both of whom scored passing grades! Similarly, there is a very substantial difference in quality between a product that is on target and one that is barely within specification.

A study at Ford Motor Company [E1] showed that transmissions with parts close to their specified target values functioned better and operated more quietly than units made with parts that are technically "within specification" but had dimensions that are spread farther away from their target values. A quantitative way is needed to express the loss that customers incur when the products perform off-target. This requires a metric that is sensitive enough to quantify quality loss for design elements that are within specifications.

Tolerances are a necessary design entity. They are useful for establishing acceptability limits when tied back to customer requirements. But how they are established and how they are used in an approach to product quality has a performance and cost impact on your products. All too often, companies have a quality strategy that allows component and performance variation with a relatively flat frequency distribution between the upper specification limit (USL) and the lower specification limit (LSL), as shown in Figure 3.2.

**Figure 3.2** Flat distribution between specification limits

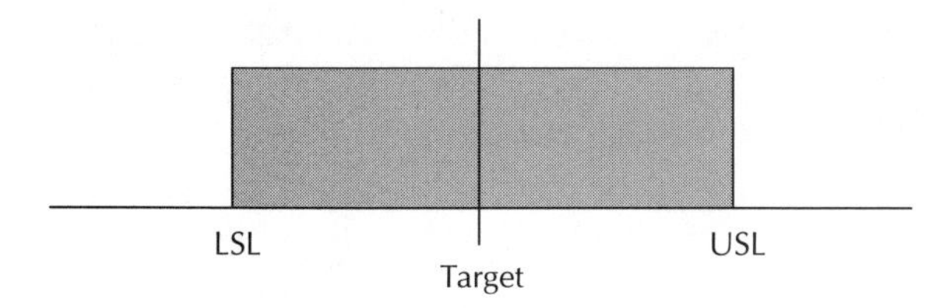

There are several types of behavior that are responsible for such a result:

1. Companies maximize productivity by shipping everything made between the limits without regard to quality differences.
2. Known process drifts are allowed to occur and manufacturing strategies are devised to maximize the output between the specification limits. For example, in machining, where tool wear causes predictable changes in a dimension, operators may "set up" at one end of the range and allow the process to produce parts until the other end is reached, thus achieving a flat distribution between the starting and ending points. Another example is in chemical processing or photofinishing, where solvent recycling causes a uniform drift from the initial "pure" conditions until the limit is reached where the solvents must be renewed.
3. Out-of-spec parts are reworked after inspection to bring them within specification. Many of the reworked parts are never put on target.
4. Variation between control limits is considered acceptable. Statistical process control (SPC) is used to distinguish between common-cause and special-cause sources of deviation from the target. Common-cause refers to unknown or uncontrolled sources of variation that are always present and accepted. Special-cause refers to sources of variation that are correctable by elimination or compensation *as they are detected*. A basic SPC strategy is to choose control limits. Any random variation within the control limits is judged to be due to common causes and is thus treated as *acceptable*. Only variations outside of these limits or lacking randomness trigger corrective action. Thus, SPC necessarily creates a distribution whose width is close to the control limits and whose shape depends on the nature of the common causes (noise factors). The control limits are set according to statistical and cost guidelines as well as quality requirements.

All of these examples share a manufacturing philosophy that aims to maximize productivity, maximize materials utilization, and minimize defects, rejects, and waste. In principle, these are all laudable goals. But there is nothing that explicitly encourages on-target manufacturing. Such companies do not *explicitly* reward or encourage their staff to work beyond good yield to produce on-target results.

Companies that focus on a quality strategy based on meeting target values exhibit a very different kind of frequency distribution, as shown in Figure 3.3.

There are several types of behaviors that are responsible for such a result:

1. A serious effort at fostering statistical literacy for everyone with product performance responsibility. From operators through engineers to management, the ability to acquire and analyze data using the statistical measures shown in Table 2.3 is a critical skill requirement.
2. The measurement and display of distributions using tables and histograms, much more so than just run charts, clearly indicate the quality of production processes.
3. The practice of Six Sigma. Achieving very high process capability and using measures such as $C_{pk}$, which penalize a process for being off target, force on-target performance. Six Sigma is discussed in detail in Chapter 18.
4. Maintaining manufacturing process capability databases. Such records assure that the correct process is specified for a required level of quality and cost performance. Without such information, there is a tendency to reduce cost *now,* resulting in quality problems *later.*
5. The use of design of experiments (DOE). Undoubtedly, this is the most powerful statistical tool for improving product quality because of its preventative nature. It is not possible to

**Figure 3.3** An improved performance distribution

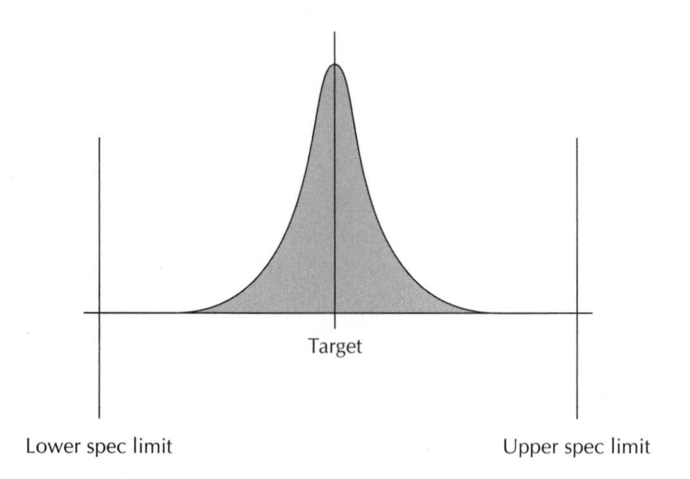


eliminate all problems purely through analysis. Therefore, the empirical methods of DOE constitute critical engineering skills. Parts II and III in this book demonstrate and explain the application of DOE to Robust Design.

6. The use of Dr. Taguchi's methods. Taguchi methods are among the most powerful engineering tools for improving product quality by the application of on-target engineering.
7. An understanding of the true cost of quality. Introduced in Chapter 1 as quality loss, the true cost of quality is poor profits, loss of customers, and loss of whole industries. The companies that have suffered such a demise do not go out of business because of out-of-spec parts. They are displaced by companies who understood that costs go down as quality improves and that quality is measured between the specification limits. Profits go up as on-target quality improves because of reduced inspection, reduced warranty and repair costs, better efficiency, and more competitive products.

All of these examples share a manufacturing philosophy that aims to maximize quality beyond simply measuring conformance to specifications. Problem prevention is greatly emphasized as being necessary to keep processes on target. The best of these companies *explicitly* reward and encourage their staff to produce on-target results.

## 3.3 The Step Function: An Inadequate Description of Quality

From the preceding discussion, a lot can be understood about a company's quality practices, reward systems, and assumptions from studying the type and width of their products' performance distributions. There is a distinct *quality culture* at companies that are practicing on-target engineering. The value system at companies that are satisfied by within specification performance can be described by Figure 3.4.

**Figure 3.4** Step function describing quality

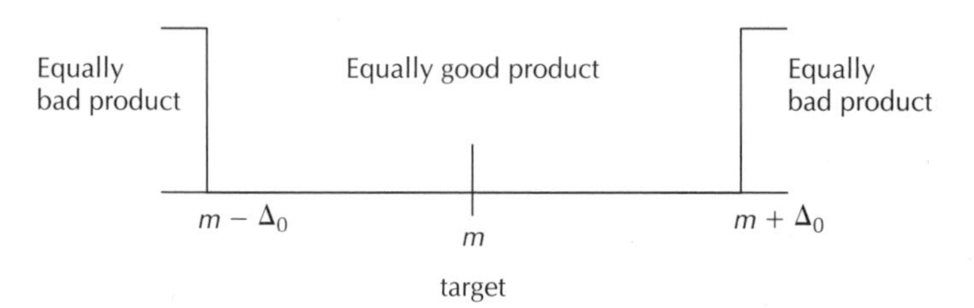



Engineering specifications often take the form of a target value, $m$, with a bilateral tolerance written as $m \pm \Delta_0$. It is an error to interpret such a specification to mean that all values in the range of $m - \Delta_0$ to $m + \Delta_0$ are of *equally good* quality for the customer, and that all values outside this range are of *equally bad* quality. The step function is an inadequate description of quality.

## 3.4 The Customer Tolerance

*Tolerances* are defined as the limit at which some economically measurable action is taken. On-target engineering does not eliminate the need for tolerances. How far from the target can the product performance get before the customer would reject the product? How far from the target can a component be before manufacturing should reject the unit? These are important quantities for deciding actions.

The *customer tolerance,* $\Delta_0$, corresponds to the point at which a significant number of customers take economic action because of off-target performance. The concept of a customer tolerance limit can be demonstrated with an example of human behavior. The parabolic curve is universally applicable to the frequency distribution of individual preferences for the level of a nominal target quality characteristic. For example, consider a hypothetical poll taken of people's inclination to adjust a thermostat that is controlling the room temperature against ambient conditions. Each person states at what ambient temperature deviation from 70°F they would become uncomfortable and adjust the thermostat to compensate. Figure 3.5 represents the typical response that a large group (>20 people) would generate.[1]

Notice that as the ambient temperature drifts farther and farther away from 70°F, more and more people are inclined to adjust the thermostat. As a result, the general form of a parabola emerges. One commonly used standard for the customer functional limit is the deviation at which 50% of the customers *take action* to correct the deviation. This is shown by the dotted lines in Figure 3.5.

## 3.5 The Quality Loss Function: A Better Description of Quality

Companies that are practicing on-target engineering have an alternative approach to the limitations the step function exhibits as a measure of quality. The quadratic Quality Loss Function (*QLF*), or simply the *quadratic loss function,* was developed by Dr. Taguchi to provide a better estimate of the monetary

1. This example was originally introduced by Peter T. Jessup, "The Value of Continuing Improvement," IEEE Conference, 1985.

**Figure 3.5** Average preference distribution

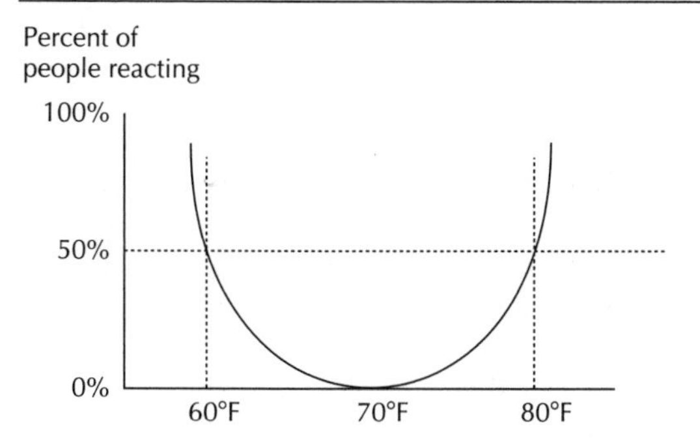


loss incurred by manufacturers and consumers as product performance deviates from its target value. The quadratic quality loss function, Equation 3.1, approximates the quality loss in a wide variety of situations.

$$L(y) = k(y - m)^2 \tag{3.1}$$

where $L(y)$ is the loss in dollars due to a deviation away from targeted performance as a function of the measured response, $y$, of the product; $m$ is the target value of the product's response; and $k$ is an economic constant called the *quality loss coefficient.*

Figure 3.6 shows the quality loss function. It should be compared to Figure 3.5. At $y = m$, the loss is zero and the loss increases the further $y$ deviates from $m$. The quality loss curve typically represents the quality loss for an average group of customers. The quality loss for a specific customer would vary depending on that customer's tolerance and usage environment. However, it is not necessary to derive an exact loss function for all situations. That would be too difficult and not generally applicable.

The quality loss function can be viewed on several levels:

1. As the unifying concept of quality and cost, allowing us to see clearly the underlying philosophy driving on-target engineering.
2. As a function that allows us to relate economic and engineering terms in one model.
3. As an equation that allows us to do detailed optimization of all costs, explicit and implicit, incurred by the firm, its customers, and society through the production and use of a product.

This empirical observation for the shape of the quality loss function can be derived formally [T8]. The value $y$ denotes the actual performance as measured or perceived by the customer. The value

**Figure 3.6** The quadratic loss function

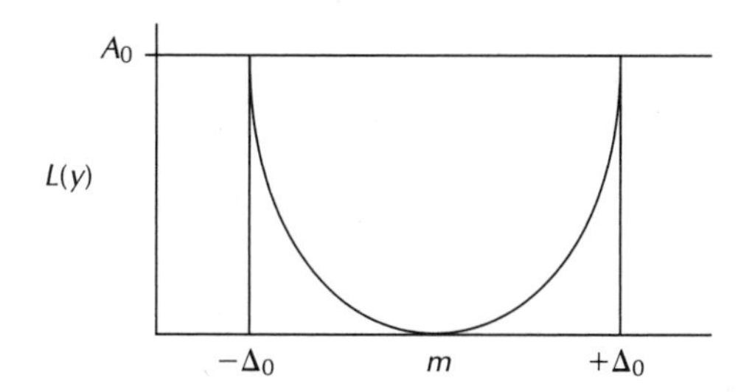

$m$ represents the target performance the customer has paid to receive. The function $L(y)$ is given by a Taylor series expansion about $y = m$:

$$L(y) = L(m) + [L'(m)/1!][y - m] + [L''(m)/2!][y - m]^2 + \dots \quad \textbf{(3.2)}$$

When the performance, $y$, is on the target, $m$, the quality loss should be zero.[2] Thus, the first term is zero: $L(m) = 0$. When the first derivative is taken at the target, the second term in the expansion is zero because the loss function is a minimum at $y = m$: $L'(m) = 0$. This leaves the third term, $[L''(m)/2!][y - m]^2$, as the leading term in the expansion. Assuming that this expansion is applied to situations where $y$ is close to $m$, the defining function for customer loss as performance, $y$, deviates from the target, $m$, is given by Equation 3.3.

$$L(y) \approx [L''(m)/2!][y - m]^2 \quad \textbf{(3.3)}$$

The higher-order terms of the expansion are inconsequential in practice, because the loss function is generally applied close to the target, where higher-order terms are relatively small. The term $[L''(m)/2!] = k$, the quality loss coefficient, is an economic constant of proportionality. Thus, we are left with the quadratic quality loss function given in Equation 3.1.

## 3.6 The Quality Loss Coefficient

The quality loss coefficient, $k$, is determined by first finding the functional limits or customer tolerance for $y$, the measured response. The *functional limits*, $m \pm \Delta_0$, are the points at which the product would fail or produce unacceptable performance in approximately half of the customer applications, as shown in Figure 3.5. These represent performance levels that are equivalent to the average customer tolerance. In Section 3.4, tolerances are defined as the limit at which some economically measurable action takes place. In this case the product has essentially failed, so the sum of the economically measurable consequences of failure, $A_o$, becomes the value of the quality loss function at $m \pm \Delta_0$, as shown in Figure 3.6. These consequences include repair, replacement, loss of use, waste, and all the other categories discussed earlier. Let the total of all the losses at $m \pm \Delta_0$ be equal to $A_0$ in dollars:

$$L(y) = A_0 \text{ at } y = m \pm \Delta_0 \quad \textbf{(3.4)}$$

Substituting the functional limits $m \pm \Delta_0$ and the total losses $A_0$ into Equation 3.1, the quality loss coefficient is found to be:

$$k = A_0/(\Delta_0)^2 \quad \textbf{(3.5)}$$

Remember that $A_0$ is the cost to replace or repair the product, including the losses incurred by the manufacturer and customer, as a consequence of off-target performance. Some typical situations that figure into repairing or replacing a product are the following:

- Loss due to lack of access to the product during repair
- The cost of parts and labor needed to make the repair
- The costs of transporting the product to a repair center

2. This assumes that the product satisfies its intended customer base when performing on-target.

Regardless of who pays for the losses—the customer, the manufacturer, or a third party—all losses should be included in $A_0$. Substituting Equation 3.5 into Equation 3.1 defines the quality loss function:

$$L(y) = \frac{A_0}{\Delta_0^2}(y - m)^2 \tag{3.6}$$

## 3.7 An Example of the Quality Loss Function

A spring (which stores and releases energy) is used in the operation of a camera shutter. The process used to wind and form the dimensions necessary to constrain the spring wire has a certain level of variability.

Variability in the forming process results in variability of the measurable parameter called the spring rate (measured in ounces of force per inch of deflection). Assume that the target spring constant is $m = 0.5$ oz./in. The shutter speed varies because of several factors—one of which is the spring rate. Assume that the functional limits for the spring constant are $m \pm 0.3$ oz./in. This means that half the customers who purchase a camera with a spring constant of $m \pm 0.3$ oz./in. would consider the camera defective because of improperly exposed pictures. Let's also assume that the average cost for repairing or replacing a camera with unacceptable shutter speed due to an off-target spring constant is \$20. Thus, the customer loss is $A_0 = \$20$. By substituting these values into the quality loss function, we find:

$$L(y) = \frac{\$20}{0.3^2}(y - 0.5)^2 = \$222(y - 0.5)^2 \tag{3.7}$$

The loss function is generally used in establishing manufacturing tolerances during the tolerance design process, as illustrated in the case study at the end of this chapter. It is also used to help define the signal-to-noise metrics used in parameter design. This is why it is thoroughly developed in this book.

Equation 3.7 can be used to calculate the expected quality loss, in dollars, for various springs. For example, the quality loss to customers who purchased a camera with a spring constant of 0.25 is:

$$L(0.25) = \$222(0.25 - 0.5)^2$$
$$L(0.25) = \$222(-0.25)^2$$
$$L(0.25) = \$13.88$$

For a camera with a spring constant of 0.435, the quality loss is:

$$L(0.435) = \$222(0.435 - 0.5)^2$$
$$L(0.435) = \$222(-0.065)^2$$
$$L(0.435) = \$0.94$$

## 3.8 The Types of Quality Loss Functions

There are four types of quality loss functions. Three are derivatives of the general case called the nominal-the-best case.

### 3.8.1 The Nominal-the-Best Case (NTB)

This is the case just used for the description of the loss attributed to the off-target spring force. The loss function is shown in Figure 3.6; Equation 3.6 is the nominal-the-best function. In the NTB case the measured response, $y$, always has a specific target value, $m$. The quality loss is equally undesirable on either side of the target.

Here are some examples of the nominal-the-best type of problem:

- Control of aerosol flow from a spray can
- Boring an engine cylinder to an aim diameter
- Controlling the diameter of a filament for a light bulb
- Controlling the texture of an ice cream product
- Controlling the gain of an op-amp
- Maintaining the part geometry of an injection-molded part
- Creating a particular hue in mixing paint

### 3.8.2 Average Quality Loss for Nominal-the-Best Cases

The nominal-the-best loss function can be applied to just one part or product or to the average loss associated with more than one unit. The concept of the average loss is central to the signal-to-noise ratio concept. Therefore, it is worth the effort of deriving the average loss and seeing how it decomposes into two parts: the contribution due to the mean being off target and the contribution due to the variance.

The average quality loss is found by defining an average of the measure of off-target performance $(y - m)^2$, which is the argument of the loss function Equation 3.1. The average off-target value is referred to as the *mean square deviation* or *MSD*. The deviation referred to is from the target $m$. The average loss, $\overline{L(y)}$, is given by

$$\overline{L(y)} = k(\text{MSD}) \tag{3.8}$$

where

$$\begin{aligned}\text{MSD} &= \frac{1}{n}\left[(y_1 - m)^2 + (y_2 - m)^2 + (y_3 - m)^2 + \ldots + (y_n - m)^2\right] \\ &= \frac{1}{n}\sum_{i=1}^{n}(y_i - m)^2 \\ &= \frac{1}{n}\sum_{i=1}^{n}(y_i^2 - 2y_i m + m^2)\end{aligned}$$

If we apply the definition of the mean (Equation 2.1) to the second term in the parentheses, then:

$$\text{MSD} = \frac{1}{n}\sum_{i=1}^{n}(y_i^2) - 2\bar{y}m + m^2$$

where $\bar{y}$ is the average value (mean) of the individual quality characteristic values, $y_i$. By adding and subtracting equal terms, the MSD can be expressed as the sum of two squares.

$$\text{MSD} = \frac{1}{n}\sum_{i=1}^{n}(y_i^2) - \bar{y}^2 + \bar{y}^2 - 2\bar{y}m + m^2$$

Using an identity (see problem 18 at the end of this chapter), the final equation is obtained:

$$\text{MSD} = \frac{1}{n}\sum_{i=1}^{n}(y_i - \bar{y})^2 + (\bar{y} - m)^2 \qquad \textbf{(3.9)}$$

Notice that the MSD is now seen to be made up first of a sum that expresses the average square deviation from the *mean,* and second of a term that expresses the deviation of the *mean* from the *target.* This decomposition is seen again and again as the mathematics of Robust Design is explored. The sum $(1/n)\sum_{i=1}^{n}(y_i - \bar{y})^2$ is the population standard deviation, $\sigma^2$. Thus:

$$\text{MSD} = \sigma^2 + (\bar{y} - m)^2 \qquad \textbf{(3.10)}$$

Equation 2.2 gives the expression for the variance, $S^2$, that is based on a sample from the total population. The variances are related by:

$$S^2 \approx \sigma^2 \text{ for } n > 30 \qquad \textbf{(3.11)}$$

Therefore, the loss function for $n$ units is

$$L(y) \approx k[S^2 + (\bar{y} - m)^2] \qquad \textbf{(3.12)}$$

Equation 3.12 can be used to calculate the average quality loss for a set of products. Returning to the spring problem, let there be two machines making the springs, one new spring winder and an older model that has a good deal of wear and tear on it. Again, the target value is $m = 0.5$ oz./in. The average loss function is:

$$L(y) = \$222[S^2 + (\bar{y} - m)^2]$$

| *Winder* | *Data* | $S^2$ | $\bar{y}$ | $(\bar{y} - m)^2$ | $L(y)$ |
|---|---|---|---|---|---|
| New | 0.37, 0.41, 0.37<br>0.43, 0.39, 0.35<br>0.40, 0.36 | 0.0007 | 0.385 | 0.0132 | 3.08 |
| Old | 0.55, 0.67, 0.70<br>0.54, 0.41, 0.32,<br>0.46, 0.66 | 0.0184 | 0.539 | 0.0015 | 4.41 |

Here it can be seen that the new machine has the lower variance but is, on average, fairly far from the target. The old machine gives the impression that it is doing pretty well by averaging around the target—but it has a lot of variability. As a result, the older machine has the greater loss. It should be possible to further reduce the loss for the new machine by adjusting the mean onto the target. It is easier to shift the tight variance onto the target than it is to tighten a widely varying distribution.

With this form of the loss function available, several key points can now be made that form the basis for a strategy that drives all our engineering optimization endeavors. To reduce loss, the MSD must be reduced. This can be done in two ways:

1. *Reduce* the variability that is causing the dispersion of the data about $\bar{y}$, thus minimizing $S^2$.
2. *Adjust* the average response $\bar{y}$ to fall on the target $m$, $(\bar{y} - m)^2 = 0$.

These concepts are used in Chapter 4 to derive the signal-to-noise ratio.

### 3.8.3 The Smaller-the-Better Case (STB)

Some responses never have a *negative* value. In addition, their targeted response is ideally *zero.* These are referred to as smaller-the-better. Because $m = 0$, $L(y) = k(y - 0)^2$. Thus, the STB quality loss function is given by

$$L(y) = k(y)^2 \qquad \textbf{(3.13)}$$

As the value of $y$ gets farther from zero, the performance gets worse and the loss starts to increase.

Here are some examples of smaller-the-better cases:

- Microwave oven radiation leakage
- Time it takes to get the first copy out of a copy machine
- Paper jams in a copier
- Defects on an image
- Background density on a text image
- Automotive exhaust pollution
- Steering column vibration
- Electromagnetic interference from consumer electronics
- Corrosion of metals

Background density in a text image is an example of the smaller-the-better case. In the copier industry, one measure of the acceptability of a copy is the amount of background toner particles that adhere to the portion of the copy that is intended to be white. Minimizing the residual toner in white areas is a smaller-the-better objective. It has been determined that approximately half of the customers will not tolerate a background level beyond the standard measure of 1.2 background units. Beyond that level, a service call is placed at a cost of $200 plus the cost of the down time of the copier—typically

**Figure 3.7** The smaller-the-better loss function

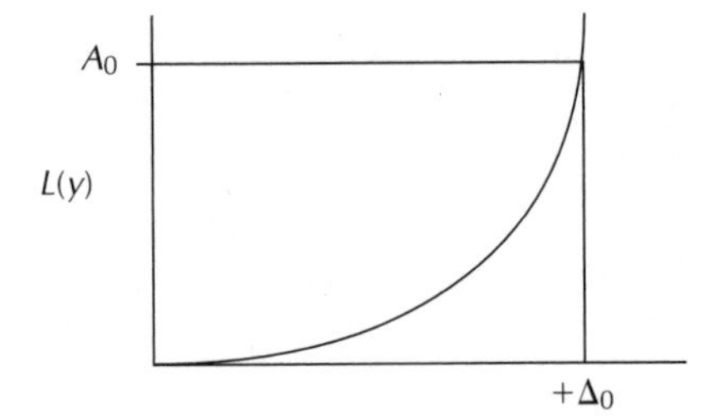

valued at approximately \$150 per hour. If the average copier down-time is 2.5 hours, the customer's loss is \$375. Thus, the total cost is $A_0$ = \$200 + \$375 = \$575.

$$k = \frac{A_0}{\Delta_0^2} = \frac{\$575}{1.2^2} = \$399.30$$

Thus, the STB loss function is given by

$$L(y) = \$399.30(y)^2$$

### 3.8.4 Average Quality Loss for Smaller-the-Better Cases

Just as the NTB case is modified for calculating the loss for more than one item, the same can be done for the STB case. Again, the mean square deviation is used:

$$\text{MSD}_{\text{STB}} = \frac{1}{n}\sum_{i=1}^{n} y_i^2 \approx [S^2 + (y)^2] \quad \textbf{(3.14)}$$

Consequently, the average loss for the STB case is

$$L(y) = k[S^2 + \bar{y}^2] \quad \textbf{(3.15)}$$

Since the target is zero in the STB case, $\bar{y}$ is the deviation of the mean from zero. The ideal function is focused on the smallest response value possible. The losses can be added to account for total product economic performance as it relates to a direct measure of physical performance.

Equation 3.15 can be used to calculate the expected quality loss, in dollars, for the STB copier background case where $k$ = \$399.30. For example, consider two machines in one print shop giving the following performance:

| *Machine* | *Data* | $S^2$ | $\bar{y}$ | $S^2 + \bar{y}^2$ | $L(y)$ |
|---|---|---|---|---|---|
| #1 | 0.64, 0.56, 0.71<br>0.55, 0.59, 0.75<br>0.64, 0.76 | 0.0068 | 0.65 | 0.4293 | \$171.41 |
| #2 | 0.55, 0.67, 0.70<br>0.94, 0.71, 0.82,<br>0.86, 0.96 | 0.0203 | 0.776 | 0.6229 | \$248.70 |

The table shows the first machine to be producing the smallest amount of loss.

### 3.8.5 The Larger-the-Better Case (LTB)

Some measured responses, while never having negative values, are better as their value gets larger. Ideally, in LTB cases, as the performance value approaches infinity the quality loss approaches zero. This loss function is simply the reciprocal of the smaller-the-better case:

$$L(y) = k[1/(y)^2] \quad \textbf{(3.16)}$$

**Figure 3.8** The larger-the-better loss function

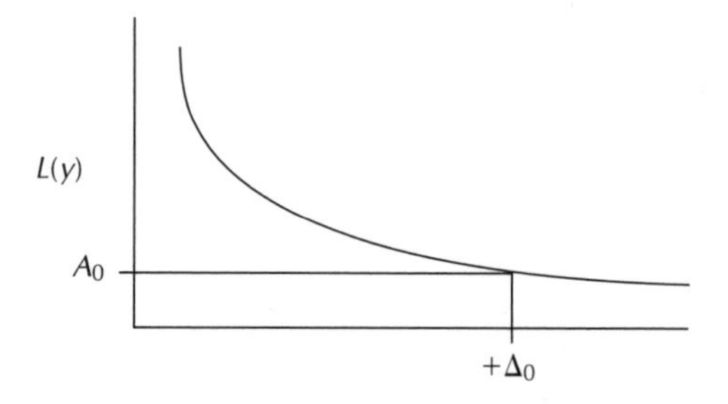


To determine $k$ for this loss function, the reciprocal relationship must be used. Thus, the loss $A_0$ is related to the functional limit $\Delta_0$ by $A_0 = k[1/\Delta_0^2]$, and the loss function coefficient is given by:

$$k = A_0\Delta_0^2 \tag{3.17}$$

Here are some examples of larger-the-better cases:

- The strength of a permanent adhesive
- Weld strength of a joint
- The traction capability of a tire
- Automobile gas mileage
- Corrosion resistance of an auto body

The seal strength of a vacuum blower housing in an office copier is an example of a larger-the-better case. The better it can run under widely varying use environments, the better it is for minimizing loss. When the blower seal fails to operate, it costs \$40 to replace: \$20 in part costs and another \$20 in installation labor costs. While the device that uses the vacuum blower sits idle, the cost to the customer is \$340 per hour. On average it takes 30 minutes to replace the blower.

$$A_0 = \$170 + \$40 = \$210$$

The seal can suffer changes in its adhesive properties as a result of vibration over time, humidity, and temperature. So, as the deterioration noises assail the adhesion of the seal, the vacuum loss grows to an unacceptable level that approximately 50% of the customers will notice. The larger the adhesive stability, the better. Seal integrity is measured by testing the seal adhesion strength (psi). The seal level at which the vacuum loss becomes objectionable is 20 psi. Thus, $\Delta_0 = 20$ psi.

The LTB loss function is given by:

$$k = A_0\Delta_0^2$$
$$k = (\$210)(20)^2$$
$$L(y) = 84{,}000[1/(y)^2]$$

In this case the seal strength of a returned machine is measured at 13 psi. Thus,

$$L(y) = 84{,}000[1/(13)^2]$$
$$L(y) = 84{,}000[0.0059]$$
$$L(y) = \$497$$

Therefore, when the seal strength drifts to a low of 13 psi, a total loss of about $500 is incurred. How can this be when the cost of repair is only $210? The consequences of seal failure include increased audible noise, loss of toner dust containment resulting in machine and office contamination, image quality degradation, and, of course, a service call. If the machine is allowed to degrade beyond the nominal functional limit, because of lack of attention or misdiagnosis of the failure mode, then additional losses are incurred.

### 3.8.6 Average Quality Loss for Larger-the-Better Cases

Just as for the nominal-the-best and smaller-the-better cases, it is useful to calculate the average LTB quality loss. The mean square deviation for the larger-the-better case is given by Equation 3.18.

$$\text{MSD} = \frac{1}{n}\sum_{i=1}^{n}(1/y_i)^2 = \frac{1}{n}[1/(y_1)^2 + 1/(y_2)^2 + 1/(y_3)^2 + \ldots + 1/(y_n)^2 \qquad \textbf{(3.18)}$$

Thus, the average LTB loss function is given by Equation 3.19:

$$L(y) = k(\text{MSD}) = k\left[\frac{1}{n}\sum_{i=1}^{n}(1/y_i)^2\right] \qquad \textbf{(3.19)}$$

It is possible to express the MSD in terms of the mean, $\bar{y}$, and the variance, $S^2$, of the quality characteristic after some algebraic manipulation [M1]. This is required because of the reciprocal in the LTB case. Thus, the MSD is given by:

$$\text{MSD} \cong \frac{1}{\bar{y}^2}\left(1 + \frac{3S^2}{\bar{y}^2} + \frac{4\hat{\mu}_3}{\bar{y}^3} + \frac{5\hat{\mu}_4}{\bar{y}^4}\right) \qquad \textbf{(3.20)}$$

where the higher-order terms are given by:

$$\hat{\mu}_3 = \frac{1}{n}\sum_{i=1}^{n}(y_i - \bar{y})^3 \text{ and } \hat{\mu}_4 = \frac{1}{n}\sum_{i=1}^{n}(y_i - \bar{y})^4 \qquad \textbf{(3.21)}$$

The higher-order terms can be neglected only if two conditions are met:

1. The distribution is not too skewed (i.e., it is close to normal or Gaussian), so that $\hat{\mu}_3 \approx 0$, by symmetry arguments.
2. $\bar{y} >> \hat{\mu}_4$ (i.e., the distribution width is not too great).

The likelihood of these conditions being met routinely is not high. Therefore, we recommend that you use Equation 3.19 when calculating the larger-the-better MSD, foregoing any calculation shortcuts.

### 3.8.7 The Asymmetric Nominal-the-Best Loss Function

There are instances when it is more harmful for a product's performance to be off-target in one direction than in the other. There are *two distinct quality loss coefficients* required for this case—one for each direction away from the target. The asymmetric loss function takes on the following form:

$$L^+(y) = k^+(y - m)^2, y > m \qquad \textbf{(3.22a)}$$

$$L^-(y) = k^-(y - m)^2, y \leq m \qquad \textbf{(3.22b)}$$

Equation 3.19 can be used to calculate the expected quality loss, in dollars, for the LTB seal adhesion problem for the vacuum blower for two separate sites. Each has eight machines operating in it. One site is in Miami, Florida, and the other is in St. Paul, Minnesota. The loss function for this case is

$$L(y) = \$84{,}000\left[\frac{1}{n}\sum_{i=1}^{n}(1/y_i)^2\right]$$

| *Site* | *Data* | *MSD* | *L(y)* |
|---|---|---|---|
| Miami | 17, 21, 30, 12<br>10, 24, 16, 27 | | |
| St. Paul | 37, 28, 30, 42<br>29, 32, 36, 25 | | |

After transforming the data by taking the reciprocal of the square of the measured seal strengths:

| *Machine* | *Data* | *MSD* | *L(y)* |
|---|---|---|---|
| Miami | 0.0035, 0.0023, 0.0011, 0.0069<br>0.0100, 0.0017, 0.0039, 0.0014 | 0.0039 | $323.40 |
| St. Paul | 0.0007, 0.0013, 0.0011, 0.0006<br>0.0012, 0.0010, 0.0008, 0.0016 | 0.0010 | $84.00 |

According to the data from the investigation of their seal strengths, the St. Paul site has the smaller loss. The vacuum seal is expected to deteriorate because of the noises of vibration, temperature, and humidity. The lower loss figures for St. Paul are a manifestation of the fact that they will last longer and result in fewer or later service calls, thus minimizing losses in real dollars.

Examples of the Asymmetric Loss Function:

- The amount of toner a copier uses to make an image. Too much toner use increases toner consumption, and too little produces weak images that are immediately unacceptable.
- Refrigerator temperature variation. More food will spoil if the temperature is above target than will spoil if the temperature is below target.

As an example, consider the case of temperature drift in a refrigerator. The standard target for most refrigerators is 40°F. Consider the consequences of being above and below this targeted temperature. When the temperature gets above 50°F, several things can happen that annoy the consumer. These include tepid food and drink that is not pleasant to the taste, and spoilage due to accelerated bacterial growth. Each of these can cause economic loss, losses due to discarding *and* replacement of food, and losses due to illness from ingestion of tainted food. When the temperature gets below 30°F, there may be some damage due to ice crystals, but there should be little food lost. The loss function constant $k^- < k^+$.

**Figure 3.9** The asymmetric nominal-the-best loss function

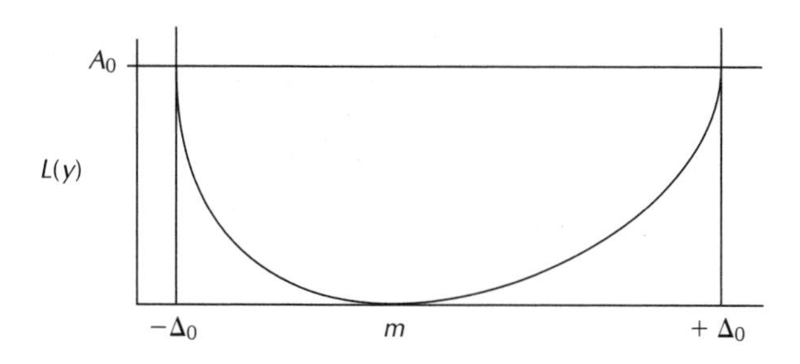


The following loss function is used for the case of drifting above 40 degrees:

$$L^{+}(y) = k^{+}(y - m)^2, y > m$$

where $k^{+} = A_0/\Delta_0^2$. As in the nominal-the-best case, $A_0$ is the monetary loss incurred by approximately 50% of customers when the temperature drifts to a value $\Delta_0$ from the target. Assume an average value of $50 for lost food replacement and $100 for the service call and parts to correct the temperature problem when the temperature goes to 50°F. Thus, $A_0 = \$150$ when $\Delta_0 = +10°F$.

$$k^{+} = \$150/(10°F)^2 = 1.5\ \$/°F^2$$

The upper loss function is

$$L^{+}(y) = 1.5(y - 40)^2$$

The following loss function is used for the case of drifting below 40 degrees:

$$L^{-}(y) = k^{-}(y - m)^2, y \le m$$

where $k^{-} = A_0/\Delta_0^2$. Again, $A_0$ is the monetary loss incurred by approximately 50% of customers when the temperature drifts to a value $\Delta_0$ from the target. Assume an average value of $10 for lost food replacement and $100 for a service call and parts to correct the temperature problem when the temperature goes to 30°F. Thus, $A_0 = \$110$ when $\Delta_0 = -10°F$.

$$k^{-} = \$110/(-10°F)^2 = 1.1\ \$/°F^2$$

The lower loss function is:

$$L^{-}(y) = 1.1(y - 40)^2$$

## 3.9 Loss Function Case Study

The following case study [F3] illustrates the use of the loss function to evaluate solutions to an engineering problem. As shown, the solution involves some trade-offs. But the quadratic loss function makes it possible to evaluate quantitatively compromises that involve cost and quality. Parameter design experiments have been used to optimize the robustness of the system. This case study illustrates the tolerance design phase where ways of improving quality that also affect unit manufacturing cost are considered. It is that kind of trade-off that you, as a practical engineer, are frequently asked to make.

**Figure 3.10** Toning gap configuration

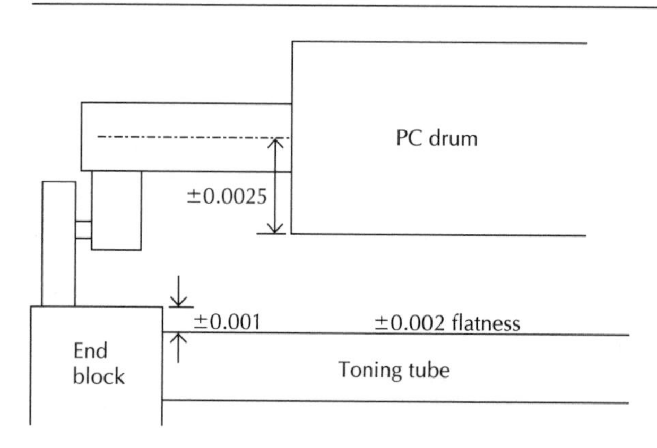


During product development for an electrophotographic printer, there is concern that variability in the hardware and materials might not be consistent with the image quality requirements. The concern is focused particularly on the toning station and the gap between the toning tube and the photoconductor (PC) cartridge. The parameter design experiments establish the strong dependence of image density uniformity on toning gap uniformity. There are doubts as to whether manufacturing would be capable of consistently producing the required toning gap uniformity at a reasonable cost. Figure 3.10 shows the toning gap configuration.

To gain the confidence needed to proceed with the product program, the following questions need to be addressed.

a. What degree of image uniformity is required for the product?
b. What is the sensitivity of print uniformity to toning gap variability?
c. What tolerance should be set for gap uniformity?
d. Can a toning station be designed and manufactured to meet this tolerance?
e. How do other sources of print nonuniformity affect the gap uniformity requirement?
f. What can be expected in terms of manufacturing reject rates and customer service calls related to print nonuniformity?

The tolerance equation based on the quadratic loss function that is used to analyze this problem is given by Equation 3.23 [T8]:

$$\Delta_i = \sqrt{\frac{A_i}{A_0}}\,\frac{\Delta_0}{\beta_i} \tag{3.23}$$

where

$\Delta_i$ = the manufacturing tolerance for the subsystem part $i$

$A_i$ = the cost to repair/replace part $i$

$A_0$ = the cost of violating the system tolerance

$\beta_i$ = the sensitivity of the system output to variation in the subsystem part $i$

$\Delta_0$ = the functional limit (customer tolerance) of the system output.

This equation is obtained from the nominal-the-best loss function, Equation 3.6, by the following derivation. The customer tolerance for image quality, $\Delta_0$, can be related to the part tolerance by considering the response of the system quality characteristic (image quality, $y$) to the subsystem quality characteristic (toning gap, $y_i$). This system sensitivity is approximated as a straight line with slope $\beta_i$ in the region of nominal performance, as shown in Figure 3.11.

Thus, the following relationships apply:

For the quality characteristics, $y = \beta_i y_i$;

For the nominal target, $m = \beta_i m_i$.

Thus, the loss function for image quality in the toning gap is given by:

$$L(y_i) = A_0/\Delta_0^2[\beta(y_i - m_i)]^2$$

The tolerance limit for the part characteristic is $y_i - m_i = \pm\Delta_i$. At the tolerance limit the loss, $L(y_i)$, equals the cost of scrap, $A_i$. Thus,

$$A_i = A_0/\Delta_0^2(\beta\Delta_i)^2$$

Solving this equation for $\Delta_i$ results in the tolerancing Equation 3.23. Relating the cost of scrapping the part to the quality loss allows us to make the following statement:

> *If part* i *is inspected and found to be outside the computed tolerance,* $\Delta_i$ *(Equation 3.23), then it is cost-effective to repair/replace the part at cost* $A_i$.

Even if there is no provision for actual inspection and repair/replacement on the assembly line, this equation may be used to set tolerances for design and procurement purposes. Note that this equation is applied to each toleranced item individually. This makes the Taguchi tolerancing approach very simple to apply, since the calculation includes no interaction among the individually toleranced items. In the Taguchi approach, subsystem part tolerances "add up" according to the cost structure, not statistically.

A refinement to Equation 3.23 may be required if significant rejection rates result from its application. Equation 3.23 also assumes that the repair/replacement part is perfect [F1]. In the present application, Equation 3.23 is sufficiently accurate and realistic for rough estimates of manufacturing and

**Figure 3.11** System sensitivity

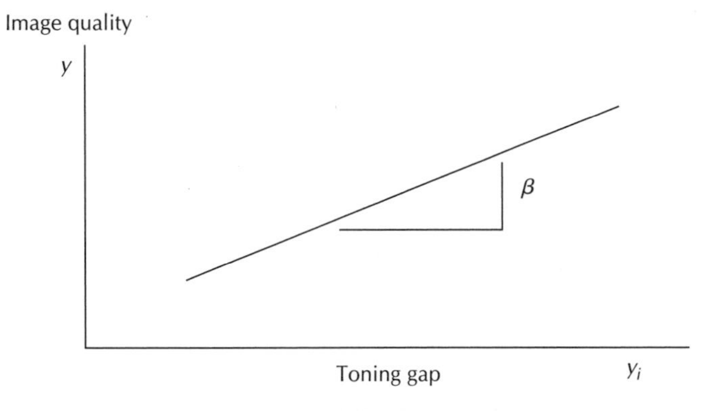

**Table 3.1** Density uniformity customer tolerance vs. nominal density and spatial frequency

| | *Spatial frequency (cycles/mm)* | | | |
|---|---|---|---|---|
| *Nominal density* | *0.3 mm$^{-1}$* | *0.2 mm$^{-1}$* | *0.013 mm$^{-1}$* | *0.0057 mm$^{-1}$* |
| 0.25 | ±0.015 | ±0.015 | ±0.022 | ±0.030 |
| 0.65 | 0.015 | 0.022 | 0.037 | 0.045 |
| 0.95 | 0.015 | 0.030 | 0.052 | 0.067 |
| 1.35 | 0.030 | 0.060 | 0.090 | 0.15 |

service implications. For the purposes of this case study, the only system-level image quality characteristic considered is image density uniformity. There are, of course, many others. The trade-offs that need to be made between competing quality characteristics are beyond the scope of this simple example.

Applying Equation 3.23 requires several inputs. The (system) image uniformity functional limit, $\Delta_0$, is taken from the product image quality (IQ) requirements, relaxed to the level at which 50% of customers would call for service. As shown in Table 3.1, this uniformity requirement depends on density level and spatial frequency. Failure at any density level is assumed to generate a service call. The low-frequency (across the page) requirements are most important since the type of toning-gap nonuniformities considered here are predominantly low-frequency. Thus, the highlighted functional limit, 0.045 (6.9%) density nonuniformity is the critical functional limit.

An estimated average service call cost of \$500, including labor, travel, and parts, is used for $A_0$. The subsystem repair/replace costs $A_i$ are taken from the preliminary bill-of-material, applying some judgment as to which components are associated with the toning gap. These costs are detailed in Table 3.2. The sensitivity of the image density uniformity to the toning gap nonuniformity is given in Table 3.3.

**Table 3.2** Estimates of the costs of corrective actions at the factory

| | |
|---|---|
| *Toning tube regrind* | \$22 |
| *Toning station* | \$100 |
| *PC drum* | \$44 |

**Table 3.3** Sensitivity data for the toning subsystem

| *Subsystem* | *Nominal density* | *Image functional limit (% nom)* | *Sensitivity* | *Subsystem functional limit (% nom)* |
|---|---|---|---|---|
| | $m$ | $\Delta_0$ | $\beta$ | $\Delta_i = \beta\Delta_0$ |
| Toning gap | 0.65 | 0.045 | 0.146 | 0.0066″ |

**Table 3.4** Toning tolerances computed using Equation 3.23 and estimated actual variability (3-sigma) expressed as a percentage of nominal

| *Subsystem component* | *Component cost* | *Computed tolerance* | *Part variability* | *Part mfg. capability* |
|---|---|---|---|---|
| | $A_i$ | $\Delta_i$ | $3\sigma$ | $\Delta_i/3\sigma$ |
| Toning gap | $144 | 0.0035 in. | 0.0034 in. | 1.03 |
| PC runout | $44 | 0.0019 in. | 0.0025 in. | 0.78 |
| Tube flatness | $22 | 0.0014 in. | 0.002 in. | 0.69 |
| End block | $100 | 0.0029 in. | 0.001 in. | 2.94 |

The subsystem tolerances can now be calculated using Equation 3.23. For the overall system tolerance, the corrective action would be replacing the entire toning-gap interface, the toning station, and the PC drum for a net cost $A_i$ = $144. We can also consider the tolerance breakdown by contributing subsystems, which are shown in Figure 3.10. Thus, replacing only the PC drum would cost $44; replacing only the toning tube would cost $22; and replacing the end blocks would cost $100 (the end blocks cannot be extracted from the toning subsystem). Using $A_0$ = $500 and $\beta\Delta_0 = 0.0066$ in., the resulting values for $\Delta_i$ are shown in Table 3.4, along with the expected 3-sigma variability levels. The resulting subsystem and subsystem component manufacturing capabilities are also shown.

Analyzing this table, we see that the overall quality of the interface, relative to its parts cost at about three standard deviations, is not too good, but not too bad either. However, when considering the contributing components to the interface, the PC runout and toning tube flatness are major contributors to the nonuniformity, again relative to their cost.

The PC drum quality is considered separately and will not be discussed further here. Several mechanical design approaches are considered for the toning station parts that define the toning gap. In the toning subsystem, the $22 corrective action involves the factory replacement of the toning tube assemblies. The toning tubes are assembled into the end blocks. The assembly is then ground flat using a slab cutter. Using an automated gap sensing fixture, these assemblies are measured at several spots along the length of the four tubes for flatness against a special reference drum. Prior to insertion in the printer, each toning tube assembly that exceeds a sort limit can be rejected to improve quality.

Inspection results in an increase of unit manufacturing cost (UMC). This increase can be evaluated along with the improvement in quality loss to find the optimum sort level. The team includes in the analysis the issues of factory inspection reject rates and field image quality failure rates, in addition to parts cost and average quality loss. To address these issues, a Monte Carlo simulation is used for the entire electrophotographic process. Critical subsystem parameters are treated as random variables, corresponding to the subsystem variability, as given in Table 3.4 for the toning subsystem.

In particular, for the toning subsystem, each dimension shown in Figure 3.10 is considered a random variable. The gap inspection procedure is simulated with toning tube assembly flatness thresholds of 1.0 mil, 2.0 mil, and no inspection. Rejection rates are tracked, along with statistics on post-inspection gap variability and IQ failures. Average quality loss is computed from the mean nonuniformity. Quality loss is divided into components for the toning tube assembly and the PC cartridge. Since an average of 10 PC cartridges would be consumed over the life of each product unit, the individual PC cartridge quality loss is multiplied by 10 in figuring life-cycle cost. IQ failures over life

**Table 3.5** Results of baseline variations simulation and total cost of quality calculations

| | *No sort* | *±2 mil sort* | *±1 mil sort* | *AC bias (no sort)* |
|---|---|---|---|---|
| Gap variability (3 sigma) | ±3.4 | ±3.3 | ±3.2 | ±3.4 |
| *Quality loss (gap)* | | | | |
| Shell assembly | $5.75 | $5.40 | $3.35 | $2.56 |
| PC cartridge | $9.05 | $9.05 | $9.05 | $4.02 |
| Lifetime | $96.25 | $95.90 | $93.85 | $42.76 |
| *UMC* | | | | |
| 1 × shell assembly | $40 | $40 | $40 | $57 |
| 10 × PC cartridge | $444 | $444 | $444 | $444 |
| *Shell assembly* | | | | |
| Reject rate | 0 | 14% | 74% | 0 |
| Prorated UMC | $0 | $3 | $16 | $0 |
| *IQ failure rate* | | | | |
| Per assy. (4 color) | 0.20% | 0.05% | 0.00% | 0.05% |
| Lifetime | 1.3% | 0.33% | 0.00% | 0.33% |
| Prorated service cost | $6 | $2 | $0 | $2 |
| *Total cost (lifetime)* | $586 | $585 | $594 | $546 |

are attributed mostly to the PC cartridges, because they are not factory-inspected, and because 10 of them are used over the product life. These simulation results are summarized in Table 3.5.

The team agrees on a 2-mil inspection sort tolerance for the toning tube assemblies. This is not greatly different from the 1.4 mil tolerance shown in Table 3.4. Inspection to a 1-mil standard reduces quality loss due to mechanical variability to the point that IQ uniformity failures in the field are virtually eliminated. But the total cost with the 1-mil standard is higher, after accounting for the 74% rejection rate. The no-inspection case had slightly higher quality loss and a significant IQ failure rate in the field. The 2-mil standard appeared to be a reasonable compromise of low quality loss, low rejection rate, and low IQ field failure. It is further noted that, with inspection at the 2-mil level, the "blame" for IQ failure would now lie predominantly with other subsystems.

Finally, the last column in Table 3.5 indicates the results obtained with a change in the toning concept. This new approach features an AC toning voltage bias rather than the DC bias currently in use. The results for the AC toning bias indicate a significant improvement in quality that justifies the increased UMC. That is, the net reduction of quality loss ($42.76 vs. $95.90) is greater than the net

**Table 3.6** Loss functions

| *Type* | *Loss function* | *Mean square deviation* | *Average loss function* |
|---|---|---|---|
| Nominal-the-best | $L(y) = \frac{A_0}{\Delta_0^2}(y - m)^2$ | $MSD = S^2 + (\bar{y} - m)^2$ | $\overline{L(y)} = \frac{A_0}{\Delta_0^2}[S^2 + (\bar{y} - m)^2]$ |
| Smaller-the-better | $L(y) = \frac{A_0}{\Delta_0^2}(y)^2$ | $MSD = S^2 + (\bar{y})^2$ | $\overline{L(y)} = \frac{A_0}{\Delta_0^2}[S^2 + (\bar{y})^2]$ |
| Larger-the-better | $L(y) = A_0\Delta_0^2(1/y)^2$ | $MSD = \frac{1}{n}\sum_{i=1}^{n}(1/y_i)^2$ | $\overline{L(y)} = A_0\Delta_0^2\left[\frac{1}{n}\sum_{i=1}^{n}(1/y_i)^2\right]$ |

increase in UMC ($57 vs. $40). For this reason, the team recommended using the AC concept for future product development.

## 3.10 Summary

The loss function is based upon the concept of on-target engineering. All deviations from the target imply some loss of quality, and that quality loss can be equated to monetary loss. The quadratic form is shown to be the simplest approximation of the actual losses that are suffered as a quality characteristic deviates from its target value. The average loss function for the nominal-the-best, smaller-the-better, and larger-the-better forms are all determined by the product of the economic constant ($k$) and the mean square deviation (MSD) from the target. Table 3.6 summarizes these quadratic loss functions.

To determine the quality loss coefficient, $k$, two pieces of information are needed:

1. The functional limit or customer tolerance, $\Delta_0$, usually defined as the level at which 50% of the customers would be dissatisfied enough to initiate corrective action.
2. The total monetary losses, $A_0$, incurred as a result of exceeding the functional limit.

These are very important and useful quantities to know, but they can be very difficult to determine exactly. To describe quality conceptually using the loss function, it is not necessary to determine $k$ accurately. For vendor evaluation, tolerance design, and economic cost of quality analysis, accurate information for $\Delta_0$ and $A_0$ is needed.

Generally, such information is not available during product development. Concept design involves new technologies or new designs for which the functional limits are yet to be determined. New product features mean that the customer requirements are not yet known. The cost of failure in the field is usually derived from experience. For evolutionary products, that information might be available. For revolutionary products that information is usually unknown. Fortunately, at the early stages in product development, relative, not absolute, quality is of interest. Good engineering requires that we make countless decisions, evaluating alternatives, trying to decide which is better. How much better is often less certain.

One other property of the loss function is that it deals with deviations from the target. In the case of the average loss function, the mean square deviation from the target is used. However, early in product development, do we really know what the target is? How often have changing requirements or system level optimization (e.g., trade-offs) made you feel as if you are riding a roller coaster, getting thrown to and fro? Building robustness into the design, by making it less sensitive to noise factors, is a consistent requirement even when the target is subject to change. For further illustrations of the use of the loss function in tolerance design, see *Quality Engineering in Production Systems* by Taguchi, Elsayed, and Hsiang [T8].

## Exercises for Chapter 3

1. Define what is meant by the term "quality characteristic" and give several physical examples of quality characteristics.
2. Propose a process for gathering loss-function data for the development of a new personal computer.
3. What are the costs that are typically included in quantifying a loss function?
4. Why is it important that the multifunctional product development team treat the loss function as an approximation of the quality loss?
5. At what point in the product development process should a loss function be constructed?
6. What phase of the off-line quality control process makes the most use of the quality loss function? For what purpose?
7. Why is the practice of driving quality efforts around having products "in spec" not advisable?
8. Explain why so many companies have made "in spec" quality processes the norm over the last 50 years, and why it is hard for them to change.
9. Explain the rationale behind the "on-target" approach to quality.
10. What is the underlying principle behind the quality loss function?
11. How is the economic constant, $k$, calculated? How would you obtain the information necessary to perform the calculation?
12. Describe several quality characteristics that fit the STB approach to quantifying quality loss.
13. Describe several quality characteristics that fit the LTB approach to quantifying quality loss.
14. Describe several quality characteristics that fit the NTB approach to quantifying quality loss.
15. Describe several quality characteristics that fit the asymmetric NTB approach to quantifying quality loss.
16. What is the MSD and why is it used?
17. Marketing teams and engineering teams need to work together to apply the loss function. Describe any cultural barriers that may prevent them from cooperating. Suggest some ways to remedy these cultural problems within modern industry.
18. Prove the identity:

$$\frac{1}{n}\sum_{i=1}^{n}(y_i^2) - \bar{y}^2 = \frac{1}{n}\sum_{i=1}^{n}(y_i - \bar{y})^2$$

CHAPTER 4

# The Signal-to-Noise Ratio

## 4.1 Properties of the S/N Ratio

The preceding chapter should make it clear that, although the quality loss function is a powerful concept for optimizing robustness, it has some deficiencies. However, the basic mathematical structure of the quality loss function can be leveraged into the key measure of robustness, the signal-to-noise ratio. Before doing so, consider the loss function and how it can be used:

1. The loss function can be used to quantify a design's quality.
2. The loss function can be used to compare the expected cost of quality relative to the manufacturing cost.
3. The loss function can be used to determine tolerances.

These are all useful, but they are retrospective. The loss function looks back at an existing design's performance. It does not necessarily predict the ultimate performance of a system. This is because the loss function is not independent of adjustment of the mean after reducing variability. That is, if a system is stable in the presence of noise but not on target, then the quality loss is high. However, a simple adjustment might put such a system on target, resulting in very low quality loss. Thus, the loss function is not a suitable metric to use for parameter design optimization where it is useful to reduce variability *independent* of putting the system on target.

The signal-to-noise metric is designed to be used to optimize the robustness of a product or process. Here is a list of the properties of the ideal signal-to-noise metric:

1. The S/N ratio reflects the variability in the response of a system caused by noise factors.
2. The S/N ratio is independent of the adjustment of the mean. In other words, the metric would be useful for predicting quality even if the target value should change.
3. The S/N ratio measures relative quality, because it is to be used for comparative purposes.
4. The S/N ratio does not induce unnecessary complications, such as control factor interactions, when the influences of many factors on product quality are analyzed.

A good signal-to-noise ratio has all of these properties. In order to achieve these properties, engineering analysis and judgment must be used to select the appropriate S/N ratio. In subsequent chapters there are many examples and further discussion to help illustrate how to use engineering analysis to select the appropriate S/N treatment. In fact, successful practitioners of Dr. Taguchi's approach to Robust Design develop custom S/N ratios that are optimized for the specific engineering task at hand. These can become important proprietary advantages for their firms.

## 4.2 Derivation of the S/N Ratio

The general procedure for creating an S/N ratio is as follows:

1. The mean square deviation (MSD) for the average loss function, which is made up of measures of variability such as the variance, is used as the basis for the S/N ratio.
2. The MSD is modified to make the S/N ratio independent of adjustment (IA) of the mean to the target.
3. The resulting expression is mathematically transformed into decibels (by the use of a logarithm), as shown in Equation 4.8. This makes the S/N ratio a measure of relative quality and helps reduce the effect of interactions between control factors.

This procedure for generating an S/N ratio can be illustrated using a nominal-the-best case study. Consider a chemical manufacturing process that consists of two control factors and a measured response that is of the NTB type. Assume that the physics of the process dictates that the response is mainly a function of temperature and time. The goal is to optimize the response against variability in spite of various sources of noise that exist in the actual production process. An additional goal is to maintain the reduced variability when the average response is put on target.

An essential practice in Robust Design, before you start to generate data, is to consider what is already known about the system so as to guide the experimental effort. In this case, the following is known about the effects of the two control factors.

1. Temperature has a definite effect on the response. It has been observed that temperature affects how much variation in the response occurs because of the various noises. Temperature also affects the reaction rate, so the amount of product is affected in some way by temperature as well. The optimum setting for the temperature in light of the amount of time and the presence of noises is not yet known.
2. Time also has a significant effect on the response. It is believed, however, that time does not interact with the noises in such a way as to mitigate their effects. Instead, time appears to act as an adjustment factor. An *adjustment factor* is one that can be used to *shift* the average response value up and down to hit the target level for the reaction.

Thus, both control factors affect the response, and at least one of them (time) has the useful property that it can adjust the average response. The target value for the response is 50 units of measure. The particular type of unit is irrelevant to the derivation.

Table 4.1 shows some data obtained by repeating an evaluation of two temperature levels run at a fixed level of time. Each run is replicated four times, and the results are summarized by the mean response, $\bar{y}$, and the standard deviation, $S$.

**Table 4.1** Data for a chemical manufacturing process run at constant time

| *Run #* | *Temp.* | *Avg.* | *Std. dev.* | *Loss* |
|---|---|---|---|---|
| 1 | 300°F | 35 | 1.2 | $226.44 |
| 2 | 350°F | 46 | 8.5 | $88.25 |

The average loss function for nominal-the-best is used to develop the derivation. The economic constant can be ignored ($k = 1$), because the relative quality is of interest. The real cost or the actual limits are not particularly important here. Therefore the economic constant is not important to the derivation.

$$L = [(\bar{y} - m)^2 + S^2] \quad \textbf{(4.1)}$$

So things look pretty bleak for setting the temperature at 300°F—with an average of 35 being so far from 50. But the lower temperature does have a benefit: a small variation (1.2) compared to the large variation (8.5) from setting the temperature at 350°F. Something about the lower temperature is making the process work better. The loss function, however, because it penalizes the 300°F condition for being off target, does not recognize the more robust condition from an economic viewpoint.

The focus on being far from the target needs to be changed, *momentarily,* to concentrating on reducing variability *first.* Later, the process can be put on target. This is why it is important to have a metric that is independent of adjustments of the mean. Let us go back to the chemical manufacturing process and introduce another control factor that can be used to adjust the average response onto the target response.

### 4.2.1 The Scaling Factor [P2]

Consider the behavior of the response when time is introduced into the experiment as a control factor. As indicated earlier, time is expected to shift the level of the response. If the reaction rate remains the same over the range of times of interest (you would know that range from your engineering experience), then the response varies proportionally with time. But what about the variability? If variation in the amount of product is caused by variation in the reaction rates, then the standard deviation should

For our first set of experimental runs, the loss is

$$L = [(35 - 50)^2 + 1.2^2]$$
$$L = \$226.44$$

and for the second set of experimental runs, the loss is

$$L = [(46 - 50)^2 + 8.5^2]$$
$$L = \$88.25$$

**Figure 4.1** Effect of a scaling factor on the response

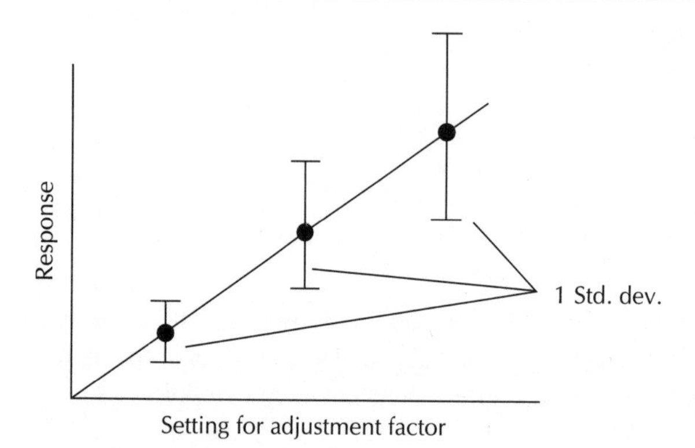


scale proportionally with time. In that case, the behavior of the control factor, time, and the response and its standard deviation can be illustrated by Figure 4.1.

By adjusting the time, any target response chosen can be hit. However, the standard deviation also grows in proportion to the increase in time. The intent is to shift the average response as needed and pay only a minimum penalty in growth of variance from doing so. When an adjustment factor has the property that the mean and standard deviation are both proportional to it, that adjustment factor is referred to as a *scaling factor.*

The scaling factor enables the two-step optimization process in this nominal-the-best problem. First, the standard deviation that would be found at any particular time is reduced. This can be done by the application of other control factors. Second, a scaling factor, such as time, is used to adjust onto the target. Although the standard deviation changes as the mean is moved onto the target, this is a necessary price to pay to minimize the quality loss, *provided that we have exhausted all other means at our disposal to reduce the variation seen in Figure 4.1.*

## 4.2.2 The Signal-to-Noise Ratio

The S/N ratio is a new metric that takes into account the effect of applying the adjustment factor. It should not change as the adjustment factor, time, is changed, but it should be sensitive to the magnitude of the standard deviation at a specific value of time. A metric that satisfies these requirements is the ratio of the mean and the standard deviation, referred to as the coefficient of variation (COV), which is given in Equation 2.5.

What happens when the effects of the adjustment factor, time, are applied to the COV, $\bar{y}$, and the Quality Loss? Table 4.2 shows the data from Table 4.1 for the chemical manufacturing example, with time added as an explicit factor.

**Table 4.2** Data for a chemical manufacturing process including time as an explicit factor

| *Run #* | *Temp.* | *Time* | *Avg.* | *Std. dev.* | *COV* | *Loss* |
|---|---|---|---|---|---|---|
| 1 | 300°F | 20 | 35 | 1.2 | 0.034 | $226.44 |
| 2 | 350°F | 20 | 46 | 8.5 | 0.185 | $88.25 |

Because of our assumption that the mean and standard deviation both scale proportionally with time, the following scaling process can be carried out.

1. Divide the target value ($m = 50$) by the actual mean from the first set of experimental runs ($\overline{y} = 35$). This is called the *scaling ratio* (*SR*).

$$\text{SR} = m/\overline{y} = 1.428 \tag{4.2}$$

2. Multiply the time ($t = 20$) by the scaling ratio to get the desired value to put the system on target after adjustment.

$$t_a = (m/\overline{y})t = 1.428 \times 20 = 28.56 \tag{4.3}$$

3. Multiply the mean ($\overline{y} = 35$) by the value of the scaling ratio. This puts the mean on the target (50) after adjustment. Since time is assumed to be a scaling factor, as shown in Figure 4.1, this is justified.

$$\overline{y}_a = (m/\overline{y})\overline{y} = 1.428 \times 35 = 50 \tag{4.4}$$

4. Multiply the standard deviation ($S = 1.2$) by the scaling ratio to get the variation after adjustment. (We assume that as the mean is adjusted, the standard deviation must also be adjusted because they scale together.)

$$S_a = (m/\overline{y})S = 1.428 \times 1.2 = 1.713 \tag{4.5}$$

In Equations 4.2 to 4.5, $t_a$ is the time after adjustment to the target and $t$ is the time before adjustment, $\overline{y}_a$ is the average response after adjustment to the target and $\overline{y}$ is the average before adjustment, and $S_a$ is the standard deviation after adjustment to the target and $S$ is the standard deviation before adjustment.

The loss after adjustment, $L_a$, and the coefficient of variation COV $= S/\overline{y}$ after adjustment are calculated using the new numbers after adjusting the mean onto the target. The ratio $S/\overline{y}$ is unchanged by the scaling factor. First, look at what happens to the loss function after adjustment:

$$L_a = [(\overline{y}_a - m)^2 + S_a^2] \tag{4.6}$$

but $(\overline{y}_a - m) = 0$ when the mean is put on target. Thus,

$$L_a = [(0)^2 + (S_a)^2]$$

$$L_a = \left[\frac{m}{\overline{y}} S\right]^2$$

$$L_a = m^2\left[\frac{S^2}{\overline{y}^2}\right] \tag{4.7}$$

Thus, the coefficient of variation reappears in the loss function *after* adjustment. Table 4.3 summarizes the situation after using the adjustment process.

A profound economic difference is displayed when the minimum variation set point is identified and its response shifted with the scaling factor to put the response on target—*even with the scaling penalty.* Now the true optimal temperature set point, 300°F, is seen clearly. This is despite the fact that the loss without adjustment implies that the 350°F setting is the optimal set point. Why did the loss function imply the wrong thing? Because the loss function itself is not a performance metric that is independent of adjustment. However, in cases involving linear adjustment, the coefficient of variation is independent of adjustment.

**Table 4.3** Data for a chemical manufacturing process including time as an explicit factor

| *Run #* | *Temp.* | *Time* | *Avg.* | *Std. dev.* | *COV* | $L_a$ |
|---|---|---|---|---|---|---|
| 1 | 300°F | 28.5 | 50 | 1.7 | 0.034 | \$2.89 |
| 2 | 350°F | 21.7 | 50 | 9.24 | 0.185 | \$85.37 |

What can be learned from this example?

1. Quality loss without adjustment is not an appropriate way of evaluating the optimum control factor setting. Mixing the location and dispersion $[(\bar{y} - m)^2 + S^2]$ response components together distorts our ability to accurately calculate the true quality loss.
2. Quality loss with adjustment provides a method that removes the location component from the loss function, revealing the component (dispersion) that *purely expresses the sensitivity to noise.* It is preferable to find adjustment factors that have a *large effect on the mean response,* but a small to moderate effect on increasing variability during adjustment. Remember, when the mean is adjusted with a scaling factor, the variance also changes.

The relationship between the mean and the variance, in this example, is a classic illustration of a correlation. As one factor changes, the other changes as well. In this case the correlation is entirely predictable based upon our engineering knowledge of the system. It is critical that engineers recognize and use that type of engineering know-how to improve the efficiency of their experimentation and analysis.

## 4.3 Defining the Signal-to-Noise Ratio from the Mean Square Deviation

The adjusted mean square deviation of the average quality loss function actually accounts for *sensitivity to noise.* This is the key concept behind the signal-to-noise ratio. The S/N ratio is designed to separate the effects of noise on the response from the mean value of the response. As stated earlier, the ideal metric should have these four properties:

1. The metric should reflect the variability in the response.
2. The metric should be independent of the adjustment of the mean.
3. The metric should measure relative quality.
4. The metric should not induce unnecessary complications, such as control factor interactions, when the influences of many factors on product quality are analyzed.

Robust Design utilizes a metric based on the adjusted quality loss function to define the signal-to-noise (S/N) ratio:

$$S/N = -10 \log(\mathrm{MSD_{IA}}) \quad \textbf{(4.8)}$$

where the $\mathrm{MSD_{IA}}$ is the mean square deviation expressed in a way that is independent of adjustment (IA). The $-10$ log(base 10 logarithm) puts the S/N ratio into *decibel* units (dB) and arranges that as the value of the $\mathrm{MSD_{IA}}$ gets smaller, the S/N ratio gets larger.

In the preceding example,

$$\mathrm{MSD_{IA}} = [S^2/\bar{y}^2] \tag{4.9}$$

As the mean squared $\bar{y}^2$ (*the signal*) increases and the variance $S^2$ (*the noise*) decreases, the performance against the effects of noise is maximized while simultaneously the loss is minimized. This is the power of this unique engineering metric. This metric has the first two properties listed earlier, i.e., it reflects the essential variability of the system in a way that is independent of the mean. The process by which this is visible to the engineer employs an analysis of the average S/N values for individual control factor (i.e., design parameter) levels, along with the analysis of the average performance values in standard engineering units (Chapter 7).

The metric is improved by applying Equation 4.8 to give the nominal-the-best S/N ratio, Equation 4.10. In Equation 4.10, the minus sign has been brought inside the argument of the logarithm to invert the $\mathrm{MSD_{IA}}$. (See box for properties of the logarithm.) This shows clearly that the S/N ratio is the ratio of the desired result, $\bar{y}$, divided by the undesired result, $S^2$.

$$S/N = 10 \log\left(\frac{\bar{y}^2}{s^2}\right) \tag{4.10}$$

However, a question arises: Why does the use of the logarithm improve the S/N metric? Why not just use Equation 4.9? This is done for several reasons to be explored in detail in the chapters to follow. If we recall the four desired properties for the S/N ratio, we see that the logarithm has several useful properties that help enhance the quality metric's properties. With this transformation, multiplicative changes in the ratio $\bar{y}^2/S^2$ are transformed to additive changes, thus making the S/N ratio proportional to relative quality (property #3). In addition, the log function makes the metric more additive in the statistical sense (property #4).

An *additive response* is one for which the influence of each control factor is relatively independent of the effects of the other control factor settings. This means that the effects of the control factors on the metric simply add, without any complicated cross terms. In other words, one can say that the log transform helps enhance the numerical independence of the control factors' measured effects from one another. Products that are intentionally engineered to have additive effects between their control factors are easier to optimize, thus increasing the efficiency of the product development process. An *additive design* is one in which the design parameters, components, subsystems, and modules work independently and harmoniously to achieve the product requirements. Products that have many requirements placed on design parameters because of their relationship (interactions) with other control factors are highly constrained and difficult to optimize. Design and analysis for additivity using the log transform as part of the S/N ratio helps clarify the control factor main effects so that the design can still be optimized efficiently. The effect of interactions between control factors is viewed as a nuisance constraint or noise. The log transform is a mathematical way to make the response less sensitive to this form of noise. The log transform effectively suppresses the numerical effect of the control factor interactivity.

## 4.4 Identifying the Scaling Factor

The preceding nominal-the-best derivation is carefully constructed to illustrate the concept of the signal-to-noise ratio. A key element of the derivation is the existence of an adjustment factor that is also used as a scaling factor, Equations 4.2–4.5. In experimental work, how do we know that there is a scaling factor? How do we know, prior to running the experiment, that the scaling relationships work and

## Properties of the Logarithm

The following algebraic relationships apply to logs:

1. $\log(A \times B) = \log(A) + \log(B)$
2. $\log(A/B) = \log(A) - \log(B)$
3. $C \times \log(A) = \log(A^c)$

For base 10 logarithms, the following quantitative relationship holds:

4. $\log(2) \approx 0.3$, $\log(0.5) \approx -0.3$
5. $\log(10) = 1.0$
   $\log(100) = 2.0$
   $\log(1000) = 3.0$, etc.

Figure 4.2 shows a plot of 10 log($x$):

**Figure 4.2** Plot of 10 log($x$)

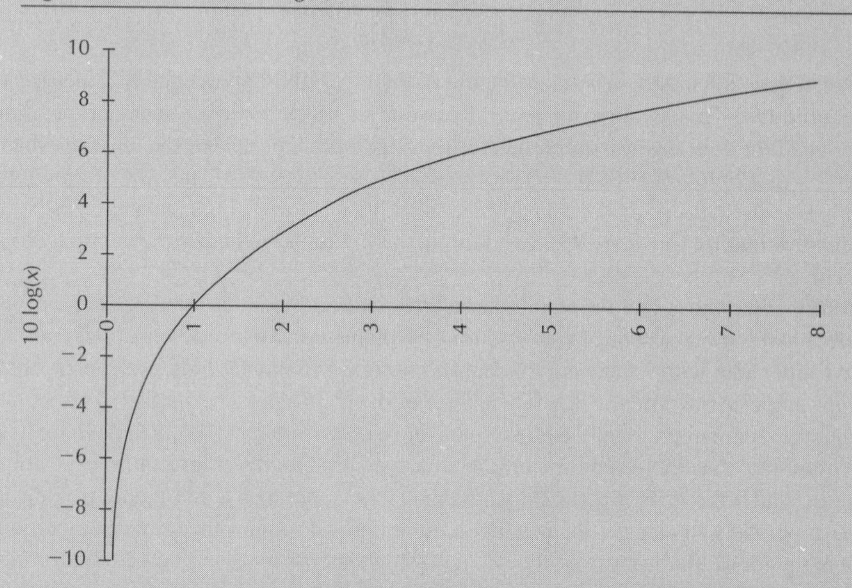


If there are moderate to mild interactive effects present between control factors, ($A \times B$) for example, it stands to reason that the log transform helps to promote additivity according to the following relationship:

$$\log A \times B = \log A + \log B$$

the S/N ratio has all the beneficial properties ascribed to it? The answer, of course, is engineering know-how. The more that is known about your system prior to conducting parameter design, the more likely the simplifying power built into the S/N ratios works to your advantage.

This means that you need to try to measure quality characteristics and use control factors that are likely to have a relationship such as that shown in Figure 4.1. In some cases there is enough engineer-

**Figure 4.3** Ideal behavior for the NTB scaling factor

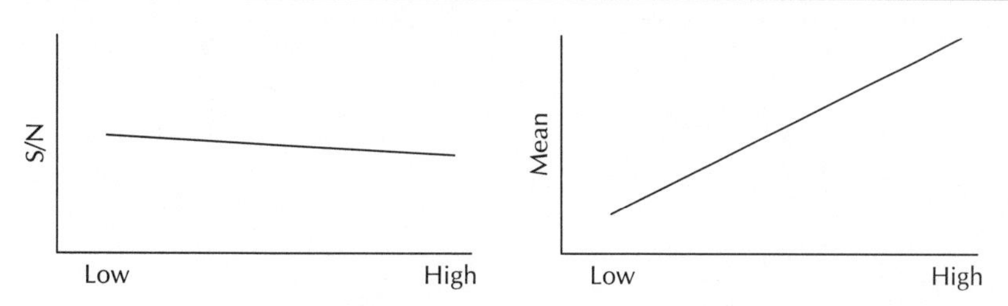


ing knowledge about the system to say that there is at least one factor that is known to behave as a scaling factor. It is common for the static nominal-the-best case to be used to hunt for scaling factors. Once this knowledge is in hand, the static approach can be replaced with the more capable dynamic approach discussed in Chapter 6.

Returning to the static problem for the moment, let us assume that an experiment has been run with several control factors and a good quality characteristic. Now what? Identifying a scaling factor is done by analyzing the response data for each control factor. One that has a significant effect on the mean response but little effect on S/N is shown in Figure 4.3. Because of the way the nominal-the-best S/N ratio is defined in Equation 4.10, such a factor must have the scaling properties.

The relatively small change in the S/N ratio, together with the large change in the mean, indicates that both the mean and standard deviation scale with this factor. This factor can be used to shift the mean onto the target while incurring only a little loss in noise sensitivity if the slope of the S/N plot is low.

A factor that has a significant effect on the signal-to-noise ratio and little influence on the mean is illustrated in Figure 4.4. Although this is not a scaling factor, it does represent a substantial opportunity to reduce variability. The large change in the S/N ratio, coupled with the small change in the mean, indicates that this factor is useful for minimizing sensitivity to noise.

A factor that affects neither the S/N ratio nor the mean response is shown in Figure 4.5. Because factors of this type have little effect on the variability or the mean, they do not need to be optimized. Rather, they can be set at their most economical levels with little impact on overall target performance. Alternatively, such factors are used for other functions in the design. Otherwise, factors that display this behavior are to be avoided in clean sheet designs. They are indicative of a factor that is not doing a lot for the design. For leveraged designs, or current product, they may be candidates for cost reduction by redesign or removal.

**Figure 4.4** Behavior of a factor used to improve robustness

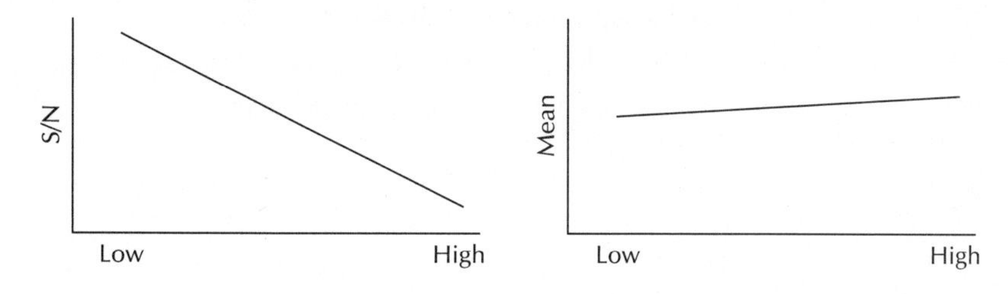

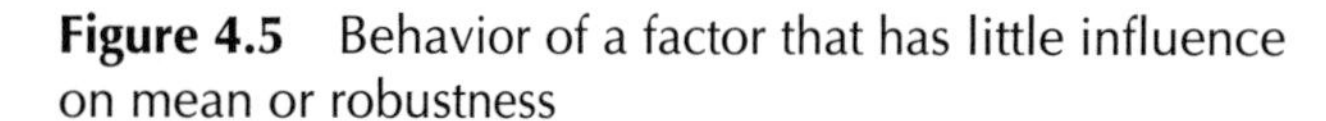

**Figure 4.5** Behavior of a factor that has little influence on mean or robustness

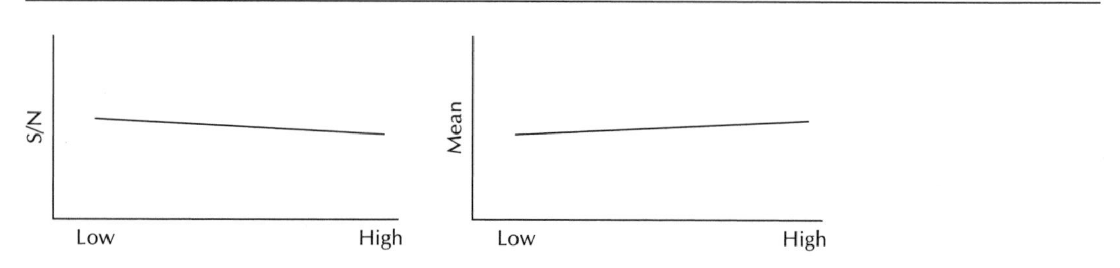


## 4.5 Summary

In this chapter, the desired properties of the S/N ratio have been described. They are listed in Table 4.4. These properties are important for the application of the two-step optimization process, described in Chapter 2, and the application of design of experiments, discussed in Chapter 7 and throughout Part II, Parameter Design.

The general procedure for developing the S/N ratio from the loss function has been described. A specific derivation is given using the nominal-the-best case. The next two chapters describe the S/N ratios for a variety of cases. Because the S/N ratio contains substantial engineering content, specific S/N ratios are defined according to the design requirements of the quality characteristic. While this may seem a bit confusing and unnecessary at first, it is, in fact, part of the power of Dr. Taguchi's signal-to-noise ratio.

The linear scaling factor concept used for the nominal-the-best problem is generalized for application to other S/N ratios in the next chapter by the definition of tuning factors. *Scaling factors* are those factors that can be used to adjust the mean without affecting the nominal-the-best S/N ratio (Equation 4.10) because both the mean and the standard deviation scale together. *Tuning factors* are those factors that can be used to adjust the mean onto target without affecting whatever S/N ratio is being used. Depending upon the type of S/N ratio, tuning factors may also be scaling factors, or tuning factors may have other properties instead.

**Table 4.4** Properties of an ideal signal-to-noise ratio

1. The S/N ratio reflects the variability in the response of a system caused by noise factors.
2. The S/N ratio is independent of the adjustment of the mean. In other words, the metric would be useful for predicting quality even if the target value should change.
3. The S/N ratio measures relative quality because it is to be used for comparative purposes.
4. The S/N ratio does not induce unnecessary complications, such as control factor interactions, when the influences of many factors on product quality are analyzed.

## Exercises for Chapter 4

1. Why is the data from an experiment mathematically transformed into a signal-to-noise ratio?
2. Sketch the relationship that often exists between a response and an adjustment factor.
3. Explain how to exploit the behavior of the variance in order to optimize robustness before putting the response on target.
4. What would you do in a situation where a control factor displays sensitivity to both S/N performance and average performance?
5. Explain the role of the coefficient of variation in defining the S/N ratio.
6. Explain the relationship between the loss function and the S/N ratio (NTB).
7. For the NTB case, what behavior is expected from the standard deviation as the value of the mean response increases?
8. Define the term *adjustment factor*.
9. What are the four ideal properties that are associated with the S/N ratio?
10. When both S/N and mean performance are weakly affected by a control factor, what criteria can be used to optimize the level of such a factor?

CHAPTER 5

# The Static Signal-to-Noise Ratios

## 5.1 Introduction, Static vs. Dynamic Analysis

There are two broad classes of signal-to-noise (S/N) ratios available. Static forms apply to cases where the quality characteristic's target has a fixed level—for example, an aim density for black in a text printer. Dynamic forms apply to cases where the quality characteristic operates over a range of values—for example, the tone scale or gray density range in continuous-tone printers or photographs. Static S/N ratio cases are primarily focused on a fixed-target, as opposed to having a designed capability to adjust to changing targets. In the static case, the robustness goal is to minimize the mean square deviation from the target. The equations necessary to calculate the S/N ratios are given in Chapters 5 and 6, but you will want to use a software package such as WinRobust for your computations.

The dynamic S/N methodology is used to optimize a design's input–output function. In other words, it is used to optimize a design's tunability rather than to optimize the response to a fixed value. Very often a design's ideal input–output function is a simple linear equation in which the input parameter may be used to adjust the mean response. The input parameter is thus an adjustment factor. For that reason, the dynamic case is used when an adjustment factor is known, or postulated, based on the physics or engineering analysis of the system. Such a parameter is also referred to as a *signal factor.* In technology development or "clean sheet" design, the dynamic method, using one or more signal factors in the design, is a very powerful technique to develop a tunable system. The dynamic case is discussed fully in Chapter 6. Let's first look at the static S/N ratios.

There are a variety of static S/N ratios. Selecting the proper static S/N ratio for a particular application is very important. This selection depends on the physics of the problem, engineering insight, and experience, as well as the ideal function for the particular case. Different methods of analysis are tied to the different S/N ratios. A number of the S/N ratios, e.g., smaller-the-better, larger-the-better, operating window, and nominal-the-best, are standard in quality engineering. Their derivation is illustrated using the concepts introduced in Chapter 4 for developing S/N ratios. New S/N ratios continue to be developed and introduced by Dr. Taguchi and others [W3]. This is an active field of research. In this chapter and the next, we introduce some of these new S/N ratios. By learning from these examples, you will be able to customize the standard S/N ratios for your applications and perhaps develop your own unique S/N ratios too.

Static S/N ratios are useful in simple cases where there is typically one response variable. The two-step optimization process, illustrated throughout this book, is applied to the nominal-the-best S/N ratio. Nominal-the-best (NTB) allows for the possibility of identifying tuning factors. The NTB S/N ratio is a good metric to use when attempting to optimize an existing design. In this situation, a scaling factor, which can put the quality characteristic on target, is sought. The smaller-the-better (STB) and larger-the-better (LTB) analyses are unique because it is usually not possible to hit the targets (zero or infinity, respectively). Instead, the mean square deviation from the "target" is used as the sole optimization criterion and the mean and variance are not treated separately. STB and LTB methodologies are useful in optimizing existing designs, but are not preferred when designing new products or processes, because of the lack of tuning capability.

Control factors, noise factors, and measured response values are utilized when working with static cases. The S/N ratio is typically calculated from at least two samples of data in each experimental run—for example, a repeated run of the control factor settings with changing levels of noise that induce significant variability in the data.[1] Noise factors are used in the experiment to force variability to occur, rather than depending on the random variation and sources of variability that may be intermittent. Robust performance is low variation in spite of the noises. The most significant noises must be carefully accounted for and included in the experiment. This engineered version of induced variability is unique to the Taguchi approach and is what promotes the optimization process. Linking this with concepts that are "pre-engineered for additivity" is a crucial strategic step in this methodology (see Chapter 8).

## 5.2 The Smaller-the-Better Type Signal-to-Noise Ratio

The smaller-the-better S/N ratio is based on the smaller-the-better loss function. The distinguishing characteristics of the smaller-the-better type problem are these:

- Response values or quality characteristics are continuous and nonnegative.
- The desired value of the response is zero.
- There is no scaling or adjustment factor; the goal is simply to minimize the mean and variance simultaneously.

Because there is no adjustment factor in these problems, the two-step optimization process is not applied. Instead, the smaller-the-better mean square deviation, Equation 3.14, is minimized, thus simultaneously reducing variation and putting the process as close to target ($m = 0$) as possible.

Minimizing the MSD in Equation 3.14 is the same as maximizing the S/N, which is defined by Equation 5.1.

$$S/N_{STB} = -10 \log[MSD] = -10 \log\left[\frac{1}{n} \sum_{i=1}^{n} y_i^2\right] \quad \textbf{(5.1)}$$

$$S/N_{STB} = -10 \log[S^2 + \bar{y}^2] \quad \textbf{(5.2)}$$

1. It *is* possible for the STB and LTB cases to base the S/N ratio calculation on one value to economize experimental effort.

In applying the S/N ratio, the raw data values, $y_i$, are influenced by the noise factor effects. These noise conditions are chosen to stress the system, producing large values, or outliers, if the system is susceptible to the effects of the noises. Thus, the S/N ratio measures the effect of noise. A closer examination of Equation 5.1 reveals why this particular mathematical expression is used.

$$S/N = -10 \log[1/n(\Sigma y_i^2)] = -10 \log[1/n(y_1^2 + y_2^2 + y_3^2 + \ldots + y_n^2)]$$

1. The individually measured response values, $y_i$, are squared to increase or bias the impact of any large $y$ values. This is done in order to be consistent with the quadratic loss function. Even without the discussion in Chapter 3, it is clear that any noise conditions that result in large $y_i$ values are of great concern. In practice, the few failures that occur because the quality characteristic is far from the target consume most of our attention.
2. The $y_i^2$ values are averaged to form the MSD. This packs the S/N ratio with the wealth of information the noise factors supply. Dealing with a single S/N ratio, rather than a set of individual responses, makes the analysis of the control factor effects much simpler, as shown in Chapter 7.
3. The logarithm of the MSD is taken to compress the data, in addition to the benefits already discussed in Chapter 4. To make the data even easier to understand and work with, a multiplicative factor, $-10$, is placed in front of the log function to make it obvious that the maximum S/N ratio represents the minimum quality loss. Therefore, the S/N for run #1, in Table 5.1, which has a smaller argument in the logarithm, is larger than the S/N for run #4 ($-7.7 > -37.7$). The factor of 10 converts the transformed number into *deci*bels. This simply magnifies the difference between the varying S/N values in the experimental region.

For example, consider the experiment shown in Table 5.1, where the total number of defects for four sets of molded parts is counted and the data span four orders of magnitude. The number of defects for each run can be analyzed by plotting the data vs. run number, as shown in Figure 5.1. The plot demonstrates that run #4 would dominate any analysis of the set points used for each of the runs. This is unfortunate because there may be useful differences among runs 1, 2, and 3 that could be exploited to reduce the number of defects and thus produce a better product.

Now consider what happens when the data are modified by taking the logarithm. The compression effect reduces several orders of magnitude to a range between 0 and $-4$. This is very useful when evaluating the effects of several control factors. While the smaller-the-better S/N ratio should depend mainly on the largest individual measurements, it is not helpful for the analysis of design options to be

**Table 5.1** Data illustrating the STB S/N ratio

| *Run* | *Mean square number of defects* | *log (MSD)* | *−10 log (MSD)* |
|---|---|---|---|
| 1 | 6 | 0.77 | −7.7 |
| 2 | 60 | 1.77 | −17.7 |
| 3 | 600 | 2.77 | −27.7 |
| 4 | 6000 | 3.77 | −37.7 |

**Figure 5.1** Number of defects vs. run number

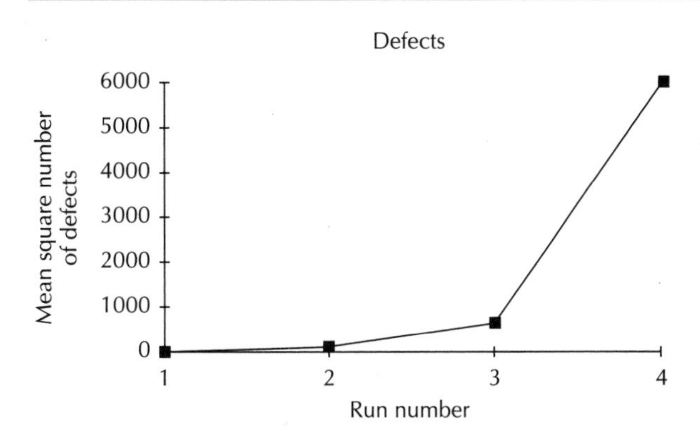


dominated or distorted by the largest S/N ratio. Rather, every possible opportunity, big and small, to improve the system's robustness should be exploited. Figure 5.2 illustrates the effect of these transformations on the sample data. Now the difference between runs 1 and 2 appears just as significant as the difference between runs 3 and 4.

The bottom line is that this transform is used to simplify data interpretation and magnify the main control factor effects. Part of this simplification occurs because the argument of the smaller-the-better S/N ratio takes both the average and the variability around the average into account, as shown in Equation 5.1. It should be noted that this simplification comes at a price: The mean square and the variance are combined (confounded) in the mean square deviation. In analyses of complex designed experiments, the simplification is valuable and not much information is lost. In some situations, such as benchmarking, where comparisons are limited in scope and as much information as possible is wanted, consider evaluating the mean, the variance, *and* the S/N ratio to get several "views" of the data.

**Figure 5.2** Transformed number of defects vs. run number

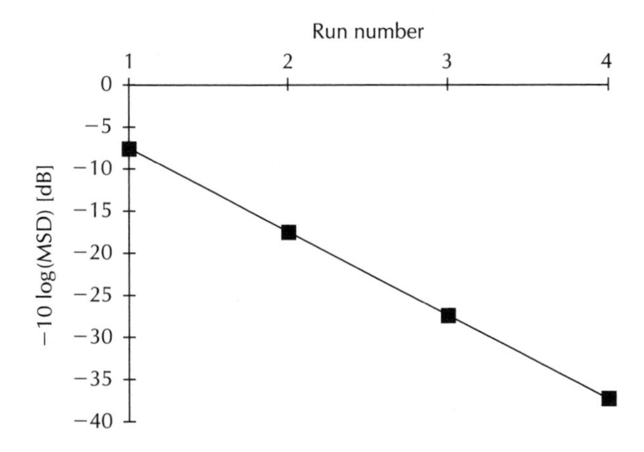

**Table 5.2** Data for the injection molding example

| *Machine* | *Noise 1* | *Noise 2* | *Noise 3* |
|---|---|---|---|
| 1 | 0.011 | 0.013 | 0.015 |
| 2 | 0.006 | 0.009 | 0.007 |
| 3 | 0.009 | 0.019 | 0.015 |
| 4 | 0.020 | 0.026 | 0.022 |

### 5.2.1 An Injection Molding Process Benchmarking Example

This example compares four different molding machines to determine which has the best performance with respect to part shrinkage. The control factors are Machines 1, 2, 3, and 4. The noise factors in this benchmarking exercise are three different types of plastic. The response in this case is part shrinkage in inches. Ideally the shrinkage = 0. A full factorial experiment based on the four different machines and the three noise factors is designed and carried out. The goal is to compute the average shrinkage $\bar{y}$, the MSD, the smaller-the-better quality loss, and the smaller-the-better S/N ratio for each machine. The loss is set to $A_0$ = \$20 at the customer tolerance $\Delta_0$ = 0.05 in., which means $k$ = 8,000 \$/in.$^2$

An evaluation of each of the four machines under each of the three noise conditions gives the data shown in Table 5.2. It appears, from this data, that machine number 2 gives the best performance (lowest shrinkage). But by how much? What if the machines have different capital cost or operating expense? To fully evaluate this case, the loss function and S/N ratios are employed. Using Equation 5.2 and, for the loss, Equation 3.15, $L(y) = k[S^2 + \bar{y}^2]$, the results shown in the first row of Table 5.3 are obtained.

As an exercise, fill in the rest of Table 5.3. From the values that you obtain you should be able to note the following about the S/N as a measure of performance:

- S/N increases as the mean decreases
- S/N increases as variability decreases
- For each S/N increase of 3 dB, the loss per piece decreases by half
- A gain of $x$ dB in S/N means the loss is reduced by $(1/2)^{x/3}$

**Table 5.3** Data and computations for the injection molding example

| *Machine* | *Noise 1* | *Noise 2* | *Noise 3* | *Mean* | *Std. dev.* | *Loss* | *S/N* |
|---|---|---|---|---|---|---|---|
| 1 | 0.011 | 0.013 | 0.015 | 0.013 | 0.002 | $1.38 | 37.62 |
| 2 | 0.006 | 0.009 | 0.007 | | | | |
| 3 | 0.009 | 0.019 | 0.015 | | | | |
| 4 | 0.020 | 0.026 | 0.022 | | | | |

## 5.3 The Larger-the-Better S/N Ratio

The distinguishing characteristics of the larger-the-better type problem are the following:

- Response values are continuous nonnegative numbers ranging from 0 to infinity
- The desired value of the response is infinity or the largest number possible
- This type of S/N has no scaling or adjustment factor
- Larger-the-better problems are the reciprocal of the smaller-the-better problems

Again, because there is no adjustment factor in these problems, the two-step optimization process is not used. Instead, the larger-the-better mean square deviation, Equation 3.18, is minimized. Minimizing the MSD is the same as maximizing the S/N, which is defined by Equation 5.3.

$$\mathrm{S/N_{LTB}} = -10\log[\mathrm{MSD}] = -10\log\left[\frac{1}{n}\sum_{i=1}^{n}\left(\frac{1}{y_i^2}\right)\right] \tag{5.3}$$

### 5.3.1 A Welding Process Example

This example involves a welding process that can be performed by four different machines. Here, the noise factor is the welding machine operator. The quality characteristic in this case is weld strength, in pounds per square inch, psi. You are to compute the average weld strength, the MSD, the Quality Loss, and the average S/N for each machine. The loss is set at $A_0$ = \$800 at the customer tolerance $\Delta_0$ = 2000 psi, which means $k = A_0\Delta_0^2 = 3.2 \times 10^9$ \$psi$^2$.

Complete the rest of Table 5.4 using the larger-the-better equations, 3.19 for the quality loss and 5.3 for the S/N ratio. You should find that machine #3 is the best and machine #2 is the worst—that much is clear from the raw data. More interesting is the comparison of machines #1 and #4. They are much closer in performance, but the MSD, loss, and S/N ratio take into account the variability of machine #1 in a way that the mean alone cannot. Note that there is about a 3 dB difference between them, which is equivalent to a 2× difference in the quality loss.

The preceding examples, the smaller-the-better injection molding study and the larger-the-better welding process study, are simple illustrations of the use of the Quality Loss Function and the S/N ratio to quantify the performance of several alternatives. Such comparative studies represent a common engineering task. These examples illustrate how to go beyond looking only at the average performance and to include variation in the engineering thought process. This is a critical robustness issue. Through years of experience, engineers have acquired a good intuitive feel for how to interpret mean responses.

**Table 5.4** Data and computations for the welding process example

| *Machine* | *Noise 1* | *Noise 2* | *Noise 3* | *Mean* | *MSD* | *Loss* | *S/N* |
|---|---|---|---|---|---|---|---|
| 1 | 9340 | 3030 | 7830 | 6733 | $4.56 \times 10^{-8}$ | \$145 | 73.4 |
| 2 | 2980 | 2450 | 3100 | | | | |
| 3 | 7890 | 9100 | 8340 | | | | |
| 4 | 6550 | 6700 | 6950 | | | | |

But high reliability and Six Sigma performance are achieved by studying variations and the distributions that describe them, and by paying attention to the outlier data points. Most engineers do not have good intuition when it comes to the behavior of distributions, simply because of a lack of experience. By measuring performance using noise factors to induce variation, engineers can experience the effects of variation. The common statistics, variance and standard deviation, are the starting point for measuring variation. The loss function provides the added insight of the mean square deviation. The S/N ratio modifies the data in a way that is necessary when design of experiments is applied for product optimization, as shown in Part II: Parameter Design.

## 5.4 The Operating Window: A Combination of STB and LTB

The smaller-the-better and larger-the-better cases can be combined to form a useful S/N ratio for solving problems that require the establishment of an operating window [C2]. Often the engineering system has a binary-type performance, i.e., the performance tends to be good or bad (1 or 0) with a rapid gradation between these performance extremes. This creates a substantial problem for efficient experimentation: how to quantify the system performance. One way is to collect data for a large number of systems or system trials and compute the ratios of bad results to total results. This results in a percentage or fraction of defective results. (The appropriate signal-to-noise treatment for that type of analysis is discussed in Section 5.5.) Such a metric, however, has two major drawbacks:

1. The experimental effort required to obtain a meaningful (i.e., statistically valid) value can be very large when the probability of failure for any one trial is low.
2. The metric is devoid of any physical information—it just adds up unfortunate events.

A much more useful analysis can be obtained by measuring the locations of the thresholds between the good and bad behaviors. An *operating window* is the range between two performance limit thresholds. The smaller-the-better analysis is applied to the lower threshold, and the larger-the-better analysis is applied to the upper threshold. The operating window requires that there exist a critical parameter upon which the threshold measurements are based. Then the discontinuous behavior can be optimized using a physically meaningful, continuous engineering parameter, referred to as the operating window signal factor. Thus, the threshold values are the quality characteristics for the operating window analysis.

The operating window signal factor behaves as illustrated in Figure 5.3. Here the two-step optimization process is applied. The robustness objective is to make the window as wide as possible. The second step is to put the nominal value of the signal factor at the center of the window. When this is done, the system is inherently robust because the width of the window is determined using noise factors to stress the system. A wide window insures that variation in the operating window signal factor does not cause performance variation.

The performance thresholds can be defined as the level of the signal factor where 50% of the tests fail, as shown in Figure 5.3. Finding the threshold level at a probability of failure, $p = 0.50$, takes a much smaller experimental effort than finding the (very small) value of $p$ near the midpoint of the operating window. The location in the probability distribution where $p = 0.5$ is a statistically robust value. In comparison, the probability of failure at some point on the distribution's tail where $p << 0.5$

**Figure 5.3** Illustration of the operating window

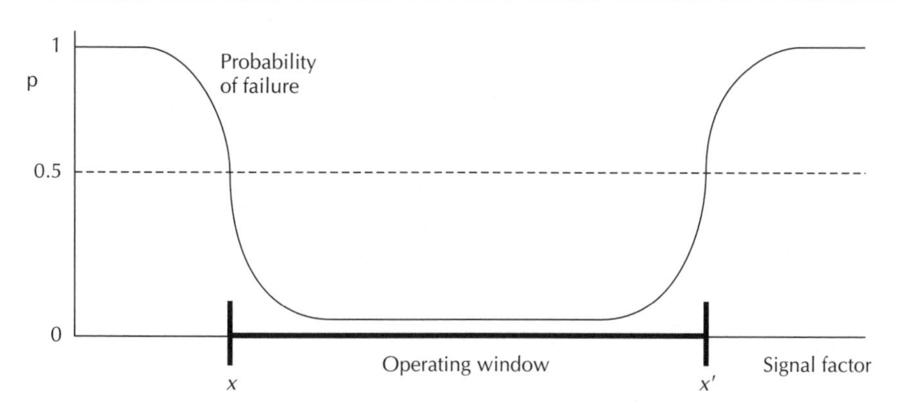


is a nonrobust[2] statistic that is difficult to measure. When the system performance is fairly reliable at the nominal level of the operating window signal factor, the probability of failure is low. In that case, finding the value of the failure probability with precision and trying to optimize it is very difficult.

The distance between $x$ and $x'$ in Figure 5.3 constitutes the operating window. The signal-to-noise is obtained from the following experimental objective:

1. Minimize $x$, treating it as a smaller-the-better case;
2. Maximize $x'$, treating it as a larger-the-better case.

Thus, the operating window S/N ratio is given by Equation 5.4.

$$\mathrm{S/N_{OW}} = \mathrm{S/N_{STB}}(x) + \mathrm{S/N_{LTB}}(x') = -10\log\left(\frac{\Sigma x^2}{n}\right) - 10\log\left(\frac{\Sigma (1/x')^2}{n}\right) \tag{5.4}$$

The following are examples of operating window problems:

1. In a computer floppy disk reader, the head-to-track alignment is the signal factor, and the levels where the error rate is 10% are the threshold values. (Note that in this case it may be appropriate to use as the threshold level 1% or 0.1%. However, measuring data transfer errors at a few parts per billion, which would be a typical error rate at nominal conditions, is exceedingly difficult.)
2. In a paper feeder, the friction force between the feed rollers and the sheet is the signal factor. The level where there is a misfeed rate of 50% is the lower threshold, and the level where there is a multifeed rate of 50% is the upper threshold.
3. In medication dosage for cold symptoms, the amount of medication is the signal factor. The level where there is no effect in half the patients is the lower threshold level, and the level where half the patients suffer some side effect is the upper threshold level.
4. In medication dosage for treating cancer cells, the medication level that kills 50% of the cancer cells in vitro is the lower threshold, and the medication level that kills 50% of the healthy cells in vitro is the upper threshold.

2. Statistically, nonrobust means that the value is strongly influenced by the shape of the distribution. In practice, this means that many more tests are required to determine the probability accurately.

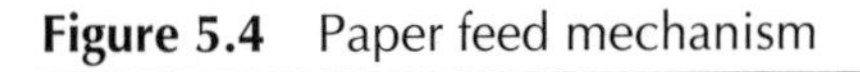

**Figure 5.4** Paper feed mechanism

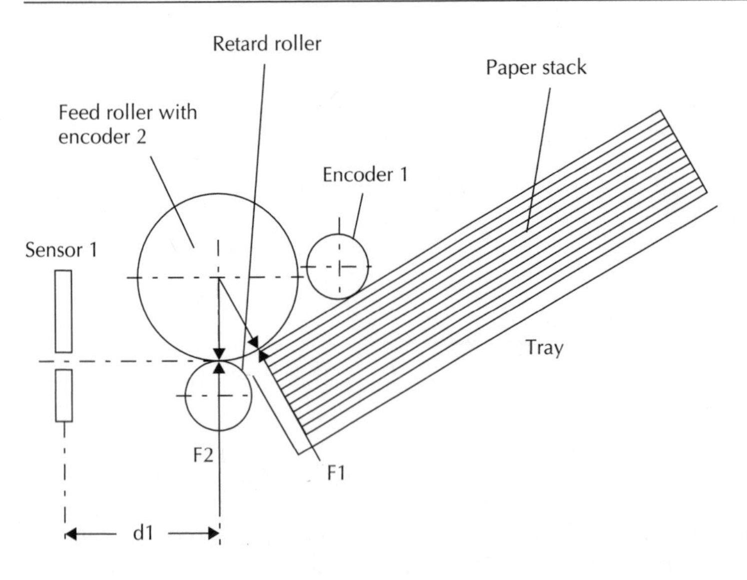


## 5.4.1 An Operating Window Case Study[3]

The following case study involves a device that is designed to feed sheets of paper through a printer, shown in Figure 5.4. The typical method for testing paper jams in a printer's paper feeder is to sample 5000 feedings and then record the number of feed failures. A better technique is to focus on what makes the feeder function. What forms of energy are going into the device? What transformation of that input energy takes place, and what is the desired form of output energy that supports the intended function?

Here are some facts about a typical paper feeding mechanism. Paper feeding is activated by a friction feed roller that is engaged by a spring-loaded mechanism. A normal force is being produced by the spring-loaded mechanism, so the potential energy of the spring induces a force, F1—physics at its simplest level! The ideal engineering function should be conceived in terms of this force.

The problem becomes one of identifying an operating window signal factor that is a manifestation of the intensity of the force F1. Two threshold values are measured for F1 in the following way. Twenty sheets are fed at different values of F1 ranging from 1.0 N to 10.0 N. Misfeeds and multifeeds are determined by sensor 1. The engineering team decided to use as a performance threshold $p = 0.20$ rather than the usual 50%. A simple regression curve is produced for each set of data and two threshold values are determined:

1. $x$ = the force at which 20% of the sheets fail to feed.
2. $x'$ = the force at which 20% of the actuations cause more than one sheet to begin to feed.

Notice how the quality characteristics $x$ and $x'$ are directly tied to the physics of the device. Counting paper misfeeds would be an acceptable statistical treatment of the problem, but would make

3. Contributed by Dr. Reinhold Weltz and Mr. Markus Weber, Eastman Kodak Company.

poor use of engineering knowledge. Such an approach does not lead to physical insight into how to improve the design. Such a data treatment would lead to unnecessary and confusing interactions in the data analysis (purely because it is not related to the physical explanation of the problem). Most, but not all, interactions in data are caused by improperly evaluating the real physics of the problem, rather than side effects or aftereffects. The operating window, as shown here, addresses the real engineering problem. The experimental objective is to minimize $x$ and maximize $x'$. The distance between $x$ and $x'$ is the operating window. The wider the mechanism's operating window, the larger is the range within which one piece of paper can be fed without jamming. For each of the experiments, various noise factors are used, including paper weight and type, environment, alignment errors, and spacing errors. The latter two represent expected assembly variations, which are important internal noise sources.

Eighteen experiments are made using different combinations of control factors and levels (see Chapter 7). We do not go through a detailed analysis here, but Figure 5.5 shows the results of the 18 experiments graphically with the bars marked Run 01 to Run 18. The threshold limits shown can be converted to a S/N ratio using Equation 5.4. The resulting S/N ratios are analyzed using analysis of means (Chapter 7) to predict which combinations of control factor set points are responsible for the good results seen in runs 4, 7, 12, 17, and 18. Once those trends are known, the set points can be combined to "design in" the wide operating window shown for run 30. Compare the final result with the original design's operating window shown at the top of the figure. There is a substantial improvement in robustness. Note that the analysis of each run allows results with wide operating windows to be combined regardless of the location of the operating window. This illustrates the power of both the S/N ratio and the two-step optimization approach.

## 5.5 A Signal-to-Noise Ratio for Probability

The operating window is one approach to deal with a problem in which the system has binary-type performance, i.e., the performance tends to be good or bad (1 or 0). The operating window requires that there exists a parameter that has a rapid transition between these performance extremes upon which threshold measurements are based. The ideal operating window signal factor is based on the physics or energy transformation of the system.

How should we treat a problem where there is no known parameter that is suitable for use in an operating window analysis? Often these are cases where the individual events are not enumerable. In such a case, the probability or percent ratio of successful events to total opportunities, often referred to as *yield,* is measured directly as a macro variable. The following are examples where yield is used as the quality characteristic:

1. Chemical reactions, where the yield represents the ratio of individual molecules that have successfully reacted to form the desired state.
2. Separation processes, such as ore extraction, where the yield represents the percent of metal mass that is successfully separated from the ore.
3. Electrostatic toner transfer in an office copier machine, where the transfer efficiency is the percent of toner particles that are transferred from the photoconductor film to the copy paper.

All these represent larger-the-better type problems where the yield (or transfer efficiency) is to be maximized.

**Figure 5.5** Results of the force limits operating window case study

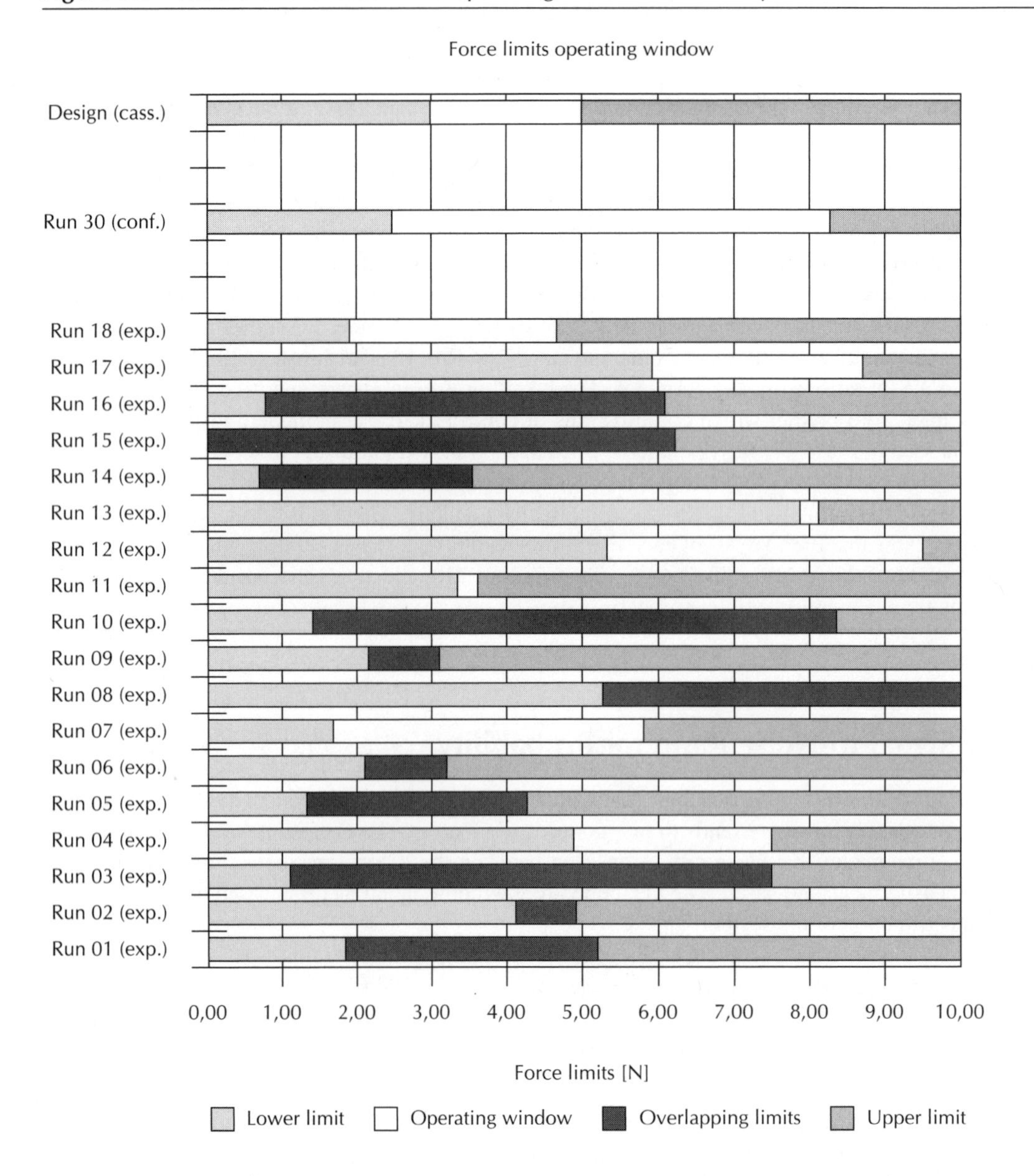


It is also possible to formulate smaller-the-better problems in a similar manner, such as the fraction defective [P2] for a tolerance-based manufacturing metric, or the probability of paper-feed stoppage in a printer. Note that these applications are measures of problems or failure modes, rather than measures that focus on a desired or target behavior. It is difficult to improve product performance based solely on reliability data. This is because there is little engineering information in measures of failure.

It is far better to identify the desired physical behavior, and then to measure how close to ideality the system can perform.

The smaller-the-better and larger-the-better cases are used when the range of values possible for the quality characteristic goes from zero to infinity. The target values within that range are zero and infinity, respectively. In the probability case discussed here, however, the range of values is bounded. That is, the quality characteristic, $p$, is confined to the range:

$$0 \leq p \leq 1.$$

Let us calculate an S/N ratio for yield, $y$, by assuming that the system under study has two possible states for each event, $y_i$, (e.g., a molecular reaction or toner particle transfer) that are represented by 0 and 1. A 1 is assigned when the event results in the desired state, and a 0 is assigned when the event results in a failure to achieve the desired state. The data can be formed into a series, $y_1, y_2, \ldots, y_n$, of zeros and ones.

The probability of success (the yield) is given by

$$p = \bar{y} = \frac{y_1 + y_2 + \ldots + y_n}{n} \tag{5.5}$$

The signal-to-noise ratio has the same form as that given in Equation 4.10, S/N $= 10 \log(\bar{y}^2/s^2)$, since the design goal is to maximize the mean while minimizing variability. Because the values of $0^2 = 0$ and $1^2 = 1$,

$$p = \frac{1}{n}\sum_{i=1}^{n} y_i \quad \text{and} \quad p = \frac{1}{n}\sum_{i=1}^{n} y_i^2 \tag{5.6}$$

The variance, $S^2$, is given by Equation 2.2,

$$S^2 = \frac{1}{n-1}\sum_{i=1}^{n}(y_i - \bar{y}_i)^2$$

After we expand the expression for the variance and use Equations 5.5 and 5.6, the S/N ratio is given by

$$\begin{aligned}
S^2 &= \frac{1}{n-1}\sum_{i=1}^{n}(y_i - \bar{y})^2 \\
&= \frac{1}{n-1}\sum_{i=1}^{n}(y_i^2 - 2y_i\bar{y} + \bar{y}^2) \\
&= \frac{1}{n-1}\left[\sum_{i=1}^{n} y_i^2 - 2\bar{y}\sum_{i=1}^{n} y_i + \sum_{i=1}^{n} \bar{y}^2\right] \\
&= \frac{1}{n-1}[np - 2np^2 + np^2] \\
&= \frac{n}{n-1}[p - p^2]
\end{aligned}$$

Since $n >> 1$ in almost all cases, the variance is given by

$$S^2 \cong p(1 - p) \tag{5.7}$$

Thus, the S/N ratio is given by

$$\text{S/N}_{p\text{-LTB}} = 10 \log\left(\frac{\bar{y}^2}{s^2}\right) = 10 \log\left(\frac{p^2}{p(1-p)}\right)$$
$$= 10 \log\left(\frac{p}{1-p}\right) \quad \textbf{(5.8)}$$

Note that Equation 5.8 applies when the engineering system is intended to maximize the probability $p$ (e.g., yield). For those cases where it is desirable to minimize the probability $p$ (e.g., fraction defective) the S/N ratio is the reciprocal of Equation 5.8:

$$\text{S/N}_{p\text{-STB}} = -10 \log\left(\frac{p}{1-p}\right) \quad \textbf{(5.9)}$$

Note the change in sign in front of the logarithm to account for the reciprocal relationship.

One of the properties of the S/N ratio listed in the preceding chapter is that it helps minimize interactions between control factors when it is used as the response in a designed experiment. The S/N ratio for the larger-the-better probability, Equation 5.8, can be used to illustrate how an S/N ratio can result in improved additivity. Remember that additivity implies that the individual effects of several parameters, such as control factors, on the quality metric can simply be added when constructing a mathematical model. In order for this to be possible, the quality characteristic itself must use the entire number line. Otherwise, the quality metric is not closed to the operation of addition.[4] For a probability, such as yield, the number line is bounded as shown in Figure 5.6.

The yield, $p$, is transformed by the ratio, $[p/(1-p)]$. This results in an expanded range on the number line as shown in Figure 5.7. Finally, the S/N ratio expression, Equation 5.8, uses the entire number line as shown in Figure 5.8, and is therefore closed to the operation of addition. This means simply that there is no a priori reason to say that an additive model cannot be applied to the S/N ratio. Unfortunately, the same cannot be said of the yield.

The use of the S/N ratio to help simplify the analysis of control factor effects is very important in designed experiments. Smaller experiments can be run, and the complexity of the resulting model is reduced. As a result, experiments using the S/N ratio are much easier to accomplish and comprehend. The preceding analysis helps illustrate how the S/N ratio accomplishes this. The following example illustrates the reduction in control factor interactions that results from the application of the S/N ratio.

### 5.5.1 A Chemical Process Yield Example

Let us now look at a quantitative example to illustrate how the S/N ratio, Equation 5.8, promotes additivity, thus simplifying the analysis. Here data are given for a simple chemical reaction describing the effect of two control factors: the amount of accelerator added, and the temperature. Physically, the effect of each factor is not negated by the other's effect, within the range of values chosen (the experimental space). Thus, the accelerator is a chemical additive that increases the speed of the reaction regardless of the temperature level. Further, the temperature is a factor that increases the speed of the

---

4. The yield is not closed to the operation of addition because it is possible to add yields $y_1$, $y_2$, $y_3$ and get a result $y = y_1 + y_2 + y_3$ that is not an allowed value on the number line for yield. (Consider $y_1 = y_2 = y_3 = 0.5$.)

**Figure 5.6** Available number line for yield

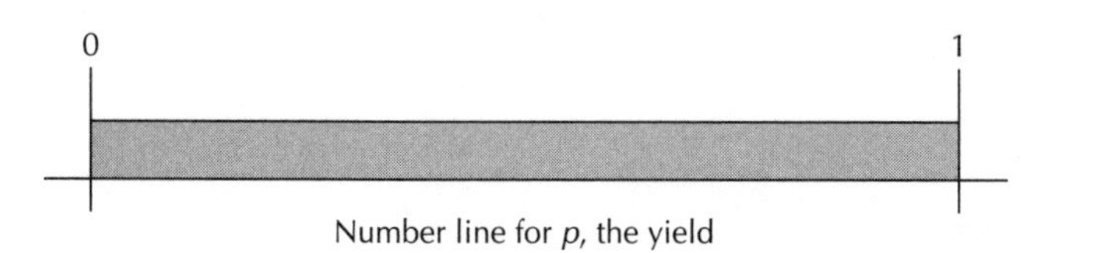


**Figure 5.7** Available number line for yield after transformation

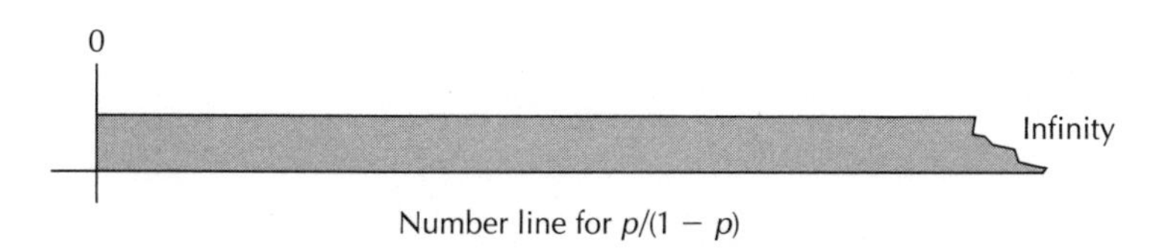


**Figure 5.8** Available number line for the $S/N_{p-\mathrm{LTB}}$

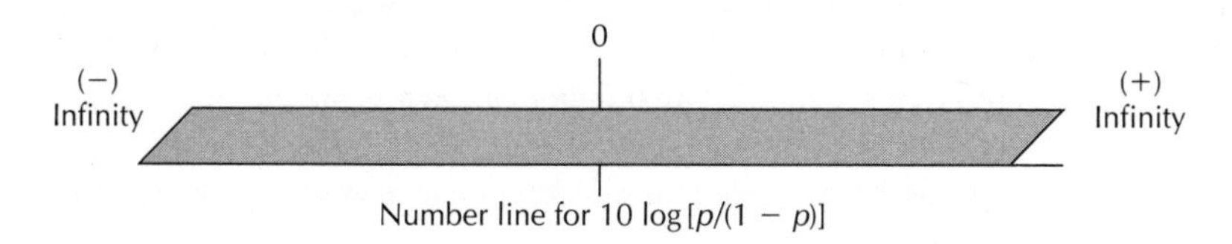


reaction regardless of the accelerator level. The yield increases as the reaction speed increases, subject to the limitation of 0 < yield < 100%. The data in Table 5.5 describe the results.

Compare the change in the yield for experiments 1 and 3 with the change in the yield for experiments 2 and 4. Increasing the accelerator concentration from a low level (L) to a high level (H) consistently increases the yield. However, the magnitude of change is not the same, indicating that the degree of improvement depends on the temperature level. Similarly, compare the change in the yield for experiments 1 and 2 with the change in the yield for experiments 3 and 4. Increasing the temperature from a low level (L) to a high level (H) consistently increases the yield. However, again the magnitude of change is not the same, indicating that the degree of improvement depends on the accelerator concentration level. When the magnitude of the effect of a factor depends on the level of another factor, that is referred to as an *interaction* between the factors. This one happens to be mild.

**Table 5.5** Data for chemical yield example

| *Experiment* | *Accelerator* | *Temperature* | *Yield* | *S/N* |
|---|---|---|---|---|
| 1 | L | L | 0.4 | −1.7 |
| 2 | L | H | 0.8 | 6.0 |
| 3 | H | L | 0.6 | 1.8 |
| 4 | H | H | 0.9 | 9.5 |

**Figure 5.9** Yield interaction plot (temperature 1 = low, 2 = high)

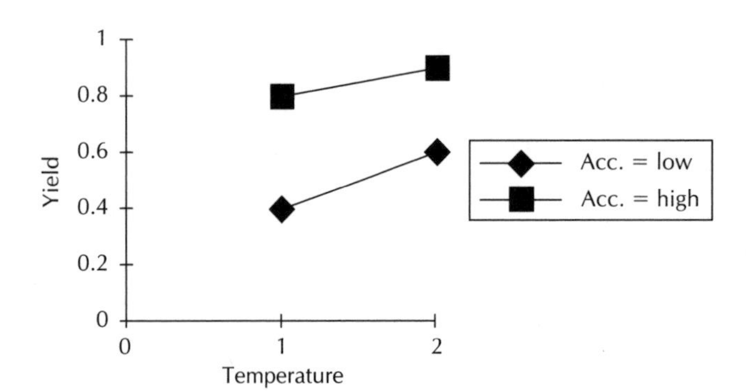


The last column, the S/N ratio, is computed by simply applying Equation 5.8 to the yield values. Now, making the same comparisons described in the previous paragraph, both the direction of change *and* the magnitude of change are consistent. This means that, using the S/N as the response, there is no interaction. Its effect has been numerically suppressed.

It is easier to see these effects by employing a simple graphing tool known as the *interaction plot.* First consider the interaction plot for the yield values, shown in Figure 5.9. There is clearly a difference in slopes for the two lines. Thus, the magnitude of the effect of temperature as it is changed from low (1) to high (2) depends upon the accelerator concentration.

Now look at a similar graph for the S/N ratios (Figure 5.10). Here the lines are parallel, thus indicating that there is no interaction. So, let's ask the question—what happened? In the first case, yield is limited. It cannot exceed 100%. As the sum of the individual effects approaches 1.0, this limit forces diminishing returns, *but not because of the physics!* Exactly the same data, after the S/N ratio transformation, show clearly that there is no fundamental interaction. That is the power of using the S/N ratio.

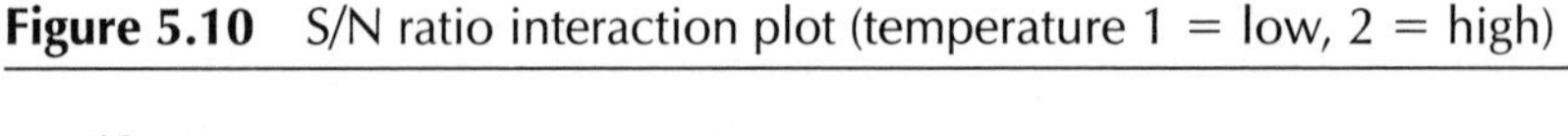

**Figure 5.10** S/N ratio interaction plot (temperature 1 = low, 2 = high)

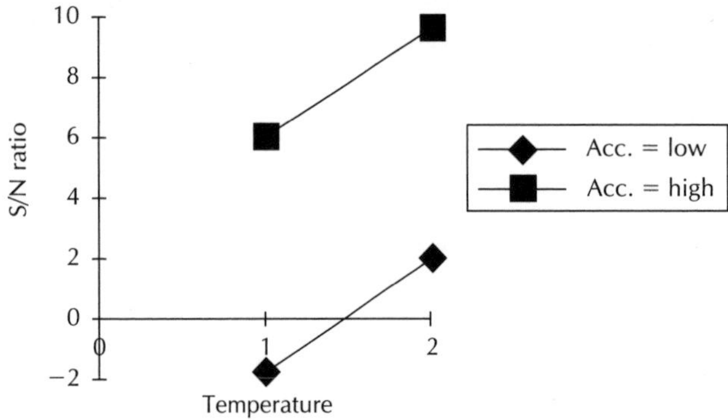

**Figure 5.11** Reaction concentrations for a simple three-stage reaction

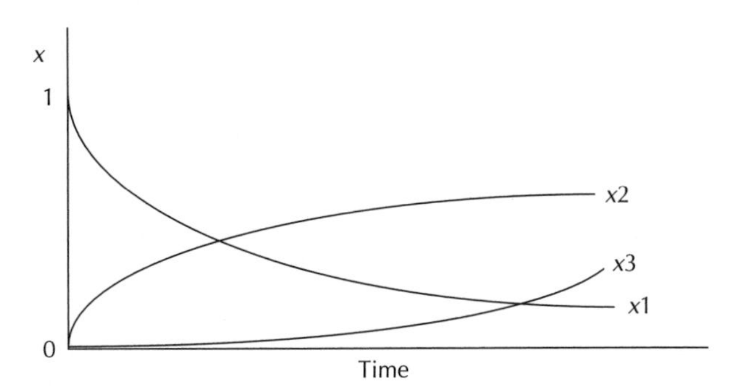


## 5.5.2 A New S/N Ratio for Consecutive Chemical Reactions

Recently, Dr. Taguchi [T10] has developed and published a new S/N ratio for a simple first-order chemical process that has three stages:

$$M1 \Rightarrow M2 \Rightarrow M3$$

where M1 is the initial substance and M2 is the desired substance being produced by the chemical reaction M1 ⇒ M2. Such a first-order reaction is often temperature-initiated, but the heat energy can cause the desired product to break down, thus producing a by-product substance caused by the chemical reaction M2 ⇒ M3.

Let the concentration fraction of the substances M1, M2, and M3 be $x_1$, $x_2$, and $x_3$, respectively. Normally, such a problem is analyzed by tracking the percent fractions of each of the materials as the reaction proceeds. A typical plot of the fractions is shown in Figure 5.11. It is clear that the behavior is complex because the relationship among $x_1$, $x_2$, and $x_3$ is not monotonic.[5]

The challenge is to produce an S/N ratio that describes the system in a way that is independent of adjustment. The yield, given by $x_2$, can be treated using Equation 5.8, but it is also possible to produce more or less M2 simply by adjusting the time of the reaction. To optimize the efficiency of the reaction, it is necessary to maximize $x_2$, while minimizing $x_1$ and $x_3$. This can be done by combining the larger-the-better and smaller-the-better S/N ratios for probability to form an operating window for this problem.

If there are no other by-products, then the following connecting relationships apply:

$$\begin{aligned} Y_1 &= x_1 \\ Y_2 &= x_1 + x_2 \\ Y_3 &= x_1 + x_2 + x_3 = 1 \end{aligned} \tag{5.10}$$

These are very useful for simplifying the problem, because their behavior *is* monotonic, as shown in Figure 5.12.

5. Monotonic is defined as a consistent inequality between parameters. For example, in Figure 5.12, $Y_3 > Y_2 > Y_1$ for all time values. The concentration fractions, $x_1$, $x_2$, and $x_3$, do not have this simple relationship.

**Figure 5.12** Reaction concentrations for the connecting functions, Equation 5.10

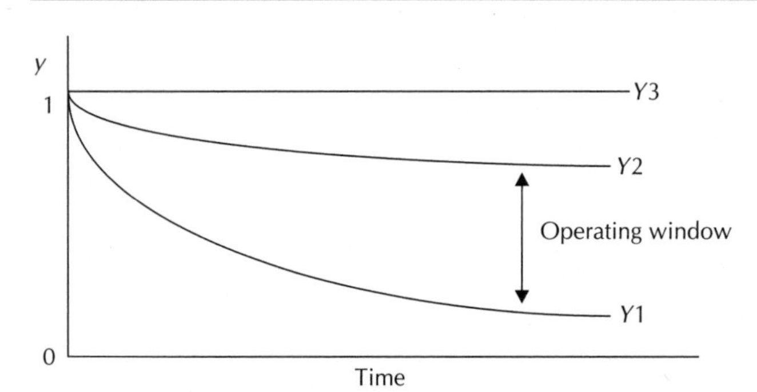


Recognizing that $Y_3$ is a constant, the system can be optimized by minimizing $Y_1$ *while maximizing* $Y_2$, as shown in Figure 5.12. Thus, there is an operating-window relationship between $Y_1$ and $Y_2$. Because $x_2$ is part of $Y_2$, optimizing the operating-window condition optimizes the reaction efficiency. Furthermore, because the operating window can be optimized using any value of time, it is independent of adjustment, thus making time the tuning factor in this problem.

Applying Equation 5.9, the *p*-STB S/N ratio for $Y_1$ is given by Equation 5.11:

$$\mathrm{S/N}_{Y_1} = -10 \log\left[\frac{Y_1}{1-Y_1}\right] \tag{5.11}$$

Applying Equation 5.8, the *p*-LTB S/N ratio for $Y_2$ is given by Equation 5.12:

$$\mathrm{S/N}_{Y_2} = 10 \log\left[\frac{Y_2}{1-Y_2}\right] \tag{5.12}$$

The operating window S/N ratio is the sum of the S/N ratios in Equations 5.11 and 5.12:

$$\begin{aligned}\mathrm{S/N}_{\mathrm{Chem}} &= 10 \log\left[\frac{Y_2}{1-Y_2}\right] - 10 \log\left[\frac{Y_1}{1-Y_1}\right] \\ &= 10 \log\left[\frac{Y_2}{Y_1}\left(\frac{1-Y_1}{1-Y_2}\right)\right]\end{aligned} \tag{5.13}$$

For example, consider the data for three reactions shown in Table 5.6. All three of these examples have the same yield, 60%. However, it appears in Case 1 that there is an excessive amount of material, $x_1$, not yet reacted. More product, $x_2$, can be produced simply by running the reaction longer. In Case 3,

**Table 5.6** Data for illustrating the chemical reaction S/N ratio

| | $x_1$ | $x_2$ | $x_3$ |
|---|---|---|---|
| *Case 1* | 0.30 | 0.60 | 0.10 |
| *Case 2* | 0.20 | 0.60 | 0.20 |
| *Case 3* | 0.05 | 0.60 | 0.35 |

**Table 5.7** Data for illustrating the chemical reaction S/N ratio

| | $Y_1$ | $Y_2$ | $Y_3$ | S/N |
|---|---|---|---|---|
| *Case 1* | 0.30 | 0.90 | 1.00 | 13.2 |
| *Case 2* | 0.20 | 0.80 | 1.00 | 12.0 |
| *Case 3* | 0.05 | 0.65 | 1.00 | 15.5 |

it appears that an excessive amount of by-product, $x_3$, has been produced. Perhaps in this case more product can be produced by reducing the reaction time. These data need to be analyzed in a way that is independent of adjustment to clarify the situation. This can be done by using the connecting relations, Equation 5.10, and the S/N ratio, 5.13. The results obtained are shown in Table 5.7.

The S/N values indicate that Case 3 will give the optimum yield when each of the cases is put on target. The operating window target is usually taken to be centered between the threshold limits. In this case, the target is to maximize the yield, $x_2$, while minimizing $x_1$ and $x_3$. Assuming that loss due to $x_1$ and $x_3$ is the same, then the optimum yield occurs when the reaction is balanced, that is, when $x_1 = x_3$. This is already true for Case 2, but what happens to the yield for the other two cases, which have higher S/N values, after tuning the time to achieve the balance condition? When $x_1 = x_3$, the connecting Equation 5.10 can be solved to give

$$Y_1 = (1 - x_2)/2 \qquad \textbf{(5.14a)}$$

$$Y_2 = (1 + x_2)/2 \qquad \textbf{(5.14b)}$$

Equations 5.14a and 5.14b can be substituted into Equation 5.13 to solve for the expected yield, $x_2$. The resulting solution is given by Equation 5.15:

$$x_2 = \frac{10^{\frac{S/N}{20}} - 1}{10^{\frac{S/N}{20}} + 1} \qquad \textbf{(5.15)}$$

Equation 5.15 is used to predict what the yield will be if the time is adjusted to the value at which $x_1 = x_3$, as shown in Figure 5.13. Applying Equation 5.16 to the three cases given earlier, we derive the expected yields:

- Case 1, $x_2 = 0.64$
- Case 2, $x_2 = 0.60$
- Case 3, $x_2 = 0.71$

Now it is clear why the S/N analysis indicated that Case 3 is superior to the other cases. But this would not have been apparent without a quality metric that is independent of adjustment. This example once again illustrates the power of the signal-to-noise ratio. Using a quality metric that is independent of adjustment along with the two-step optimization procedure, with time as the tuning factor, allows engineers to use all the information they possess to optimize their system. This example also illustrates how transformations, such as those given by the connecting relationships, Equation 5.10, are used to reformulate a complex problem in a simple monotonic manner. This approach is explored further in Part II, Parameter Design.

**Figure 5.13** Reaction concentrations for a simple three-stage reaction showing the optimum reaction time value

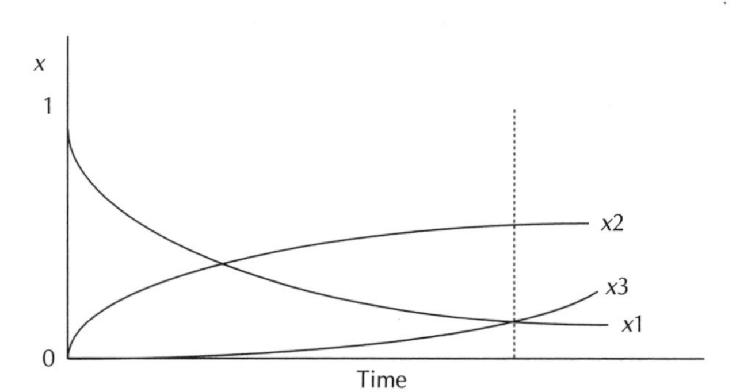


## 5.6 The Nominal-the-Best Signal-to-Noise Ratios

We now look at the two S/N ratios that are associated with the nominal-the-best loss function, Equation 3.12. The reason for multiple forms of the S/N ratios is that a lot of engineering knowledge is packed into the S/N ratio. Thus, there are unique S/N ratios for different situations. As before, some of the key aspects of each problem are described, from an engineering perspective, that suggest which S/N ratio is ideally suited for the problem analysis.

Regardless of the type of S/N ratio, there are a few common features. One is that the S/N ratio is always maximized. Another is that the S/N ratio is based on the average Quality Loss Function or the mean square deviation. The MSD for the nominal-the-best type is given in Equation 3.10. If we substitute the sample variance, $S^2$, for the population variance $\sigma^2$, the nominal-the-best MSD is given by

$$\text{MSD}_{\text{NTB}} = [S^2 + (\bar{y} - m)^2] \tag{5.16}$$

But Equation 5.16 describes the MSD in a way that is not independent of adjustment. There are two different nominal-the-best cases that require modifying the MSD in different ways to get the form most likely to be independent of adjustment.

### 5.6.1 Type I, Where the Mean and Standard Deviation Scale Together

The following characteristics distinguish the type I nominal-the-best problem:

- Response values are continuous and nonnegative, ranging from zero to infinity
- This type of problem has a nonzero target value and zero variance when the mean response is at zero

Some examples of the type I nominal-the-best include the following:

- Image density
- Projectile range

**Figure 5.14** Scaling assumption for the type I NTB

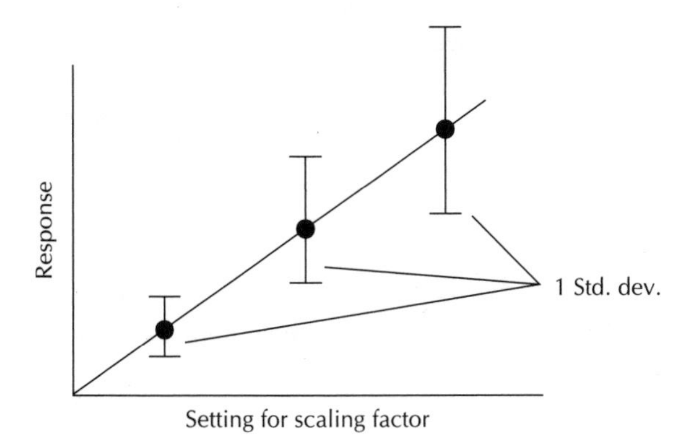


- Elastomer part dimension (variation due to temperature and humidity)
- Carbon resistor values

These are all examples of quality characteristics whose variation typically grows as their nominal magnitude grows. For example, in the case of 10% resistors, the absolute value for the resistance variation is 1000× for a 1 MΩ resistor compared to a 1 kΩ resistor. Figure 5.14 illustrates how these characteristics would look when graphed with their variability indicated by the error bars.

This case has a scaling factor to adjust the mean onto the target. In Chapter 4, a nominal-the-best S/N ratio is derived from the quality loss function using the scaling factor assumption. Adjustment factors can be used to adjust the mean, but a scaling factor is a special adjustment factor with the additional property that the standard deviation and the mean scale proportionally, as shown in Figure 5.15. If this relationship holds true, then the type I nominal-the-best S/N ratio has the desirable property of being a quality metric that is independent of adjustments to the mean.

The required condition, that there be no nonnegative values for the quality characteristic, insures

**Figure 5.15** Ideal relationship between the standard deviation and the mean for the type I NTB

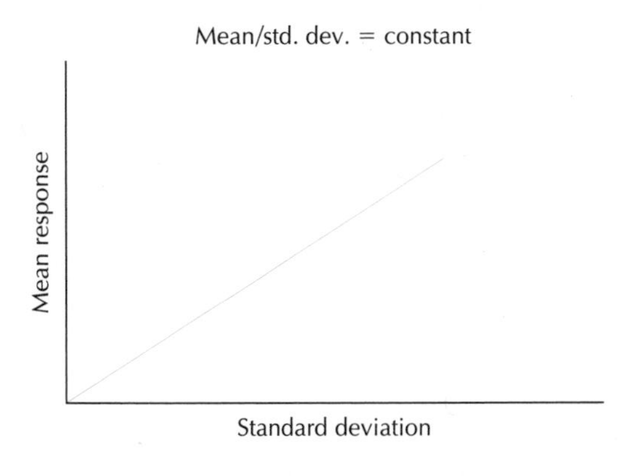

that the quality characteristic has an absolute zero. If this is the case, then the standard deviation *must* get smaller as the mean approaches zero, as shown in Figure 5.14. This makes it much more likely that a scaling factor relationship exists. It is not known if there is a linear relationship, nor is it known how far from the origin the relationship holds. The experimental results indicate the range of validity of these assumptions. For that reason, the experiment, in part, has the goal of finding a scaling factor. A good experimental plan includes control factors that have the potential for being scaling factors. During technology development, scaling relationships should be investigated and understood and used for nominal-the-best and dynamic analyses. Otherwise, the nominal-the-best approach is used to uncover such a factor in an existing design.

The nominal-the-best S/N ratio is based on the MSD given in Equation 5.16. Where the scaling assumption is applied, the MSD given in Equation 4.9 is independent of adjustment (IA) as discussed in Chapter 4. Applying the equation $S/N = -10\log(MSD_{IA})$,

$$S/N_{TypeI-NTB} = 10\log\left(\frac{\bar{y}^2}{S^2}\right) \qquad \textbf{(5.17)}$$

## 5.6.2 Interpreting the Type I Nominal-the-Best S/N Ratio

Let's apply the type I nominal-the-best S/N ratio to a table of simple numbers in order to better understand how this particular metric works relative to changes in the mean and variance. The calculations are done using Equation 5.17 where $\bar{y}$ and $S^2$ given in Equations 2.1 and 2.2, respectively.

Table 5.8 shows the type I NTB S/N ratio values as a function of both the mean and the variance. The highlighted values show how the mean and standard deviation can change together while the S/N remains constant. Remember that the ratio $S/\bar{y}$ is defined as the coefficient of variation. This chart shows that the COV and the type I NTB S/N ratio are very similar. If the mean grows and the variance stays constant, as highlighted in Table 5.9, the S/N increases as shown. This is important to notice, as it can matter a great deal when attempting to interpret the S/N ratio. If the mean stays constant and the variance decreases, as highlighted in Table 5.10, the S/N increases as shown. This is also important to notice when interpreting the S/N ratio. Each S/N increase of 3 dB results in a loss per piece that decreases by half. Table 5.11 gives an indication of how the use of logarithms translates into the interpretation of the S/N analysis [B4].

**Table 5.8** Nominal-the-best analysis highlighting constant $S/\bar{y}$

| $y_1$ | $y_2$ | $y_3$ | *Mean* | *Std. dev.* | $S^2$ | *S/N* |
|---|---|---|---|---|---|---|
| 9 | 10 | 11 | 10 | 1 | 1 | 20 |
| 90 | 100 | 110 | 100 | 10 | 100 | 20 |
| 99 | 100 | 101 | 100 | 1 | 1 | 40 |
| 900 | 1000 | 1100 | 1000 | 100 | 10,000 | 20 |
| 990 | 1000 | 1010 | 1000 | 10 | 100 | 40 |
| 999 | 1000 | 1001 | 1000 | 1 | 1 | 60 |

**Table 5.9** Nominal-the-best analysis highlighting constant standard deviation

| $y_1$ | $y_2$ | $y_3$ | Mean | S | $S^2$ | S/N |
|---|---|---|---|---|---|---|
| 9 | 10 | 11 | 10 | 1 | 1 | 20 |
| 99 | 100 | 101 | 100 | 1 | 1 | 40 |
| 999 | 1000 | 1001 | 1000 | 1 | 1 | 60 |

**Table 5.10** Nominal-the-best analysis highlighting changing standard deviation

| $y_1$ | $y_2$ | $y_3$ | Mean | S | $S^2$ | S/N |
|---|---|---|---|---|---|---|
| 900 | 1000 | 1100 | 1000 | 100 | 10,000 | 20 |
| 990 | 1000 | 1010 | 1000 | 10 | 100 | 40 |
| 999 | 1000 | 1001 | 1000 | 1 | 1 | 60 |

Let's take a look at what a 3 dB and 6 dB improvement in the type I NTB S/N means in terms of the variance and standard deviation.

$$\text{S/N} = 10\log(\bar{y}^2/S^2)$$

Suppose there is a 3 dB improvement in the S/N ratio due to a design change:

$$\begin{aligned}3\text{ dB} &= \text{S/N}_1 - \text{S/N}_0\\ 3 &= 10\,[\log(\bar{y}_1^2/S_1^2) - \log(\bar{y}_0^2/S_0^2)]\\ 0.3 &= \log\bar{y}_1^2 - \log S_1^2 - \log\bar{y}_0^2 + \log S_0^2\end{aligned}$$

Setting the mean on target, $\bar{y}_0 = \bar{y}_1$. Thus,

$$\begin{aligned}0.3 &= \log S_0^2 - \log S_1^2\\ 0.3 &= \log(S_0^2/S_1^2)\text{. Now take the antilogarithm.}\\ 10^{0.3} &= (S_0^2/S_1^2)\\ 10^{0.3} &\approx 2 = S_0^2/S_1^2\\ 2 &= S_0^2/S_1^2\\ S_0^2 &= 2S_1^2\\ S_0 &= 1.4S_1\end{aligned}$$

Similarly, if there is an S/N improvement of 6 dB:

$$\begin{aligned}6\text{ dB} &= \text{S/N}_1 - \text{S/N}_0\\ 0.6 &= \log S_0^2 - \log S_1^2\\ 10^{0.6} &= (S_0^2/S_1^2)\\ 4 &= S_0^2/S_1^2\\ S_0 &= 2S_1\end{aligned}$$

This indicates that a 6 dB improvement results in a 2× change in distribution width or, for example, *capability index* (Chapter 18). A 3 dB improvement results in a 2× change in the MSD or quality loss.

**Table 5.11** Percent change in the variance or quality loss with changes in the S/N ratio [B4]

| *Change in S/N* | 0.3 | 0.5 | 1 | 2 | 3 | 6 | 10 |
|---|---|---|---|---|---|---|---|
| *Change in* $S^2$ | 1.07 | 1.12 | 1.26 | 1.59 | 2 | 4 | 10 |
| *Percent change* | 7.2 | 12.2 | 25.9 | 58.5 | 100 | 400 | 1000 |

### 5.6.3 Type II, Where the Standard Deviation Is Independent of the Mean

The following characteristics distinguish the type II nominal-the-best problem:

- Response values are continuous and can take on either positive or negative values.
- For this type of problem, the target can be zero.

The type II NTB S/N ratio is referred to by Phadke [P2] as the Signed Target S/N ratio. The type II case differs from the type I case in that there is no reason, given the distinguishing characteristics, to assume that the standard deviation scales with the mean. In addition, the targeted response can be and often is zero, which is not desirable with the type I S/N ratio because $\log(0) = -(\text{infinity})$!

Some examples of the type II nominal-the-best include the following:

- Temperature controller errors around nominal
- Offset voltage for a power supply [P2]
- Skew in image to paper registration in an office copier

These are examples where the target is zero and the variation is largely independent of how close to the target the mean value is. Examples where the mean value target is not zero but the variation does not scale with the mean include the following:

- Machined part dimensions
- Casting dimensions
- Molded part dimensions

Figure 5.16 illustrates how these characteristics would look for the type II case, when graphed with their variability indicated by the error bars. Recall that the mean and standard deviation are both proportional to a scaling factor. In this case there is no scaling assumption. However, the two-step optimization process does apply to type II problems. This means that an adjustment factor is required to put the mean onto the target. Maximizing the S/N ratio requires the control factor set points that produce the least amount of variance. Ideally, for the type II NTB S/N ratio, there exists a factor that can adjust the mean onto the target without affecting the S/N ratio. Such an adjustment factor is referred to as a *tuning factor.* For this case, the existence of a tuning factor requires that the mean and standard deviation be independent, as Figure 5.17 shows.

Such behavior need not be evidenced by all adjustment factors. Only one tuning factor is required to accomplish the two-step optimization process. The condition that the quality characteristic values can be positive or negative means that there is no a priori reason in this case to assume that scaling applies, as shown in Figure 5.16. It is important that a true zero value for

**Figure 5.16** Adjustment assumption for the type II NTB

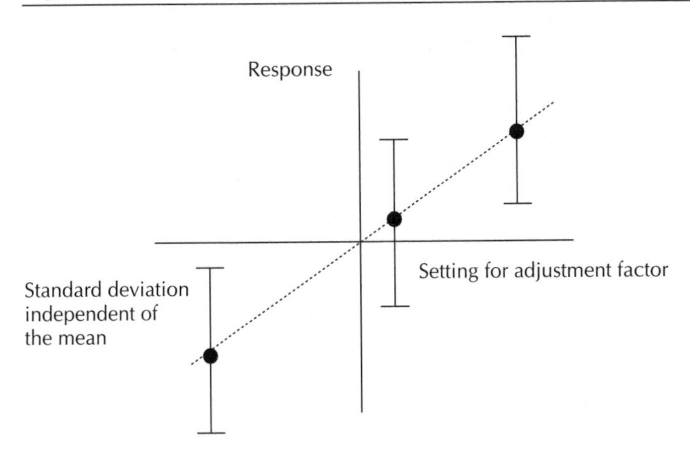


the quality characteristic exists, as opposed to a zero value obtained by applying an offset to what would otherwise be a type I nominal-the-best problem. (This is illustrated in Chapter 8 using a catapult example.) The experimental results indicate the range of validity of these assumptions. For this reason, the experiment, in part, has the goal of finding a tuning factor. A good experimental plan should include control factors that have the potential for being tuning factors.

Recall that the nominal-the-best MSD $= [S^2 + (\bar{y} - m)^2]$, Equation 5.16. If the standard deviation is not related to the mean, the form of the MSD that is independent of adjustment (IA) is given simply by $\mathrm{MSD_{IA}} = S^2$. Thus, the type II nominal-the-best S/N ratio takes the following form:

$$\mathrm{S/N_{TypeII\text{-}NTB}} = -10\log(S^2) \qquad \textbf{(5.18)}$$

**Figure 5.17** Ideal relationship between the standard deviation and the mean for the type II NTB

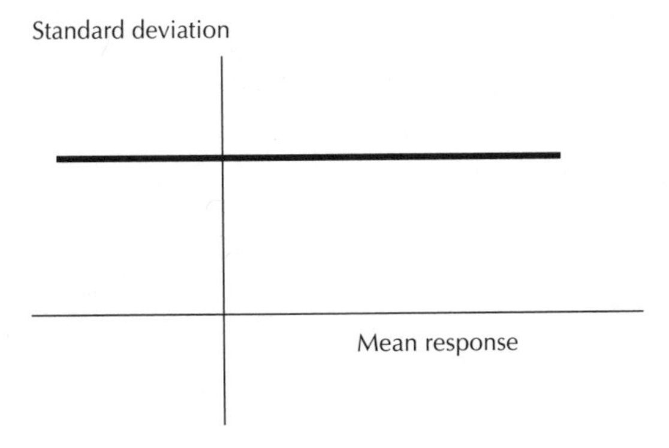

Let's take a look at what a 3 dB and 6 dB improvement in the type II NTB S/N means in terms of the variance and standard deviation. Setting the mean on target, $\bar{y}_0 = \bar{y}_1$, thus:

$$\text{S/N} = -10\log(S^2)$$

Suppose there is a 3 dB improvement in the S/N ratio due to a design change:

$$\begin{aligned}
3\text{ dB} &= \text{S/N}_1 - \text{S/N}_0 \\
3 &= 10[-\log(S_0^2) + \log(S_1^2)] \\
0.3 &= \log S_0^2 - \log S_1^2 = \log(S_0^2/S_1^2) \\
10^{0.3} &= (S_0^2/S_1^2) \\
10^{0.3} &\approx 2 = S_0^2/S_1^2 \\
2 &= S_0^2/S_1^2 \\
S_0{}^2 &= 2S_1^2 \\
S_0 &= 1.4S_1
\end{aligned}$$

Similarly, if there is a S/N improvement of 6 dB:

$$\begin{aligned}
6\text{ dB} &= \text{S/N}_1 - \text{S/N}_0 \\
0.6 &= -\log S_1^2 + \log S_0^2 \\
10^{0.6} &= (S_0^2/S_1^2) \\
4 &= S_0^2/S_1^2 \\
S_0 &= 2S_1
\end{aligned}$$

This indicates that a 6 dB improvement results in a 2× change in distribution width or, for example, *Capability Index* (Chapter 18). A 3 dB improvement results in a 2× change in the MSD or quality loss. These results are the same as seen with the type I NTB S/N ratio.

## 5.7 Two-Step Optimization

The operating window and the type I and type II nominal-the-best problems occur frequently in engineering design. As mentioned several times, the two-step optimization process introduced in Chapter 2 is used.

**Step 1:** Maximize S/N to minimize sensitivity to the effects of noise. To accomplish this, first identify the control factors that have significant slope in their S/N plots, and then select control factors that correspond to the maximum S/N values.

**Step 2:** Adjust the mean response onto the target response. To accomplish this, identify the control factor that has both the greatest slope from the mean plots and the smallest slope in the S/N plots. This control factor is used to adjust the mean.

The following two plots, Figures 5.18 and 5.19, demonstrate how to choose factors that are good for minimizing the effects of noise and shifting the mean. These plots are compared to identify control factors that minimize variability, which are shown in bold in Figure 5.18. They are also compared to

**Figure 5.18** Plot of the control factor effects on the S/N ratio

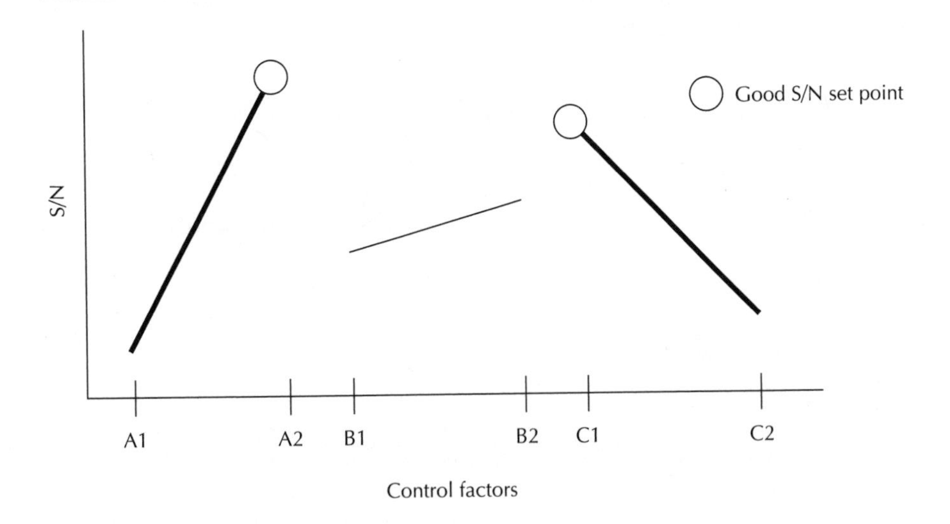


identify a scaling or tuning factor, which is shown in bold in Figure 5.19. The key is that factors whose effect on the S/N ratio is *relatively* strong should be used to reduce variability. Factors whose effect on the mean is *relatively* strong should be used as scaling factors (type I NTB) or as tuning factors (type II NTB, or other S/N ratios).

It takes some engineering judgment to determine the best scaling or tuning factor from several possibilities. By analyzing the response plots for S/N and mean, the control factors can be classified into one of four types:

**Figure 5.19** Plot of the control factor effects on the average response

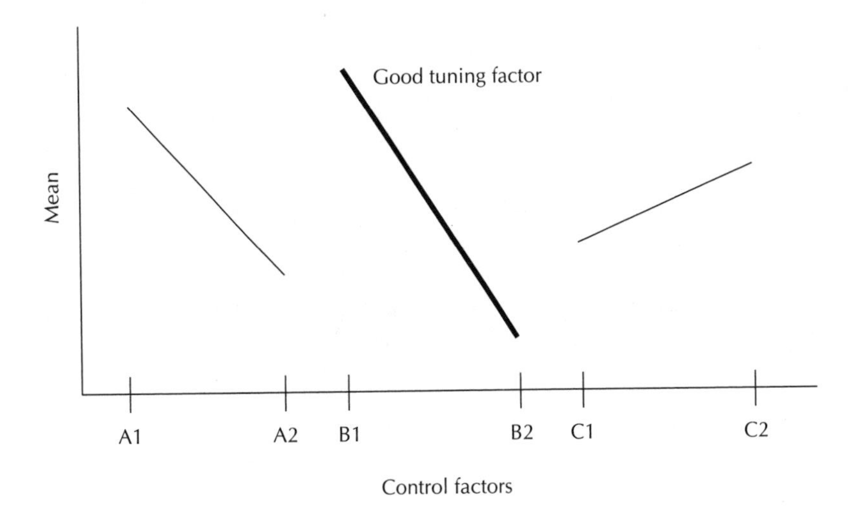

1. Those that affect both the mean and the S/N. The control factors that affect both metrics are typically used to minimize the effects of noise.
2. Those that affect the mean but not the S/N. A true scaling or tuning factor is one that has very little slope in the S/N plot, but a relatively large effect on the mean. A factor that has too much sensitivity to noise leads to difficulties when it is used to shift the mean, because it also increases variability.
3. Those that affect the S/N and not the mean. The control factors that affect only the S/N ratio should be set at the levels that give the highest S/N ratio values.
4. Those that affect neither the S/N or the mean. Typically those control factors that do not show significant effects on the S/N *or* mean response are set at the most economical levels in order to keep costs low.

## 5.8 A Comparative Analysis of Type I NTB and Type II NTB

These two S/N ratios, Equations 5.17 and 5.18, measure different aspects of variability. It is important to understand exactly what each measure is capable of telling about design parameters and their levels. The *Type I NTB S/N* measures variability in relation to the actual mean and is expressed as plus or minus a certain percentage (%) of the mean. This illustrates its roots in the COV. The *Type II NTB S/N* measures the variability directly and thus is given in terms of plus or minus absolute units.

This set of general recommendations is intended to help discriminate between the subtleties of the different measures of variability. Your team's insight into the physical situation and the complexities of the technology being manipulated in the design process will be your guide during S/N ratio selection. It is highly recommended that some preliminary experiments be done to get a quantitative feel for how variability is acting in a specific case. Having a general map of the nature of the design's path of variation with respect to the mean response is of great value in the selection of metrics. The best way to do this is to develop an *energy flow map* of the process or design. This divides the paths of energy into productive and nonproductive work. The variability is dependent, to a large extent, on the nature of the energy transformations and the efficiency with which the design facilitates the transformations into productive work.

## 5.9 A Note on Notation

One warning before we move on to the next chapter on dynamic S/N ratios. The notation for the statistics and the S/N ratios has not been standardized as of yet. Differing uses of mathematical symbols have arisen in the Taguchi method's literature over the years. We want you to be conversant in any book to which you may choose to refer. Some authors, by mixing up the symbols for population and sample statistics, have not been as precise as we have tried to be about the statistical notation. Thus, the same formulas used in this text are found in other texts by substituting $\mu$ for $\bar{y}$, and $\sigma^2$ for $S^2$. Dr. Taguchi and the other Japanese authors have standardized their own notation. The variance, $S^2$, is given by $V_e$. The nominal-the-best S/N ratio is defined using a new statistic, the sum of squares for the mean, $S_m$, often called the sensitivity, which is defined by

$$S_m = \frac{\left(\sum_{i=1}^{n} y_i\right)^2}{n} \tag{5.19}$$

**Table 5.12** Static S/N ratios

| *Case* | *This book* | *Japanese notation* |
|---|---|---|
| Smaller-the-better | $S/N = -10 \log\left[\frac{1}{n}\sum_{i=1}^{n} y_i^2\right]$ | Same |
| Larger-the-better | $S/N = -10 \log\left[\frac{1}{n}\sum_{i=1}^{n} \frac{1}{y_i^2}\right]$ | Same |
| Operating window | $S/N_{OW} = -10 \log\left(\frac{\Sigma x^2}{n}\right) - 10 \log\left(\frac{\Sigma (1/x')^2}{n}\right)$ <br> $x$ is the lower threshold, and $x'$ is the upper threshold. | Same |
| Probability | $S/N_{p-STB} = -10 \log[p/(1-p)]$ <br> $S/N_{p-LTB} = 10 \log[p/(1-p)]$ | Referred to as the omega transformation |
| Nominal-the-best, type I | $S/N_{TypeI-NTB} = 10 \log\left(\frac{\bar{y}^2}{S^2}\right)$ | $\eta = 10 \log\left[\frac{1}{n}\frac{S_m - V_e}{V_e}\right]$ |
| Nominal-the-best, type II | $S/N_{TypeII-NTB} = -10 \log(S^2)$ | $\eta = -10 \log [V_e]$ |

With these definitions, the nominal-the-best S/N ratios, as they are found in the Japanese literature, are shown in Table 5.12. Also shown there are the S/N ratio formulas defined in this chapter.

## 5.10 Summary

The static S/N ratios are the most commonly used metrics in Robust Design. We have tried to show how each of the S/N ratios discussed in this chapter satisfies the ideal S/N ratio properties described in Chapter 4. They relate directly to the quality loss function's mean square deviation, and their use is consistent with the two-step optimization process. They are ideal for benchmark studies, where noise is used to induce variability to get a better indication of performance. They are most important, however, for designed experiments, where their use significantly simplifies the analysis. You should become very familiar with their use, how to compute them, and, most importantly, when to use them. The variety of types of S/N ratios is limitless. The noise factors and mathematical manipulation of the data form unique *S/N treatments* that are customized for the system being studied.

## Exercises for Chapter 5

1. Why is the nominal-the-best (I) S/N ratio preferred over the smaller-the-better and larger-the-better S/N ratios?
2. Under what conditions is it advisable to use the probability form of the S/N ratio?
3. Explain why the STB and LTB S/N ratios have no scaling factor associated with their use.
4. What is meant, specifically, by the term "static" S/N ratio?
5. Name four cases that would require the application of the STB S/N approach to optimization.
6. Name four cases that would require the application of the LTB S/N approach to optimization.

7. Name four cases that would require the application of the NTB (I) S/N approach to optimization.
8. Name four cases that would require the application of the NTB (II) S/N approach to optimization.
9. Name two cases that would require the application of the Operating Window approach to optimization.
10. What are the four possible effects a control factor can have on the S/N and mean performance in an experiment? Sketch these effects.

CHAPTER 6

# The Dynamic Signal-to-Noise Methods and Metrics

## 6.1 Introduction

The dynamic methods are, without exception, the most powerful and widely applicable methods we know of for rapid technology development. Good product quality requires stable, robust, on-target performance. But in technology development, what is the target? That depends on the application of the technology. For example, a film manufacturing process may produce polyester (Mylar) films of thicknesses ranging from less than 0.001 in. for a Kodak Colorease PS printer dye donor substrate to 0.007 in. for an Ektaprint photoconductor loop. In each case, consistency is very important. Also important is the ability of the manufacturing engineer to adjust the film thickness. What is not desirable, but is often the practice, is to reoptimize the entire process setup every time the product requirements change. From film manufacturing to plastic injection molding to die casting, the product quality depends on the skill of the operator as she "tweaks" the process every time the product specs change. The dynamic methods address this issue by allowing the manufacturing engineer to *optimize the system around a function, not a number.* In other words, the method optimizes the system over a range of output values while providing an independent parameter to adjust the output.

The goal in the dynamic method is to optimize a system's quality characteristic as a function that generates a *range* of outputs. Each of the static cases discussed in the previous chapter has only one value for the output optimization goal. A dynamic case has *multiple* values for the output optimization goals. Where necessary for the two-step optimization, the static problems require an adjustment factor. Ideally, the adjustment factor is a tuning or scaling factor so that the process can be put on target without affecting the S/N ratio. The dynamic approach accommodates a range of input signals that have an ideal relationship with the output response. Thus, the dynamic factor is an adjustment factor chosen *prior to running the experiment,* based upon engineering knowledge. This is a key difference between the dynamic case and the static nominal-the-best case. The dynamic method is closely related to multivariate analysis, but avoids much of the mathematical complexity of more statistically rigorous analysis techniques.

Applications of the dynamic method are found in technology development where the final target for the technology may not be known or may depend upon the application. Similarly, dynamic applications are found in process development where a manufacturing engineer, for example, may have a process, such as the coating process described earlier, that must be adjustable over a range. These methods enable engineers to invest their company's resources efficiently and strategically in building a family of products or processes from a single development activity.

Another application of the dynamic method is in product development where a feature of the product requires a functional relationship—for example, the steering of an automobile, where the driver interacts dynamically with the car. Lastly, the dynamic method is applied to measurement engineering where the output of a measurement system is a function of the true value being measured.

Several types of functional relationships exist, the nature of which determines the type of dynamic problem. The linear form, where the signal factor is a scaling factor as well, is by far the most common and powerful. There are two types of linear forms: the *zero-point proportional* form and the *reference-point proportional* form. As explained later, nonlinear ideal functions are not as desirable but can be analyzed easily after linearization. In general, the dynamic method can be described using the P-diagram [P2] shown in Figure 6.1.

In addition to the noise factors and control factors, a signal factor is shown that is used to adjust the response, according to the form of the ideal function. Thus, the two-step optimization process, with one signal factor, is done as follows: First, the variability around the linear function is reduced. Second, the process is put on target or controlled dynamically by adjusting the signal factor. Here the primary purpose of the control factors is to negate the influence of the noise factors.

It may be desirable to adjust the sensitivity of the response to the signal value (the slope in a linear case) in order to achieve the desired output range or hit a target sensitivity such as 1:1 in a measurement system. If this is necessary, one of the control factors can be used to adjust the slope (sensitivity) of the linear relationship between the response and the signal factor. This is another form of the two-step optimization process. First, the variability around the linear function is reduced. Second, the slope of the linear function is put on target. In such a case, the *function* is adjusted to the

**Figure 6.1** P-diagram for the dynamic case with one signal factor

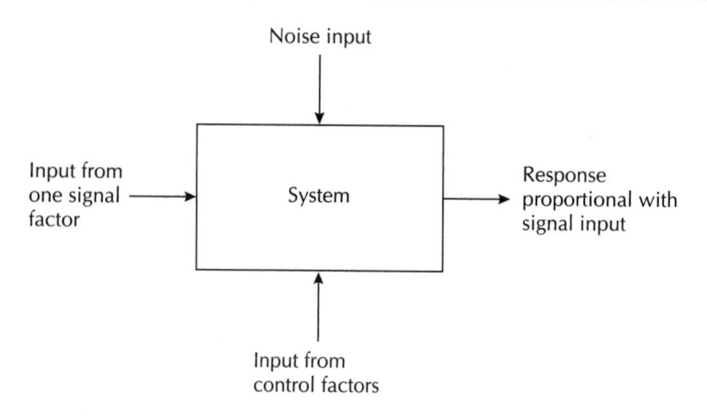

**Table 6.1** Examples of problems that can be analyzed using the dynamic method

| *System* | *Objective* | *Input* | *Noise* | *Output* |
|---|---|---|---|---|
| Golfing | Control distance | Club type (force) | Wind & mental state | Distance |
| Skiing | Change direction | Shift of weight (position of feet) | Slope & surface condition | Turning radius |
| Car | Change direction | Wheel angle | Road condition, tire pressure | Turning radius |
| Car | Change speed | Gas pedal position (gear) | Road condition, # passengers | Speed |
| Automatic transmission | Change gear | Engine RPM | Part wear | Wheel RPM |
| Mill | Control dimension | Tool position (feed rate) | Tool wear, material hardness | Dimension |
| Controller | Control flow | Air pressure on diaphragm | Fluid viscosity | Valve position |
| Injection molding | Adjust dimension | Mold dimension (inj. press.) | Plastic type | Dimension |
| Metal plating | Control thickness | Plating time (current) | Part location in vat | Plating thickness |
| Seismograph | Measure earthquake | Vibration of ground | Part variability | Richter scale data plot |

target. A dynamic method for accomplishing this two-step optimization process using two signal factors is discussed in the "double dynamic" section of this chapter.

Table 6.1 shows a few examples of problems set up in the dynamic format.[1]

## 6.2 The Zero-Point Proportional Case

The zero-point proportional case uses a simple linear relationship between the response, $y$, and the signal factor, $M$:

$$y = \beta M \tag{6.1}$$

This is the equation for a straight line having the general form $y = mx + b$. The response is linearly related to the signal factor—hence the name "proportional." In addition, the line is assumed to go through zero, i.e., the $y$ intercept $b = 0$; hence the name "zero-point." The ideal function for the zero-point proportional case is shown in Figure 6.2.

The signal factor is a control factor known to be capable of adjusting the output response in a linearly proportional fashion. Such knowledge derives from known engineering input–output relationships. These are some examples:

1. Table 6.1 used courtesy of the American Supplier Institute.

**Figure 6.2** Ideal function for the zero-point proportional case

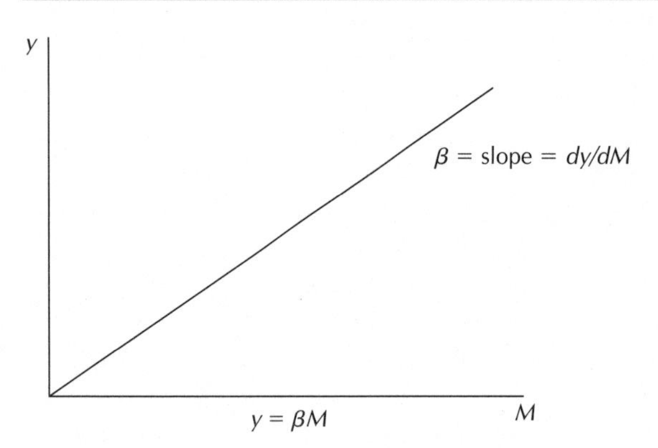


1. The relationship between the input force applied to a spring and the extension or compression of the spring: $F = kx$.
2. The relationship between current passing through a resistor and the output voltage: $V = IR$.
3. The relationship between instantaneous velocity and angular velocity in circular motion: $v = r\omega$.
4. The relationship between the stress and the strain in elastic materials: $\sigma = E\epsilon$.

These are all well-known examples; there are many more that can be used as the foundation for a dynamic analysis.

The zero-point proportional equation is useful for dynamic relationships. For instance, the ability to adjust the output $y$ by changing the signal factor $M$ requires that there is some nonzero slope (power of proportionality). However, there is always the possibility of variation in the signal factor, e.g., noise! If the slope of the line is constant (i.e., no curvature), then the transmitted error is constant. Optimizing the system to minimize that error results in a robust system independent of adjustment. The zero intercept insures that additional error due to the variation in the intercept is avoided. Optimizing the system close to the intercept insures that the system can be adjusted down to very small values of $M$ and $y$.

If there is a nonlinear relationship, in order to use the mathematical treatment developed for the S/N ratio, the data must be mathematically transformed to act in a linear fashion. It is also possible to develop nonlinear forms of dynamic analysis. We discuss that approach after looking at the simpler forms.

The value of this approach is that it can act just like a series of nominal-the-best cases. Each level of the signal factor is a local optimization point for all the control factors against the noise. Each signal factor level has a mean and variance for the response variable associated with it. These data are fitted to a line using the method of least squares. The slope of this line is $\beta$. Figure 6.3 illustrates how data are fitted using the zero-point proportional form.

There is no dynamic form of the quality loss function to use as a guide for defining the signal-to-noise ratio. The dynamic S/N ratio is closely related to the nominal-the-best case. In Chapter 4, it is shown that the nominal-the-best S/N ratio conceptually takes on the following form:

$$S/N_{NTB} = \frac{\text{Power of the mean}}{\text{Power of the variability around the mean}} = 10 \log \frac{\bar{y}^2}{S^2}$$

**Figure 6.3** Data fit to the zero-point proportional form

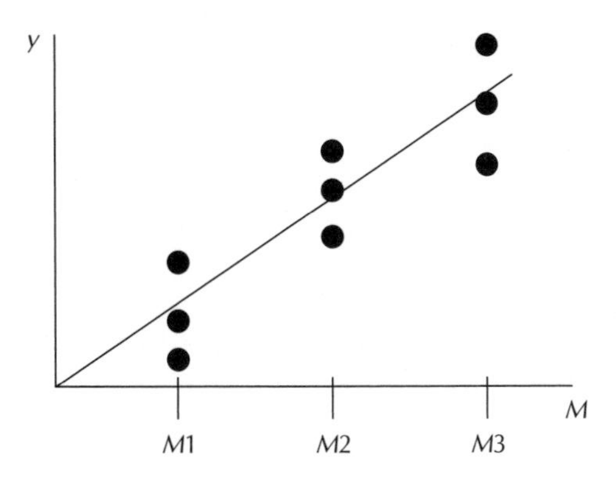


Similarly for the dynamic case, the S/N ratio can be defined conceptually by

$$S/N_{\text{Dynamic}} = \frac{\text{Power of proportionality between } M \text{ and } y}{\text{Power of variability around the proportionality}}$$

For the linear cases, the mathematical form of the S/N ratio is given by

$$S/N = 10 \log \frac{\text{Slope}^2}{\text{MSE}} \tag{6.2}$$

The mean square error (MSE) is the average of the square of the distances (residuals) from the measured responses to the best fit line (Equation 6.9). A higher S/N means better ability to meet the design's quality objective.

### 6.2.1 Reflection Densitometer Example

Here is an example of how the dynamic signal-to-noise ratio can be used to assess the quality of measurement systems. This application is useful for benchmarking different measurement devices against one another. Measurement systems require an exact 1:1 relationship between the true value being measured (the signal factor) and the system output (the measured response).

Measurement systems can be characterized by the following three qualities:

1. **Linearity**—how closely the input–output relationship is described by a straight line
2. **Sensitivity**—the magnitude of the slope of the line
3. **Variability**—deviations from the line caused by noise factors

These characteristics are illustrated using the example of an image analysis instrument called a reflection densitometer. The function of a densitometer is to determine how a particular area of an image reflects light. For simplicity, only black-and-white imaging is considered here. The amount of white light that is reflected back into the sensor carries information about how much of the incident light is absorbed by gray and black portions of the image.

Table 6.2 gives a set of sample data for four different densitometers. These data are used to compare the performance of the densitometers using the dynamic method. The data consist of six density measurements for each densitometer. Because the accuracy and linearity are performance characteris-

**Table 6.2** Data for the densitometer example

| *Densitometers* | $M_1 = 0.20$ | $M_2 = 0.60$ | $M_3 = 1.00$ |
|---|---|---|---|
| Densitometer 1 | 0.14, 0.16 | 0.59, 0.61 | 1.04, 1.06 |
| D #1 Average | 0.15 | 0.60 | 1.05 |
| Std. Deviation 1 | 0.0141 | 0.0141 | 0.0141 |
| Densitometer 2 | 0.09, 0.11 | 0.69, 0.71 | 0.84, 0.86 |
| D #2 Average | 0.10 | 0.70 | 0.85 |
| Std. Deviation 2 | 0.0141 | 0.0141 | 0.0141 |
| Densitometer 3 | 0.31, 0.33 | 0.61, 0.63 | 0.91, 0.93 |
| D #3 Average | 0.32 | 0.62 | 0.92 |
| Std. Deviation 3 | 0.0141 | 0.0141 | 0.0141 |
| Densitometer 4 | 0.05, 0.25 | 0.50, 0.70 | 1.0, 1.2 |
| D #4 Average | 0.15 | 0.60 | 1.10 |
| Std. Deviation 4 | 0.1414 | 0.1414 | 0.1414 |

tics to be examined in this dynamic analysis, the data are collected at three different input signal factor levels. The signal factor is the image density of the test sample. The levels chosen are:

$M_1 = 0.20$, a low-density image

$M_2 = 0.60$, a medium-density image

$M_3 = 1.00$, a moderately high-density image

The densitometer is expected to give good linearity over this range of densities.

Notice that for each densitometer there are two readings at each signal factor level. The first reading is taken at the end of the day on a Friday, and the second reading is taken on the next Monday morning by a different lab technician. This introduces noise into the readings, because different technicians on different days may have imperceptible but important differences in technique. Thus, the data are based on known noise factors. Figures 6.4–6.7 show the plots of the raw data.

### 6.2.2 S/N Calculation

The metrics used by the dynamic method are the S/N ratio and slope $\beta$ for each densitometer. Table 6.3 shows the data, clearly separating the two compound noise conditions:

N1 = Friday + afternoon + technician 1,

N2 = Monday + morning + technician 2.

**Figure 6.4** Densitometer 1: Good linearity, slope $\beta$ = 1.03, and variability $S$ = 0.014

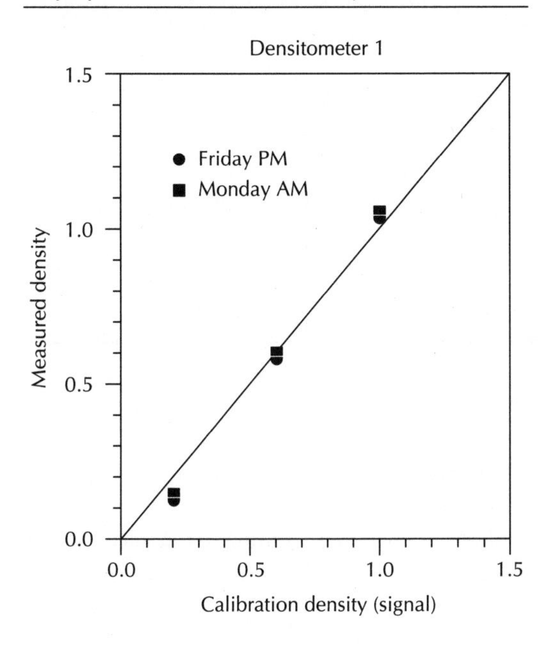


**Figure 6.5** Densitometer 2: Poor linearity, slope $\beta$ = 0.92, and variability $S$ = 0.014

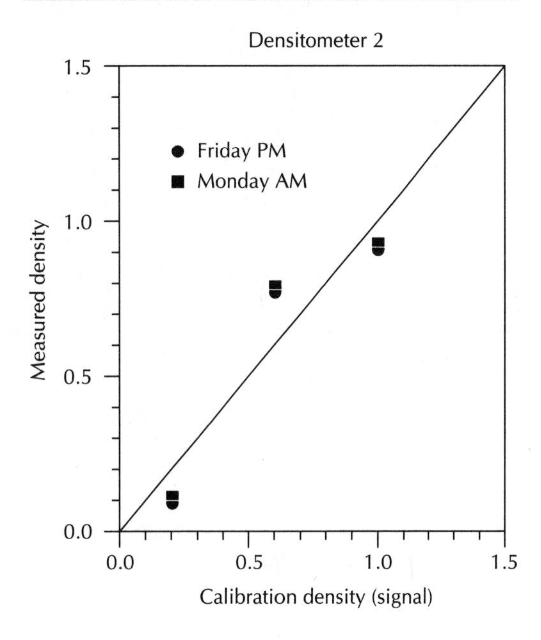


**Figure 6.6** Densitometer 3: Good linearity but nonzero intercept, slope $\beta$ = 0.97, and variability $S$ = 0.014

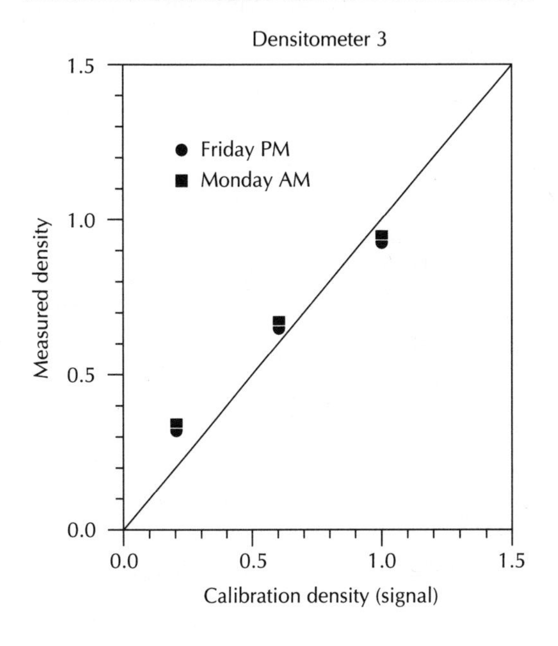


**Figure 6.7** Densitometer 4: Good linearity, slope $\beta$ = 1.06 , and variability $S$ = 0.14

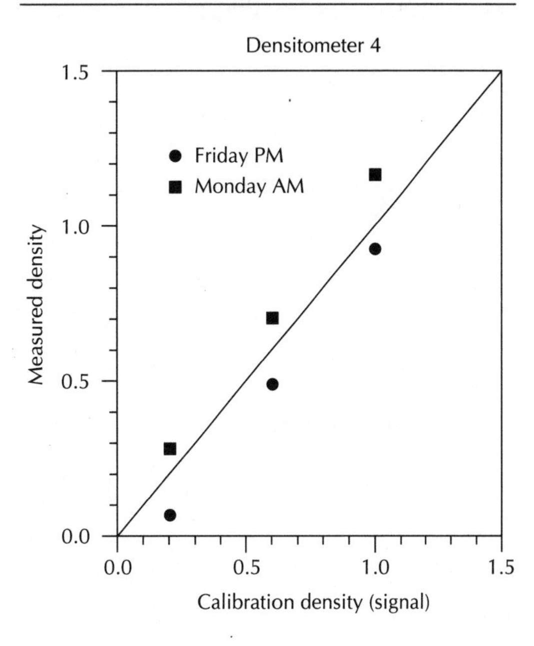

**Table 6.3** Data for the densitometer example showing linear combinations

| *Densitometers* | *Noise cond.* | $M_1 = 0.20$ | $M_2 = 0.60$ | $M_3 = 1.00$ | *Linear comb.* |
|---|---|---|---|---|---|
| Densitometer 1 | N1 | 0.14 | 0.59 | 1.04 | 1.422 |
| Densitometer 1 | N2 | 0.16 | 0.61 | 1.06 | 1.458 |
| Densitometer 2 | N1 | 0.09 | 0.69 | 0.84 | 1.272 |
| Densitometer 2 | N2 | 0.11 | 0.71 | 0.86 | 1.308 |
| Densitometer 3 | N1 | 0.31 | 0.61 | 0.91 | 1.338 |
| Densitometer 3 | N2 | 0.33 | 0.63 | 0.93 | 1.374 |
| Densitometer 4 | N1 | 0.05 | 0.50 | 1.0 | 1.310 |
| Densitometer 4 | N2 | 0.25 | 0.70 | 1.2 | 1.670 |

Also added to the table are linear combinations of the raw data (defined in Equation 6.7) that are useful in the calculation.

The method of least squares can now be used to fit lines to the data according to the linear model obtained from the zero-point proportional case, Equation 6.1:

$$y_{ij} = \beta M_i \tag{6.3}$$

For the signal factors, $i = 1, 2, \ldots, k$, where $k = 3$ because there are three $M_i$ values. For the noise replicates, $j = 1, 2, \ldots, r_0$ where $r_0 = 2$ because there are two $N_j$ values.

The slope, $\beta$, is determined by the least squares method, which minimizes the sum of the squares of the data around a "best fit" line. This is done by minimizing $SS_V$, the sum of the squared variation around the line,

$$SS_V = \sum_{i=1}^{k} \sum_{j=1}^{r_0} (y_{i,j} - \beta M_i)^2 \tag{6.4}$$

by solving $dSS_V/d\beta = 0$, the value of $\beta$ that minimizes the sum is found. Thus,

$$\frac{dSS_V}{d\beta} = -2 \sum_{i=1}^{k} \sum_{j=1}^{r_0} M_i(y_{i,j} - \beta M_i) = 0$$

$$\sum_{i=1}^{k} \sum_{j=1}^{r_0} M_i y_{i,j} - \beta \sum_{i=1}^{k} \sum_{j=1}^{r_0} M_i^2 = 0$$

Solving for the slope, $\beta$,

$$\beta = \frac{\sum_{i=1}^{k} \sum_{j=1}^{r_0} M_i y_{i,j}}{\sum_{i=1}^{k} \sum_{j=1}^{r_0} M_i^2} = \frac{\sum_{i=1}^{k} \sum_{j=1}^{r_0} M_i y_{i,j}}{r_0 \sum_{i=1}^{k} M_i^2} \tag{6.5}$$

It is useful to define shorthand notations for some of the frequently seen sums in the calculations. That way, as the calculations are performed, these sums can be calculated once, then stored and reused as needed. Therefore, let

$$r = r_0 \sum_{i=1}^{k} M_i^2 \tag{6.6}$$

and

$$L_j = \sum_{i=1}^{k} M_i y_{i,j}. \tag{6.7}$$

The latter sum, $L_j$, is the *linear combination* that is given for each row in Table 6.3. The quantity $r$ is simply a function of the signal factors, and is therefore the same for each of the densitometers.

$$r = 2(0.20^2 + 0.60^2 + 1.00^2)$$
$$r = 2.80$$

If we combine Equations 6.5 and 6.7, the slope, $\beta$ is given by

$$\beta = \frac{\sum_{j=1}^{r_0} L_j}{r} \tag{6.8}$$

$$\beta_1 = [1.422 + 1.458]/2.80$$

$$\beta_1 = 2.88/2.80 = 1.03$$
$$\beta_2 = 2.58/2.80 = 0.92$$
$$\beta_3 = 2.71/2.80 = 0.97$$
$$\beta_4 = 2.98/2.80 = 1.06$$

The S/N ratio is defined by Equation 6.2. The slope is given by Equation 6.8. To complete the calculation of the S/N ratio, the mean square error, MSE, must be determined. Now that the individual $\beta$'s are determined, the MSE can be found directly by calculating

$$\text{MSE} = \frac{\text{SS}_\text{V}}{r_0 k - 1} = \frac{1}{r_0 k - 1} \sum_{i=1}^{k} \sum_{j=1}^{r_0} (y_{i,j} - \beta M_i)^2 \tag{6.9}$$

The values for $y_{i,j}$ and $\beta$ determined for each densitometer can be substituted directly into Equation 6.9 to find the MSE. This can be done easily in a computer program. To make it easier to code a spreadsheet and avoid having to iterate for the solution, the calculation can be simplified and generalized by solving the MSE algebraically using Equation 6.5 for $\beta$.

First, expand the square term in the sum:

$$\text{SS}_\text{V} = \sum_{i=1}^{k} \sum_{j=1}^{r_0} (y_{i,j}^2 - 2\beta y_{i,j} M_i + \beta^2 M_i^2) \tag{6.10}$$

The first term in Equation 6.10 is the total sum of squares, $SS_T$:

$$SS_T = \sum_{i=1}^{k} \sum_{j=1}^{r_0} y_{i,j}^2 \tag{6.11}$$

Thus,

$$SS_V = SS_T - \sum_{i=1}^{k} \sum_{j=1}^{r_0} 2\beta y_{i,j} M_i + \sum_{i=1}^{k} \sum_{j=1}^{r_0} \beta^2 M_i^2$$

Now the slope $\beta$ can be substituted in the equation to give

$$SS_V = SS_T - 2 \sum_{i=1}^{k} \sum_{j=1}^{r_0} \left( \frac{\sum_{i=1}^{k} \sum_{j=1}^{r_0} M_i y_{i,j}}{r_0 \sum_{i=1}^{k} M_i^2} \right) y_{i,j} M_i + \sum_{i=1}^{k} \sum_{j=1}^{r_0} \left( \frac{\sum_{i=1}^{k} \sum_{j=1}^{r_0} M_i y_{i,j}}{r_0 \sum_{i=1}^{k} M_i^2} \right)^2 M_i^2$$

$$= SS_T - 2 \left( \frac{\sum_{i=1}^{k} \sum_{j=1}^{r_0} M_i y_{i,j}}{r} \right) \sum_{i=1}^{k} \sum_{j=1}^{r_0} y_{i,j} M_i + \left( \frac{\sum_{i=1}^{k} \sum_{j=1}^{r_0} M_i y_{i,j}}{r_0 \sum_{i=1}^{k} M_i^2} \right)^2 \sum_{i=1}^{k} \sum_{j=1}^{r_0} M_i^2$$

$$= SS_T - \frac{\left( \sum_{i=1}^{k} \sum_{j=1}^{r_0} M_i y_{i,j} \right)^2}{r}$$

This expression can be simplified by defining the sum of squares for the slope:

$$SS_\beta = \frac{\left( \sum_{i=1}^{k} \sum_{j=1}^{r_0} M_i y_{i,j} \right)^2}{r} = \frac{\left( \sum_{j=1}^{r_0} L_j \right)^2}{r} \tag{6.12}$$

Thus,

$$SS_V = SS_T - SS_\beta \tag{6.13}$$

Applying Equation 6.11, the total sums of squares for the densitometers are:

$$SS_{T1} = 0.14^2 + 0.16^2 + 0.59^2 + 0.61^2 + 1.04^2 + 1.06^2 = 2.9706$$

$SS_{T1} = 2.9706$
$SS_{T2} = 2.4456$
$SS_{T3} = 2.6670$
$SS_{T4} = 3.2450$

After we apply Equation 6.12, the sums of squares due to the slope for the densitometers are:

$$SS_{\beta 1} = [1.422 + 1.458]^2/r = (2.88)^2/2.8 = 2.9623$$

$$SS_{\beta 1} = 2.9623$$
$$SS_{\beta 2} = 2.3773$$
$$SS_{\beta 3} = 2.6229$$
$$SS_{\beta 4} = 3.1716$$

Combining the sums of squares, as given by Equation 6.13, the sums of squares due to variation from the best fit line for the densitometers are

$$SS_{V1} = 2.9706 - 2.9623 = 0.0083$$

$$SS_{V1} = 0.0083$$
$$SS_{V2} = 0.0683$$
$$SS_{V3} = 0.0441$$
$$SS_{V4} = 0.0734$$

Now, the mean square error can be determined by applying Equation 6.9:

$$MSE_1 = \frac{0.0083}{(6 - 1)} = 0.0017$$

$$MSE_1 = 0.0017$$
$$MSE_2 = 0.0137$$
$$MSE_3 = 0.0088$$
$$MSE_4 = 0.0147$$

The S/N ratio can now be determined by applying Equation 6.2, $S/N = 10 \log(\beta^2/MSE)$. Table 6.4 summarizes the calculations made for the four densitometers and gives the S/N ratio values obtained. The dynamic S/N ratio analysis indicates that densitometer 1 is significantly better than the others. (Recall that 3 dB is a 2× improvement in quality, so 10 dB is a lot!) This conclusion should have been anticipated from examining Figures 6.4–6.7. The graphical analysis, in conjunction with the S/N analysis, can teach a lot about the four densitometers and the three main issues of interest in quantifying the quality of a measurement system: linearity (or deviation from linearity), sensitivity (slope), and variability (due to noise).

**Table 6.4** Results of the densitometer S/N calculation

| *Meter #* | $SS_T$ | $SS_\beta$ | $SS_V$ | *MSE* | $\beta$ | *S/N* |
|---|---|---|---|---|---|---|
| 1 | 2.9706 | 2.9623 | 0.0083 | 0.0017 | 1.03 | 32.41 |
| 2 | 2.4456 | 2.3773 | 0.0683 | 0.0137 | 0.92 | 22.37 |
| 3 | 2.6670 | 2.6229 | 0.0441 | 0.0088 | 0.97 | 24.73 |
| 4 | 3.2450 | 3.1716 | 0.0734 | 0.0147 | 1.06 | 23.32 |

**Meter 1** has the highest S/N ratio and, from its plotted output, it appears to be the best performer according to the three criteria just mentioned. It is the benchmark; the other densitometers are compared to it.

**Meter 2** has severe nonlinear performance. The nonlinear performance results in a large MSE. Nonlinear behavior is as bad as sensitivity to noise in this case and results in a substantially reduced S/N ratio.

**Meter 3** has linear performance, but does not follow the zero-point proportional curve. In fact, a two-parameter best-fit curve would have an intercept of about 0.15. For that reason, its MSE is much larger than that of meter 1, although the variation due to noise is similar to that of meter 1. The impact of the intercept is to reduce the S/N ratio. This is because the ideal behavior for a measurement system is not satisfied when the intercept requirement is violated. A nonzero intercept results in poor calibration and difficulty in measuring samples having small values..

**Meter 4** is linear and has a relatively high slope, although not as high as that of meter 1. Its real problem shows up in its sensitivity to noise, which spreads the data from the best fit line, resulting in a large mean square error. In this case the S/N ratio is reduced by the impact of the noise factors.

Thus, the zero-point proportional S/N ratio neatly summarizes the salient quality issues into a single metric. With data and analysis like this, decisions on how to spend corporate dollars for equipment purchases or subsystem purchases from other suppliers to incorporate into a product can be quantitatively justified. This is how Taguchi methods put real power into an activity such as benchmarking.

### 6.2.3 Decomposing the Sum of Squares Due to Variation

Note that the S/N ratio treats all sources of deviation from the ideal function line equally—both variation due to noise and failure to follow the ideal zero-point proportional relationship. This trait is characteristic of the S/N ratio. It confounds all the sources of quality loss in an effort to simplify the analysis. Further analysis is required if one wishes to sort out the factors that contribute to the variation. In the case of the dynamic S/N analysis, there are always two possible sources of variation:

1. Deviation due to sensitivity to noise factors
2. Deviation due to lack of fit or nonlinearity (in the zero-point proportional case)

The following analysis shows how to decompose the $SS_V$ into two sums of squares, $SS_{LOF}$, the sum of squares due to lack of fit, and the $SS_N$, the sum of squares due to the noise factors. The decomposition is given by the following equation:

$$SS_V = SS_{LOF} + SS_N \tag{6.14}$$

Continuing the densitometer example, the sum of squares due to noise can be found, using the linear combinations, Equation 6.7:

$$SS_N = \frac{r_0 \sum_{j=1}^{r_0} L_j^2}{r} - \frac{\left(\sum_{j=1}^{r_0} L_j^2\right)^2}{r}$$

$$= \frac{1}{r}\left[2(L_1^2 + L_2^2) - (L_1^2 + 2L_1L_2 + L_2^2)\right]$$

$$= \frac{1}{r}(L_1^2 - 2L_1L_2 - L_2^2) = \frac{(L_1 - L_2)^2}{r} \quad \textbf{(6.15)}$$

The values of the linear combinations for the four densitometers are given in Table 6.3, and $r$ = 2.80. The remaining variation due to lack of fit, $SS_{LOF}$, is found by solving Equation 6.14. After we apply these equations to the four densitometers, we obtain the results shown in Table 6.5.

Table 6.5 shows that the total variation from the line for Densitometers 2 and 3 are primarily due to lack of fit (LOF), while for densitometer 4 it is largely due to sensitivity to noise. The decomposition of sums of squares is referred to as Analysis of Variance, which is discussed in Chapter 17.

## 6.3 The Reference-Point Proportional Case

The zero-point proportional ideal function is the foundation of the dynamic method. Whenever possible it should be used as the ideal function, *even if the data do not follow the zero-point form!* The reason for this is that the system performance should be measured against the ideal behavior in order to force significant robustness improvement. For the reasons described earlier, there are good arguments to show that the zero-point proportional form gives the most robust results. Remember, S/N ratios are used for comparison purposes, either for benchmarking of systems or in an orthogonal array experiment where the calculations are meant to describe the variations due to the parameters in the matrix. Thus, if all the runs have similar intercepts, that contribution to the comparative calculations drops out.

Despite the preceding argument, there are cases where it just makes good engineering sense to depart from the zero-point case. For example, focusing on the linear equation only in the range where the results are generated (which may be far from the zero point) allows the sensitivity of the analysis to be greatly enhanced. Otherwise, the variation in a group of data points away from the origin is overwhelmed by the effect of having the line go through the origin in a zero-point calculation. There are also cases where the ideal linear relationship simply does not go through zero. These situations are

**Table 6.5** Results of the densitometer sum of squares analysis

| *Meter #* | $SS_T$ | $SS_\beta$ | $SS_V$ | $SS_N$ | $SS_{LOF}$ | *S/N* |
|---|---|---|---|---|---|---|
| 1 | 2.9706 | 2.9623 | 0.0083 | 0.00046 | 0.00784 | 32.41 |
| 2 | 2.4456 | 2.3773 | 0.0683 | 0.00046 | 0.06784 | 22.37 |
| 3 | 2.6670 | 2.6229 | 0.0441 | 0.00046 | 0.04364 | 24.73 |
| 4 | 3.2450 | 3.1716 | 0.0734 | 0.04629 | 0.02711 | 23.32 |

**Figure 6.8** Reference-point proportional ideal function

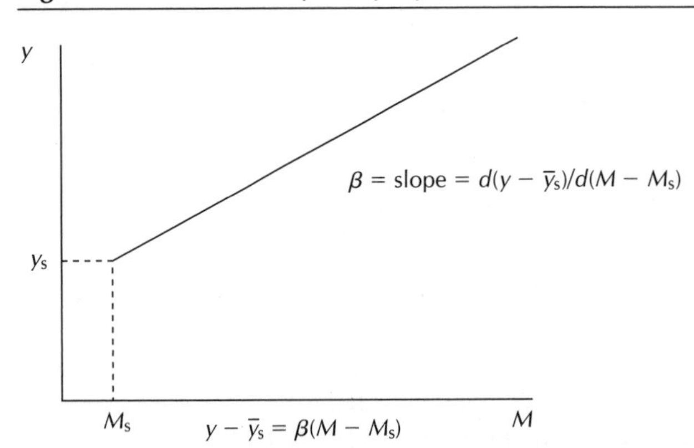


referred to as *reference-point proportional cases.* The reference-point proportional case can be represented graphically as shown in Figure 6.8.

The model that this case is based on takes the following form:

$$y - \bar{y}_s = \beta(M - M_s) \tag{6.16}$$

where $y$ is a response value, $\bar{y}_s$ is the reference response, $\beta$ is the slope of the line representing the relationship, $M$ is the signal factor value, and $M_s$ is the value of the reference signal factor.

The best-fit line does not go through zero. Therefore, a reference signal factor and reference response are established. Because this case references a point other than zero, the point chosen as the reference can be any point along the line. Often the smallest signal factor level is chosen. Typically, the reference response value, $\bar{y}_s$, is the average of the responses obtained at the reference signal factor, $M_s$. In other cases, there is a known reference response value obtained from the ideal function. Equation 6.16 is equivalent to a shift in the axes, as shown in Figure 6.9.

**Figure 6.9** Shift in axes used to define reference-point proportional ideal function

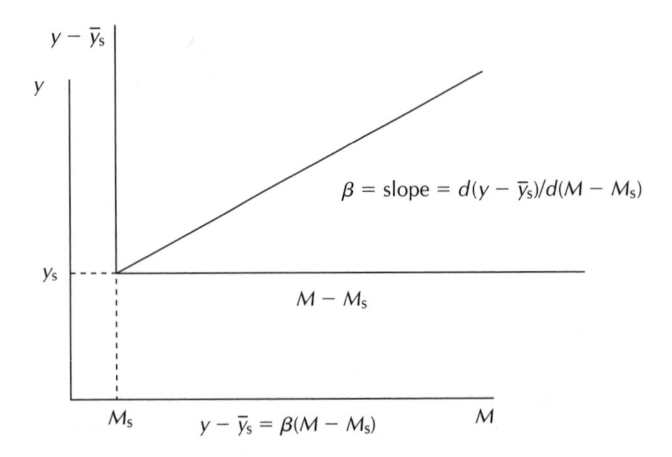

Thus, the reference-point proportional problem is essentially equivalent to the zero-point proportional problem, after the transformations are applied. Therefore, the same equations can be used as before with the appropriately transformed data. This greatly simplifies the analysis and confirms the zero-point analysis as the foundation of the dynamic analysis. Other linear forms can be similarly transformed, as discussed elsewhere [T5].

## 6.4 Nonlinear Dynamic Problems

The question of what to do with situations where the underlying physical mechanism is inherently nonlinear is a bit more problematic. This is because nonlinearity carries with it a fundamental penalty in robustness. This can be seen by considering the transmission of variation of the signal factor illustrated in Figure 6.10.

This plot for a simple nonlinear ideal function shows that as the slope gets steeper, the same amount of variation in the signal factor causes much greater variation in the response. This violates a condition used in the two-step optimization: that the problem of reducing variation is treated separately from the problem of putting the process on target. Nevertheless, there are cases where the known physics of the system dictate a nonlinear response. Therefore, it may be necessary to transform the data so as to linearize the ideal function. Some useful transforms are given in Table 6.6.

These transformations are all examples of a general procedure that can be applied to a great many nonlinear problems [H6, pp. 91 and 92]. When doing so, plot the data using the transformed variables for the axes to see clearly the resulting linear behavior. Transform all the signal factors and response values, and then apply the zero-point equations. (Note that some of the transformations give results in reference point forms.) A case study illustrating this approach is given in Section 6.4.1.

Because exponential relationships are so ubiquitous in engineering, the logarithmic transformations (last two in Table 6.6) are extremely important. The exponential form is frequently used in analyzing physical and chemical phenomena. For example, consider the ideal thermal function of a simple coffee cup. Is it to maintain the temperature of the hot drink indefinitely, suggesting a static nominal-the-best analysis? That might be appropriate for a Thermos bottle, but a coffee cup should allow the

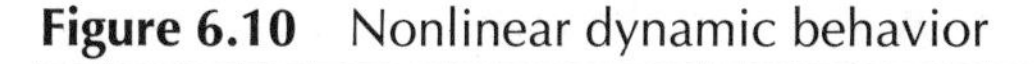

**Figure 6.10** Nonlinear dynamic behavior

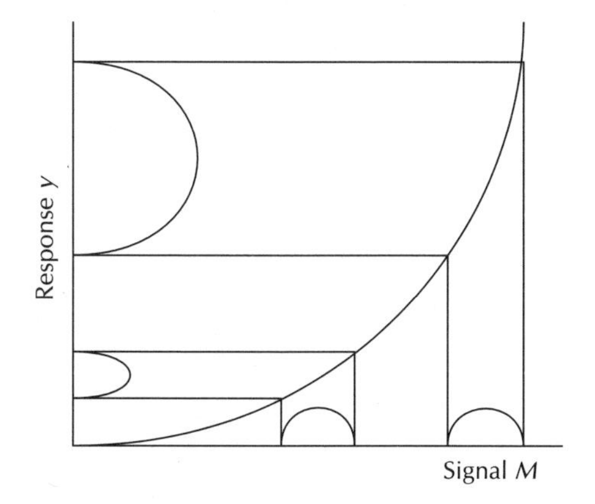

**Table 6.6** Transformations useful for linearizing nonlinear forms

| *Ideal function* | *Transformation to linearize* |
|---|---|
| $y = ax^b$ | $y' = y^{1/b}$, $M = x$<br>$y' = (a^{1/b})\,M$ |
| $y = x/(a + bx)$ | $y' = 1/y$, $M = 1/x$<br>$y' = aM + b$ |
| $y = a + b/x$ | $y' = y$, $M = 1/x$<br>$y = bM + a$ |
| $y = ae^{bx}$ | $y' = \log y$, $M = x$<br>$y' = bM + \log a$ |
| $y = 1 - e^{-bx}$ | $y' = \log[1/(1 - y)]$, $M = x$<br>$y' = bM$ |

beverage to cool down gradually. In fact, the ideal behavior is a dynamic relationship of temperature, $T$, with time, $t$, given by the exponential equation

$$T = T_0 \exp(-\alpha t)$$

where $T_0$ is the initial temperature and $\alpha$ is the thermal heat loss coefficient. This equation can be linearized using the transformation given in Table 6.6, and the resulting data can be analyzed using the dynamic S/N ratio. Such expressions are typical of many physical processes and chemical reactions. Thus, a wide range of problems can be analyzed using the dynamic S/N ratio. The following case study illustrates this with a manufacturing process optimization study.

### 6.4.1 Dynamic Case Study: Coating Uniformity

This case study[2] illustrates the application of the dynamic S/N ratio using the last transformation listed in Table 6.6. The process to be optimized is a roll coating method known as ring coating. In this method, a roller is pulled through a gasket containing a mixture of solvents and solids. Some of the material transfers by the mechanism of surface tension. A sketch of the process is shown in Figure 6.11. This method is very efficient because it requires very little material to be available (exposed

**Figure 6.11** Ring coating

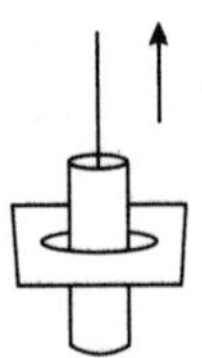

2. Contributed by Marc Bermel, Eastman Kodak Company.

**Figure 6.12** Axial run-out nonuniformity in coating process

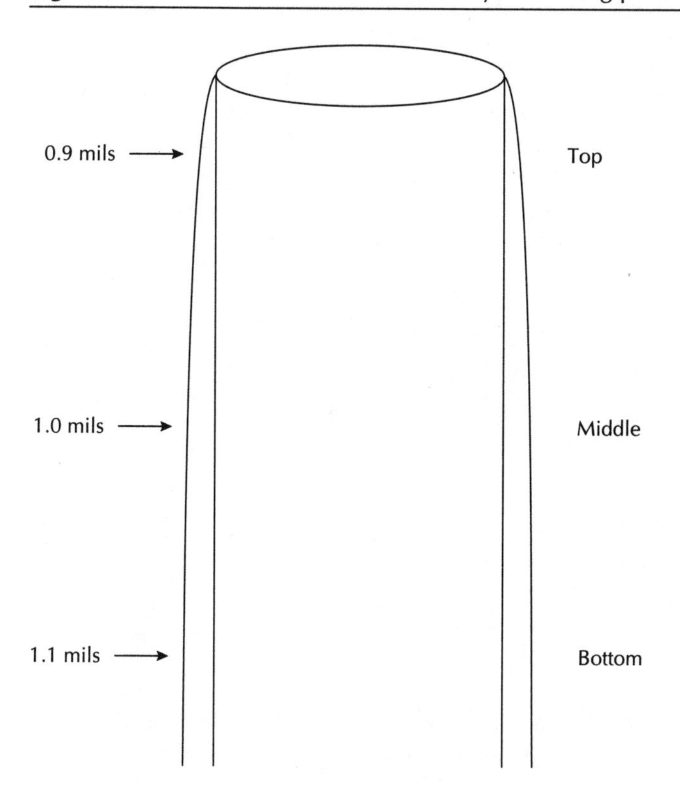


to the environment) during the coating process. By this method, rollers coated with polymeric materials may be fabricated. The rollers are used to fix toner images, by the application of heat and pressure, in a desktop printer. The coated material acts as a release surface, to prevent the molten toner from sticking. The coating must be thin and uniform to insure even heating throughout the image.

Initial results for the roller show a slight nonuniformity of the coating thickness, as shown in Figure 6.12. The $\pm 10\%$ variation in thickness is within the process tolerance. The variation grows dramatically, however, if the solids concentration is increased to produce a thicker coating. This effect is shown by the error bars in Figure 6.13. The sensitivity of the coating uniformity at high loading levels can be viewed in two ways:

1. Although the current target is 0.001 in. (1 mil), a change in requirements to a thicker coating (2 mil) would create a manufacturability issue.
2. The variability at high loading levels is an opportunity to improve the quality of the coating thickness at nominal loading levels. By improving the coating thickness uniformity at 0.002 in. thickness, using a dynamic analysis, the coating uniformity at 0.001 in. thickness is improved as well. There is much more information content in experiments performed at a higher thickness where the noise effects are magnified.

Note that the dynamic relationship is nonlinear. In fact, the physics of the coating process dictates that the thickness, $y$, follows an exponential relationship with the solids concentration, $C$, up to the

**Figure 6.13** Coating thickness vs. solids concentration

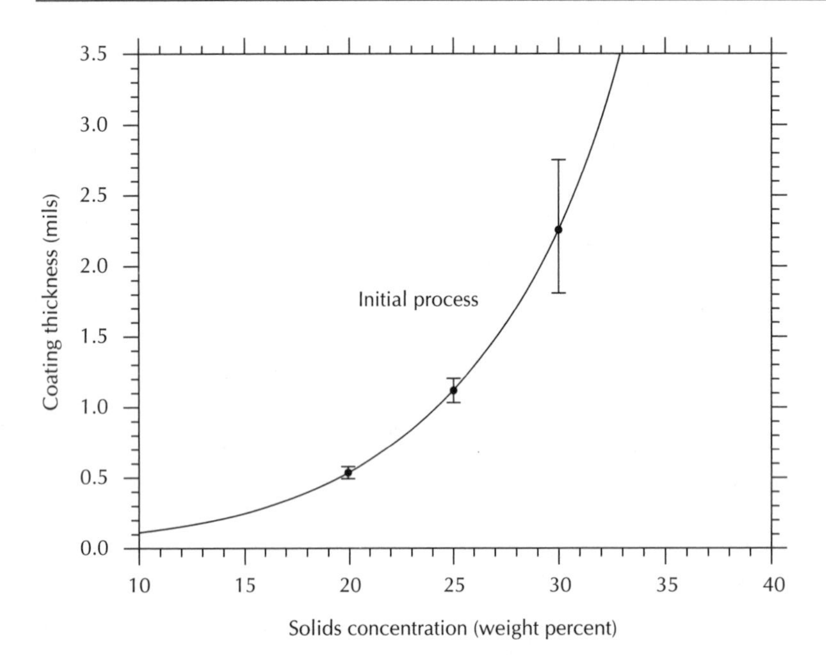


the point where the coating fails due to sagging and dripping. This is described by Equation 6.17, where $y$ is measured in mils (0.001 in.) and $\alpha$ and $t_0$ are empirically determined constants.

$$y = t_0[\exp(\alpha C) - 1] \tag{6.17}$$

Data such as those shown in Figure 6.13 can be linearized using the following transformation:

$$y' = \ln(1 + y/t_0) = \alpha C$$

The resulting data are analyzed using the zero-point proportional form. The zero-point proportional analysis focuses on minimizing the sensitivity to noise (location) and increasing sensitivity (slope) if possible. The dynamic analysis also focuses on the ability of the system to follow the ideal behavior, Equation 6.17.

The results of optimization experiments in this case show that the critical factor in determining the coating uniformity is the speed at which the roller is pulled through the gasket. A lower speed gives an improvement of 7 dB compared to the previous level. The resulting variation in roller thickness is about $\pm 2\%$. As a result of this study, a relatively small experiment, the process capability is significantly improved and axial thickness variability is reduced.

## 6.5 The Double-Dynamic Signal-to-Noise Ratio

The dynamic method represents a generalization of the nominal-the-best case in two respects. First, the dynamic analysis is done by performing a series of static problems, but at different signal factor levels. Rather than analyzing each signal factor level experiment separately, the power of the dynamic method

**Figure 6.14** P-diagram for the dynamic case with two signal factors

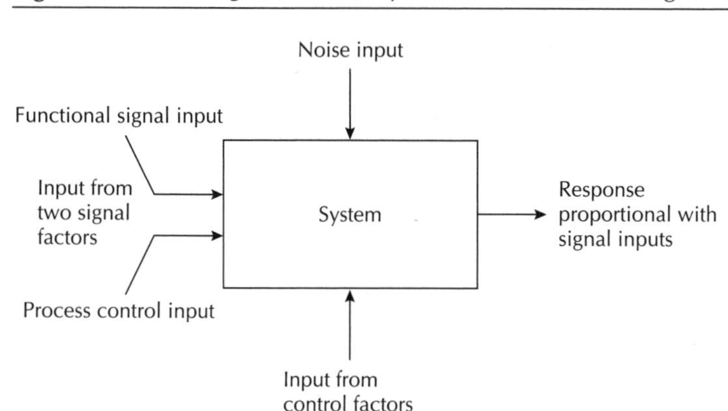


is used to pull all the data together in a single analysis. Second, the adjustment factor necessary to put the nominal-the-best response on target is incorporated into the dynamic analysis. This is done by using engineering knowledge to select an appropriate signal factor when setting up the experiment.

The double-dynamic case extends this progression one step further by considering the slope of the dynamic zero-point proportional case, $\beta$, itself to be a nominal-the-best type of parameter. Generally, when analyzing the dynamic problem, the slope is treated as a response to be studied. Control factors are categorized according to how they affect the S/N ratio and how they affect the slope. The goal is to find at least one factor that can be used to adjust the slope onto the target.

We can build upon the basic dynamic case by introducing a second signal factor, intended to adjust the slope of the fundamental zero-point relationship. The two signal factors are referred to as the *functional input signal* and the *process control signal.* Functional input refers to that which drives or causes the basic functional transformation in the design or process. Process control refers to that which modifies or adjusts the transformation but does not cause the function itself. As a result, the P-diagram shown in Figure 6.1 is modified as shown in Figure 6.14.

## 6.5.1 The Double Dynamic Signal Factors

The signal factors used for the double-dynamic case are formally defined as follows:

*Functional Signal Factor (FSF):* An input factor that has a linear relationship with the response and is fundamentally causing the energy transformation and consequently driving the functional output of the design or process. The FSF is identified as $M$.

*Process Signal Factor (PSF):* An adjustment factor that has the property of modifying the linear relationship of the FSF with the response. The PSF is identified as $M^*$.

Usually, the PSF is selected from one of the control factors. This way, the slope of the dynamic relationship can be controlled. Another option available with the PSF is to use a critical noise factor. In this case the ideal function's response to a range of critical noise is tested for the express purpose of identifying a compensation strategy. Such a compensation strategy can be used to keep the functional relationship on target even over a range of anticipated noise factor levels.

The double-dynamic case can be represented graphically as shown in Figure 6.15. The ideal function in this case is a family of zero-point proportional lines. The way in which the PSF modifies the slope of the lines determines the classification of the double-dynamic problem. The model for each

**Figure 6.15** Double-dynamic zero-point proportional ideal function

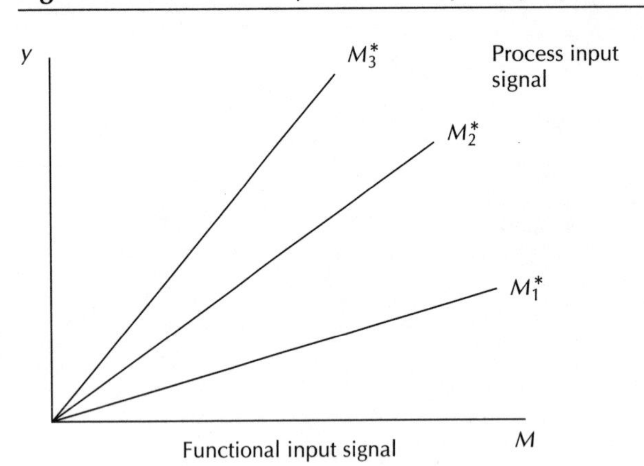


form is established based on the ideal function the design is intended to produce. The physical relationship between the signal factors and the response determines the form selected. Wherever possible, the problem is transformed back to a zero-point proportional form for the purpose of calculating the S/N ratio.

### 6.5.2 Cases 1 and 2: $y = \beta MM^*$ and $y = \beta M/M^*$

If the slope is linearly related to the PSF, the ideal function is given by

$$y = \beta MM^* \tag{6.18}$$

The product of the two signal factors is linearly related to the response. Plotting the response, $y$, vs. $MM^*$ clearly shows that the ideal function reverts to the zero-point proportional form (Figure 6.16).

**Figure 6.16** Double-dynamic ideal function, $y = \beta(MM^*)$

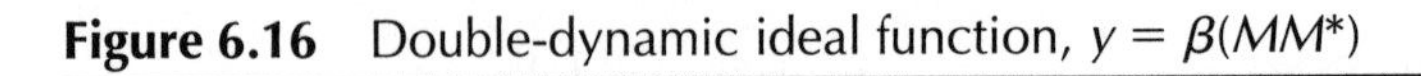

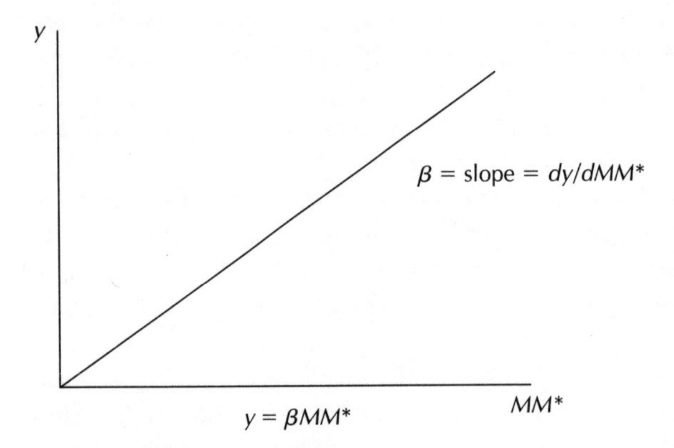

**Figure 6.17** Double-dynamic ideal function, $y = \beta(M/M^*)$

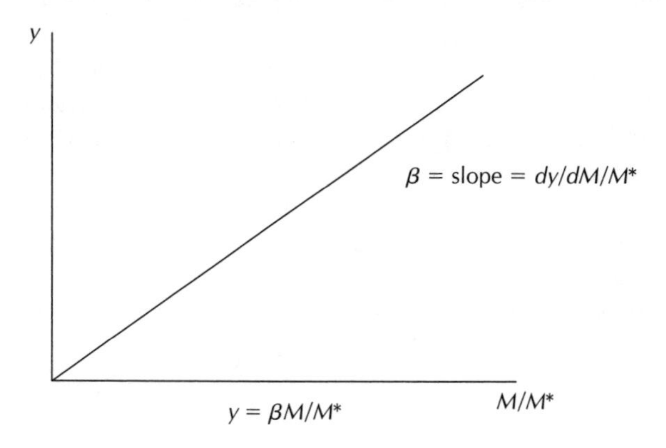


Some examples for this linear form:

1. Ohm's Law for a resistor, $V = IR = (\rho d/A)I$, where $V$ is the voltage across the resistor, $\rho$ is the volume resistivity of the resistor's material, $d/A$ is the geometry factor (length divided by area), and $I$ is the current passing through the resistor. Thus, $M = I$ and $M^* = d/A$.
2. Hooke's Law for an elastomeric material: $F = kx$, $k = EA/l$, where $F$ is the applied force, $x$ is the resulting displacement, $E$ is the modulus of elasticity, $A$ is the area, and $l$ is the width of the sample. Thus, $M = x$ and $M^* = A/l$.

If the slope is inversely related to the PSF, the ideal function is given by

$$y = \beta M/M^* \tag{6.19}$$

The ratio of the two signal factors is linearly related to the response. Plotting the response, $y$, vs. $M/M^*$, we see that the ideal function reverts to the zero-point proportional form (Figure 6.17).

Some examples for this linear form:

1. The deflection, $y$, of a cantilevered beam, which is a function of the applied load, $M$, and the moment of inertia, $M^*$. Thus, $y = \beta M/M^*$.
2. The flow rate through a fuel pump, $y = \beta M/M^*$, where $M$ is the pump power and $M^*$ is the system back pressure [C3].

Altering the problem to a form that is similar to the zero-point proportional case allows the use of the analysis previously developed to calculate the S/N ratio and the slope, $\beta$. Note that there is no longer any need for adjustment to hit the target. The PSF, $M^*$, is used to tune the dynamic relationship, and the FSF is used to hit the target. The friction wheel example in Section 6.5.4 shows how this is done.

### 6.5.3 Case 3: $y = [\beta + \beta^*(M^* - M^*_{ref})]M$

A linear relationship is the most generic form for the relationship between the signal factors for the ideal function. Thus, the ideal equation is given by

$$y = [\beta + \beta^*(M^* - M^*_{ref})]M \tag{6.20}$$

A good example of this form is an application to describe a blood oxygen sensor [B8]. In that case, the FSF, $M$, is the $O_2$ level in the serum, and the PSF, $M^*$, is the blood temperature. The reference temperature $M^*_{\mathrm{ref}}$ can be chosen a priori or set equal to the average of the $M^*$ values. The parameter $\beta$ is the slope between the response and the FSF. The parameter $\beta^*$ is the slope of the relationship between $\beta$ and $M^*$. As described in the paper by Bires, surgeons can choose different metabolism rates for their patients by modifying the patient's body temperature. This causes a different response of the measurement device. One of the uses of the double-dynamic relationship is to characterize the slope change due to a significant noise factor, in this case temperature. This allows the process control engineer to add compensation to maintain accuracy at various temperature levels. The compensation process is facilitated by the robustness of the open-loop behavior.

Equation 6.19 is transformed to the reference-point proportional form by dividing by $M$. The resulting equation is

$$y/M = \beta + \beta^*(M^* - M^*_{\mathrm{ref}}) \tag{6.21a}$$

or

$$y/M - \beta = \beta^*(M^* - M^*_{\mathrm{ref}}) \tag{6.21b}$$

If we set $y' = y/M - \beta$ and $M' = (M^* - M^*_{\mathrm{ref}})$, then the S/N ratio can be calculated, as shown earlier using the zero-point proportional form:

$$y' = \beta^* M' \tag{6.22}$$

## 6.5.4 Friction Wheel Example

As a simple illustration of the double-dynamic concept, consider the friction wheel transmission shown in Figure 6.18. The energy transformation in this case is the transfer of torque from the drive shaft to the load shaft. The ideal function is $\omega_L = \omega_D$, where $\omega_L$ is the rotational velocity of the load

**Figure 6.18** Friction wheel transmission

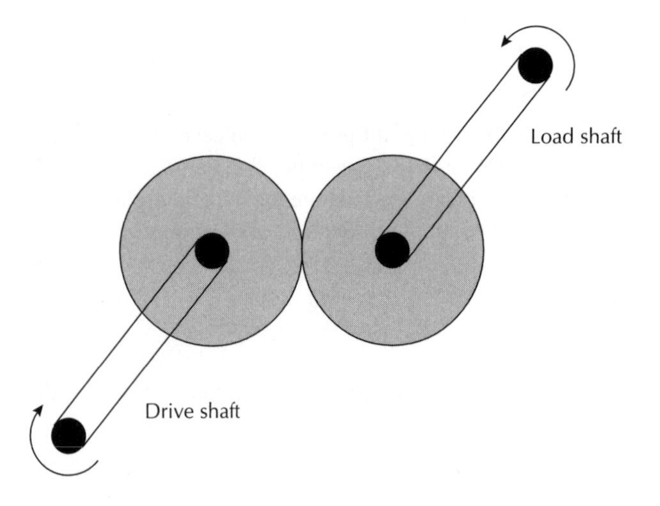

shaft and $\omega_D$ is the rotational velocity of the drive shaft. There is a double-dynamic relationship here that can be described by the P-diagram in Figure 6.19.

The nature of the double-dynamic relationship is clear from the physics of the drive system shown. The functional signal factor (FSF) is $M = \omega_D$. The process signal factor is the friction wheel diameter ratio, $M^* = r_L/r_D$, where $2r_L$ is the load shaft friction wheel diameter and $2r_D$ is the drive shaft friction wheel diameter. Thus, the ideal function is given by

$$\omega_L = \beta MM^* = \beta\omega_D(r_L/r_D) \tag{6.23}$$

Before doing the calculation, consider how the double-dynamic experiment is actually run. The experiment consists of an orthogonal array (Chapter 7), loaded with the control factors and their respective levels. Each experimental run from the array is performed at the appropriate functional ($M$) and process signal factor ($M^*$) levels. For each of the $M$ and $M^*$ levels, the specified noise replicates are run as the experimental plan dictates. Thus, each experimental run in the orthogonal array is repeated several times depending on the number of $M$, $M^*$, and noise factors. If there are three $M$'s, three $M^*$'s, and two noise levels, then there are 18 repeats. Figure 11.3 illustrates the double-dynamic layout.

For the friction wheel example, assume that there are three functional signal factors, $M1$, $M2$, and $M3$; three process signal factors, $M1^*$, $M2^*$, and $M3^*$; and four noise factors, $N1$, $N2$, $N3$, and $N4$. A full factorial study of these factors results in 36 measurements of the response, $\omega_L$. The resulting data set is shown in Table 6.7.

Figure 6.20 is a plot of the raw data showing the characteristic double-dynamic behavior described in Figure 6.15.

Calculating the signal-to-noise ratio is accomplished by combining $M$ and $M^*$ into a single value. This product can then be used as the basis of a zero-point proportional calculation. Table 6.8 shows this procedure. The column labeled $MM^*$ gives the product of the two signal factors, which is used as the signal factor in the calculation to follow. Figure 6.21 shows the raw data plotted as a function of the *product* $MM^*$, indicating the resulting zero-point proportional form. Figure 6.21 illustrates the concept underlying Figure 6.16.

**Figure 6.19** P-diagram for friction wheel transmission

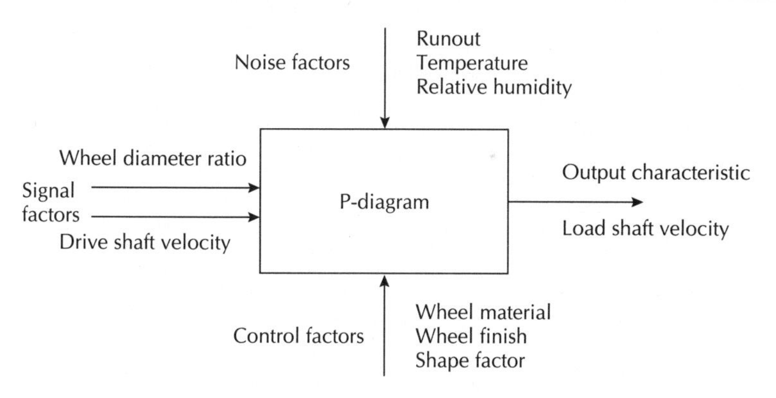

**Table 6.7** Data for the friction wheel example

| | | *M* (revolutions per minute) | | | | | |
|---|---|---|---|---|---|---|---|
| | | *25* | | *50* | | *75* | |
| *M** | 2.0 | 48 | 50.5 | 99 | 103 | 150 | 155 |
| | | 47 | 55 | 97 | 110 | 151 | 170 |
| | 1.0 | 24 | 25.5 | 47 | 51 | 71 | 77 |
| | | 24 | 28 | 48 | 55 | 69 | 81 |
| | 0.5 | 12 | 12.5 | 24 | 25 | 35 | 37 |
| | | 12 | 13 | 24 | 26 | 36 | 40 |

**Figure 6.20** Plot of raw data vs. *M* and *M**

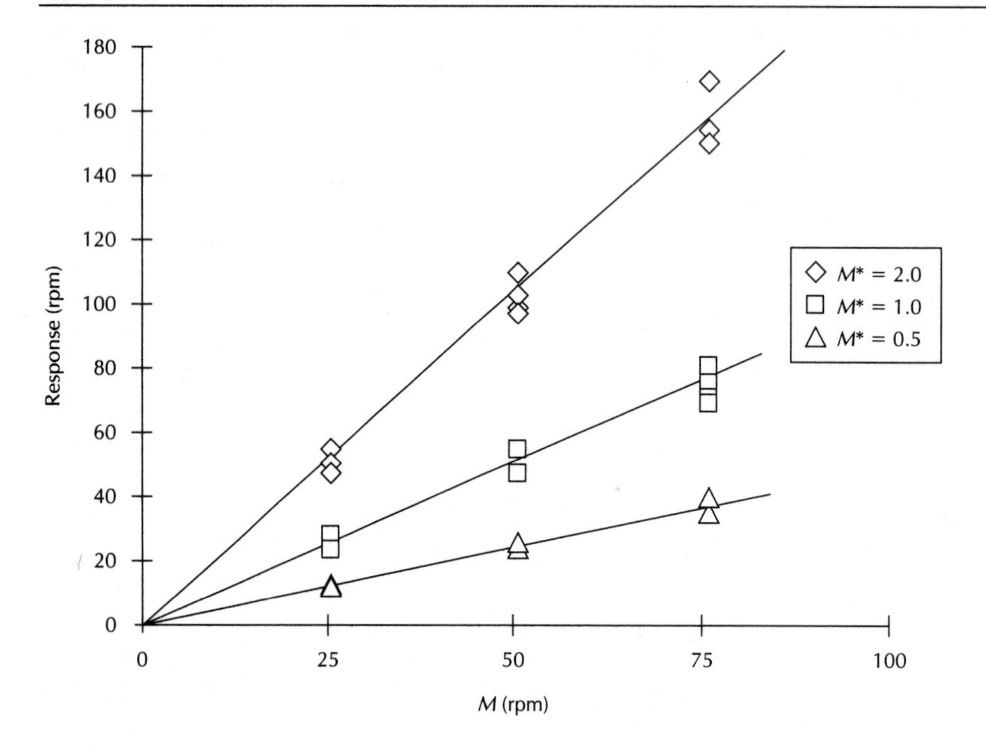

**Figure 6.21** Plot of raw data vs. $MM^*$

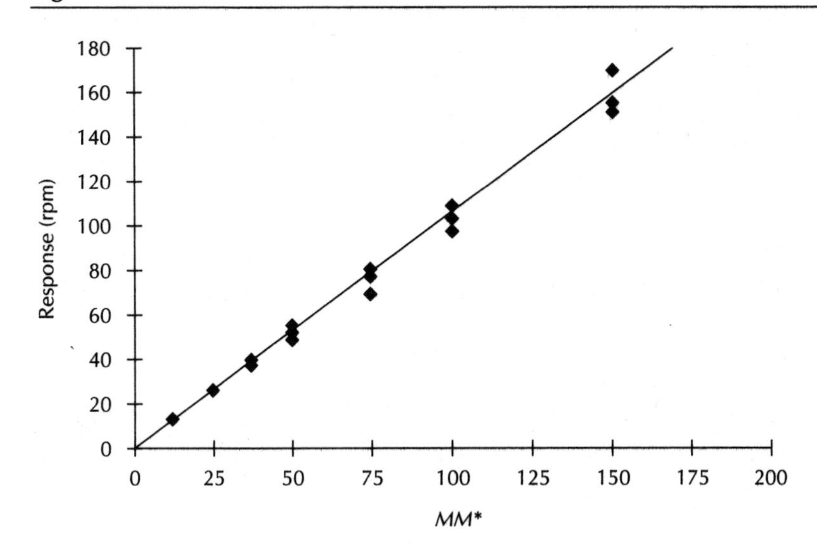


**Table 6.8** Data for the friction wheel example as a function of the signal and noise factors

| $M^*$ | $M$ | $N$ | $MM^*$ | $\omega_L$ | $M^*$ | $M$ | $N$ | $MM^*$ | $\omega_L$ | $M^*$ | $M$ | $N$ | $MM^*$ | $\omega_L$ |
|---|---|---|---|---|---|---|---|---|---|---|---|---|---|---|
| 2 | 75 | 1 | 150 | 150 | 1 | 75 | 1 | 75 | 71 | 0.5 | 75 | 1 | 37.5 | 35 |
| 2 | 75 | 2 | 150 | 151 | 1 | 75 | 2 | 75 | 69 | 0.5 | 75 | 2 | 37.5 | 36 |
| 2 | 75 | 3 | 150 | 155 | 1 | 75 | 3 | 75 | 77 | 0.5 | 75 | 3 | 37.5 | 37 |
| 2 | 75 | 4 | 150 | 170 | 1 | 75 | 4 | 75 | 81 | 0.5 | 75 | 4 | 37.5 | 40 |
| 2 | 50 | 1 | 100 | 99 | 1 | 50 | 1 | 50 | 47 | 0.5 | 50 | 1 | 25 | 24 |
| 2 | 50 | 2 | 100 | 97 | 1 | 50 | 2 | 50 | 48 | 0.5 | 50 | 2 | 25 | 24 |
| 2 | 50 | 3 | 100 | 103 | 1 | 50 | 3 | 50 | 51 | 0.5 | 50 | 3 | 25 | 25 |
| 2 | 50 | 4 | 100 | 110 | 1 | 50 | 4 | 50 | 55 | 0.5 | 50 | 4 | 25 | 26 |
| 2 | 25 | 1 | 50 | 48 | 1 | 25 | 1 | 25 | 24 | 0.5 | 25 | 1 | 12.5 | 12 |
| 2 | 25 | 2 | 50 | 47 | 1 | 25 | 2 | 25 | 24 | 0.5 | 25 | 2 | 12.5 | 12 |
| 2 | 25 | 3 | 50 | 50.5 | 1 | 25 | 3 | 25 | 25.5 | 0.5 | 25 | 3 | 12.5 | 12.5 |
| 2 | 25 | 4 | 50 | 55 | 1 | 25 | 4 | 25 | 28 | 0.5 | 25 | 4 | 12.5 | 13 |

The calculation of the slope and S/N ratio is done as before, with appropriate modifications made to account for the fact that the number of runs at each combined signal factor level, $MM^*$, is not always the same. For example, there are twice as many runs for $MM^* = 50$ and $MM^* = 25$ as there are for the other levels of $MM^*$. That is not a problem if close attention is paid to the original equations given earlier in this chapter. After substitution of $(MM^*)_i$ for the signal factor (recall Equation 6.5), the double-dynamic slope is given by

$$\beta = \frac{\sum_{i=1}^{k} \sum_{j=1}^{r_0} (MM^*)_i y_{i,j}}{\sum_{i=1}^{k} \sum_{j=1}^{r_0} (MM^*)_i^2} \tag{6.24}$$

The zero-point proportional calculation done earlier assumes that there is an equal number of noise replicates for each level of the signal factor. Here, as noted earlier, that is not the case. The summation limit, $r_0$ (the number of noise factors), is effectively a function of the level of $i$; $r_0 = 4$ except when $MM^* = 50$ or $MM^* = 25$, in which case $r_0 = 8$. Compare the following definition of $r$ with Equation 6.6:

$$r = \sum_{i=1}^{k} \sum_{j=1}^{r_0(i)} (MM^*)_i \tag{6.25}$$

Equation 6.25 can be used whenever there is an unequal number of replicates for the signal factors. If $r_0$ is the same at each signal factor level, then the summation over $j$ can simply be replaced by $r_0$ as is done in Equation 6.6.

For the friction wheel case, $r = 183750$. Completing the calculation for the slope:

$$\beta = 188406.3/183750 = 1.025$$

The mean square error is found by applying Equation 6.13 to find $SS_V$, where $SS_T$ is given by Equation 6.11 and $SS_\beta$ is given by Equation 6.12.

$$SS_V = \sum_{i=1}^{k} \sum_{j=1}^{r_0} y_{i,j}^2 - \frac{\left(\sum_{i=1}^{k} \sum_{j=1}^{r_0} (MM^*)_i y_{i,j}\right)^2}{r} = 193806.8 - \frac{(188406.3)^2}{183750} = 626.26$$

The number of degrees of freedom for this experiment (Chapter 7) is given by the total number of runs minus 1. Thus, the mean square error is

$$MSE = \frac{SS_V}{r_0 k - 1} = \frac{626.26}{36 - 1} = 17.893$$

The S/N ratio is then given by

$$S/N = 10 \log \frac{\beta^2}{MSE} = -12.3$$

The slope of 1.025 indicates that, as expected, the velocity of the driven shaft is essentially a function of $MM^*$, $\omega_L = M^*\omega_D$. If the 2.5% error is real (it is probably not statistically significant), it could be due to loading or overdrive. The negative value of the S/N ratio is a reflection of the spread in the data seen in Figure 6.21. By itself, it is not possible to know whether that is a problem. In a designed experiment, however, where control factors such as machining process, wheel material, and loading force are being evaluated, the S/N values are used as a measure of each test treatment's performance.

This case study is primarily for the purpose of illustrating the calculations required for the double-dynamic analysis. All sources of deviation from the ideal relationship, Equation 6.23, are included in the S/N analysis. Thus, sensitivity to noise factors such as temperature or runout that affect the wheel diameters, $r_L$ or $r_D$, in addition to load variations or friction are captured in the S/N analysis. But the most powerful aspect of the double-signal analysis is that the friction wheel transmission is treated as a system, and any combination of wheel diameters can be included in a single analysis.

## 6.6 Summary

The dynamic S/N metrics represent an important extension to the static S/N metrics. The dynamic method is especially useful because it incorporates the signal factor for adjusting to target, a feature that is critical for on-target engineering. The primary dynamic form is the zero-point proportional case, $y = \beta M$, where $y$ is the quality characteristic, $M$ is the signal factor, and $\beta$ is the slope of the linear relationship. The S/N ratio includes the slope, a measure of the dynamic sensitivity, and the mean square error, a measure of the variation from the ideal function. The S/N ratio is given by

$$\text{S/N} = 10 \log \frac{\text{slope}^2}{\text{MSE}}$$

where the MSE is given by

$$\text{MSE} = \frac{1}{r_0 k - 1} \sum_{i=1}^{k} \sum_{j=1}^{r_0} (y_{i,j} - \beta M_i)^2$$

Dynamic analysis requires one to optimize the *function* rather than just a single result. It is a principle of the dynamic method that the ideal function can usually be defined in terms of the zero-point equation. It requires engineering analysis to identify the ideal function that is to be optimized. Identifying the ideal function is discussed in Part II, Parameter Design, especially Chapter 8. The significance of the zero-point relationship is that it offers the most robust adjustment capability for the quality characteristic possible. All other functional forms compromise the robustness somewhat. However, it is possible to optimize any function by applying an appropriate linearizing transformation.

The dynamic method automatically incorporates the means for adjusting the quality characteristic onto target using the signal factor. In some cases it is desirable to tune the function, or slope $\beta$, onto target. For those cases, the double-dynamic analysis is used. The process signal factor, $M^*$, is used to tune the slope of the functional relationship.

Because of the importance of the two-step optimization process, where the variability is reduced first and the process is put on target second, a number of adjustment parameters have been defined in Part I, Quality Engineering Metrics. They are gathered here and defined to clarify their distinction:

- **Adjustment Factor:** This is any factor that is useful for adjusting a quality characteristic onto the target. Adjustment factors may have an effect on the S/N ratio.
- **Tuning Factor:** This is an adjustment factor that can put the mean quality characteristic on target without affecting the S/N ratio.

- **Scaling Factor:** This is a tuning factor that is appropriate for the type I nominal-the-best S/N ratio. Both the mean and the standard deviation change proportionally to a scaling factor, thus leaving the S/N ratio constant.
- **Signal Factor:** This is an adjustment factor (preferably a scaling factor for the zero-point proportional case) chosen for a dynamic analysis. The assumption here is that the system is understood well enough that an appropriate scaling factor has already been identified. In double-dynamic analyses, it is referred to as the Functional Signal Factor to distinguish it from the Process Signal Factor.
- **Process Signal Factor:** This is the tuning factor chosen for adjusting the slope of the dynamic relationship onto target.

## Exercises for Chapter 6

1. Explain the difference between a static case and a dynamic case.
2. Explain the difference between a single-dynamic and a double-dynamic case.
3. Define what is meant by a signal factor, and provide three examples of products or processes that have signal factors associated with them.
4. What three properties are used to define the quality of dynamic measurement systems? Define what each term means.
5. Explain the difference between applying dynamic methods "actively" and "passively."
6. Provide several examples of a functional signal factor as used in the context of a double-dynamic experiment.
7. Provide several examples of a process signal factor as used in the context of a double-dynamic experiment.
8. Define the math models that underlie the zero-point and reference-point proportional methods for single-dynamic cases.
9. Why is it so important to focus on the ideal function for setting up a dynamic case? What are the consequences of setting up a dynamic case that has a signal not linearly related to the response?
10. What recourse do we have if a nonlinear relationship exists between the signal and the response?
11. What is the strategic importance of applying dynamic methods early in the technology development or commercialization process?
12. What is $\beta$ used for in a dynamic optimization process?
13. Provide an example of when to use a zero-point proportional case and when to use a reference-point proportional case.
14. Provide an example of when you would apply a double-dynamic case.

# PART II

# The Parameter Design Process

We have discussed the quality metrics used for robust design: the mean and variance, the loss function, and the S/N ratio. For the remainder of this book, the focus is on how to use design of experiments (DOE) and the S/N ratio to improve product and process designs. After some background on DOE, the elements of the Parameter Design process are presented in their natural sequence. Several examples—cooling a drink with ice, a paper gyrocopter, a catapult, and a belt-drive system—are threaded throughout Parts II and III of the book to illustrate aspects of Robust Design.

Remember, from the discussion of the loss function, robustness has two goals: low variation and on-target behavior. The key to a successful and efficient optimization of the design set points is the disciplined use of engineering knowledge in planning the experiment. It should be clear from our discussion of the metrics of Robust Design that variability is the enemy in engineering [H2]. The ideal system is one that is without loss, without variability. As any experienced engineer knows, in all real processes, variability is present. The loss function and the S/N ratio are used to measure how far from the ideal state the real process is. If these metrics are optimized, the system approaches perfection, significantly improving customer satisfaction.

The system or subsystem to be optimized is schematically represented by an analysis tool introduced by Madhav Phadke, called the "P-Diagram" (Parameter Diagram) [P2]. Parameter design uses the control factors, noise factors, signal factor, and the measured output response shown in the P-diagram. The P-diagram, shown in Figure II.1, is a useful tool for organizing the engineering factors in parameter optimization.

The experimental design methodology used for studying how a number of parameters impact a measured response is introduced in Chapter 7. The orthogonal array, in the form of fractional factorials as well as full factorials, facilitates the study of many factors in one highly efficient experiment. The results of such an experiment are evaluated using analysis of means, a simple yet powerful method to characterize how the experimental parameters influence the response. The ice water example, a smaller-the-better problem, is used to illustrate the DOE process. Chapter 7 also introduces WinRobust Lite, the software package included with this book.

**Figure II.1** The P-diagram

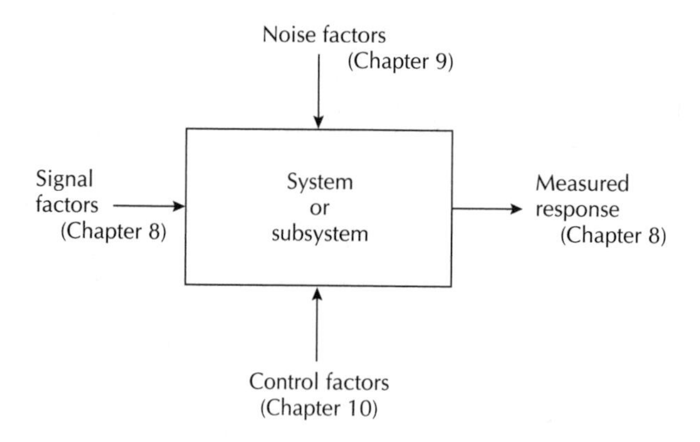


Successful parameter optimization requires a meaningful quality characteristic as the response to be optimized. Chapter 8 shows how to choose quality characteristics that are related to the main function and energy transformations that occur within a design. Putting the quality characteristics on target is important. Dynamic analysis, used to optimize the ideal function with the use of a signal factor, is critical to the efficiency and effectiveness of the parameter design process. The signal factor can be the one factor necessary for tuning. All the other controllable parameters are used to satisfy the engineering requirements of high quality at low cost. The catapult, a device capable of throwing a projectile over various distances, is introduced in Chapter 8 to illustrate how to select the proper quality characteristic.

The noise factors used in the parameter optimization experiment are critical to the success of the design. An appropriate set of noise factors stresses the system in a way that exposes how the system deviates from the ideal function. Noise factors, their selection, and experimentation to test the noise factor effects are discussed in Chapter 9. Noise factors for the ice water and the catapult are used as examples. The chapter concludes with a case study employing noise factors to certify the performance of a film feeding device.

The parameters that need to be optimized to improve the robustness of the system are the control factors. These are the design parameters or formulation variables that determine the system's performance. The selection of control factors and their levels is discussed in Chapter 10. Both the catapult and the gyrocopter, a device similar to a paper airplane that spins rapidly while floating slowly to the ground, are analyzed using a unique control factor matrix to design for additivity.

In Chapter 11, the layouts for several types of parameter optimization experiments are discussed. There it is shown how to apply the DOE methodology, using separate treatments for control factors, signal factors, and noise factors, to efficiently evaluate the control factors' ability to make the design approach the ideal function. The data analysis of the parameter optimization experiment, using the methods covered in Chapter 7, is then discussed in Chapter 12. Finding the optimum levels for control factors and then verifying the optimum is ultimately the goal in Robust Design. Prediction of the optimum performance using the additive model is used to verify the performance improvement and to determine for additivity. A belt drive example is used to illustrate the process and calculations.

The Parameter Design process is demonstrated in Chapter 13 with the three examples—cooling water with ice, the catapult, and the gyrocopter—that are introduced in Chapters 8–12. The latter two examples use the dynamic analysis. The ideal function for the catapult is a linear relationship between the distance a projectile travels and the amount of lever-arm pull-back. This dynamic relationship can be used to hit a target at any point in a range of approximately 2 to 12 feet from the catapult. The ideal function for the gyrocopter is a linear relationship between time of descent and height, with the largest possible slope. These are all simple experiments that you can actually do yourself.

Part II concludes with Chapter 14 by looking at three examples derived from actual case studies performed at Kodak. As you might expect, "real life" studies are both interesting and "messy." We have chosen *not* to sanitize them; they are presented basically as they were accomplished so that you can learn some of the "tricks of the trade" that experienced practitioners use. WinRobust Lite software is used to do the analyses in the following chapters. As a result, we hope that you will see how to apply these techniques to improve the quality and quantity of your engineering output.

Here is an overview of strategies discussed in Part II for use in Parameter Design:

1. Choose quality characteristics that reflect the essential function.
2. Use appropriate arithmetic transformations such as the S/N ratio.
3. Choose noise levels that are large enough to overwhelm experimental error and put the focus on robustness. Use noise experiments to determine how the noises should be grouped so as to stress the system in the most efficient manner.
4. Choose control factors and levels that (a) have effects that are easily distinguishable and (b) can be logically grouped in energy terms. Where appropriate, use sliding levels to improve experimental efficiency and reduce interactions.

These strategies each require the use of engineering experience and disciplined experimental design techniques. The following pages provide engineering methods for robust product design.

CHAPTER 7

# Introduction to Designed Experiments

## 7.1 Experimental Approaches

This chapter introduces one of the most powerful statistical tools developed in the 20th century. Because it is such a powerful tool, many books have been written on the subject of experimental design. These include classic works by the inventor of DOE, Sir Ronald Fisher [F2], and works by contemporary authors such as Box, Hunter, and Hunter [B10]. A number of additional books are listed in the bibliography. Dr. Taguchi is the author of several books on this topic [T2, T5, T6]. There are also numerous classes taught in universities, industry, and by consultants. It is not our intention to compete with these luminaries. Rather, in keeping with the engineering spirit throughout this text, we seek to explain the power of this technique using the simplest of approaches. Where necessary, we discuss the statistical assumptions underlying the method and show where and how to bend the "rules" to accomplish the product optimization task effectively without complication. We respect the diversity of opinion that exists in modern quality methods and provide an open-minded representation of the subject at hand.

Consider the following question: How should a team design an effective set of experiments to determine the optimal levels of a set of control factors (design parameters)? This is the engineering objective. Building complex mathematical models that can be illustrated by contour plots and polynomial equations is not the ultimate goal of the experimental study. The key role of the design parameters is to minimize the product's sensitivity to noise. The engineer's task is to specify the design parameters.

Experiments are used, by engineers, to study the effects of parameters as they are set at various levels. The experimental boundaries are established by which factors are included in an experiment and the set points (levels) at which they are evaluated. These boundaries are referred to as the *experimental design space.* The following four approaches have been devised to study specific experimental design spaces:

1. Build–test–fix
2. One-factor-at-a-time experiments

3. Full factorial experiments
4. Orthogonal array experiments (fractional factorials)

Build–test–fix is an ineffective and inefficient method that necessarily leads to long cycle times and poor reproducibility; it is strongly dependent on the skill of the experimenter. It has worked at times, and will at times in the future, but *even a stopped clock is right twice a day!*

Build–test–fix is completely inadequate for two reasons:

1. It is impossible to know if a true optimum is achieved, because a build–test–fix cycle is considered complete as soon as functionality is achieved. But this approach is indicative of within-specification thinking. There is no information on how close to ideal the resulting design is. Any result within the tolerance limits is considered equal as a result of build–test–fix.
2. Build–test–fix is consistently slow because it requires luck and intuitive skill, and because it leads to rework, since there is always a need to improve performance. On-target performance is not achieved, so reoptimization and fire-fighting regularly occur following build–test–fix.

Let us look at the other three approaches in detail, using as an example the study of seven two-level factors, i.e., parameters that can be set to either of two values. This example is chosen because it is very common to have several parameters to specify in a design. The factors need to work harmoniously to accomplish the ultimate goal of customer satisfaction. The goal of the study is to find the optimal level for each of the seven factors.

### 7.1.1 One-Factor-at-a-Time Experiments

This is the traditional method used by scientists and engineers in academia and industry. How is it practiced in research organizations? The first factor is thoroughly studied under fixed conditions. Once it is well characterized, the experimentalist moves on to study another factor thoroughly, until all seven factors are well characterized. This approach has been successful in developing a scientific understanding of the effect of a parameter. One-factor-at-a-time can be used to test many levels of a factor under precision (low-noise) conditions. It can also be used to develop physical models.

The biggest weakness of this approach, as it is commonly practiced, is that it is slow. In an engineering application, time is critical. Table 7.1 shows how this method might look for seven factors in as few runs as possible.

Table 7.1 describes the level settings for each of the seven factors that is to be used for the eight runs. The runs are referred to as *treatment combinations.* The experiment consists of the entire eight runs. Each column corresponds to one of the factors; each row corresponds to one of the runs in the experiment. For each run, a response or quality characteristic is measured. The table indicates that for the first treatment combination all the factors are set to level 1. A test is performed on the design and the result, $R_1$, is measured and recorded. This result is referred to as the measured *response.* Now, for the second treatment combination, factor A is changed to its level 2, while holding all the other factors constant. The response, $R_2$, is measured. The analysis of this part of the experiment is done by comparing $R_1$ to $R_2$. Depending on which run gives a better result for the response, the optimum level for factor A is chosen. Assuming that the optimum level is $R_2$, factor A is set at level 2 for the remainder of the experiment. The next pair of results, $R_2$ and $R_3$, is used to compare $B_1$ and $B_2$ (B at its

**Table 7.1** An experimental design for one-factor-at-a-time

| *Run* | *A* | *B* | *C* | *D* | *E* | *F* | *G* |
|---|---|---|---|---|---|---|---|
| 1 | 1 | 1 | 1 | 1 | 1 | 1 | 1 |
| 2 | 2 | 1 | 1 | 1 | 1 | 1 | 1 |
| | | | | *Comparison of $A_1$ vs $A_2$* | | | |
| 3 | 2 | 2 | 1 | 1 | 1 | 1 | 1 |
| 4 | 2 | 2 | 2 | 1 | 1 | 1 | 1 |
| 5 | 2 | 2 | 2 | 2 | 1 | 1 | 1 |
| 6 | 2 | 2 | 2 | 2 | 2 | 1 | 1 |
| 7 | 2 | 2 | 2 | 2 | 2 | 2 | 1 |
| 8 | 2 | 2 | 2 | 2 | 2 | 2 | 2 |

level 1 and B at its level 2, respectively). This process is continued until each factor has been evaluated by a pairwise comparison.

Each pairwise comparison reflects the effect of the factor that is being changed. This is because of the assumption that all other sources that could cause variation of the response are held constant. Thus, the only effect that could cause the response to vary is the control factor that is being changed. Such pairwise comparisons are common in designed experiments and are referred to as *factor effects.*

Does this approach fix the problems with one-factor-at-a-time? It does not give information about how the effect of a factor changes when the other factors change. While one factor is being changed, all the others are held constant. In this example, the comparison of $A_1$ to $A_2$ is done with all the other factors set at level 1. By the end of the tests, the other factors may be all set at level 2. Is it known that $A_2$ is still better than $A_1$, now that all the other factors have changed? Certainly not. If the effect of factor A depends upon the setting of another factor, this is referred to as an *interaction.*

### Interactions between Factors

To illustrate this concept of interactions, imagine that a test is done to find the conditions that will give the fewest waste emissions from an oxidation reaction. Factor A is the reaction temperature (at two levels) and factor B is the reaction pressure (at two levels). The emission levels for all four combinations possible for factors A and B are shown in Figure 7.1. The way the experiment has been planned, using one-factor-at-a-time, a sub-optimal result of 9% emissions will be found, rather than the optimum of 5%. This is because of an interaction between factors A and B, which results in $A_2$ being better than $A_1$ only when factor B is at level 1.

(*continued*)

**Figure 7.1** Result of one-factor-at-a-time in the presence of an interaction

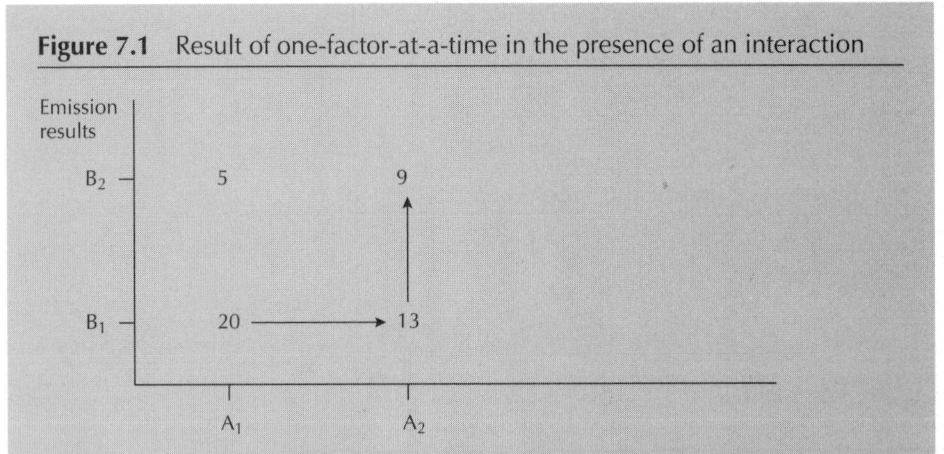


These results can be plotted in another way, referred to as the interaction plot. Figure 7.2 is a plot of the response vs. factor A. But there are two data points, corresponding to the two levels for factor B, shown for each level of factor A. The data points for the same level of factor B are connected by a line for clarity. The lines are not parallel when there is a significant interaction. If the factors are independent in their effects on the response, then the lines are parallel, thus providing a visual representation of noninteractive behavior between parameters.

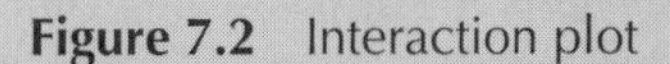

**Figure 7.2** Interaction plot

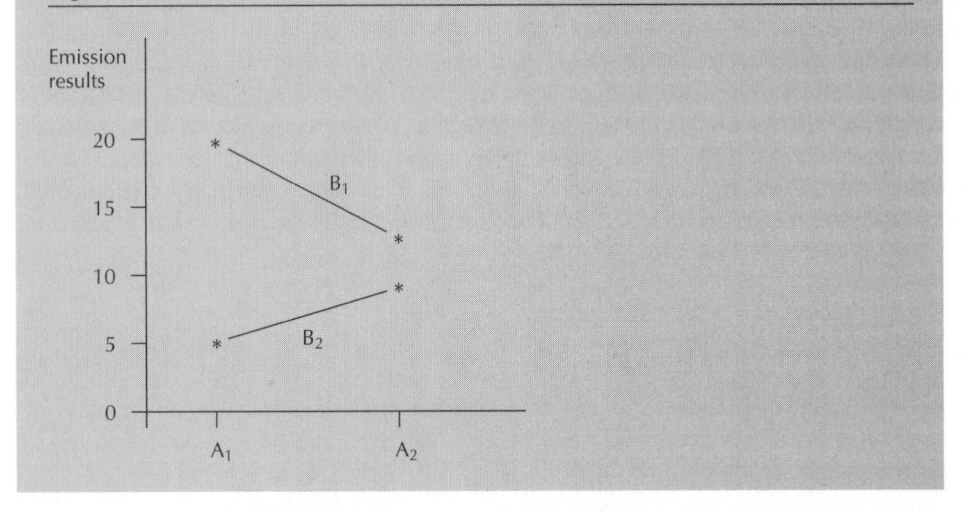


Frequently, tests that lead to conclusions about factors never get rechecked as time goes on and other factors have been changed. Unfortunately, when the early experiments finally do get repeated, the result for the optimum level of the factor often changes because it depends on what level the other factors are set to. They, too, may have changed. Thus, the whole optimization process goes into a loop that doesn't always converge. (Although the money *does* always run out eventually.)

Another problem with the original experimental plan shown in Table 7.1 is that there is an unequal number of replicates for each factor level. $A_1$ was only tested one time, $A_2$ was tested seven times. Thus, the experiment is not balanced.

### 7.1.2 Full Factorial Experiments

A full factorial experiment investigates all possible combinations of all factor levels. Table 7.2 shows a few of all the combinations that are possible with seven two-level factors.

The total number of combinations that are possible can be calculated using the formula given in Equation 7.1.

$$\text{\# of combinations} = y^x \tag{7.1}$$

where $x$ = # of factors that have $y$ levels. Even for the simple example of seven factors and two levels for each factor, there are 128 combinations ($2^7$).

The full factorial approach investigates all possible combinations, maximizing the possibility of finding a favorable result. (Pick-the-winner!) All that is needed is to examine the 128 responses to find the most favorable. The factor levels that produced that result are assumed to be the optimum combination. As in the previous example, this assumes that all other factors, outside of the seven control parameters, are held constant so that the variation in the responses is due only to the factor effects. This is a highly questionable assumption when dealing with 128 separate runs.

The biggest weakness of this approach is that too many experiments are used for the amount of information needed to understand the factor effects [B3, B9]. In fact, full factorial experiments are only practical with small numbers of factors and levels. For example, 13 factors with 3 levels of each factor have $3^{13}$ (1,594,323) combinations! Therefore, statistical techniques are required to generate enough information in a small number of tests to understand what the optimum levels for the control factors are and what the resulting robustness of the system is. It is not necessary to investigate every factor combination to find the optimum set. A simple statistical analysis allows us to infer the optimum level for each factor by looking at the factor's average effect on the response.

The results of the full factorial experiment can be analyzed to develop an empirical math model for the factor effects. The simplest available method to do so, the *analysis of means (ANOM),* Section 7.2, is used in this book. Assume that there are 128 response values, $R_i$, and that each one is a quantitative measure of the design's performance. (For example, the response could be a critical part

**Table 7.2** An experimental design for a full factorial

| *Run* | *A* | *B* | *C* | *D* | *E* | *F* | *G* |
|---|---|---|---|---|---|---|---|
| 1 | 1 | 1 | 1 | 1 | 1 | 1 | 1 |
| 2 | 1 | 1 | 1 | 1 | 1 | 1 | 2 |
| 3 | 1 | 1 | 1 | 1 | 1 | 2 | 1 |
| . | . | . | . | . | . | . | . |
| . | . | . | . | . | . | . | . |
| 126 | 2 | 2 | 2 | 2 | 2 | 1 | 2 |
| 127 | 2 | 2 | 2 | 2 | 2 | 2 | 1 |
| 128 | 2 | 2 | 2 | 2 | 2 | 2 | 2 |

dimension, the amount of material produced, the yield of a reaction, the density of an image, etc.) Runs 1 through 64 are done with factor A at level 1, and runs 65 through 128 are done with A at level 2. Thus, the factor effect can be found by taking the average or mean response of all the runs with $A_1$ and comparing that value with the mean response for $A_2$.

$$\text{Effect of } A_1 = (R_1 + R_2 + \ldots + R_{63} + R_{64})/64$$
$$\text{Effect of } A_2 = (R_{65} + R_{66} + \ldots + R_{127} + R_{128})/64 \quad \textbf{(7.2)}$$

In a similar fashion, the other factor effects can be found by averaging the respective runs at level 1 and at level 2. For example, the factor effect for B can be found by averaging the runs at $B_1$ (1–32 and 65–96) and the runs at $B_2$ (33–64 and 97–128). Clearly, 64 data points are not required to establish a measure of performance when A is at level 1. The only reason for running a full factorial experiment is that only a few factors are being studied and there is a need to comprehensively account for all factor effects, including *every* possible factor interaction.

### 7.1.3 Orthogonal Array Experiments

The *orthogonal array* is a method of setting up experiments that only requires a fraction of the full factorial combinations. The treatment combinations are chosen to provide sufficient information to determine the factor effects using the analysis of means. The orthogonal array imposes an order on the way the experiment is carried out. Orthogonal refers to the balance of the various combinations of factors so that no one factor is given more or less weight in the experiment than the other factors. Orthogonal also refers to the fact that the effect of each factor can be mathematically assessed independently of the effects of the other factors.

Table 7.3 is an example of an orthogonal array that would accommodate the seven-factor example. As before, this array describes the level settings for each of the seven factors that is to be used for the eight runs. In this case, the treatment combinations are chosen to uniformly span the experimental space represented by the 128-run full factorial. For the first run, all factors are set at level 1. For the second run, factors D, E, F, and G are changed to level 2; A, B, and C remain at level 1. The remaining runs are set up as coded in the matrix until all eight runs are complete. Notice that this array cannot provide all factor interaction effects.

**Table 7.3** An orthogonal array design for a fractional factorial experiment

| *Run* | *A* | *B* | *C* | *D* | *E* | *F* | *G* |
|---|---|---|---|---|---|---|---|
| 1 | 1 | 1 | 1 | 1 | 1 | 1 | 1 |
| 2 | 1 | 1 | 1 | 2 | 2 | 2 | 2 |
| 3 | 1 | 2 | 2 | 1 | 1 | 2 | 2 |
| 4 | 1 | 2 | 2 | 2 | 2 | 1 | 1 |
| 5 | 2 | 1 | 2 | 1 | 2 | 1 | 2 |
| 6 | 2 | 1 | 2 | 2 | 1 | 2 | 1 |
| 7 | 2 | 2 | 1 | 1 | 2 | 2 | 1 |
| 8 | 2 | 2 | 1 | 2 | 1 | 1 | 2 |

**Table 7.4** Illustration of balance in an orthogonal array

| *Run* | *A* | *B* | *C* | *D* | *E* | *F* | *G* |
|---|---|---|---|---|---|---|---|
| 1 | 1 | 1 | 1 | 1 | 1 | 1 | 1 |
| 2 | 1 | 1 | 1 | 2 | 2 | 2 | 2 |
| 3 | 1 | 2 | 2 | 1 | 1 | 2 | 2 |
| 4 | 1 | 2 | 2 | 2 | 2 | 1 | 1 |
| 5 | 2 | 1 | 2 | 1 | 2 | 1 | 2 |
| 6 | 2 | 1 | 2 | 2 | 1 | 2 | 1 |
| 7 | 2 | 2 | 1 | 1 | 2 | 2 | 1 |
| 8 | 2 | 2 | 1 | 2 | 1 | 1 | 2 |

This array can be used to illustrate the concept of orthogonality in a matrix experiment. Level 1 and level 2 occur the same number of times in each column in the array. Furthermore, for the four rows with level 1 in column A, two rows have level 1 in column B and two rows have level 2 in column B. The same can be said for the four rows with level 2 in column A. In fact, the same balance of factor levels can be found for every pair of columns in the array. This balance is illustrated for the first two columns in Table 7.4. When we compare $A_1$ to $A_2$ in a response table, the effect of B in $A_1$ is the same as the effect of B in $A_2$.[1] Orthogonality is a pairwise property of the columns in the array. This means that columns can be left empty without destroying the orthogonality of the array. However, *all* the treatment combinations must be run to maintain the balance condition and preserve orthogonality!

The analysis of means is done on the eight response values according to the pattern of ones and twos in each column. The columns are set up to give seven independent factor effects from the eight responses. There are no other patterns available that would be balanced. This combination of balance and independence is what makes this array *orthogonal.*

## 7.2 The Analysis of Means (ANOM)

Let us now look at the details of the analysis of means using a simple example that the reader can actually replicate at home or work. The analysis of three parameters that can affect the cooling of one-half cup of water is done using the software package, WinRobust Lite, provided with this book.[2] The experiment described was actually run by engineering technology students in the College of Applied Science and Technology at the Rochester Institute of Technology. The materials used in this experiment are easily obtained at a local grocery store. The response for this problem is the temperature of the water, *y*, after specific amounts of time. This is a quality characteristic because cooler water is more palatable

1. Factor B can only affect the analysis of factor A through an interaction or due to experimental error.
2. Note that the analysis here is done using an L8 option, which is available in the fully featured version of WinRobust, but omitted from the book package. However, you can still do this experiment and the ANOM analysis by using the L12 option. Simply assign the four factors to any of the four columns in the L12 and run the indicated treatment combinations. This will be an excellent practice exercise. You should come to very similar conclusions to those found in this section.

to a customer in a setting such as a restaurant or cafeteria; therefore, the water's temperature is a direct measure of the product's quality.

The basic system consists of one-half cup of water, with ice added. The temperature is measured after the ice is added. The following parameters are studied:

1. **$H_2O$:** The volume of water being cooled. This represents variability in filling the cup. The levels chosen for the study are 2 and 4 extra tablespoons added to the nominal one-half cup of water.
2. **Thermo:** The physical location of the thermometer in the cup. The levels chosen are bottom and top surface measurement locations. This represents the two modes of customer use—sipping from the top of the cup or using a straw to drink from the bottom.
3. **Time:** The time elapsed before measuring the temperature. This represents how quickly the customer is served and then waits before taking the first sip. The levels chosen are 10- and 30-second delay times before the temperature measurement is recorded.

**Figure 7.3** L8, with factor levels shown, in WinRobust

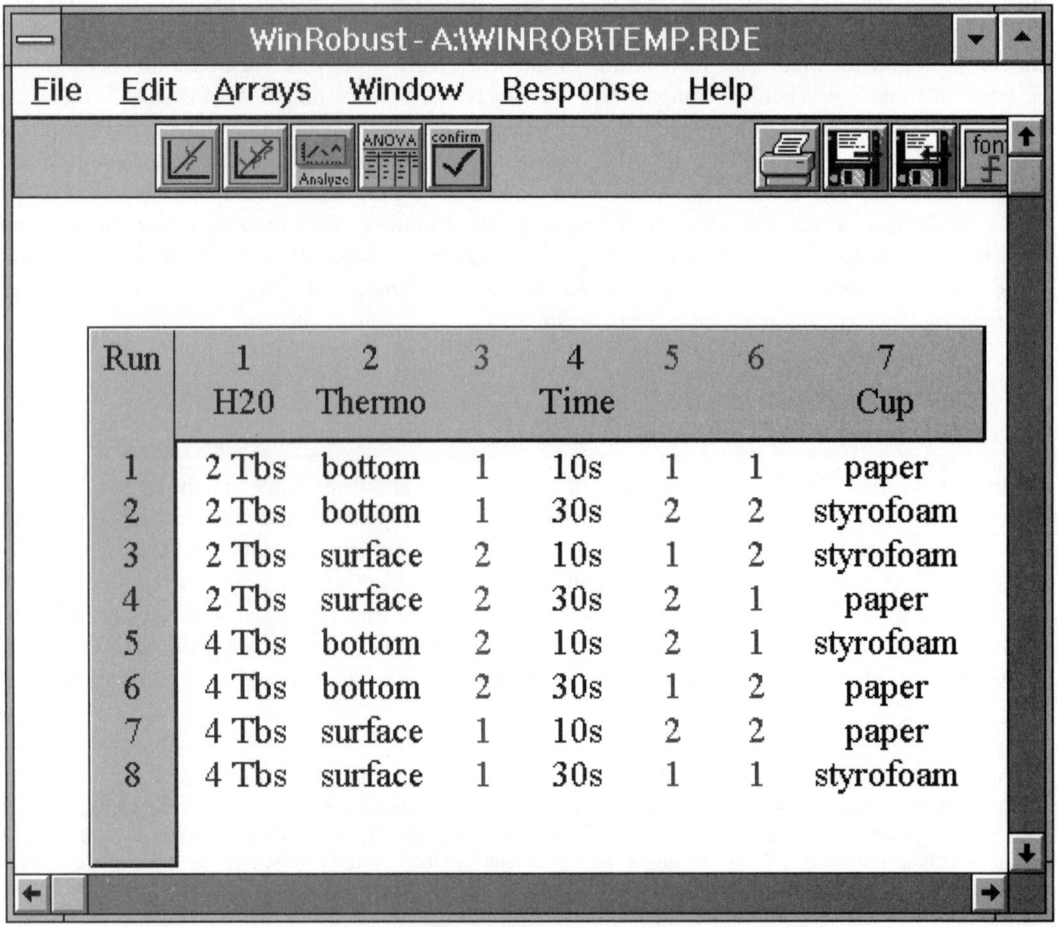

WinRobust - A:\WINROB\TEMP.RDE

File Edit Arrays Window Response Help

| Run | 1<br>H20 | 2<br>Thermo | 3 | 4<br>Time | 5 | 6 | 7<br>Cup |
|---|---|---|---|---|---|---|---|
| 1 | 2 Tbs | bottom | 1 | 10s | 1 | 1 | paper |
| 2 | 2 Tbs | bottom | 1 | 30s | 2 | 2 | styrofoam |
| 3 | 2 Tbs | surface | 2 | 10s | 1 | 2 | styrofoam |
| 4 | 2 Tbs | surface | 2 | 30s | 2 | 1 | paper |
| 5 | 4 Tbs | bottom | 2 | 10s | 2 | 1 | styrofoam |
| 6 | 4 Tbs | bottom | 2 | 30s | 1 | 2 | paper |
| 7 | 4 Tbs | surface | 1 | 10s | 2 | 2 | paper |
| 8 | 4 Tbs | surface | 1 | 30s | 1 | 1 | styrofoam |

4. **Cup:** The type of cup used to hold the water. This represents different containers commonly in use. The levels chosen are paper and Styrofoam.

The WinRobust window in Figure 7.3 shows the parameters entered into the eight-run or L8 orthogonal array. Each parameter or factor is assigned to a column. There are only four factors, so only four columns are shown with parameter labels. The remaining three columns are ignored in this analysis.

The WinRobust default is to calculate one of the standard S/N ratios discussed in the previous chapters. It is possible to override the S/N calculation in the *Response* menu and analyze a set of response values that are entered by the user by choosing *Select Static S/N Ratio* and then picking *Custom (None)* for the response. With this choice, no transformation is done on the data; they remain in the units in which they were entered.

The experiment was done in eight runs. Each run corresponds to the treatment combinations given in a row of the L8 array. Thus, for each run, different combinations of the parameters $H_2O$, Thermometer, Time, and Cup are chosen. The resulting data, after entry into WinRobust, are shown in Figure 7.4.

The analysis of means (ANOM) is done by taking averages of the temperatures that correspond with the factor levels. For example, the first four results, $y_1$, $y_2$, $y_3$, and $y_4$, correspond to the

**Figure 7.4** Data input to WinRobust

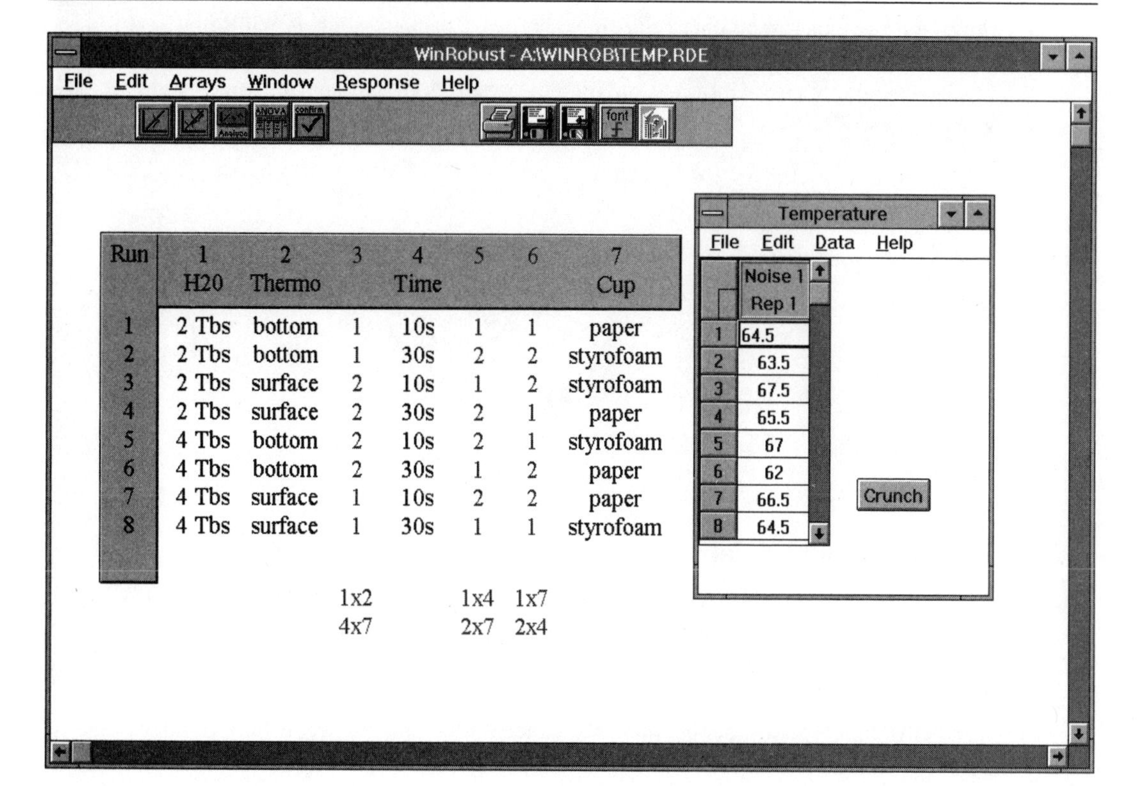

low level for $H_2O$, 2 Tbs. The last four results, $y_5$, $y_6$, $y_7$, and $y_8$, correspond to the high level for $H_2O$, 4 Tbs. Thus,

$$\bar{y}_{H2O(2Tbs)} = (y_1 + y_2 + y_3 + y_4)/4 = (64.5 + 63.5 + 67.5 + 65.5)/4 = 65.25$$
$$\bar{y}_{H2O(4Tbs)} = (y_5 + y_6 + y_7 + y_8)/4 = (67 + 62 + 66.5 + 64.5)/4 = 65$$

Similarly, for the other factors,

$$\bar{y}_{Thermo(bot)} = (y_1 + y_2 + y_5 + y_7)/4 = (64.5 + 63.5 + 67 + 62)/4 = 64.25$$
$$\bar{y}_{Thermo(surf)} = (y_3 + y_4 + y_7 + y_8)/4 = (67.5 + 65.5 + 66.5 + 64.5)/4 = 66.0$$

$$\bar{y}_{Time(10s)} = (y_1 + y_3 + y_5 + y_7)/4 = (64.5 + 67.5 + 67 + 66.5)/4 = 66.375$$
$$\bar{y}_{Time(30s)} = (y_2 + y_4 + y_6 + y_8)/4 = (63.5 + 65.5 + 62 + 64.5)/4 = 63.875$$

$$\bar{y}_{Cup(Styro)} = (y_1 + y_4 + y_6 + y_7)/4 = (64.5 + 65.5 + 62 + 66.5)/4 = 64.625$$
$$\bar{y}_{Cup(Paper)} = (y_2 + y_3 + y_5 + y_8)/4 = (63.5 + 67.5 + 67 + 64.5)/4 = 65.625$$

These results are summarized in Table 7.5, generated by WinRobust. It is also helpful to look at a plot of these results, shown in Figure 7.5. The results can be interpreted qualitatively based on the graphical representation shown in Figure 7.5. Thermometer location and time are the most significant factors; cup and water level are less important. In Chapter 9, these results are used to compound these factors for use as a noise array in an optimization experiment.

The ANOM results can be obtained using a software package such as WinRobust or any spreadsheet, or they can be calculated by hand. A good check on the results is that the average for any pair of low and high factor levels should give the same result. This is because the same eight values are used in each calculation. For example,

$$[\bar{y}_{H2O(2Tbs)} + \bar{y}_{H2O(4Tbs)}]/2 = [(y_1 + y_2 + y_3 + y_4) + (y_5 + y_6 + y_7 + y_8)]/8$$

Similarly, for the other factors,

$$[\bar{y}_{Thermo(bot)} + \bar{y}_{Thermo(surf)}]/2 = [(y_1 + y_2 + y_5 + y_7) + (y_3 + y_4 + y_7 + y_8)]/8$$
$$[\bar{y}_{Time(10s)} + \bar{y}_{Time(30s)}]/2 = [(y_1 + y_3 + y_5 + y_7) + (y_2 + y_4 + y_6 + y_8)]/8$$
$$[\bar{y}_{H2O(2Tbs)} + \bar{y}_{H2O(4Tbs)}]/2 = [(y_1 + y_4 + y_6 + y_7) + (y_2 + y_3 + y_5 + y_8)]/8$$

## 7.2.1 The Predictive Model

A predictive model based on the ANOM results can now be formulated. The predictive model is formed by considering each factor effect's optimum level contribution to the deviation from the overall mean value for the experiment. Interactions can be included in the predictive model, as shown in Chapter 16; here they are ignored.

The predictive model states that the quality characteristic, temperature, is the overall average, $\bar{y}$,

**Table 7.5** Factor effects table output by WinRobust

| *Factor* | *Level* | *Temperature (S/N)* |
|---|---|---|
| $H_2O$ | 2 Tbs | 65.25 |
| | 4 Tbs | 65.00 |
| Thermo | Bottom | 64.25 |
| | Surface | 66.00 |
| Time | 10s | 66.38 |
| | 30s | 63.88 |
| Cup | Paper | 64.62 |
| | Styrofoam | 65.62 |

plus the deviations from the average due to each factor level effect. Thus, the predicted value for the quality characteristic $y$ is given by Equation 7.3:

$$y(A,B,C,D) = \bar{y} + (\bar{y}_A - \bar{y}) + (\bar{y}_B - \bar{y}) + (\bar{y}_C - \bar{y}) + (\bar{y}_D - \bar{y}) \quad \textbf{(7.3)}$$

The values of factor effects $\bar{y}_A$, $\bar{y}_B$, $\bar{y}_C$, and $\bar{y}_D$ depend upon the levels of A, B, C, and D, respectively. For the ice water data, the overall mean is given by

$$\begin{aligned}\bar{y} &= (y_1 + y_2 + y_3 + y_4 + y_5 + y_6 + y_7 + y_8)/8 \\ &= (64.5 + 63.5 + 67.5 + 65.5 + 67 + 62 + 66.5 + 64.5)/8 = 65.0625\end{aligned}$$

The lowest temperature can be obtained by the settings $H_2O$ = 4 Tbs, Thermometer = bottom, Time = 30s, and Cup = paper. This treatment combination corresponds to run #6 in the experiment, which did have the lowest temperature. However, $H_2O$ and Cup do not seem to be significant factors

**Figure 7.5** Factor effects plot output by WinRobust

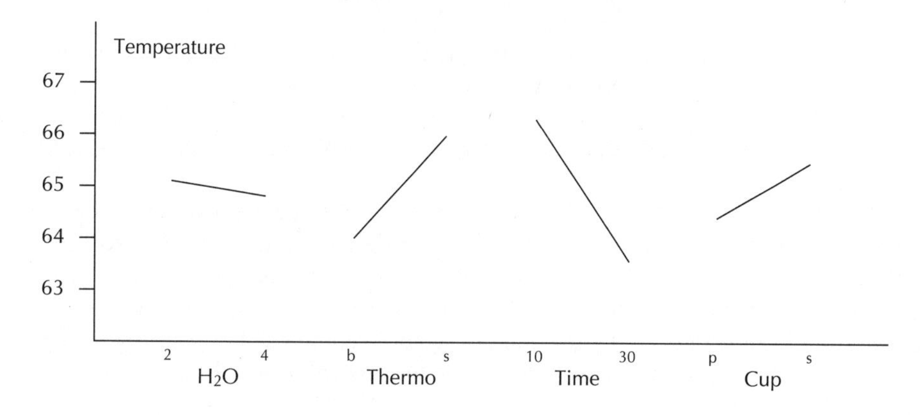

Figure 7.6 Additive model in WinRobust

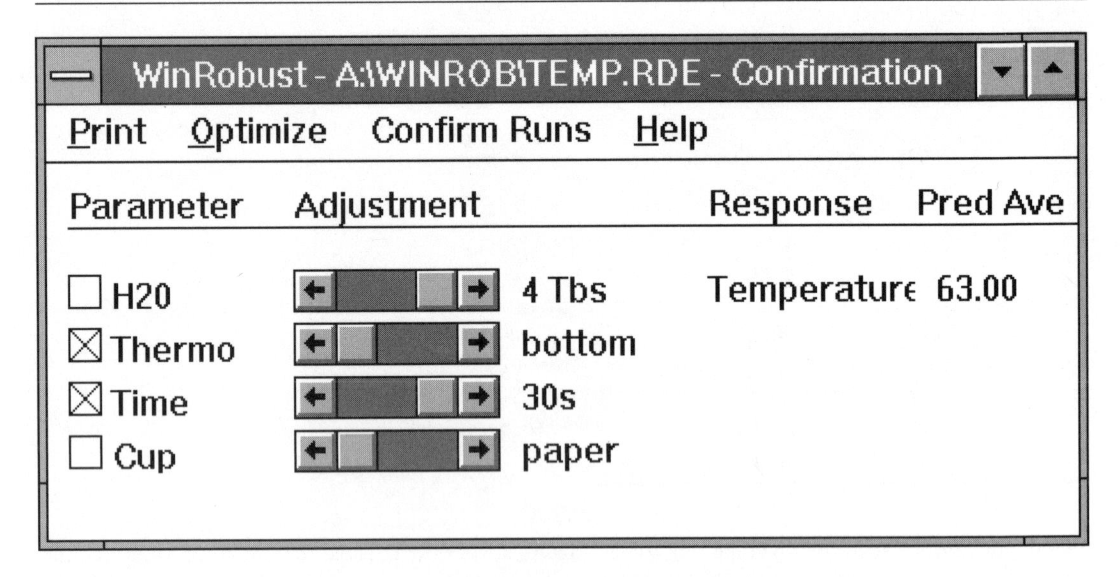


and can be omitted from the model. Thus, ignoring interactions, the model predicts that the temperature using Thermometer = bottom, Time = 30s is

$$\begin{aligned}\text{Temperature} &= \bar{y} + (\bar{y}_{\text{Thermo(bot)}} - \bar{y}) + (\bar{y}_{\text{Time(30s)}} - \bar{y}) \\ &= 65.06 - (64.25 - 65.06) + (63.88 - 65.06) \\ &= 65.06 - 0.81 - 1.18 = 63.07\end{aligned}$$

This calculation can be done using WinRobust by clicking on *Confirmation and Prediction* in the *Window* menu. The parameters can be set for the predictive equation, and by clicking on the boxes before the parameters, the contribution from the insignificant factors can be eliminated, resulting in the response shown in Figure 7.6.

## 7.2.2 Verification

It is important in doing designed experiments to verify the predictive model by testing the modeled treatment combination and comparing the results to the prediction. Such a verification can also be used to check on the reproducibility of results and give an estimate of the variation due to experimental error (Chapter 17). Here, five runs were made at the optimum conditions, always starting with a fresh cup, water, and ice to make them true replicates.

The results obtained are 63.0, 62.5, 62.0, 62.5, and 62.0. The average of these results is given by $\bar{y} = 62.4$, and the standard deviation is given by $S_y = 0.4$. This shows that the average value from the verification test differs from the prediction by an amount of 0.6, only 1.5 standard deviations. The optimum condition can also be compared to the previous value for the same treatment combination, run #6 = 62.0. In this case the difference is only 0.4, indicating good reproducibility. This is a strong confirmation which gives confidence in both the experiment and the predictive model.

## 7.3 Degrees of Freedom

Degrees of freedom is a concept that is useful to describe how big an experiment must be and how much information can be extracted from the experiment. The number of degrees of freedom of a matrix experiment is 1 less than the number of runs in the experiment:

$$DOF_{exp} = \#\text{ runs} - 1 \tag{7.4}$$

The degrees of freedom needed to describe a factor effect is 1 less than the number of levels tested for that factor:

$$DOF_f = \#\text{ levels} - 1 \tag{7.5}$$

The problem of solving a set of simultaneous equations for a set of unknowns is a good mathematical analogy for this process. The number of equations is analogous to the degrees of freedom of a matrix experiment. The number of unknowns is analogous to the total degrees of freedom for the factor effects:

$$\text{Total } DOF_f = (\#\text{ factors})(DOF_f) \tag{7.6}$$

How does the degrees of freedom analysis look for the cases we have seen so far?

*The one-factor-at-a-time (Table 7.1):* In the array shown in Table 7.1, the $DOF_{exp} = 8 - 1 = 7$. The degrees of freedom required to calculate the main effects of each factor are Total $DOF_f = 7(2 - 1) = 7$. However, the array is not orthogonal, because the balance property is violated. This makes this experiment highly susceptible to problems with interactions.

*The full factorial (Table 7.2):* In the array shown in Table 7.2, the $DOF_{exp} = 128 - 1 = 127$. If only the main effects of the factors A through G are being studied, then the Total $DOF_f = 7$; there is a waste of 120 degrees of freedom!! This is why the full factorial is so inefficient. It generates much more information than is actually used. Even if all the two-way interactions (discussed in Chapter 16) are to be modeled, the total $DOF_f$ needed is still only 28!

*The fractional factorial orthogonal array (Table 7.3):* In the array shown in Table 7.3, the $DOF_{exp} = 8 - 1 = 7$. The degrees of freedom required to calculate the main effects of each factor are Total $DOF_f = 7(2 - 1) = 7$. In addition, the experiment is balanced!

## 7.4 Full Factorial Arrays

Any experiment in which all of the possible combinations of factor levels are tested is, by definition, a full factorial. One common example is the $2^2$ full factorial shown in Table 7.6.

The orthogonal array shown in Table 7.6 provides three DOF but only requires two DOF for the factor effects. Let us define a metric, the experimental efficiency $X$, where

$$X = \frac{\text{Total } DOF_f}{DOF_{exp}}(100\%) \tag{7.7}$$

For the $2^2$ full factorial experiment, the efficiency $X = 67\%$.

Another common example of a full factorial is the $3^2$ full factorial shown in Table 7.7. It provides eight DOF but only requires four DOF for the factor effects; $X = 50\%$ efficiency.

Table 7.8 shows a $2^3$ full factorial. It provides seven DOF but only requires three DOF for the factor effects; $X = 43\%$ efficiency.

**Table 7.6** A two-factor, two-level full factorial

| *Run* | *A* | *B* |
|---|---|---|
| 1 | 1 | 1 |
| 2 | 1 | 2 |
| 3 | 2 | 1 |
| 4 | 2 | 2 |

**Table 7.7** A two-factor, three-level full factorial

| *Run* | *A* | *B* |
|---|---|---|
| 1 | 1 | 1 |
| 2 | 1 | 2 |
| 3 | 1 | 3 |
| 4 | 2 | 1 |
| 5 | 2 | 2 |
| 6 | 2 | 3 |
| 7 | 3 | 1 |
| 8 | 3 | 2 |
| 9 | 3 | 3 |

**Table 7.8** A three-factor, two-level full factorial

| *Run* | *A* | *B* | *C* |
|---|---|---|---|
| 1 | 1 | 1 | 1 |
| 2 | 1 | 1 | 2 |
| 3 | 1 | 2 | 1 |
| 4 | 1 | 2 | 2 |
| 5 | 2 | 1 | 1 |
| 6 | 2 | 1 | 2 |
| 7 | 2 | 2 | 1 |
| 8 | 2 | 2 | 2 |

**Table 7.9** A one-factor, two-level by two-factor, three-level full factorial

| *Run* | *A* | *B* | *C* |
|---|---|---|---|
| 1 | 1 | 1 | 1 |
| 2 | 1 | 1 | 2 |
| 3 | 1 | 1 | 3 |
| 4 | 1 | 2 | 1 |
| 5 | 1 | 2 | 2 |
| 6 | 1 | 2 | 3 |
| 7 | 1 | 3 | 1 |
| 8 | 1 | 3 | 2 |
| 9 | 1 | 3 | 3 |
| 10 | 2 | 1 | 1 |
| 11 | 2 | 1 | 2 |
| 12 | 2 | 1 | 3 |
| 13 | 2 | 2 | 1 |
| 14 | 2 | 2 | 2 |
| 15 | 2 | 2 | 3 |
| 16 | 2 | 3 | 1 |
| 17 | 2 | 3 | 2 |
| 18 | 2 | 3 | 3 |

**Table 7.10** A two-factor, four-level full factorial

| *Run* | *A* | *B* |
|---|---|---|
| 1 | 1 | 1 |
| 2 | 1 | 2 |
| 3 | 1 | 3 |
| 4 | 1 | 4 |
| 5 | 2 | 1 |
| 6 | 2 | 2 |
| 7 | 2 | 3 |
| 8 | 2 | 4 |
| 9 | 3 | 1 |
| 10 | 3 | 2 |
| 11 | 3 | 3 |
| 12 | 3 | 4 |
| 13 | 4 | 1 |
| 14 | 4 | 2 |
| 15 | 4 | 3 |
| 16 | 4 | 4 |

Table 7.9 shows a $2 \times 3^2$ full factorial. It provides 17 DOF but only requires five DOF for the factor effects; $X = 29\%$ efficiency.

Table 7.10 shows a $2^4$ full factorial. It provides 15 degrees of freedom but only requires six DOF for the factor effects; $X = 40\%$ efficiency.

The message here is that full factorials make sense only for a small number of factors. However, the extra DOF can be used to your advantage to study interactions.

## 7.5 Fractional Factorial Orthogonal Arrays

The convention for naming the fractional factorial orthogonal arrays is

$$La\ (b^c)$$

where

$a$ = the number of experimental runs

$b$ = the number of levels for each factor

$c$ = the number of columns in each array

### 7.5.1 Two-Level Orthogonal Arrays

The L4 is shown in Table 7.11. The L4 can handle three factors at two levels. It is the smallest of the two-level orthogonal arrays. The L4 $DOF_{exp} = 3$. The first two columns of the L4 are the $2^2$ full factorial.

The L8 is shown in Table 7.12. The L8 can handle seven factors at two levels. The L8 $DOF_{exp} = 7$. The first, second, and fourth columns of the L8 are the $2^3$ full factorial.

The L12 is shown in Table 7.13. The L12 can handle 11 factors at two levels. The L12 $DOF_{exp} = 11$. The L12 as well as the L18 (shown in Table 7.16) have some special properties with respect to interactions that make them uniquely useful in Robust Design. For that reason, these two arrays are featured in the WinRobust Lite disk included with this book (see Appendices B and C).

**Table 7.11** L4 ($2^3$) orthogonal array

| *Run* | *1* | *2* | *3* |
|---|---|---|---|
| 1 | 1 | 1 | 1 |
| 2 | 1 | 2 | 2 |
| 3 | 2 | 1 | 2 |
| 4 | 2 | 2 | 1 |

**Table 7.12** L8 ($2^7$) orthogonal array

| *Run* | *1* | *2* | *3* | *4* | *5* | *6* | *7* |
|---|---|---|---|---|---|---|---|
| 1 | 1 | 1 | 1 | 1 | 1 | 1 | 1 |
| 2 | 1 | 1 | 1 | 2 | 2 | 2 | 2 |
| 3 | 1 | 2 | 2 | 1 | 1 | 2 | 2 |
| 4 | 1 | 2 | 2 | 2 | 2 | 1 | 1 |
| 5 | 2 | 1 | 2 | 1 | 2 | 1 | 2 |
| 6 | 2 | 1 | 2 | 2 | 1 | 2 | 1 |
| 7 | 2 | 2 | 1 | 1 | 2 | 2 | 1 |
| 8 | 2 | 2 | 1 | 2 | 1 | 1 | 2 |

**Table 7.13** L12 ($2^{11}$) orthogonal array

| *Run* | *1* | *2* | *3* | *4* | *5* | *6* | *7* | *8* | *9* | *10* | *11* |
|---|---|---|---|---|---|---|---|---|---|---|---|
| 1 | 1 | 1 | 1 | 1 | 1 | 1 | 1 | 1 | 1 | 1 | 1 |
| 2 | 1 | 1 | 1 | 1 | 1 | 2 | 2 | 2 | 2 | 2 | 2 |
| 3 | 1 | 1 | 2 | 2 | 2 | 1 | 1 | 1 | 2 | 2 | 2 |
| 4 | 1 | 2 | 1 | 2 | 2 | 1 | 2 | 2 | 1 | 1 | 2 |
| 5 | 1 | 2 | 2 | 1 | 2 | 2 | 1 | 2 | 1 | 2 | 1 |
| 6 | 1 | 2 | 2 | 2 | 1 | 2 | 2 | 1 | 2 | 1 | 1 |
| 7 | 2 | 1 | 2 | 2 | 1 | 1 | 2 | 2 | 1 | 2 | 1 |
| 8 | 2 | 1 | 2 | 1 | 2 | 2 | 2 | 1 | 1 | 1 | 2 |
| 9 | 2 | 1 | 1 | 2 | 2 | 2 | 1 | 2 | 2 | 1 | 1 |
| 10 | 2 | 2 | 2 | 1 | 1 | 1 | 1 | 2 | 2 | 1 | 2 |
| 11 | 2 | 2 | 1 | 2 | 1 | 2 | 1 | 1 | 1 | 2 | 2 |
| 12 | 2 | 2 | 1 | 1 | 2 | 1 | 2 | 1 | 2 | 2 | 1 |

The L12 is a unique array with respect to the effect of 2-factor interactions on its analysis. Interactions are defined in this chapter, in the box in Section 7.1.1, using a simple example. In general, an interaction between factors means that the effect of one factor depends upon the level of another factor. Thus, in Figure 7.2, factor A has a negative slope when factor B is at level 1 and factor A has a positive slope when factor B is at level 2. Such an interaction will also affect the analysis of means. In most of the fractional factorial orthogonal arrays, the mathematical effect of an interaction between two factors is blended with one other factor effect. This is referred to as confounding, because the ANOM result is a numerical sum of the factor effect blended with the interaction effect. These effects cannot be separately evaluated in an array where all of the columns are used. (Techniques for separately evaluating the interactions and main effects are discussed in Chapter 16.)

Consider how the subsytems and components in a physical design interact with each other. Rarely does one interaction put all of its energy into a single place. In fact, complex designs and processes can have a myriad of interactions, some large and some small, all working at once. When they work together harmoniously, the effect can be very pleasing, but when the interactive effects are harmful, they can interfere with the main functionality or effect of one or several other design elements.

The interactive effects within the L12 array are distributed evenly so that they all partially confound with one another, as is true in a physical design. We refer to the L12, the L18, and the L36 arrays (the L18 and L36 have similar interaction properties) as the *engineering arrays.* They are not suitable for modeling main effects *and* interactions. On the other hand, they give the most realistic simulation of the actual interplay of parameters that occur in engineering. The distributed effects of the interactions act in a manner similar to noise; after all, undesired interactions in a design are a nuisance. The main effects that emerge from a properly prepared study using the engineering arrays are strong enough to rely on despite noise and uncontrolled interactions. For that reason, the L12 is an excellent experimental design for optimizing physical designs that are engineered for additivity (a topic devel-

**Table 7.14** L16 ($2^{15}$) orthogonal array

| *Run* | *1* | *2* | *3* | *4* | *5* | *6* | *7* | *8* | *9* | *10* | *11* | *12* | *13* | *14* | *15* |
|---|---|---|---|---|---|---|---|---|---|---|---|---|---|---|---|
| 1 | 1 | 1 | 1 | 1 | 1 | 1 | 1 | 1 | 1 | 1 | 1 | 1 | 1 | 1 | 1 |
| 2 | 1 | 1 | 1 | 1 | 1 | 1 | 1 | 2 | 2 | 2 | 2 | 2 | 2 | 2 | 2 |
| 3 | 1 | 1 | 1 | 2 | 2 | 2 | 2 | 1 | 1 | 1 | 1 | 2 | 2 | 2 | 2 |
| 4 | 1 | 1 | 1 | 2 | 2 | 2 | 2 | 2 | 2 | 2 | 2 | 1 | 1 | 1 | 1 |
| 5 | 1 | 2 | 2 | 1 | 1 | 2 | 2 | 1 | 1 | 2 | 2 | 1 | 1 | 2 | 2 |
| 6 | 1 | 2 | 2 | 1 | 1 | 2 | 2 | 2 | 2 | 1 | 1 | 2 | 2 | 1 | 1 |
| 7 | 1 | 2 | 2 | 2 | 2 | 1 | 1 | 1 | 1 | 2 | 2 | 2 | 2 | 1 | 1 |
| 8 | 1 | 2 | 2 | 2 | 2 | 1 | 1 | 2 | 2 | 1 | 1 | 1 | 1 | 2 | 2 |
| 9 | 2 | 1 | 2 | 1 | 2 | 1 | 2 | 1 | 2 | 1 | 2 | 1 | 2 | 1 | 2 |
| 10 | 2 | 1 | 2 | 1 | 2 | 1 | 2 | 2 | 1 | 2 | 1 | 2 | 1 | 2 | 1 |
| 11 | 2 | 1 | 2 | 2 | 1 | 2 | 1 | 1 | 2 | 1 | 2 | 2 | 1 | 2 | 1 |
| 12 | 2 | 1 | 2 | 2 | 1 | 2 | 1 | 2 | 1 | 2 | 1 | 1 | 2 | 1 | 2 |
| 13 | 2 | 2 | 1 | 1 | 2 | 2 | 1 | 1 | 2 | 2 | 1 | 1 | 2 | 2 | 1 |
| 14 | 2 | 2 | 1 | 1 | 2 | 2 | 1 | 2 | 1 | 1 | 2 | 2 | 1 | 1 | 2 |
| 15 | 2 | 2 | 1 | 2 | 1 | 1 | 2 | 1 | 2 | 2 | 1 | 2 | 1 | 1 | 2 |
| 16 | 2 | 2 | 1 | 2 | 1 | 1 | 2 | 2 | 1 | 1 | 2 | 1 | 2 | 2 | 1 |

oped in Chapters 8 and 10). It is particularly suitable for doing experimental evaluations involving noise factors where two-level factors are studied (Chapter 9).

The L16 is shown in Table 7.14. The L16 can handle 15 factors at two levels. The L16 $DOF_{exp}$ = 15. Columns 1, 2, 4, and 8 of the L16 make up the $2^4$ full factorial.

### 7.5.2 Three-Level Orthogonal Arrays

Three-level series orthogonal arrays allow us to investigate three factor levels. The L9 is shown in Table 7.15. The L9 can handle four factors at three levels. The L9 $DOF_{exp} = 8$. The eight DOF can be broken down into two DOF per column. This is because the $DOF_f = (3 - 1) = 2$. Columns 1 and 2 of the L9 make up the $3^2$ full factorial.

The L18, the other engineering array included in WinRobust Lite, is shown in Table 7.16. It is similar to the L12 in its properties with respect to interactions. The first two columns (1 and 2) in the L18 are free to evaluate an interaction between the factors placed within them. Interactions between all the other columns are evenly distributed with one another. The L18 can handle one factor at two levels and seven factors at three levels. The ability to handle three-level factors makes the L18 ideal for studying control factor effects where optimization is paramount. The L18 has $DOF_{exp} = 17$. The Total $DOF_f = 1(2 - 1) + 7(3 - 2) = 15$. Columns 1, 2, and 3 of the L18 make up the $2 \times 3^2$ full factorial. Because two fewer degrees of freedom are required for factor effects than are available in the experiment, it is possible to modify the L18 to handle one six-level factor. This is done by combining the first

Table 7.15 L9 ($3^4$) orthogonal array

| Run | 1 | 2 | 3 | 4 |
|---|---|---|---|---|
| 1 | 1 | 1 | 1 | 1 |
| 2 | 1 | 2 | 2 | 2 |
| 3 | 1 | 3 | 3 | 3 |
| 4 | 2 | 1 | 2 | 3 |
| 5 | 2 | 2 | 3 | 1 |
| 6 | 2 | 3 | 1 | 2 |
| 7 | 3 | 1 | 3 | 2 |
| 8 | 3 | 2 | 1 | 3 |
| 9 | 3 | 3 | 2 | 1 |

two columns in a manner similar to that described in Section 15.3 (Upgrading a Column). The result is shown here in Table 7.17. For more information on the L18, refer to Chapter 16.

The L27 is shown in Table 7.18. The L27 $DOF_{exp} = 26$. The L27 can handle 13 factors at three levels. The Total $DOF_f = 13(3 - 1) = 26$. Columns 1, 2, and 5 of the L27 make up the $3^3$ full factorial. Larger three-level arrays and some four-level arrays are listed in the back of this book.

Table 7.16 L18 ($2^1 \times 3^7$) orthogonal array

| Run | 1 | 2 | 3 | 4 | 5 | 6 | 7 | 8 |
|---|---|---|---|---|---|---|---|---|
| 1 | 1 | 1 | 1 | 1 | 1 | 1 | 1 | 1 |
| 2 | 1 | 1 | 2 | 2 | 2 | 2 | 2 | 2 |
| 3 | 1 | 1 | 3 | 3 | 3 | 3 | 3 | 3 |
| 4 | 1 | 2 | 1 | 1 | 2 | 2 | 3 | 3 |
| 5 | 1 | 2 | 2 | 2 | 3 | 3 | 1 | 1 |
| 6 | 1 | 2 | 3 | 3 | 1 | 1 | 2 | 2 |
| 7 | 1 | 3 | 1 | 2 | 1 | 3 | 2 | 3 |
| 8 | 1 | 3 | 2 | 3 | 2 | 1 | 3 | 1 |
| 9 | 1 | 3 | 3 | 1 | 3 | 2 | 1 | 2 |
| 10 | 2 | 1 | 1 | 3 | 3 | 2 | 2 | 1 |
| 11 | 2 | 1 | 2 | 1 | 1 | 3 | 3 | 2 |
| 12 | 2 | 1 | 3 | 2 | 2 | 1 | 1 | 3 |
| 13 | 2 | 2 | 1 | 2 | 3 | 1 | 3 | 2 |
| 14 | 2 | 2 | 2 | 3 | 1 | 2 | 1 | 3 |
| 15 | 2 | 2 | 3 | 1 | 2 | 3 | 2 | 1 |
| 16 | 2 | 3 | 1 | 3 | 2 | 3 | 1 | 2 |
| 17 | 2 | 3 | 2 | 1 | 3 | 1 | 2 | 3 |
| 18 | 2 | 3 | 3 | 2 | 1 | 2 | 3 | 1 |

**Table 7.17** Modified L18 ($6^1 \times 3^6$) orthogonal array

| *Run* | *1–2* | *3* | *4* | *5* | *6* | *7* | *8* |
|---|---|---|---|---|---|---|---|
| 1 | 1 | 1 | 1 | 1 | 1 | 1 | 1 |
| 2 | 1 | 2 | 2 | 2 | 2 | 2 | 2 |
| 3 | 1 | 3 | 3 | 3 | 3 | 3 | 3 |
| 4 | 2 | 1 | 1 | 2 | 2 | 3 | 3 |
| 5 | 2 | 2 | 2 | 3 | 3 | 1 | 1 |
| 6 | 2 | 3 | 3 | 1 | 1 | 2 | 2 |
| 7 | 3 | 1 | 2 | 1 | 3 | 2 | 3 |
| 8 | 3 | 2 | 3 | 2 | 1 | 3 | 1 |
| 9 | 3 | 3 | 1 | 3 | 2 | 1 | 2 |
| 10 | 4 | 1 | 3 | 3 | 2 | 2 | 1 |
| 11 | 4 | 2 | 1 | 1 | 3 | 3 | 2 |
| 12 | 4 | 3 | 2 | 2 | 1 | 1 | 3 |
| 13 | 5 | 1 | 2 | 3 | 1 | 3 | 2 |
| 14 | 5 | 2 | 3 | 1 | 2 | 1 | 3 |
| 15 | 5 | 3 | 1 | 2 | 3 | 2 | 1 |
| 16 | 6 | 1 | 3 | 2 | 3 | 1 | 2 |
| 17 | 6 | 2 | 1 | 3 | 1 | 2 | 3 |
| 18 | 6 | 3 | 2 | 1 | 2 | 3 | 1 |

## 7.6 Summary of Chapter 7

Several experimental approaches for design optimization are considered in this chapter. One-factor-at-a-time and full factorials are not very effective approaches, although they are much better than build–test–fix methods. One-factor-at-a-time is very slow and is most appropriate in scientific research. It can work in engineering optimization, but only if interactions are not present or are very mild. Here, too, it is not possible to demonstrate that a true on-target optimum has been achieved. One-factor-at-a-time methods can require rework in the form of additional experiments, and they frequently cause schedule delays. Full factorials are in principle effective, but for more than a few parameters they are notoriously inefficient. Because of their inefficiencies, full factorial experiments are difficult to execute properly and, as a practical matter, are limited to no more than four factors at two or three levels.

Orthogonal arrays are used to run fractional factorial experiments. This chapter illustrates the application of an orthogonal array to a simple four-parameter study. The analysis technique chosen is the analysis of means (ANOM). It employs a simple averaging of the responses to find the various factor effects. The orthogonality property of the array (balance) allows the factor effects to be found independent of each other. The discussion of interactions is deferred until Chapter 16.

Degrees of freedom is a useful concept for understanding orthogonal arrays. They are defined here and are used to find the efficiency of various full-factorial and orthogonal-array experiments. Degrees of freedom is also used in Chapter 16 for understanding interactions.

**Table 7.18** L27 ( $3^{13}$) orthogonal array

| *Run* | *1* | *2* | *3* | *4* | *5* | *6* | *7* | *8* | *9* | *10* | *11* | *12* | *13* |
|---|---|---|---|---|---|---|---|---|---|---|---|---|---|
| 1 | 1 | 1 | 1 | 1 | 1 | 1 | 1 | 1 | 1 | 1 | 1 | 1 | 1 |
| 2 | 1 | 1 | 1 | 1 | 2 | 2 | 2 | 2 | 2 | 2 | 2 | 2 | 2 |
| 3 | 1 | 1 | 1 | 1 | 3 | 3 | 3 | 3 | 3 | 3 | 3 | 3 | 3 |
| 4 | 1 | 2 | 2 | 2 | 1 | 1 | 1 | 2 | 2 | 2 | 3 | 3 | 3 |
| 5 | 1 | 2 | 2 | 2 | 2 | 2 | 2 | 3 | 3 | 3 | 1 | 1 | 1 |
| 6 | 1 | 2 | 2 | 2 | 3 | 3 | 3 | 1 | 1 | 1 | 2 | 2 | 2 |
| 7 | 1 | 3 | 3 | 3 | 1 | 1 | 1 | 3 | 3 | 3 | 2 | 2 | 2 |
| 8 | 1 | 3 | 3 | 3 | 2 | 2 | 2 | 1 | 1 | 1 | 3 | 3 | 3 |
| 9 | 1 | 3 | 3 | 3 | 3 | 3 | 3 | 2 | 2 | 2 | 1 | 1 | 1 |
| 10 | 2 | 1 | 2 | 3 | 1 | 2 | 3 | 1 | 2 | 3 | 1 | 2 | 3 |
| 11 | 2 | 1 | 2 | 3 | 2 | 3 | 1 | 2 | 3 | 1 | 2 | 3 | 1 |
| 12 | 2 | 1 | 2 | 3 | 3 | 1 | 2 | 3 | 1 | 2 | 3 | 1 | 2 |
| 13 | 2 | 2 | 3 | 1 | 1 | 2 | 3 | 2 | 3 | 1 | 3 | 1 | 2 |
| 14 | 2 | 2 | 3 | 1 | 2 | 3 | 1 | 3 | 1 | 2 | 1 | 2 | 3 |
| 15 | 2 | 2 | 3 | 1 | 3 | 1 | 2 | 1 | 2 | 3 | 2 | 3 | 1 |
| 16 | 2 | 3 | 1 | 2 | 1 | 2 | 3 | 3 | 1 | 2 | 2 | 3 | 1 |
| 17 | 2 | 3 | 1 | 2 | 2 | 3 | 1 | 1 | 2 | 3 | 3 | 1 | 2 |
| 18 | 2 | 3 | 1 | 2 | 3 | 1 | 2 | 2 | 3 | 1 | 1 | 2 | 3 |
| 19 | 3 | 1 | 3 | 2 | 1 | 3 | 2 | 1 | 3 | 2 | 1 | 3 | 2 |
| 20 | 3 | 1 | 3 | 2 | 2 | 1 | 3 | 2 | 1 | 3 | 2 | 1 | 3 |
| 21 | 3 | 1 | 3 | 2 | 3 | 2 | 1 | 3 | 2 | 1 | 3 | 2 | 1 |
| 22 | 3 | 2 | 1 | 3 | 1 | 3 | 2 | 2 | 1 | 3 | 3 | 2 | 1 |
| 23 | 3 | 2 | 1 | 3 | 2 | 1 | 3 | 3 | 2 | 1 | 1 | 3 | 2 |
| 24 | 3 | 2 | 1 | 3 | 3 | 2 | 1 | 1 | 3 | 2 | 2 | 1 | 3 |
| 25 | 3 | 3 | 2 | 1 | 1 | 3 | 2 | 3 | 2 | 1 | 2 | 1 | 3 |
| 26 | 3 | 3 | 2 | 1 | 2 | 1 | 3 | 1 | 3 | 2 | 3 | 2 | 1 |
| 27 | 3 | 3 | 2 | 1 | 3 | 2 | 1 | 2 | 1 | 3 | 1 | 3 | 2 |

A simple predictive model based on the factor effects can be constructed. An important part of any orthogonal array experiment is the verification experiment. The verification experiment result is compared to the prediction to verify the success of the experiment.

Lastly, a number of common orthogonal arrays and their properties are listed in this chapter. These arrays are used in various combinations to test control factor and noise factor effects in the following chapters. It is not necessary to use every column in running an orthogonal array experiment. Recommendations for which columns to use when considering interactive effects are given in Chapter 16. Simple techniques for modifying the arrays to fit the factor-level requirements are given in Chapter 15.

## Exercises for Chapter 7

1. What is an orthogonal array?
2. What is meant by the term "experimental design space"?
3. Why do we prefer three-level experiments when conducting a parameter design experiment?
4. Why do we have limitations on how many factors can be studied in full factorial experiments?
5. What unique properties does the L12 array possess?
6. What is the difference between a full factorial experiment and a fractional factorial experiment?
7. Why is a designed experiment used during the optimization process?
8. What are some of the consequences one might expect from using the build–test–fix approach to product development?
9. When would you be justified in using a full factorial experiment?
10. Why is it important that there be balance between the control factors in an experiment?
11. What is a degree of freedom?
12. Define what is meant by "degrees of freedom" for (a) an orthogonal array and (b) a control factor.
13. What is a fractional factorial experiment?
14. What does it mean when two control factors are dependent on one another?

CHAPTER 8

# Selection of the Quality Characteristics

*In parameter design, the most important job of the engineer is to select an effective characteristic to measure as data. . . . We should measure data that relate to the function itself and not the symptoms of variability. . . . Quality problems take place because of variability in the energy transformation. Considering the energy transformation helps to recognize the function of the system [N2, p. 138].*—SHIN TAGUCHI *(Dr. Genichi Taguchi's son)*

*The efficiency of research will drop if it is not possible to find characteristics that reflect the effects of the individual factors regardless of the influence of other factors [T6, p. 61].*—DR. GENICHI TAGUCHI

## 8.1 Introduction

The measured response of a design, in Robust Design, is referred to as the *quality characteristic.* The term quality is included in the definition of the measured response to highlight the importance of the principal characteristic that customers universally desire. The quality characteristic is the response that is measured at each treatment combination of the control factors and noise factors. The S/N ratio acts as a statistic that summarizes the effects of the noise factors on the quality characteristic. It is calculated from the quality characteristic values obtained at several noise factor combinations. Noise is typically introduced into each experimental run either through compounded noise factors or through a small outer array (see Chapter 11).

This chapter provides details for defining, selecting, and measuring quality characteristics. It is useful to employ your knowledge of the physics and/or chemistry of the system to anticipate and avoid interactions. Dr. Taguchi and others [L1, P2, S2] have shown that the response used to measure the system's quality can have a large effect on how much interactivity is found when the control factor effects are analyzed. Some interactions are purely a result of using a response that mixes up various energy transformations. This does not necessarily make those responses bad. Many measures of customer-observable quality are inherently prone to interactions. Customer-observable responses that are not directly connected to the energy transformations in a product or process are inefficient for parameter optimization experiments. Do not measure quality directly, as the customer defines it. Rather, seek quality characteristics that describe the critical system processes contributing to the customer's perception of quality. When the internal processes are stable and tuned to their target, the result is dependable external (observable) behavior that drives greater customer satisfaction.

In the previous chapter, analysis of means (ANOM) is used to produce a predictive equation of how each parameter affects the response in a designed experiment. This modeling approach is applied in parameter optimization to predict how the control factors affect the S/N ratios that describe the variability of the quality characteristic. By minimizing interactive effects, the simplest possible behavior is produced. With negligible interactions between control factors, the S/N effects can be modeled by simply adding the main effects from each control factor. This is referred to as an *additive model*, which takes on the following form:

$$y = f_1(A) + f_2(B) + f_3(C) + \ldots + \text{error} \tag{8.1}$$

where $y$ is the response, the quality characteristic, or its S/N ratio, and A, B, C . . . are the control factors.[1] The error term is the difference between the actual response on the left and the predicted response based on the additive model. If the error terms are small, then there is additivity in the design. This implies weak interactions, because significant interactions would be one source of deviation. There are many reasons why additivity is desirable—e.g., smaller experiments because interactions can be neglected with easily interpreted results, designs that can withstand changes in any of the design parameters without an undue effect on the quality, and designs that are easy to implement and optimize because there are few constraints imposed by interactions.

## 8.2 Engineering Analysis in the Planning Stage

The key to minimizing interactions[2] so that the additive model can be used is effective experimental planning. The goal is to understand, measure, and control the underlying fundamental physical actions that drive system functionality.

### 8.2.1 The Ideal Function of a Design

If we know what the ideal behavior of the function in a design is, two critical issues can be clarified.

1. What is the target? Robustness, as presented in the discussion of the loss function, is inherently target-based. "Good enough," or "within specification," are not concepts that are compatible with parameter design. After the optimization stage is complete, by all means go ahead and measure the performance using these customer-oriented measures. For parameter design, however, choose a quality characteristic that is based on knowledge of the desired target. The quality characteristic must be capable of measuring the system's deviation from the target.
2. What is the intended behavior of the system? Defining the ideal state helps the intended behavior of the system be better understood. In some cases it is not easy to define the ideal function. When this occurs, it is an indication of an incomplete understanding or framing of

1. Note that the factors themselves must be linearly independent.
2. The authors mention interactions frequently in this text. The reader is cautioned not to interpret this as a call to automatically study them. Rather it is a call to engineer well so as to avoid their having any dominant effect on the robustness of the product or process.

the engineering problem. It is unfortunate, but true, that some experiments are run before the objective is properly defined. This is extremely inefficient in large-scale engineering, where rework can literally cost millions of dollars and months of time.

The ideal function is understood by considering how the design would perform if it were free of any physical constraints limiting its performance. This is not a proposal for building a perpetual motion machine, but for considering how the design would function if every part was built exactly to target, if there were no heat or frictional losses, if there was no sensitivity to customer noises. Nature's inevitable attack by noise factors causes deviations that are measured with the quality characteristic. By first considering the system's performance in the absence of any noise causing variation, the deviations that occur when the real system is built are exposed. A clear understanding of the deviations from the ideal state that occur due to noise is necessary for making the system robust.

### 8.2.2 Efficiency of Energy Transformation

All physical devices and processes that accomplish work do so via transforming energy (and material). A major goal in Robust Design is to convert as much of the input energy as possible into targeted output performance. The design should be optimally *efficient* with regard to the required energy transformations. The design should be made *insensitive* to the remaining inefficiencies in the required energy transformations, as well as to external sources of energy that tend to degrade performance.

For example, consider a radio receiver. Its antenna should efficiently capture the carrier signal from the electromagnetic field. Losses as the signal is transmitted to the amplifier reduce the signal-to-noise ratio. Signals from interfering broadcasts or cross-talk with signals at different frequencies also reduce the signal-to-noise ratio. Whether the problem is loss of the desired energy transformation or interference from other energy sources, the result is a reduction in the S/N ratio.

This is not strictly an electrical engineering phenomenon. In chemical engineering, there is often a concern about the production of unwanted by-products in a chemical reaction. There is also concern about the influence of contaminants on the reaction. In mechanical engineering, there are problems such as the production of unwanted wear and heat due to frictional losses in an engine or motor. There is also consideration of the effect of contaminated lubricants or of vibration due to tolerance stack-up.

These same issues apply on a system level. It is not desirable to have one portion of a design convert energy such that it promotes the ideal function of the system, only to have another element of the design degrade or even reverse the energy transformation. Each subsystem and each control factor should independently and harmoniously contribute to the overall response. This is the *essence* of additivity. For example, it is not a good practice to step on the accelerator and brake pedals in a car at the same time. Similarly, motor controllers that limit the speed with a brake while applying full torque are highly inefficient. Besides wasting energy, this solution is susceptible to noises.

The situation is summarized by Figures 8.1 and 8.2. Figure 8.1 shows the ideal system, where the only inflow is the intended input energy, and the only outflow is useful output energy. Figure 8.2 shows

**Figure 8.1** An ideal system (no variability)

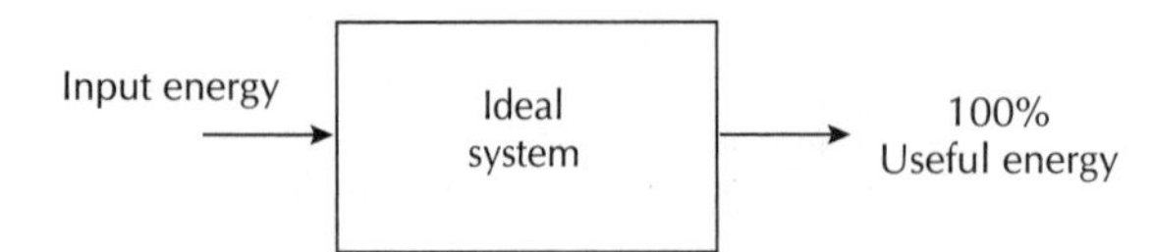

**Figure 8.2** A real system (with variability)

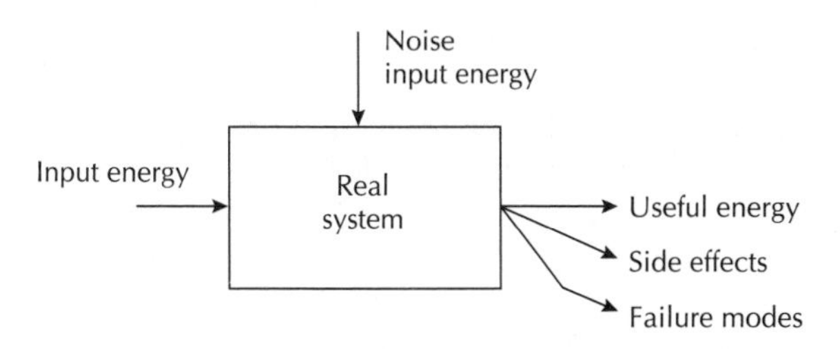


a real system, where some of the input energy goes to other, nonuseful results. In addition, other energy inputs may induce undesirable results.

### 8.2.3 Measurability of Functional Response

When selecting the quality characteristic related to an ideal function performed by a system, the engineering team must consider the transducers and meters that will be used to quantify the response. The quality characteristic should be a direct physical measure of functional phenomena. Measurements are made in *engineering units* to get at the internal functional mechanisms that actually make the design work. Measurement capability is critical to successfully developing robust designs. It is necessary to define both the main function and the means to measure it. Use of good reference books on transducers can help your team to define a reasonable quality characteristic.

The golden rule in choosing a quality characteristic is:

> *The closer you are to measuring the simplest act of physics that occurs in your design, the better.*

The most basic levels of the design are studied by defining the ideal function and using good measurement capability. The reason that high-level, customer-observable measures are so often used for quality characteristics is that they are easy to identify and they are obviously related to customer needs. However, they almost always lead to complicated, hard-to-exploit results when the data analysis is done. Running experiments that include only easy-to-measure quality characteristics can significantly limit the ability to produce breakthrough improvements. Resisting the pressure of management to measure reliability instead of physics is an issue in most industries. It is helpful to become adept at converting physical measures from engineering units into the language of management. Relating quality to loss (cost $) is an important step in this direction. It is also useful to define empirical relationships between variability and reliability. From these, S/N ratio improvements can be related to reliability improvement [P2, p. 255].

## 8.3 The Ideal Function of the Design

### 8.3.1 Relating High-Level Attributes to Ideal Functions

A useful step towards finding the ideal function and the proper quality characteristic to describe it is to distinguish the ideal function from the side effects or failure modes of the system being optimized, as shown in Figure 8.2. Focus on the efficient, useful portion of the energy transformation. The ideal function should be expressed as a dynamic relationship, using the desired output of the system as if it had no losses and no noise factor effects and the appropriate signal factor. The output and the signal factor should be physically related by the energy transformations that are required by the system to produce the results the customer desires.

The translation of the "voice of the customer" into technical terms that describe to the engineering team what the customer wants is a good starting point for defining the ideal function. The ideal function could be the physical function performed during the energy transformation process. Often the team knows the customer expectation for some high-level, observable attribute of performance. This may have to be related to "low-level" functional attributes in which the customer has little or no interest. For example, in the photographic industry, customers specify that they do not like overexposed or underexposed pictures. The engineer translates this high-level attribute into physical parameters that can be controlled, such as shutter speed variability. This is an engineering unit of measure (a low-level attribute hidden from the customer's view) that is readily measured as a continuous variable.

### 8.3.2 The Side Effects and Failure Modes of a Design

What inefficient energy paths occur in the design? What amount and forms of energy go into nonproductive work? What elements of the design can fail, and under what circumstances do they fail? What effect will this failure have on the customer? The side effects or failure modes are the undesirable outputs of the system. These are commonly observed as failures or reliability issues. These widely used performance indicators are just symptoms of the underlying physical problems. Always focus on measuring the desired physics, not the aftermath.

Knowing what the failure modes are helps establish reasonable bounds on the expectations for the design. Understanding the physics of the design's failure modes provides the insight needed to identify the noises that are most likely to contribute to the failures. Knowing the side effects of a subsystem's energy transformations provides insight into internal noises that the design can impose on itself, as well as noises that the subsystem can impose on other, neighboring, subsystems (proximity noise). Dr. Taguchi summarizes the situation with this recommendation:

- *Don't measure reliability to get reliability.*

### 8.3.3 Relating the Ideal Function to the P-Diagram

Figure 8.3 is the backbone of the P-diagram. Compare it to Figure 8.1 and Figure II.1 (Introduction to Part II). When the ideal function of the system is understood clearly, then the key dynamic relationship should be revealed. The input energy represents the signal factor. The output energy represents the response or quality characteristic. It is usually not necessary to measure the energy transformations directly. In many cases, it is not practical. But the energy transformations in the system should be

**Figure 8.3** The ideal function (part of the P-diagram)

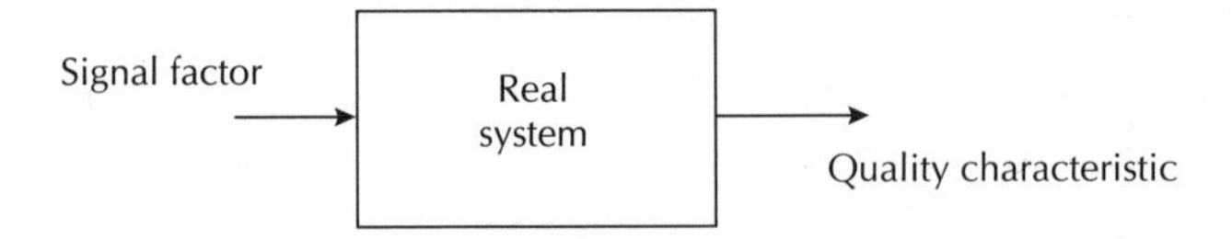

widely known and understood. From that information, appropriate signal factors and quality characteristics that are related to the energy transformations can be selected.

This approach applies even in the case of a static problem. In fact, most static problems are simply special cases of underlying dynamic problems, where the signal factor is held constant. For example, in manufacturing molded parts (e.g., plastic, metal, clay, or concrete), the part dimension is dynamically related to the mold dimension, as shown in Figure 8.4. Once the mold has been built, the signal factor is fixed and the problem appears static. Nevertheless, the underlying ideal function is that the part dimension is proportional to the mold dimension, as shown in Figure 8.5.

Train yourself to find the dynamic relationships in the system. You will find that this provides a fresh way to look at issues such as controlling the nominal value of a quality characteristic and the role of control factors. The ideal dynamic relationship is a zero-point proportional. In Chapter 6, this and other functional dynamic forms are discussed. If a design concept is difficult to define in this dynamic manner, then keep working until one is developed. Fitting a dynamic relationship to the design concepts will pay off in the long run in improved robustness and tunability.

### 8.3.4 Some Examples of Ideal Functions and Side Effects

Let us consider some simple systems and identify their ideal function and side effects to illustrate the thought process discussed herein.

**Example #1: Painting Process**

A spray painting process, shown in Figure 8.6, is used to finish a complex piece of equipment. The process could use an air gun sprayer or an electrostatic sprayer. In either case, the customer desire is for a pleasing finish. Qualities such as uniform coverage, durability, gloss, and color are important. The gloss and color are determined by the paint properties. Similarly, durability is a function of the chem-

**Figure 8.4** Dynamic relationship for molding parts

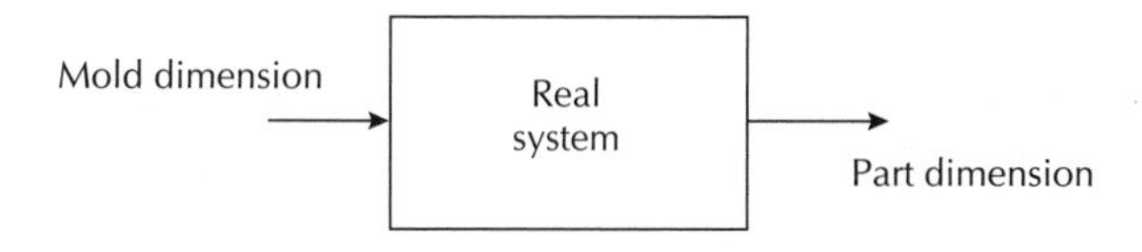


**Figure 8.5** Ideal function for molding parts

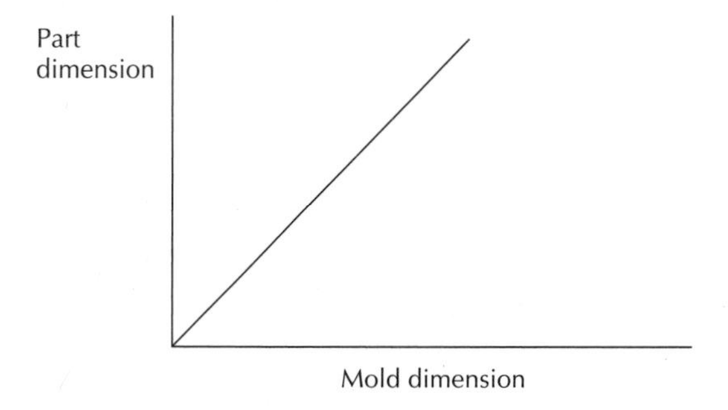

**Figure 8.6** Spray process for applying a finish

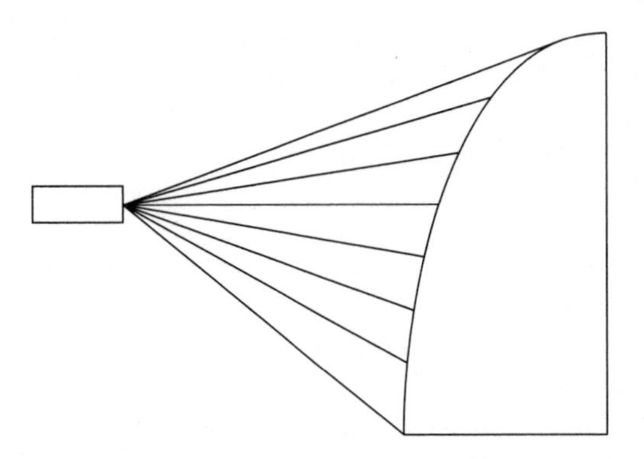

istry of paint and primer. Here the focus is on coverage. A finish that includes sags or uncoated areas (drop-outs) would be objectionable. These are considered to be failures. The ideal function is a uniform paint thickness everywhere on the part being painted. Sags and drips are phenomena that are related to gravity, which is not one of the intended energy transformations in the painting process.

The transfer of paint from the nozzle to the object is related to the energy transformation being optimized. The measure of the paint transfer is the thickness ($y$). The signal factor, if the flow rate is constant, is spraying time. In fact, this problem is a good example of a double-dynamic problem (see Chapter 6) where paint flow rate is the functional signal factor ($M$) and amount of time is the process signal factor ($M^*$). The ideal function is

$$y = \beta M M^*$$

The side effects, sags and drop-outs, suggest that location is a key (surrogate) noise. It is possible to quantify the quality of the actual process, comparing it to the ideal process, by measuring the paint thickness at several spots on the finished part. The P-diagram (Figure 8.7) shows the dynamic relationship and the surrogate noise factor, location.

The ideal function is a double-dynamic zero-point proportional case, where the paint thickness is the quality characteristic. Variability due to the noise factor, location, causes deviations from the ideal function in the actual data.

**Figure 8.7** P-diagram for a paint spray finishing process

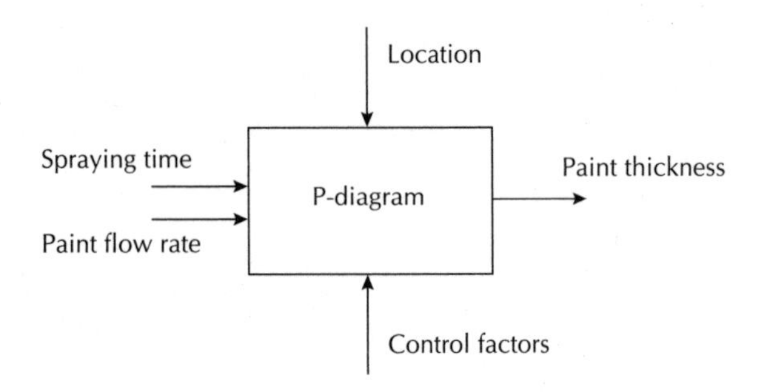

**Figure 8.8** Belt drive system

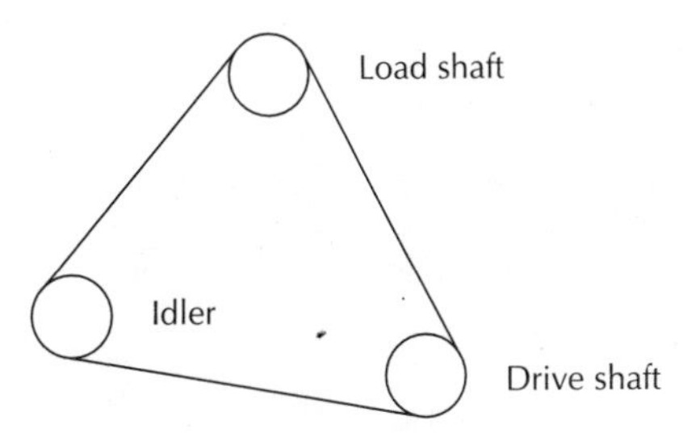


### Example #2: Belt Drive System

The system shown in Figure 8.8 transfers rotational energy from the drive shaft to the load shaft. The customer desires a smooth drive that translates into low acoustic noise and low vibration. Torque spikes, belt slippage, or acoustic noise are all examples of side effects. None of them are descriptions of the desired energy transformation. The energy transformation is best described kinematically, by relating the load shaft motion to the drive shaft motion, or kinetically, by relating the load torque to the drive torque. This is fundamentally a motion control issue. The drive shaft rotational velocity, $\omega_d$, is the signal factor, and the load shaft rotational velocity, $\omega_L$, is the response. Thus, the ideal function is

$$\omega_L = \beta \omega_d$$

The value of the slope ($\beta$) depends on the ratio of the pulley diameters, which is typically 1.0. The side effects each cause deviations from the ideal function. There are a number of noises possible in this problem: low belt tension, belt wear, load variation, acceleration, belt stretch, etc. Important measurement issues include encoder accuracy, sampling time (long sample time averages out some motion deviations), and relative phase angle measurement (to account for motion delays). Measuring the ideal function correctly is clearly nontrivial yet important.

### Example #3: Imaging

There are many applications of imaging besides photography [M6]. An important industrial application is photolithography, the process of developing a pattern onto a photoresist for circuit board application and microelectronics manufacturing. In imaging, there is usually a desire to reproduce faithfully the information in the original pattern onto the film or substrate. This process is energy-transformation-rich. One example is the transfer of light from the imaged object to the photolithography substrate. Broken lines, blurry edges, and thick or thin lines are all symptoms of variation in the energy transformation. The dimension of the pattern developed on the substrate should be proportional to the pattern dimensions in the original. Thus, the ideal function is a mapping of the original pattern to the substrate, as indicated in Figure 8.9.

The ideal function is defined by comparing the dimensions, $y_i$, of key features in the developed substrate to the corresponding dimensions, $M_i$, in the original pattern:

$$y_i = \beta M_i$$

The ideal function is a dynamic zero-point proportional case, where the measured dimensions are the quality characteristics. If variable magnification is provided, then the ideal function is best treated as a double-dynamic relationship, with $M^*$ representing the magnification factor. Variability

**Figure 8.9** Image transfer system

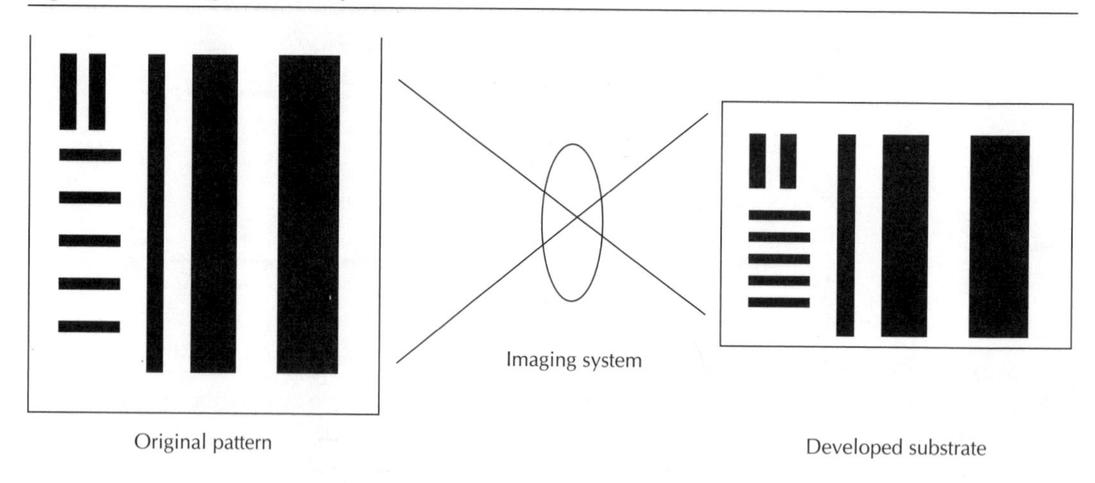


due to location in the image, loss of small dimensions, line broadening, or thinning are all nonlinearities in the actual data.

### Example #4: An Electrical Resistor

The passive element in an electrical circuit, the resistor, is ubiquitous in electronics applications. Still, the tolerance for the lowest-cost applications is a whopping 10%. This is indicative of the fact that producing the physical device is nontrivial. Besides manufacturing tolerance, there is variability due to temperature. Resistive heating is a common source of heat rise in a resistor. This is an obvious example of a side effect's undesired influence on the energy transformation. The ideal function for a resistor is Ohm's Law. In other words, the resistor is a device that transforms current, $I$, into electric potential, $\xi$. The ideal function is

$$\xi = RI$$

The ideal function is a dynamic zero-point proportional case, where the electric potential is the quality characteristic. Variability due to resistive ($I^2R$) heating is a nonlinearity in the actual data.

## 8.4 Guidelines for Choosing the Quality Characteristic

Here, then, are some desired properties for the quality characteristic that should be satisfied.

- Characteristics should be **continuous, quantitative,** and **easy to measure.**
- Characteristics should have an **absolute zero.**
- Characteristics should be **additive** (noninteractive) or at least **monotonic** (consistent) relative to their factor effects.
- Characteristics should be **complete** (cover all dimensions of the ideal function). A complete response provides all the information required to describe the ideal function.
- Characteristics should be **fundamental** (related to basic physical functions).

### 8.4.1 Continuous, Quantitative, and Easy to Measure

To properly account for the flow of energy through a design, select a continuous variable as a measure of performance. Quality characteristics that reflect energy transformations should always be expressed in engineering units (scalar or vector quantities). In other words, they should directly measure the ideal function. Do not seek indirect measures that are typical of customer-observable behavior.

Measure parameters such as these:

- Force
- Distance
- Velocity
- Acceleration
- Pressure
- Time

Avoid using indirect measures such as these:

- Yield
- Faults
- Voids
- Pass/fail
- Number of defects
- Reliability (meantime between failures)
- Appearance
- Fraction defective

Quality characteristics that focus on these types of measures are highly susceptible to non-monotonic behavior.

Quality characteristics need to be quantitative and easy to measure. Quantitative measures allow discrimination between small improvements in quality. Characteristics that are easy to measure make the experiments more practical to perform. Prototypes, referred to as robustness fixtures, should be specially built for robustness testing. They should include accurate and precise transducers and adequate control-factor and noise-factor adjustability.

### 8.4.2 Absolute Zero

Quality characteristics should have an absolute zero. An absolute zero means that there can be no negative values for the quality characteristic. The Kelvin temperature scale has an absolute zero; Celsius and Fahrenheit do not. Other examples of quality characteristics that have an absolute zero are part dimensions, mass or weight, chemical concentration, and image density. This property is sought when using the zero-point signal-to-noise ratio for dynamic problems or the nominal-the-best (type I) signal-to-noise ratio for static problems. These S/N ratios typically give the best results because they eliminate interactions between the mean and standard deviation (see Chapters 5 and 6).

### 8.4.3 Additive or Monotonic

The complexity or directness of the relationship between the quality characteristics and the control factors determines whether the system is additive, monotonic, or interactive. An additive quality characteristic follows an additive model for the control factor effects. The effects of the design parameters in such a system are independent of each other. Thus, the consequences of internal noise (e.g., manufacturing variation and control factor deterioration), design changes, or changes in the application of a technology or subsystem are minimized. With an interactive quality characteristic, the effect of a change in any one parameter has a surprising result because of interactions with other parameters. This results in a need to reoptimize a system every time anything in the system is changed or the application of the design changes.

**Figure 8.10** Additivity: Interaction plots illustrating no interactions

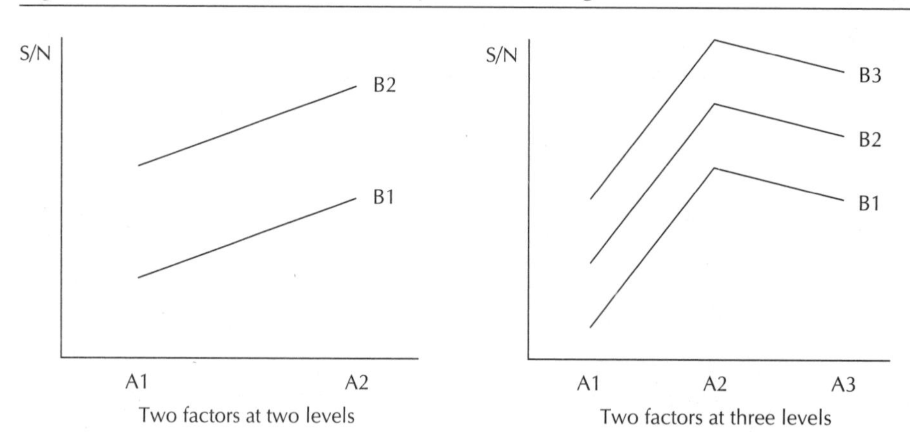


Figure 8.10 illustrates responses that are free of interactions between factors A and B. In this case, the effect on the response of changing levels for either factor is completely independent of the set points for the other factor. Compare Figure 8.10 to Figures 7.2 and 8.11, which illustrate interactive responses.

The next two figures illustrate the monotonic and interactive cases. A monotonic quality characteristic may not be totally free of interactions. For monotonic behavior, the interaction magnitude is small enough so that the control factor effects on the quality characteristic are consistent (directionally) regardless of other control factor set-points. This means that the quality characteristic always changes in the same direction (either positive or negative) when one of its factors is changed, *even if* the other control factors have changed their levels. So monotonicity refers here to the quality characteristic's directional behavior with respect to changes in levels of one or several of its other control factors. Monotonicity is not a measure of parallelism; that is what additivity measures.

Figure 8.11 illustrates responses that have a mild interaction. A perfectly accurate math model for an S/N ratio response would require the inclusion of interaction terms. The factor levels that maximize the S/N ratio are consistent, although the magnitude of their effect changes slightly due to interactions. Thus, in the two-level plot on the left, factor A's optimal level is A2, regardless of the level of factor B, and factor B's optimal level is B2, regardless of the level of factor A. Similarly, in the three-level plot on the right, factor A's optimal level is A2, regardless of factor B's level, and factor B's optimal level is B3, regardless of factor A's level. The interaction in these examples is often referred to as synergistic, because the effect of setting A2 and B2 together for the two-level factors and A2 and B3 for the three-level factors is a beneficial one.

Figure 8.12 illustrates responses that have a strong, antisynergistic interaction. From an engineering point of view, this situation is of great concern. The factor levels that maximize the S/N ratio are not consistent: They depend in a critical way on the other factor levels. Thus, in the two-level plot on the left, factor A's optimal level is A1 when factor B is at B1, but the optimal level for A is A2 when factor B is at B2. Similarly, the optimal level for factor B depends on the set point for factor A. Whenever the interaction plots cross, it is clear that the interaction is antisynergistic. In the three-level factor plot on the right, the lines do not cross, but here, too, the optimal level for one factor changes depending on the other factor. Factor A's optimal level is A2 when factor B is at B1 or B2, but the optimal level for A is A1 when factor B is at B3. Thus, the response is not monotonic with respect to factor

**Figure 8.11** Monotonicity: Interaction plots illustrating weak interactions

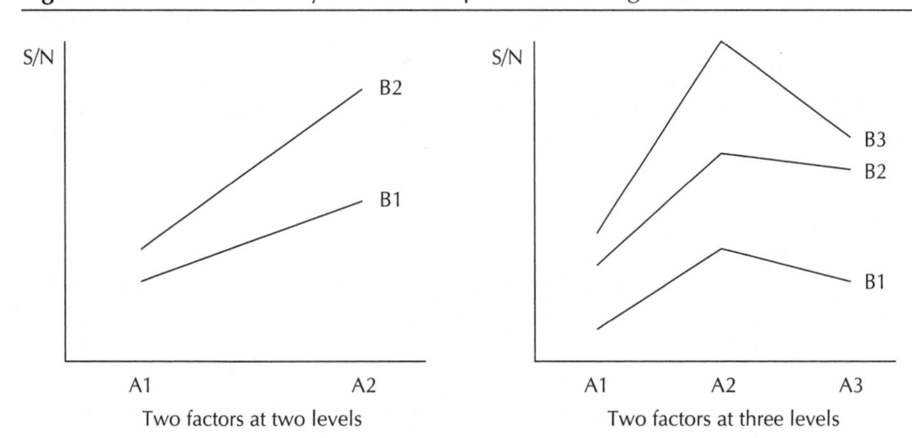


A (the *direction* of the slope has changed). On the other hand, Factor B's optimal level is B3, regardless of factor A; the response is monotonic in that case.

The nonmonotonic case is much more difficult to manage and control. Factor levels change because of manufacturing variation and factor deterioration. Interactions with other factors can magnify the influence of the variation, resulting in an excessive loss of quality. These effects are hard to visualize when there are many factors involved, and even harder to explain to team members and management. Factor levels also change when a design is altered to meet new requirements or because of new applications of a given design. In the nonmonotonic case, this requires reoptimization of the system.

Whether a system is additive or monotonic is a function of the design concept, the design parameters, and the quality characteristic. Some simple principles can be applied to find monotonic quality characteristics. Because the difference between additivity and monotonicity is a matter of degree, the same principles apply to finding an additive quality characteristic. However, it requires orthogonal

**Figure 8.12** Nonmonotonic: Interaction plot illustrating strong interactions

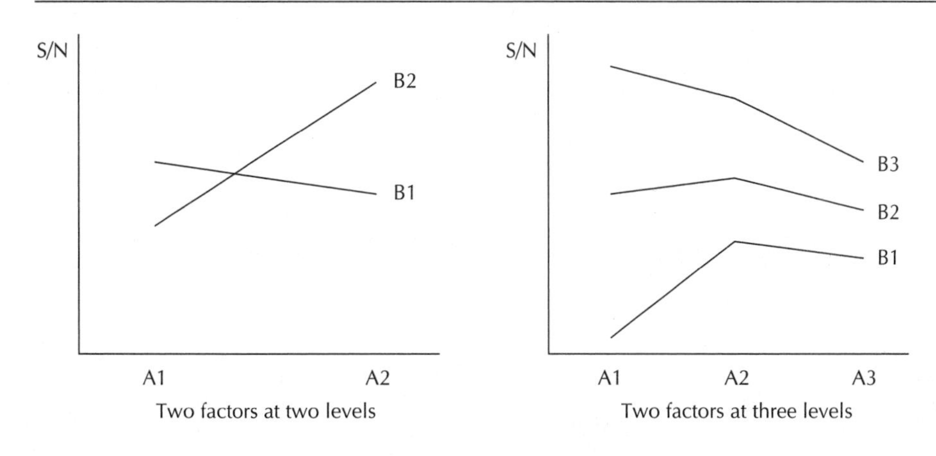

array experimentation and verification testing to demonstrate additivity. Monotonicity, on the other hand, can be analyzed using simple physical models to predict behavior, because a qualitative understanding of the factor effects is sufficient to demonstrate consistency. The quality characteristic should be examined for any obvious signs of interactions that can be anticipated. If there are any, then try to change the quality characteristic to avoid the interactive behavior. Let us consider some examples to illustrate this concept.

## Example #1: Baking Cookies

A baker plans an experiment to improve his cookie-making process. The baker is interested in increasing the number of acceptable cookies produced. The baker first tries using yield as a larger-the-better quality characteristic. The baking team performs tests using time and temperature as control factors. They generate the data shown in Figure 8.13. Clearly this quality characteristic is not monotonic. The direction of change for yield as time goes from 20 to 35 minutes depends on the level of temperature. In addition, yield is not directly related to the energy transformation. It is a reflection of customer preference for the degree of doneness of the cookies.

Consider another quality characteristic that is directly related to the transfer of energy in the baking process. The energy transfer is from the hot oven to the cool dough. An appropriate quality characteristic could be cookie temperature, cookie hardness, cookie color, or any other measurable physical value (expressed in engineering units) that is a function of the degree of doneness of the cookie dough caused by the application of heat energy.

Anyone who has done much baking will immediately recognize that cookie color is easy to use, since it requires no complex measurement capability.[3] The baking team measures the darkness of the cookies for the same time and temperature test just presented. They generated the new interaction plots shown in Figure 8.14, using the same levels for time and temperature as before.

This characteristic is monotonic. The optimum levels for A and B must be chosen based on the customer preference for degree of doneness, but the analysis of this experiment provides a consistent result.

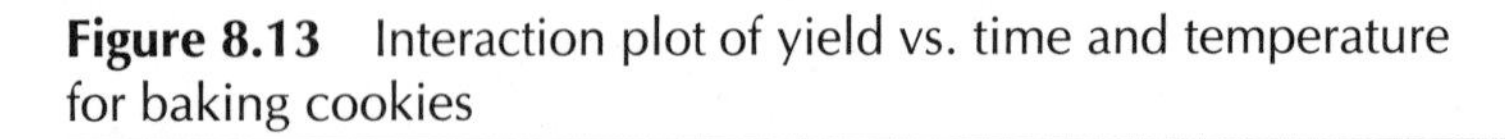
**Figure 8.13** Interaction plot of yield vs. time and temperature for baking cookies

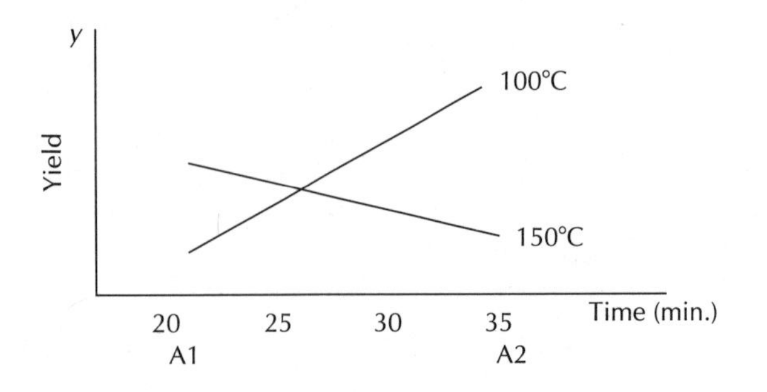


3. In fact, there is a British shade gauge reference that is used by engineers optimizing cooking processes.

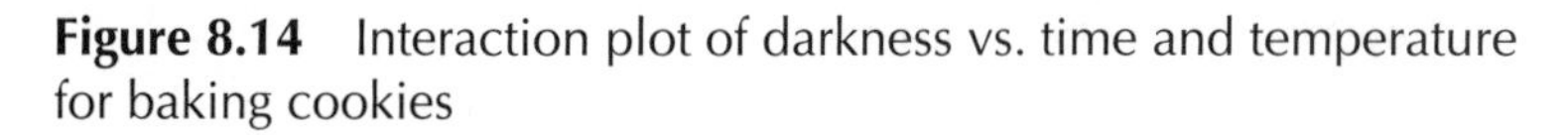

**Figure 8.14** Interaction plot of darkness vs. time and temperature for baking cookies

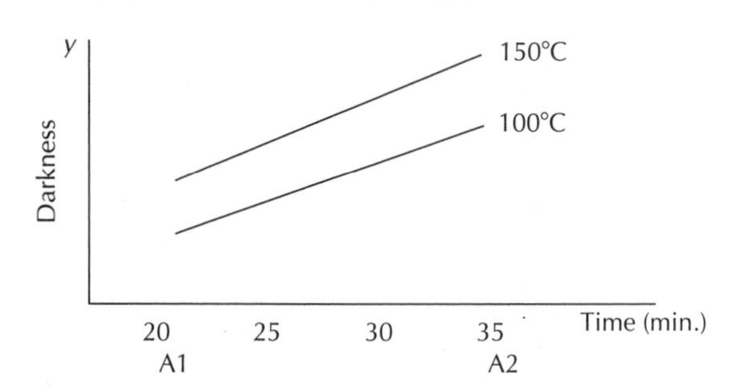


## Example #2: The Catapult

The catapult, also referred to as the statapult, is a machine (shown in Figure 10.2) that is used in the classroom by various consultants and instructors to illustrate statistical principles. The authors use this device when teaching classes at Eastman Kodak Company and at the Rochester Institute of Technology. A detailed description of this device is not needed at this point, we simply focus on the response here. Consider the possible quality characteristics to be used when trying to optimize the catapult to hit a target with a projectile such as a golf ball or small rubber ball. There are at least three options for the response describing the firing of projectiles from the catapult. The responses considered here only take into account the projectile motion in line with the target. Sideways deviations are ignored in this discussion.

*Option 1:* Measure the distance from the target to the projectile impact point using positive values for impacts beyond the target and negative values for impacts before the target. This is shown in Figure 8.15.

*Option 2:* Measure the absolute value of the distance the projectile impacts from the target. This is shown in Figure 8.16.

*Option 3*: Measure the value of the distance the projectile travels from the base of the catapult mechanism. This is shown in Figure 8.17.

Remember, when considering monotonicity, look for the interaction or dependence of one factor effect on the others. Of course it is not, in general, known a priori what all the interactions are. Nevertheless, by applying some common-sense engineering analysis to the device, *some* of the results can be foreseen. In this case, the kinematic equations of projectile motion state that the range depends on the angle and velocity of release, as well as the height of release. Focusing on angle and velocity, it can be assumed that the spring force affects the velocity in a way that is independent of the factors determining release angle. With that in mind, evaluate the three quality characteristics.

For option 1, if the angle of release is set and then the spring force is increased, the quality characteristic increases monotonically as shown in Figure 8.18. Assume there is an angle of release, $\alpha_1$, such that at a spring force, $f_1$, the ball lands at $-1$ m from the target. Also assume that there is another angle of release, $\alpha_2$, such that for the same spring force, the ball lands at $+1$ m from the target. Then the interaction plot shown in Figure 8.19 can be anticipated when the two angles are used at an increased spring force of $f_2$. The parallel lines indicate that this quality characteristic exhibits no interactive behavior with respect to this analysis. Even if the slopes are not identical, as long as $f_2 > f_1$ and

**Figure 8.15** *Option 1* for the catapult quality characteristic

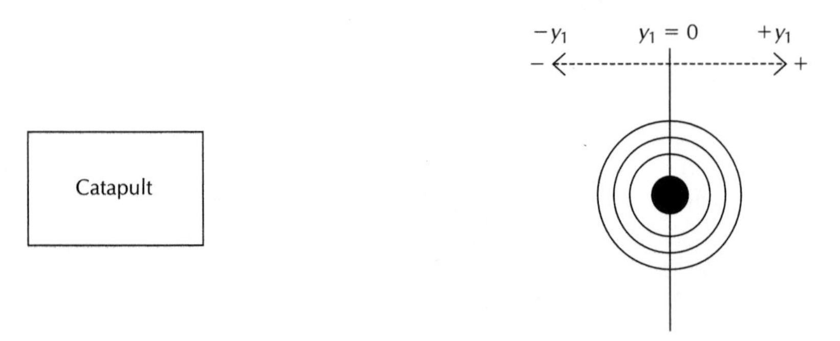


**Figure 8.16** *Option 2* for the catapult quality characteristic

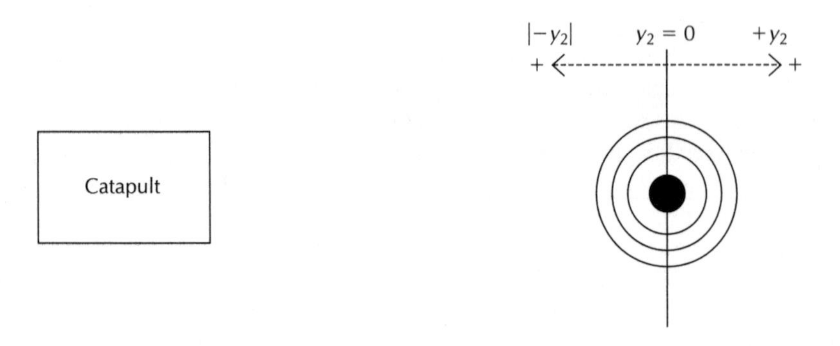


**Figure 8.17** *Option 3* for the catapult quality characteristic

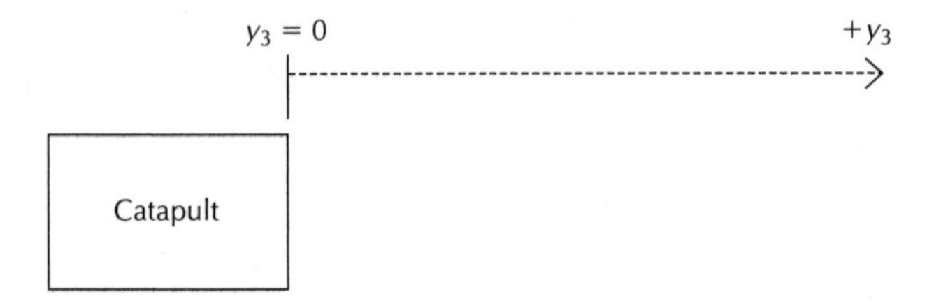


the quality characteristic is monotonic with force as shown in Figure 8.18, then the quality characteristic is also monotonic with respect to interactions.

For option 2, if the angle of release is set and then the spring force is increased, the quality characteristic does not behave monotonically, as shown in Figure 8.20. It is not monotonic because the value decreases with increasing spring force as short falls approach the target. Then the value increases with increasing spring force as the overshoots move farther away from the target. Now repeat the previous analysis using the angle of release, $\alpha_1$, so that at a spring force $f_1$, the ball lands 1 m short of the target. The value of $y$ for this result is $y = +1$. Then, repeat this analysis at the angle of release $\alpha_2$, so

**Figure 8.18** *Option 1* quality characteristic vs. spring force

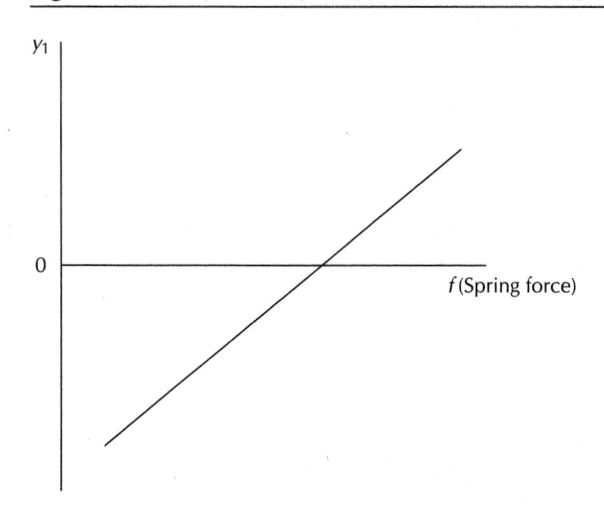


**Figure 8.19** *Option 1* interaction plot for spring force and release angle

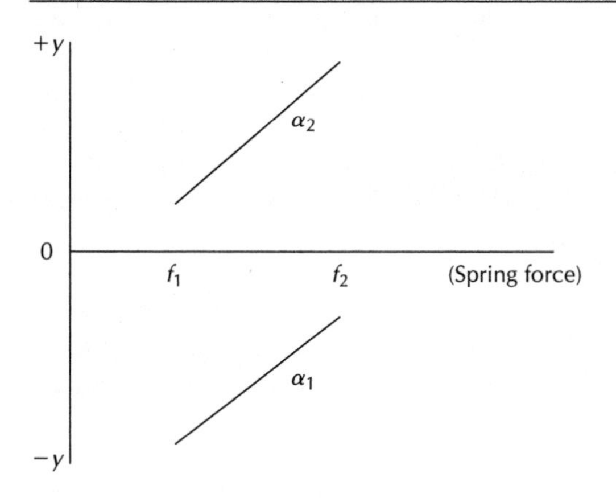


**Figure 8.20** *Option 2* quality characteristic vs. spring force

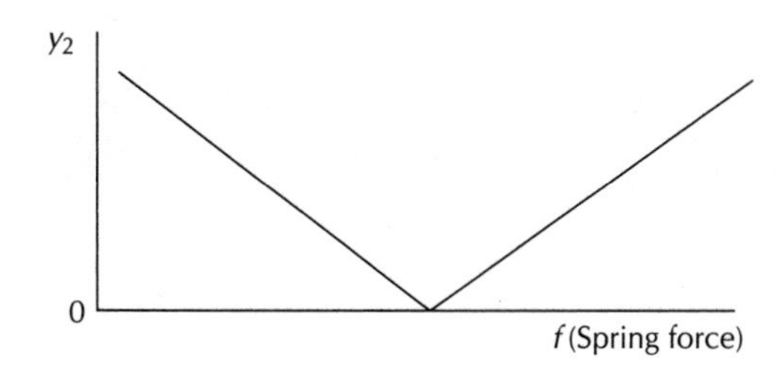

**Figure 8.21** *Option 2* interaction plot for spring force and release angle

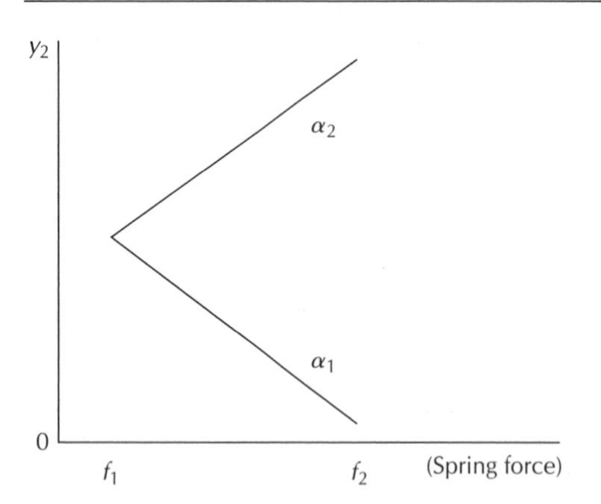


that the ball lands 1 m beyond the target. The quality characteristic is again $y = +1$. Thus, the results shown in Figure 8.21 are anticipated when the two angles are used at an increased spring force of $f_2$.

This quality characteristic clearly has a strong interaction, similar to that in Figure 8.12, as shown by this analysis. This option for the quality characteristic is also the one that is probably most closely tied to the customer desire to hit the target with a projectile. The customer would judge the results by how close the impact is to the target, with little regard to the direction of miss. So why is there an interaction here, but not in option 1? Ignoring the direction of a miss loses important information that is directly related to the energy transformation, in this case from the spring potential energy to the ball's kinetic energy. The customer does not care, *but we do!*

For option 3, if the angle of release is set and then the spring force is increased, the quality characteristic increases monotonically, as shown in Figure 8.22. Now repeat the previous analysis using the angle of release, $\alpha_1$. Assume that the target is 8 m from the catapult, so that $y_3 = 7$ m. Then, if the angle of release is changed to $\alpha_2$, so that the ball lands at 1 m beyond the target, $y_3 = 9$ m. Thus, the results shown in Figure 8.23 are anticipated when the two angles are used at an increased spring force of $f_2$.

This quality characteristic exhibits no interactive behavior with respect to this analysis. Option 3

**Figure 8.22** *Option 3* quality characteristic vs. spring force

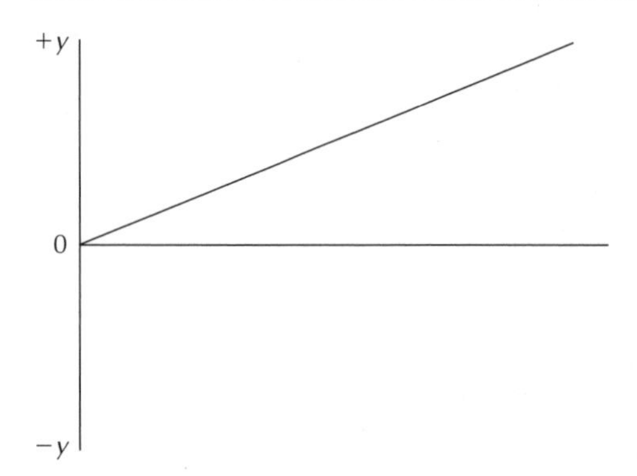

**Figure 8.23** *Option 3* interaction plot for spring force and release angle

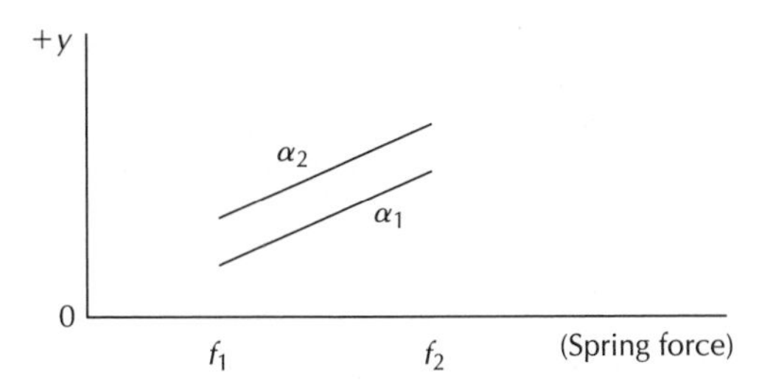


has an additional property, desirable for a quality characteristic, that clearly makes it the best option: Option 3 has an absolute zero, while option 1 lacks this property. For option 3, negative values are not possible if the catapult can only project forward.

### 8.4.4 Complete

Select characteristics that are complete (cover all dimensions of the ideal function). A quality characteristic is complete if it provides all the information needed to describe the ideal function. Trying to co-optimize a number of quality characteristics simultaneously leads to confusion and inefficiency. Co-optimizing the mean and standard deviation of a single quality characteristic is an example of a difficult, constrained optimization problem [P2]. Trying to do the same for a number of quality characteristics is even harder. The S/N ratio helps reduce the complexity, because optimizing its value improves the quality independent of the mean.

Co-optimizing several quality characteristics is typically necessary when focusing on symptoms rather than the ideal function. One quality characteristic should be chosen as the defining measure for each ideal function. Focusing on more than one quality characteristic can be a problem. The following example illustrates this concept.

The paper gyrocopter (shown in Figure 13.9) is a simple machine that you can design, build, and test using paper, scissors, and a stop watch. By experimenting with wing and body shape and dimensions, you can build a device that is a pretty good simulation of the maple tree's winged seed, the samara. It is possible, however, to construct other paper shapes that float to the ground slowly, but without the gyration feature that is intended. In fact, when running their design treatment combinations, classroom teams often see one or two runs fail to autogyrate. The usual description of the customer want for the gyrocopter is a long flight time. Based on this quality characteristic, the analysis of the experiment treats gyrocopters and nongyrocopters similarly.

The problem is that time of flight is not a complete measure. The spinning behavior of autogyration is also important. The solution, to achieve completeness, is to measure the number of revolutions the gyrocopter makes before it lands. This quality characteristic is complete because it takes into account all the dimensions of the ideal function. Gyrocopters that fall fast or that don't spin are penalized equally. Designs that spin rapidly and float slowly to the ground are much more pleasing. The problem, typically, is measurement. To count the number of revolutions requires a slow-motion video analysis. Lacking that equipment, measurements are limited to flight time.

If it is difficult to find just one quality characteristic to adequately express the function of the design, it may be necessary to subdivide the design for simplification of functions. Frequently a design is made up from any number of subsystems or components. Break the design down into more manageable subsystems that form functional subsets contributing to the overall energy transformation controlling the function of the product. Optimization of these subsystems, influenced by the noise from subsystem *and* system variability, must be done prior to system optimization. If there is more than one quality characteristic for a single subsystem, the optimization process becomes somewhat more complex. The problem then becomes one of co-optimization, where trade-offs and compromises dominate the process.

### 8.4.5 Fundamental

Characteristics should be fundamental [L1]. A response is fundamental if it does not mix mechanisms together and is not influenced by factors outside the process being optimized. Clearly, quality characteristics whose values depend on the customer tolerance or expectations are influenced by factors external to the physics in the design. Such characteristics are not fundamental. When a characteristic is not fundamental, it is almost always interactive with respect to the control factors. The customer may always be right, but customers don't generally understand the detailed physics and energy flows in the products that engineers design. Engineers cannot focus directly on the measures customers use to define quality when they are involved in the engineering of a product. They must focus on accurate *translations* of the customer's terms (Voice of the Customer) that are related to the underlying physics and chemistry, including the variability associated with these technical areas that really drive product quality.

Options 1 and 2 from the catapult discussion are examples of nonfundamental quality characteristics. This is because they depend upon the distance between the catapult and the target. That distance is entirely independent of the physics. If the catapult is fixed and the target is moved some distance, then the experimental results will change. Option 3 from the catapult discussion is fundamental. The target is not included in the definition of the quality characteristic. If the target is moved and the experiment is repeated, the same results are obtained.

Consider the quality characteristic for the surface mount assembly of an electronic component. A robot-assisted production process for circuit board manufacture places the surface-mount components on the circuit board, as shown in Figure 8.24. The circuit performs reliably only if the components are precisely located on the board. Overall circuit board yield is not a good characteristic—it is not fundamental, because there are many other steps in the manufacture of a board.

The error (distance), $r$, between the location of the component and the correct location on the board, shown in Figure 8.25, seems to be the obvious choice for a quality characteristic. However, this quality characteristic is not fundamental, because it mixes mechanisms together.

Consider the energy transformations and ideal functions. The circuit board is transported on a conveyer belt. Ideally it moves only in the direction $y$ with no $x$ error and no skew. The robot transports the component to a location *that is independent of the circuit board.* A reference system relative to the robot base is required to analyze this problem. Now the problem breaks up into two fundamental aspects: the location of the board relative to the robot when it stops to receive the component, and the location of the component when it is set down by the robot arm.

Note that in the case of the circuit board, $x$ and $y$ errors are clearly distinct and should be analyzed separately. The appropriate coordinate system for the robot depends on the direction of motion as controlled by the conveyer mechanism. By measuring the fundamental mechanisms, not the resulting quality, a much better set of quality characteristics is obtained. There is no constrained co-optimization problem here. Each of the quality characteristics measures a separate function of the board and robot

**Figure 8.24** Robotic placement of circuit board components

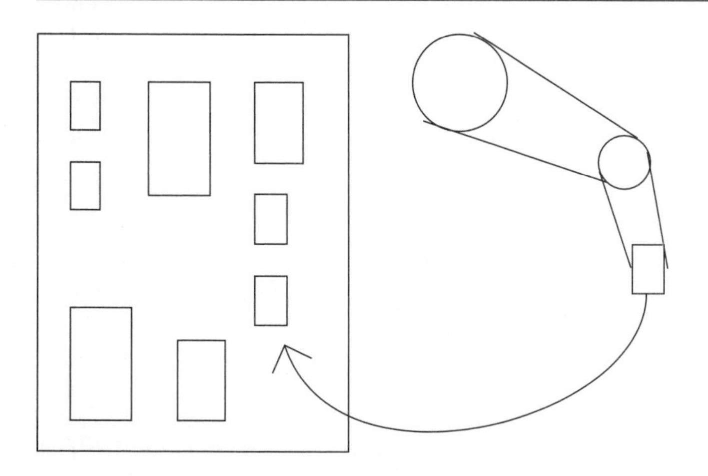

**Figure 8.25** Component placement error

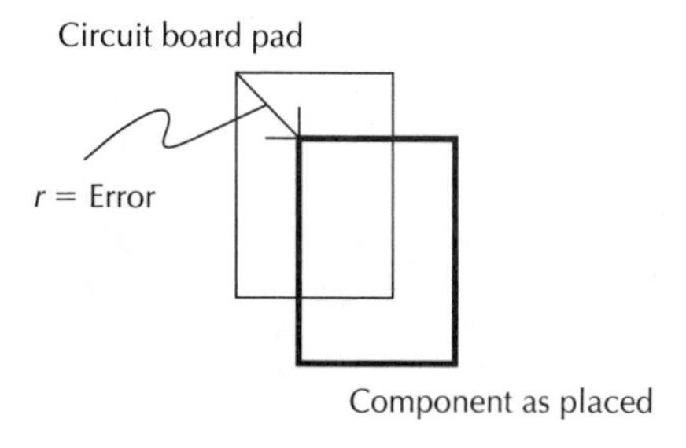


transport devices. Only when they are all, independently, on target and robust will the aggregate quality characteristic, board yield, be satisfactory.

## 8.5 Summary: The P-Diagram

At this point in the process, the backbone of the P-diagram has been developed as shown in Figure 8.26. The identification of the ideal function and the quality characteristics used to quantify it are important parts of the robustness process. These are also the most difficult parts, and often require some help from an experienced practitioner who has been trained to look for the underlying energy transformation–driven processes rather than the aftermath of deviations in these processes. If such help is not available, focus on the physics. Remember the golden rule in choosing a quality characteristic:

> *The closer you are to measuring the simplest act of physics that occurs in your design, the better.*

**Figure 8.26** The backbone of the P-diagram

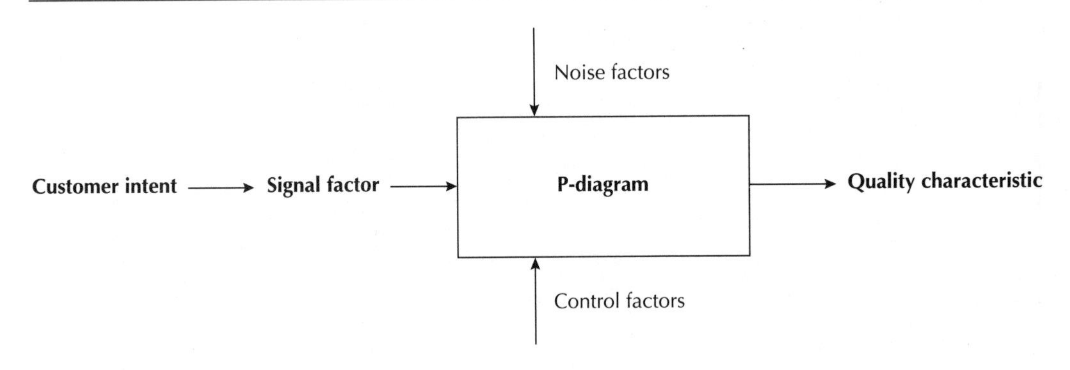


Consider the results of discussions and analyses and see if an ideal behavior has been found. Does the quality characteristic describe what is necessary for the system to do, or does it describe what is to be avoided in the system? Be particularly wary of smaller-the-better quality characteristics; they are often symptoms, not targets. Do not simply measure reliability; measure the physical processes that dictate the ultimate reliability.

## Exercises for Chapter 8

1. What is meant by the term "energy transformation"? Why are energy transformations valuable to understand and employ in the Robustness process?
2. What does it mean when we find a design has *four* distinct quality characteristics in a single-parameter design experiment?
3. Define what is meant by the additive model and explain why we seek additivity.
4. What is meant by the *ideal function* of a design?
5. What are some of the benefits of thinking about a design's performance based on energy transformations?
6. Why is it highly recommended that you design products and their constituent parts so their most basic physical functions can be measured?
7. What is a P-diagram, and how is it used?
8. What is the difference between a main effect, a side effect, and a failure?
9. Define the terms *absolute zero, monotonic, complete,* and *fundamental* as they relate to quality characteristics.
10. Why do we avoid taking measurements such as yield, number of defects, and fraction defective?

CHAPTER 9

# The Selection and Testing of Noise Factors

## 9.1 Introduction

The selection of noise factors is critical to the success of the robustness effort. It is important to know what the causes of variability are, because these are the sources of customer *dissatisfaction.* Noise factors are defined in Chapter 1 as factors that can influence the performance of a system and are not under our control during the intended use of the product. Noise factors can be external to the design. Generally, these are controlled by customers. Noise factors can also be internal to the design. These are design parameters whose nominal level may be specified, but manufacturing variation and deterioration result in deviation from the intended target.

In this chapter, the process for selecting and testing of noise factors to determine which are the most significant and how to combine them for the parameter optimization experiment is discussed. The study of variation and failure induced by the application of the noise factors in the lab is used to improve the product quality. Any laboratory study must also include measurement noise. The purposeful application of noise factors to simulate field variation in the lab allows one to clearly distinguish between noise that affects the product in its intended use and laboratory noise, which merely affects the results of an experiment.

The primary goal of the noise experiment is to identify the few noise factors that cause most of the variability in the system being optimized. This goal for the noise factor experiment is accomplished by decomposing the data from the noise factor runs using analysis of means (ANOM) to analyze the magnitude and directionality of the noise factor main effects. Further analysis to predict performance can be accomplished using analysis of variance (ANOVA), but that is deferred to Chapter 17.

The significant noise factors are combined and used in the parameter optimization experiment to test the robustness performance of the control factor combinations using the process discussed in Chapter 13. The goal is to compound the noises for the parameter optimization experiment into two groups using the directionality of the noise factor effects. Those noise factor levels that cause an increase in the response are grouped together into $CNF_+$, and those noise factor levels that cause a reduction in the response are grouped together into $CNF_-$. These two groups, $CNF_+$ and $CNF_-$, are used for each treatment combination of the control factor array. For an L18 array, this results in a total of 36 runs for the parameter optimization experiment. Robustness should not be tested using random variation. That approach is inefficient because it takes several runs to get a good estimate of the stochastic (random) variation, which is weak compared to the *real* noise. Induced variation, on the other hand, can be quantified in just two runs using compound noise factors. Remember, for the parameter optimization experiment, that the amount of variation must be evaluated for each control factor treatment combination, so efficiency is extremely important.

Other goals for the noise experiment—benchmarking performance, perfecting the experimental procedure, and testing the magnitude of measurement error—are accomplished as well. The overall system performance can be characterized by calculating a signal-to-noise ratio using the data from all the runs of the noise array. After parameter optimization is complete, the entire noise array can be retested with the new control factor set points to demonstrate the overall S/N improvement.

The experimental procedure used for the noise experiment should be the same as that which is to be used for the parameter optimization experiment—with one important difference. In the main experiment the noises are compounded. But the nature and range of the noises used, the measurement devices, and the quality characteristics are all the same for the noise experiment and parameter optimization experiment. If there are problems with the measurement system or quality characteristics, that should become apparent in the noise study. In their classic work [B10], Box, Hunter, and Hunter recommend, when starting a large experiment, that the experimentalist stop after about 10% of the work is done and do some analysis of the results to ensure that the experiment is sound. The total experimental effort (excluding confirmation runs) for the L12 noise experiment and the L18 $\times$ 2 CNF main experiment consists of 48 runs (12 + 36). The noise factor experiment represents 25% of the total effort. Thus, the approach recommended here is reasonably consistent with BHH's advice.

## 9.2 The Role of Noise Factor—Control Factor Interactions

The relationship between the control factors and noise factors in an experiment is of great interest. Robustness is achieved through the interactions between the control factors and the noise factors. There are two types of CF$\times$NF interactions, one associated with external noise and the other with internal noise. For external noise factors, the type of interaction that can lead to improved robustness is shown in Figure 9.1. Here, the noise factor has an effect on the response. This is undesirable. However, the magnitude of the effect of the noise factor on $y$ changes when the control factor shifts from level A1 to level A2. The response is far less sensitive to changes in noise levels when control factor A is set at its second level. The graph shows that there is an interaction taking place between the control factor and the noise factor.

In the case of internal noise, the noise factor is derived from the same parameter as one of the control factors. The inevitable variation of the control factor around its nominal level, due to deterioration or manufacturing variation, is considered a noise source. Figure 9.2 shows a response whose control factor effect shows curvature. A nominal Level 2 for the factor shown results in much less variation in the response, $y$, than a nominal Level 1. This can be seen from the slope of the control factor response at the two nominal levels.

Nonlinearity or curvature of the response with changes in the control factor is an opportunity for improving robustness. Therefore, the type of interaction that can lead to improved robustness is a self-

**Figure 9.1** Control factor–noise factor interaction plot

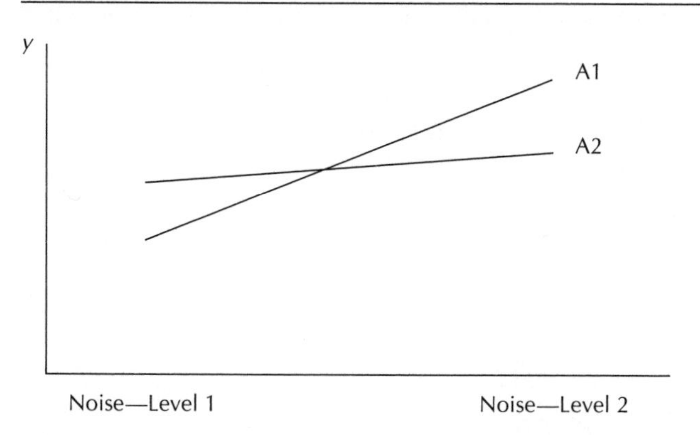


**Figure 9.2** Response curve for a control factor

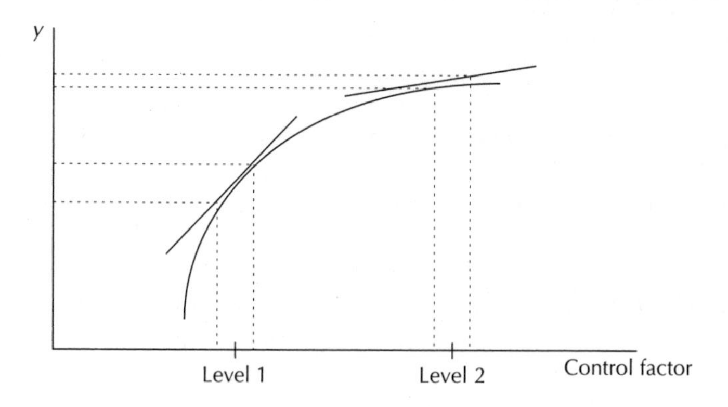


interaction or curvature. The same conclusion can also be obtained from the CF×NF interaction plot that the tangent lines in Figure 9.2 represent. In this case, the noise factor would be a small amount of variation around the nominal for the control factor, indicated by the dotted lines in Figure 9.2.

Engineers seek control factors whose interaction with the significant noise factors is sufficient to improve robustness. These effects should exceed the magnitude of experimental error. It is possible to plan and execute experiments that directly measure and quantify the CF×NF interactions [S2, S3, B6]. An alternative to directly modeling these interactions is the use of the signal-to-noise ratio to quantify the CF×NF interaction by measuring the variation in the quality characteristic when the noise factors are applied. This is usually simpler and easier to understand than the analysis of CF×NF interactions.

## 9.3 Experimental Error and Induced Noise

Experimental error is referred to by several other names. Statisticians often refer to it as the "residual" of the experimental data, meaning that portion of the data variation that is not attributable to the effect of any of the main factors in the experiment. Variation in data is quantified using the variance, which is discussed in Chapter 2. It is given by the mean square deviation from the overall experimental average. Experimental error is the variation in the data that occurs due to any of the following, uncontrolled causes:

- Measurement noise, i.e., variation due to meter error or transducer error
- Experimental noise due to variation in setting the factor levels for a test treatment
- Unaccounted for and uncontrolled factors that can influence the response
- Interactions that are not included in the analysis
- Human error

Any cause of variation that is not related to the factors assigned to the columns in the orthogonal array could be included here. The variation due to experimental error is characterized by the error variance, $S_e^2$.

The error variance can be estimated directly by replication. *Replication* means setting up the system to a specified treatment combination, then altering the factor levels before returning to the same treatment combination for another measurement. Thus, replication includes error contributions from errors in the factor settings, measurement error, and any other factors external to the orthogonal array parameters that may have varied over time. The replicate runs can be made with any particular combination of the controllable factors. The key is to reset the factors each time the replicate experiment is run.

*Repetition* means making repeated measurements on a treatment combination, one after another. This usually only includes measurement error and is not recommended for estimating experimental error. However, repetition is a good way to minimize the effect of measurement error by allowing for some averaging to be done. Repetition means that the factors are not reset before taking the repetitive measurement of the response.

Replication tells how much variability occurs when no factors are intentionally changed to a new level. The error variation can be compared to the variation that occurs when a factor is changed from low to high. If the magnitude of the resulting change is small compared to the standard deviation due to replication, then the factor is not considered *statistically* significant. If, on the other hand, the magnitude of the change is much larger than the standard deviation due to replication, then the factor is considered statistically significant. This can be expressed in a ratio (known as the $F$ ratio), using the square of the variation due to the factor and the error variance:

$$F = MS/S_e^2 \qquad \textbf{(9.1)}$$

where the MS is the mean square of the variation due to the factor. The F ratio, the mean square, and other methods of estimating the error variance are discussed more fully in Chapter 17.

A fundamental assumption of classical DOE is that the same experimental error distribution applies for the entire experimental space. The situation can be characterized by Figure 9.3, which shows the addition of experimental error as independent of the factor levels underlying the functional relationship.

Robustness is achieved by an approach that has several critical differences from the preceding discussion.

1. Noise factors are used to induce a controllable level of system variation that is separate and distinct from the experimental error discussed so far.
2. Control factors that are found to be significant in improving robustness reduce the level of variation due to noise factors. But noise factors are similar in their nature to many of the

**Figure 9.3** Experimental error independent of the factor effects

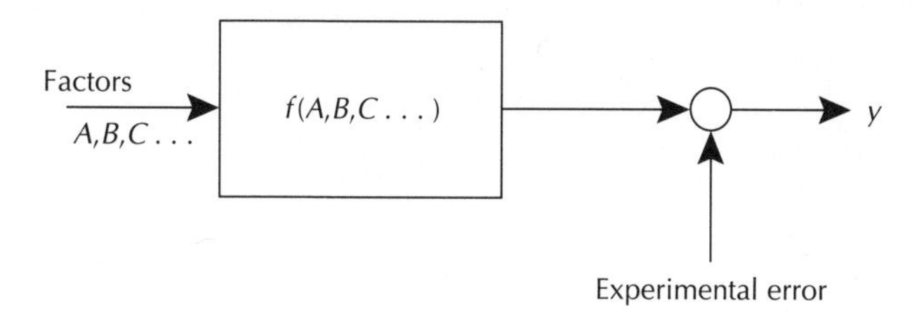


causes of experimental error. Thus, the experimental error is not always independent of the control factor levels (i.e., the control factors interact to some degree with the level of experimental error).

3. Traditionally, in designed experiments, the experimental error is assumed to be larger than many of the factors in the experiment. This assumption, to some extent, is a result of the historical development of DOE in agriculture, where it is impossible to control all the possible sources of error. This assumption is the basis for many analyses used to estimate factor significance such as pooling and half-normal plots. However, the assumption does not necessarily hold true for many laboratory experiments, especially those involving equipment and electronics. It is quite possible to run extensive experiments in which the variation found in replication (due to all sources of experimental noise) is, in fact, entirely due to measurement error. As a result, many of the factor effects exceed the error, particularly if the effect of measurement error is reduced by repetition and averaging. It is a good idea to try and ensure such conditions by using noise factors that induce variation at a level that is significantly greater than all the sources of experimental error. When this is done for each run of a parameter design experiment, it makes the effect of nuisance factors negligible.

One goal of the noise experiment is to know what the level of experimental error is and, if possible, to find a realistic noise treatment that can induce variation at a much greater level. Consider Figure 9.4, which compares field variation to laboratory variation (experimental error).

**Figure 9.4** Comparison of field distribution and lab distribution

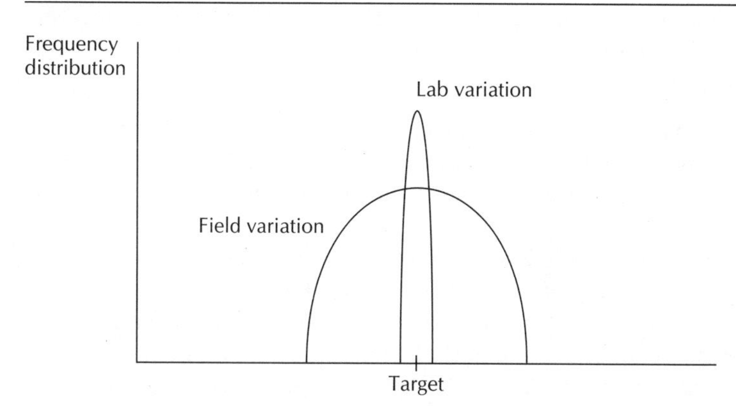

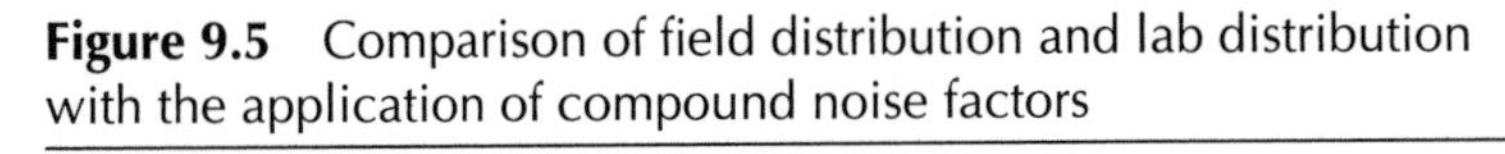
**Figure 9.5** Comparison of field distribution and lab distribution with the application of compound noise factors

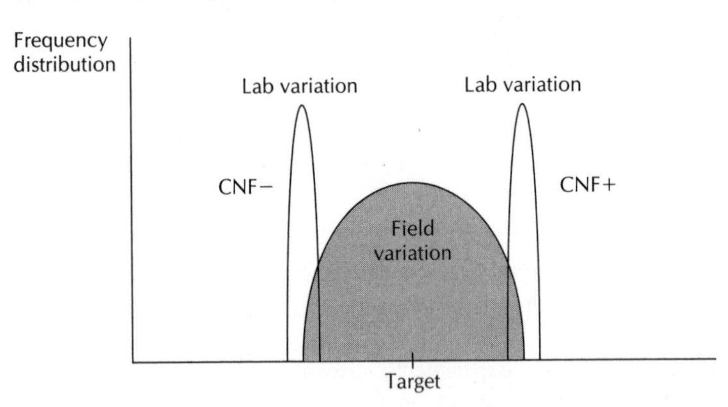


In this typical situation, it is possible to choose noise conditions that cause deviations from the target that are comparable to the range of field variation. Thus, the noise factor effects greatly exceed the experimental error, so the contribution of error to the measured variation is insignificant. The goal of the noise experiment is to find the noise levels that produce the experimental condition pictured in Figure 9.5.

## 9.4 Noise Factors

Noise factors can be grouped into four categories. The first category is *external noise,* which refers to causes of system variability that are external to the system, i.e., completely outside of our control when the product is used as intended. For example:

- Environment, e.g., temperature, relative humidity, and altitude
- Duty cycle or load, i.e., how much and how often the customer uses the product
- Setup conditions, i.e., what optional choices in a complex device are selected by the customer
- Consumables chosen by the customer for input to the system, e.g., different papers used in a copier, different fuels or lubricants or coolants used in an engine
- Foreseeable misuse and abuse

These are all examples of factors that can be controlled and simulated in the lab, but in actual use, cannot be controlled without adversely impacting customer satisfaction or adding cost. (Consider the cost impact of sensitivity to environment for a mainframe computer that has resulted in requirements for climate-controlled site placement.)

The subsystems in a complex system such as office copier machines and automobiles have another form of external noise, referred to as *proximity noise.* In this case, the noises are external to the subsystem, i.e., the unwanted effect of one subsystem on another. For example, the heat emitted from the subsystem in a high-speed copier that fuses toner to paper by the application of heat and pressure can influence the performance of neighboring subsystems. While not external to the system, the proximity effect of other subsystems is important for each subsystem team to consider when developing a list of noises.

The second category is *unit-to-unit variation,* an internal noise often mislabeled as tolerance variation.[1] This is the inevitable variation caused by noises in production. There is always some variation in parts and manufacturing processes that can cause the performance of the final product to vary. Some examples include the following:

- Dimensional variation in mechanical parts, which can stack up in assembly
- Electrical component variation
- Process parameter variation (e.g., temperature, speed, time, pressure) in setting up a manufacturing device such as a mold injector or chemical reaction
- Variation in rheological properties of polymers used in a product
- Variations in product-controlled parameters such as hydraulic pressure in a machine or vacuum level or ignition timing in an engine

These are all parameters that vary from unit to unit, often in a way that can be described probabilistically by some statistical distribution. Generally, the range of variation cannot be reduced without adding cost, in the form of tighter tolerances (more inspection and rejection) or process control (more complex design).

The third category is *deterioration noise,* also an internal noise, which refers to the inevitable degradation of equipment components or materials over time. The following are some examples:

- Corrosion, fatigue, wear, and physical degradation that occurs over time
- Buildup of impurities in recycled materials, such as oil in an engine, or chemicals (including mold injection materials) in a manufacturing process
- Loss of volatile materials over time, such as plasticizers in an automobile dashboard, or low molecular weight fractions in a polymer
- The disordering effects of entropy, such as loss of magnetization in an audio tape or video tape recording

These are all sources of variation that increase over time. We can sometimes influence the rate of degradation, but eventually all systems suffer from these types of losses.

The fourth category is *surrogate noise* [P2], which can represent external or internal noises. Surrogate noise refers to the measurement of variability without necessarily controlling the cause of the variation. This is similar to experimental error, but does not refer to measurement error. Rather, for surrogate noises, the process data are varying, but in a manner that is not directly controlled in the experiment. For example, surrogate noises could include the following:

- Repeat or replicate measurement over time during a process run
- Sampling from different locations in a cell
- Picking off the high/low values from a strip-chart recording
- Measuring image density on a copy at several different locations
- Measuring nip width between two rollers at several locations

In each case, it may be difficult to specify the noise factor, but reducing the variability would help improve quality.

1. Tolerance is not a measure of variation; it is a measure of how much variation is allowed before a part is rejected.

Surrogate noises can be classified into three groups [B7]. Such classification can be useful when trying to be thorough in noise analysis and brainstorming. These are the three groups:

1. Location, e.g., unit to unit, within a unit, operator to operator, machine to machine, plant to plant.
2. Replication, e.g., lot to lot or batch to batch, including hysteresis effects.
3. Time, e.g., hour to hour, shift to shift, day to day, week to week.

These classifications can be used as guidelines to help in thinking about additional sources of variation.

## 9.5 Choosing the Noise Factors

The choice of the noise factor treatment is a major part of the planning process. The design is made robust only against the types of noises that are chosen and noises that are related to those chosen. If there are sources of variability that affect the system in ways that are not part of the experiment, the design may be susceptible, even after optimization, to problems from the unaccounted-for noises. In our experience, this is the most common reason for disappointing results, when they occur.

The first step a multidisciplinary, multifunctional team should take is to identify all the noise factors that can introduce variability in the function of the design. This is best done through brainstorming sessions that thoroughly define and prioritize the noises [P1, Chapter 3]. The goal is to pick a set of noises that are representative of all the sources of variation that can affect the design's performance. The team needs to interview customers, service personnel, suppliers, manufacturing representatives, and assembly workers. A good brainstorming session generates a long list of noises. Remember . . . *There are many sources of variability.* Try to identify the various energy transformations in the system. Make sure you have noises that act as sources of variability for each of them. Consider noise surrogates such as time, location, and replication. Ideally, your brainstorming will result in several dozen or more noise factors.

Use energy considerations to guide your selection. The ideal function of the design should have been identified and discussed. Try to consider any possible energy transformations that may contribute to the end result (a flow chart is useful here). Consider all the factors affecting the energy transformations that occur in the system. Consider any possible sources of variation in each transformation. Any that can cause deviations from the ideal are candidate noise factors. Ask what other effects can be superimposed on the desired transformation, thus introducing variability. Variation may be a result of an inefficient energy transformation, but it could also be a result of an undesired secondary transformation that is superimposed on the primary transformation.

Be sure to consider internal processes as well as customer-observable effects. For example, film handling is accomplished by pushing a film strip with friction rollers. The energy transformation is from the roller (rotational kinetic energy) to the film (linear kinetic energy). Variation in the coefficient of friction or roller hardness or nip pressure can all affect this energy transformation.

Consider the four primary categories of noise. When the initial discussion begins to wane and nobody is producing new candidate noises, step back and look at the list. Classify the noises and see if all the categories are represented. Do this with the group—it usually gets people thinking again and additional noises will emerge. In the ice water example, introduced in Chapter 7, the customer may sip from the top of the drink or, using a straw, from the bottom. This is an external noise that is represented by the location of the thermometer. The amount of liquid in the cup is controllable in principle, but in practice there is some variation in the amount the server actually pours. This is a unit-to-unit noise that is represented by increasing or decreasing the fluid amount by two tablespoons.

Look for surrogate noise factors, ways to find or measure variability other than directly controlling the sources of variability. Surrogate noises can be very useful and are a practical way of capturing variation when the noise factors are hard to control in the lab. There are many examples of surrogate noises in the literature. The surrogate noise should be chosen to allow measured variation with little additional experimental effort. Avoid using surrogate noises such as replication that result in additional experimental runs. This is not efficient. Use noise factors to replace random variation with induced variation. Focus on forcing what can happen to happen.

## 9.6 The Noise Factor Experiment

### 9.6.1 Goals of the Noise Factor Experiment

The decision to study noise factors is nontrivial when one is faced with intense pressure to release a design or when spending limited resources on testing and experimentation to correct a known problem. Often, considerable time has already been expended planning and analyzing the ideal function and choosing your quality characteristics and noise factors. There is an urge to rush into testing possible solutions. Having been there, we know that it is hard to justify an effort that does not appear to be leading directly to a solution.

Consider the consequences of using experience, intuition, and skill to speculate for a quick solution that can be built and tested. This approach is not systematic: It depends on luck to succeed. It does give the appearance of quick response, but even when successful, there is absolutely no way to know if the solution is optimum or merely adequate. Even more critical is the issue of keeping problems from recurring. Rework, in the form of having to solve the same basic problems over again, is a significant cause of schedule delay in product development. In fact, it is not uncommon to observe that many of the problems troubling a product at shipping approval have been known and worked on ("solved") throughout the product development process. The use of inefficient approaches such as build–test–fix to solve problems is a major cause of the inability to eliminate problems completely. Inefficient approaches are ones that require an inordinately large number of tests to improve the reliability of the product when the conclusions prove to be inconsistent or fleeting.

The situation is similar with respect to noise factors. Brainstorming to produce a list of noise factors relies on prior knowledge and lore to predict what the major sources of variability are. While this is a good start, it is not adequate. It is not feasible, because of cost or lack of control, to consider all the noise factors. Carefully designed noise experiments, which are designed to provide the right information in as few tests as possible, are invaluable in preventing oversights and checking assumptions. Only by the application of a consistent experimental process can reoptimization (due to noises that were overlooked) be avoided.

The noise factor experiment has at least four goals:

1. Identify the important few noise factors that cause most of the variability. Minimize effort in the parameter optimization experiment by eliminating unimportant noises.
2. Benchmark the performance of the baseline or starting design. (This can also be applied to concept selection.)

3. Perfect the experimental procedure by trying it out during the noise experiment. This results in a smoother parameter optimization experiment, which is the more critical experiment.
4. Test the magnitude of uncontrollable noise, i.e., measurement error. Find out whether replicates are needed.

The goals of the experiment do not require actually modeling the dependence of the quality characteristics on the noises. Rather, they seek to find the relative significance of the noise factors and the directionality of their effect. The selection of noise factors and levels for the experiment should be guided by the following rule: Be sure *not* to test what you know, but *do* test what you don't know for sure.

### 9.6.2 Refining the Noise Factor List

The brainstorming session can result in a large list of noise factors. It is simply not feasible to run even a fractional factorial (orthogonal array) experiment on a very large number of factors. However, existing knowledge of the system may be sufficient to allow some prioritization and grouping of noise factors. Once an exhaustive list of noises exists, the team should rank-order the list. The initial list will surely contain some redundancy, separately named parameters that, in fact, affect the system in a similar manner.

Factors with known interrelationships can be compounded. The compounding is done by grouping noise factors that are likely to have a similar effect on the system (Figure 9.6). For example, if temperature, material, and pressure can all affect the critical dimension of a manufactured part, then those noises may not need to be tested separately. All that is needed is a *qualitative* knowledge of how they affect the dimension. If high temperature, polymer 1, and high pressure are all known to produce oversized parts, then test those levels together. For undersized parts, test low temperature, polymer 2, and low pressure together. Thus, the following compound noise factor (CNF), with two levels, can be defined:

$$\text{CNF1 (High)} = \text{Temperature (high)} + \text{Polymer (1)} + \text{Pressure (high)}$$

$$\text{CNF2 (Low)} = \text{Temperature (low)} + \text{Polymer (2)} + \text{Pressure (low)}$$

**Figure 9.6** Compounding noises

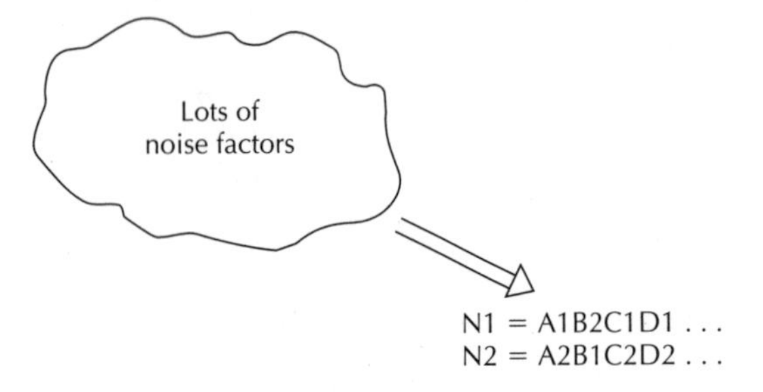

### 9.6.3 Setting up and Running the Noise Experiment

The question of how to examine the effects of noise is best approached by first realizing that the control factors must be set at their nominal values during the entire noise factor experimentation process. The noise factor experiment begins with a list of noise factors obtained by applying the process discussed in the previous section. The results of the brainstorming should be collected to produce a candidate list of up to 11 compound noises. Too many more puts too much experimental effort into studying noises; too few, and there is a risk of neglecting important sources of variability. The 11 noise factors can be studied in just 12 experiments using the Plackett–Burmann array, known as the L12 Orthogonal Array. This array is shown in Table 9.1.

The quantitative effect of any two-way interaction between the factors tested in an L12 is uniformly distributed to all columns as described in Chapter 7. This property makes the L12 array ideal for a study of the main effects of the parameters studied. The uniform distribution of the interactions increases the ability of the main effects to stand out without the need to worry about confounding in particular columns. This acts as a powerful interaction countermeasure because the main effect of any one noise, which was chosen for its presumedly large effect on the system under study, is very likely to overwhelm the interactive effects from the other noises. Thus, the noise experiment can be analyzed simply by finding the average response value corresponding to each column used in the array, as described in Chapter 7.

The noise factors chosen by the team following the process described in the previous section are assigned to the L12, one to each column. Since only 11 columns are available, the maximum number of distinct factors to be tested is 11. If necessary, some of the noise factors can be compounded and put into a single column, as described in Section 9.5. Thus, it is actually possible to have more than 11 factors in the experiment. Compounding results in a loss of information about the individual noise factor effects. For that reason, it is important to choose groupings on the basis of common physical effects so that the factors included in a compound noise are synergistic in their effects. It is, of course, always

**Table 9.1** The L12 array used for a noise experiment

| *Run* | *1* | *2* | *3* | *4* | *5* | *6* | *7* | *8* | *9* | *10* | *11* |
|---|---|---|---|---|---|---|---|---|---|---|---|
| 1 | 1 | 1 | 1 | 1 | 1 | 1 | 1 | 1 | 1 | 1 | 1 |
| 2 | 1 | 1 | 1 | 1 | 1 | 2 | 2 | 2 | 2 | 2 | 2 |
| 3 | 1 | 1 | 2 | 2 | 2 | 1 | 1 | 1 | 2 | 2 | 2 |
| 4 | 1 | 2 | 1 | 2 | 2 | 1 | 2 | 2 | 1 | 1 | 2 |
| 5 | 1 | 2 | 2 | 1 | 2 | 2 | 1 | 2 | 1 | 2 | 1 |
| 6 | 1 | 2 | 2 | 2 | 1 | 2 | 2 | 1 | 2 | 1 | 1 |
| 7 | 2 | 1 | 2 | 2 | 1 | 1 | 2 | 2 | 1 | 2 | 1 |
| 8 | 2 | 1 | 2 | 1 | 2 | 2 | 2 | 1 | 1 | 1 | 2 |
| 9 | 2 | 1 | 1 | 2 | 2 | 2 | 1 | 2 | 2 | 1 | 1 |
| 10 | 2 | 2 | 2 | 1 | 1 | 1 | 1 | 2 | 2 | 1 | 2 |
| 11 | 2 | 2 | 1 | 2 | 1 | 2 | 1 | 1 | 1 | 2 | 2 |
| 12 | 2 | 2 | 1 | 1 | 2 | 1 | 2 | 1 | 2 | 2 | 1 |

possible to leave some columns empty and use fewer than 11 factors, as illustrated in the following examples in Sections 9.8 and 9.10. For each noise factor, a high level and low level must be chosen. Three levels are not necessary because curvature effects are not required. The goal is to find the magnitude of the influence of a noise factor on the quality characteristics and the direction of that influence. For this goal, linear effects are adequate.

The team must also consider how the noise factors are to be controlled in the lab. Noise factors should represent conditions that are not controlled in the actual product usage but which are controllable in lab testing. This is very important for obtaining reproducible results with a minimum of experimental effort. For example, operating environment is a common uncontrolled external noise. In lab tests, however, environment can be controlled in an environmental chamber. In this way, extreme combinations of humidity and temperature can be used, which stress the system and simulate the field variation, but do so in a controlled and reproducible manner. The set points for the control factors should be chosen and left unchanged throughout the experiment. Use the nominal setup that is expected to produce the best possible results. It is possible that some control factor set-point combinations may produce a system whose robustness is inadequate to complete the noise factor test because of system failure. If this happens, you may need to adjust the noise factor levels to reduce the stress. Again, the engineering team's knowledge plays a major role in planning the experiment, and the team will learn a lot from the results of the study.

## 9.7 Analysis of Means for Noise Experiments

The noise factor effects from the L12 experiment are studied using the analysis of means. This reveals which noise factors have a large effect on the value of the quality characteristic, which noise factor levels make that value go high, and which make that value go low. The results from the ANOM can be shown in a factor effect plot like that shown in Figure 9.7.

**Figure 9.7** Compounding of noise factors by factor effects plot

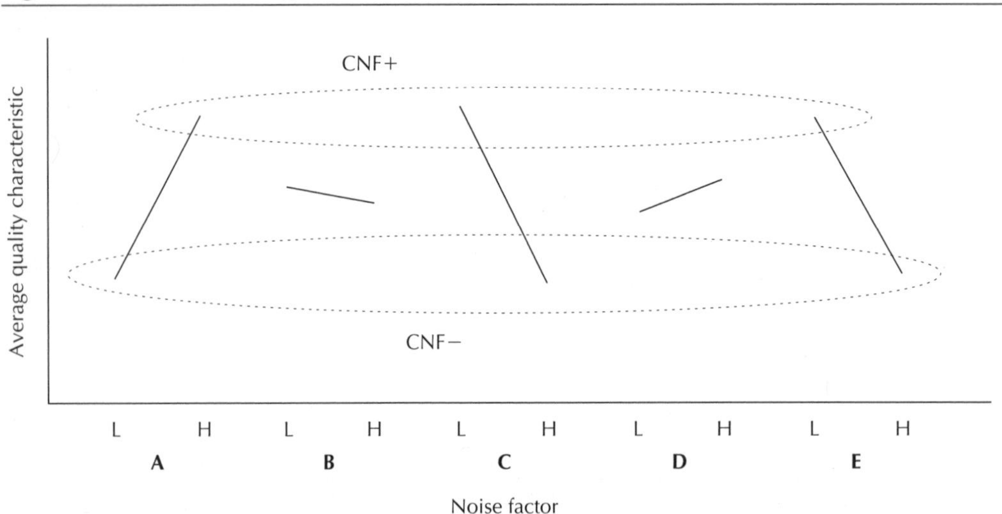

The ANOM results can be interpreted by inspection of the factor effects plot. The low slope of the factor effects for noise factors B and D indicate that they are not very significant. Factors A, C, and E, however, appear to be important. The compounding is determined by the direction of the factor effect slope. Thus, in this case, we see that for $CNF_+$ factor levels A High, C Low, and E Low (written more compactly as $CNF_+ = A2C1E1$) all work together to cause the quality characteristic to deviate in the positive direction. Similarly, for $CNF_-$ factor levels A Low, C High, and E High ($CNF_- = A1C2E2$) all work together.

The additive model can be used to predict the level of deviation expected from the compound noise factors. Thus, for the case illustrated here, the predictive equations would be given by

$$y_+ = \bar{y} + (\bar{y}_{A2} - \bar{y}) + (\bar{y}_{C1} - \bar{y}) + (\bar{y}_{E1} - \bar{y})$$

$$y_- = \bar{y} + (\bar{y}_{A1} - \bar{y}) + (\bar{y}_{C2} - \bar{y}) + (\bar{y}_{E2} - \bar{y}) \quad \textbf{(9.2)}$$

### 9.7.1 Verification

It is important to verify the results of the compounding analysis by testing the two compound noise factors. This is because of the possibility that significant interactions are neglected in the additive model. The L12 array does act as a countermeasure. But in general, direct analysis of a quality characteristic is more prone to interactions than is the analysis of the S/N ratio. Nevertheless, if a large number of noise factors are chosen initially, the compounding is likely to be very effective. Remember, the noise factors are chosen because the engineering team predicts that these factors have a significant effect all by themselves. The likelihood of a few interactive effects overwhelming several main effects compounded together is low.

The verification test is done by running the same nominal system used for the L12 experiment at the compound noise factor combinations. Occasionally the combined effect of all the noises compounded together causes the system to crash. This is some cause for concern, since it is impossible to analyze the main experiment if several runs do not produce data. If this happens, one approach is to eliminate or reduce the level of some of the noises to reduce the overall stress and then test again. Be sure to test the system again at the *original* compound noise level after parameter optimization is complete. Since the compound noises represent possible combinations that your system will encounter in actual use, it is important to know if the optimized system has achieved a satisfactory level of robustness. If the optimized design cannot handle the noise, then there will be a quality issue. Either the design concept must be altered, or the product requirements must be reconsidered.

Usually, the verification test does produce useful data where the quality characteristic level for $CNF_+$ is significantly higher than that for $CNF_-$. That result alone is adequate, but the values should be checked against the predictive equation. The comparison need not be exact, but if the results compare favorably with the prediction, then you can proceed with confidence knowing that interactions between the noise factors are not significant.

## 9.8 Examples

### 9.8.1 The Ice Water Example

The ice water example was introduced in Chapter 7. The parameters studied there are **$H_2O$** amount, **Thermometer** location, **Time** from adding the ice until the temperature is read, and **Cup** type. These are all examples of noise factors; water and cup are internal noises, thermometer and time are external

**Figure 9.8** Noise factor effects plot for ice water example

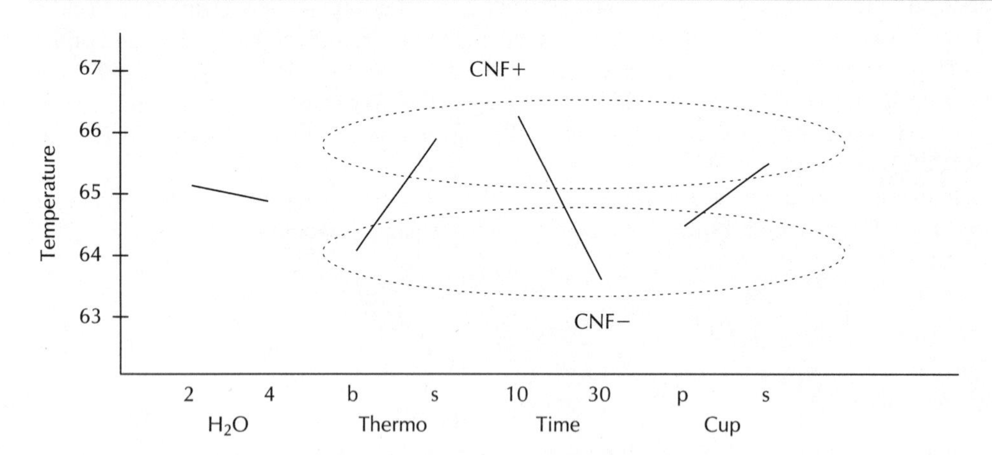


noises. Based on the results of that study, the noises can now be grouped in anticipation of the parameter optimization experiment to follow.

The results of the study are given in Table 7.5 and Figure 7.5. The factor effects plot is shown here in Figure 9.8. From this plot, it is apparent that the $H_2O$ amount is a weak noise. The other noises are compounded as indicated in the figure.

$$CNF_{+} = \text{Thermometer (surface), Time (10s), Cup (Styrofoam)}$$

$$CNF_{-} = \text{Thermometer (bottom), Time (30s), Cup (paper)}$$

The predicted value for the quality characteristic with these compound noise factors is given by the additive model. Recalling that the experimental average $\overline{y} = 65.125$, the additive models for the two noises are

$$y_{+} = 65.125 + (66.00 - 65.125) + (66.38 - 65.125) + (65.62 - 65.125) = 67.75$$

$$y_{-} = 65.125 + (64.25 - 65.125) + (63.88 - 65.125) + (64.62 - 65.125) = 62.50$$

The verification experiment given in Chapter 7, which consisted of the $CNF_{-}$ conditions, gave a result of 63.07 with a standard deviation of 0.4. Thus, the compound noise effects are indeed much stronger ($67.75 - 62.50 = 5.25$ vs. 0.4) than the random error, as expected. These compound noise conditions can be used for the parameter optimization experiment, which is given in Chapter 11.

## 9.8.2 The Catapult Experiment

The catapult experiment is introduced in Chapter 8. Consider the P-diagram, shown in Figure 9.9, for the catapult experiment. Here, the quality characteristic is the distance the projectile, a small ball, travels from the catapult. This is a function of the pull-back angle, which can be used to control the range in a dynamic experiment. The noise factors listed are all examples of sources of variation in the dis-

**Figure 9.9** P-diagram for the catapult experiment showing noise factors

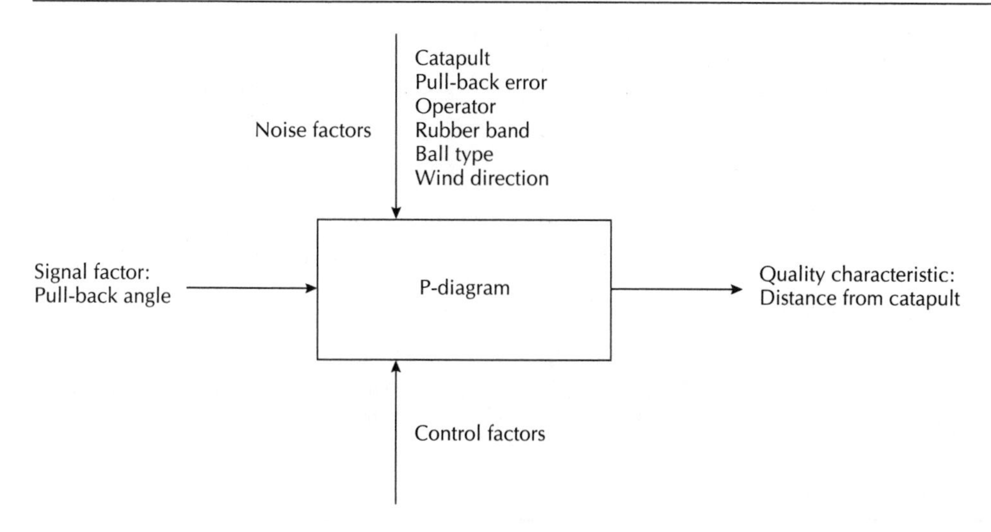


tance the ball travels. There is, of course, the possibility that simply replicating the same throw results in some variation. But the variation due to the influence of the noise factors listed should be much greater than the random variation due to replication. This is tested in the noise experiment.

The noise factors listed are described in Table 9.2. Note that only the in-line distance from the catapult is measured. Pointing errors and cross winds, which would result in lateral deviation from a target, are ignored here to keep things simple. In actuality, lateral deviation could be treated as another quality characteristic (because it represents another ideal function and we want to avoid mixing quality characteristics, which may result in interactions). There would then be additional noise factors included in the study.

**Table 9.2** Noise factor levels for catapult noise experiment

| *Noise factor* | *Level 1* | *Level 2* | *Description* |
|---|---|---|---|
| Catapult | 1 | 2 | Machine–machine variation |
| Pull-back error | −3 degrees | +3 degrees | Deviation from the nominal pull-back angle |
| Operator | Neil | Hannah | Operator variation |
| Rubber band | 1 | 2 | Unit-to-unit variation |
| Ball type | Golf | Foam | Outdoor golf ball and indoor (foam) golf balls, differ in mass and air resistance (compound noise factor) |
| Wind | Against | With | Fan is blowing against/with projectile throw |

**Table 9.3** Noise factor array for catapult experiment

| Run | *1* *Catapult* | *2* *PullBackError* | *3* *Operator* | *4* *RubberBand* | *5* *BallType* | *6* *Wind* |
|---|---|---|---|---|---|---|
| 1 | 1 | −3 deg. | Neil | 1 | Golf | Against |
| 2 | 1 | −3 deg. | Neil | 1 | Golf | With |
| 3 | 1 | −3 deg. | Hannah | 2 | Foam | Against |
| 4 | 1 | +3 deg. | Neil | 2 | Foam | Against |
| 5 | 1 | +3 deg. | Hannah | 1 | Foam | With |
| 6 | 1 | +3 deg. | Hannah | 2 | Golf | With |
| 7 | 2 | −3 deg. | Hannah | 2 | Golf | Against |
| 8 | 2 | −3 deg. | Hannah | 1 | Foam | With |
| 9 | 2 | −3 deg. | Neil | 2 | Foam | With |
| 10 | 2 | +3 deg. | Hannah | 1 | Golf | Against |
| 11 | 2 | +3 deg. | Neil | 2 | Golf | With |
| 12 | 2 | +3 deg. | Neil | 1 | Foam | Against |

These six factors are assigned to the first six columns in the L12 array, resulting in the treatment combinations shown in Table 9.3.

The noise factor test consists of throwing balls using the factor levels indicated for the 12-treatment combinations shown. In a static mode of testing, one pull-back angle is used throughout the test. All the other catapult control factors are fixed as well. The response is the distance the ball travels; the resulting data are analyzed as shown in the next chapter. The test can also be done in the dynamic mode with three throws at different pull-back angles for each row of the L12. In that case, the response analyzed is the slope of the zero-point proportional best fit line.

Generally, no signal-to-noise ratio analysis is done for the noise experiment because there are no other significant noise factors separate from those in the L12 array. The exception to this rule is if surrogate noise factors are included, separate from the L12. An example of this would be replication. If three throws are made for each treatment combination, all at the same pull-back angle, the resulting variation could be analyzed using an S/N ratio. This approach is discussed further in Section 9.10.

## 9.9 Other Approaches to Studying Noise Factors

The orthogonal array is a powerful tool for studying a limited number of noise factors in just a few experiments. But what should you do if you suspect a large number of noise factors are active, but there is insufficient knowledge to prioritize and compound them? A technique for testing the compounding and isolating the key noises from a large set is taught by the quality consultant Dorian Shainin. These techniques, which make use of extensive compounding, are referred to as Component

Search and Variable Search [B7]. Both of these techniques infer the noise effects from observations of variation in products. Rather than controlling noises to induce variation, systems that are exhibiting variation are analyzed to determine what noises are causing the variation. These approaches are not as systematic as an orthogonal array experiment, but they can, nevertheless, be used effectively in many situations.

Component Search is applied to equipment assembly operations. In this method, a group of machines exhibiting good behavior is contrasted with a group exhibiting bad behavior. The group size can be as small as three machines each. By disassembly and reassembly, components in the two groups are swapped and the results are observed. If the relative behavior flip-flops, then the swapped components are judged to be important sources of variation. If the behavior change is small, so that the superior group maintains its superiority, then the swapped components are judged to be insignificant factors. Since many components can be swapped at once (effectively compounding them), a small series of pairwise tests can rapidly converge on the critical few noise factors.

Variable Search is a generalization of Component Search, which is not restricted to assembled equipment already exhibiting variation in performance. In this case, any potential noise factor that can be controlled for the purposes of a pairwise test is a candidate variable. The method is not restricted to components that can be physically swapped. All the potential noise factors are assigned two levels according to the expectation for better or worse system performance. The first pairwise test is a comparison of the system setup with all the best levels for the noise factors vs. the system setup with all the worst levels. Usually this produces a great disparity in performance. Factors, or small groups of factors, are then changed from worst to best and best to worst, while holding all the other factors constant. Again, when the system performance is seen to correlate strongly with the variable level change, those variables are identified as significant noises. Dr. Taguchi briefly mentions these techniques in his book *Systems of Experimental Design* [T6, pp. 985–986].

## 9.10 Case Study: Noise Experiment on a Film Feeding Device

The goal of this study [N3] is to investigate the variability in transport speed of a 35-mm film strip (4 to 36 frames long) by an automated drive mechanism, to be used in conjunction with a photo-finishing processor. A sketch of the design concept is shown in Figure 9.10. The film drive mechanism consists of four pairs of polymeric rollers. Each pair is mounted on a shaft; the shafts are driven in unison by a timing belt and stepper motor. The drive motor is driven by a computer controller. One of the functions includes scanning the film by an imaging sensor. The scan must be made at a well-controlled velocity in order to avoid introducing image defects. The velocity of the film is measured using an encoder, which is also shown in Figure 9.10.

The team holds a brainstorming session to help determine the most likely contributors to drive motion nonuniformity. The manufacturing variability of the drive rollers, skew of the shafts, belt tension, and contamination are selected as the most likely contributors. As shown in Figure 9.10, there are actually four shafts and eight rollers, so there are 14 (including belt tension and contamination) possible noise factors to consider in this study. Instead, however, the variation of each pair of rollers is compounded, e.g., each pair is oversized or undersized together, reducing the total number of factors actually tested to 10. The potential film steering that could occur when one roller was oversized while

**Figure 9.10** Sketch of film drive device (top view)

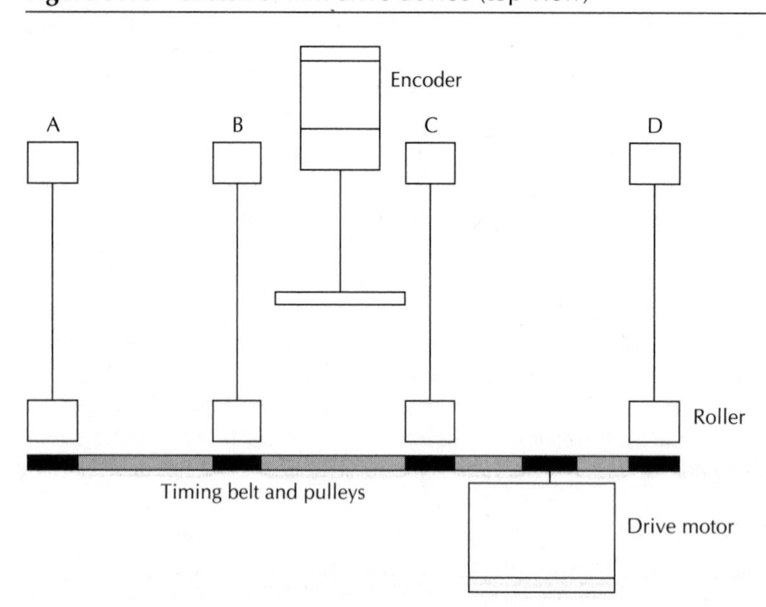


the other roller was undersized is accounted for by the noise factor skew. Table 9.4 shows the tolerance levels that were chosen for the actual tests.

The design chosen for the experiment is the L12 orthogonal array. The array with its factors is shown in Table 9.5. Note that there are 11 columns, but only 10 noise factors after compounding. This is not a problem: The last column in the array is left empty. This design causes the quality characteristic to vary due to the noise factors. The noise factor levels chosen were one-third of their tolerance limits. The tolerance limits were assumed to represent a $3\sigma$ level of variability. Therefore, the resulting variation expected in the quality characteristic is $1\sigma$.

The quality characteristic, film velocity, is measured for approximately 700 ms at the beginning, middle, and end of a 36-frame film strip. The three locations are, in effect, examples of surrogate noise factors that look for variations due to the length of the film strip already processed. The encoder con-

**Table 9.4** Levels of noise factors chosen for film feeder experiment

| *Factor* | *Min.* | *Max.* |
|---|---|---|
| Roller diameter | 0.368″ | 0.372″ |
| Skew | −0.007″ | +0.007″ |
| Belt tension | 22.5 g | 45 g |
| Dirt | Clean | Film dust |

**Table 9.5** Noise array for film feeder experiment

| Run | 1 DiaA | 2 DiaB | 3 DiaC | 4 DiaD | 5 SkewA | 6 SkewB | 7 SkewC | 8 SkewD | 9 Tension | 10 Dirt | 11 |
|---|---|---|---|---|---|---|---|---|---|---|---|
| 1 | .368 | .368 | .368 | .368 | −.007 | −.007 | −.007 | −.007 | 22.5 | Low | 1 |
| 2 | .368 | .368 | .368 | .368 | −.007 | .007 | .007 | .007 | 45 | High | 2 |
| 3 | .368 | .368 | .372 | .372 | .007 | −.007 | −.007 | −.007 | 45 | High | 2 |
| 4 | .368 | .372 | .368 | .372 | .007 | −.007 | .007 | .007 | 22.5 | Low | 2 |
| 5 | .368 | .372 | .372 | .368 | .007 | .007 | −.007 | .007 | 22.5 | High | 1 |
| 6 | .368 | .372 | .372 | .372 | −.007 | .007 | .007 | −.007 | 45 | Low | 1 |
| 7 | .372 | .368 | .372 | .372 | −.007 | −.007 | .007 | .007 | 22.5 | High | 1 |
| 8 | .372 | .368 | .372 | .368 | .007 | .007 | .007 | −.007 | 22.5 | Low | 2 |
| 9 | .372 | .368 | .368 | .372 | .007 | .007 | −.007 | .007 | 45 | Low | 1 |
| 10 | .372 | .372 | .372 | .368 | −.007 | −.007 | −.007 | .007 | 45 | Low | 2 |
| 11 | .372 | .372 | .368 | .372 | −.007 | .007 | −.007 | −.007 | 22.5 | High | 2 |
| 12 | .372 | .372 | .368 | .368 | .007 | −.007 | .007 | −.007 | 45 | High | 1 |

verts the speed into a frequency signal that is converted to a voltage signal by a digital oscilloscope. The resulting voltage signal can be analyzed to give the average velocity and standard deviation during the 700 ms data acquisition period. The results are shown in Table 9.6 for each of the 12 runs.

Here, only the variation of the average speed is considered. The "within a measurement" standard deviation is revisited in Chapter 17. The mean speed for each noise factor combination is given by

**Table 9.6** Data from film feeder noise experiment

| | Start | | Middle | | End | |
|---|---|---|---|---|---|---|
| Experiment | Avg. speed | Std. dev. | Avg. speed | Std. dev. | Avg. speed | Std. dev. |
| 1 | 4.232 | 0.0322 | 4.248 | 0.0515 | 4.177 | 0.0306 |
| 2 | 4.254 | 0.0304 | 4.245 | 0.031 | 4.204 | 0.0296 |
| 3 | 4.258 | 0.0378 | 4.23 | 0.0344 | 4.26 | 0.0353 |
| 4 | 4.238 | 0.0405 | 4.231 | 0.0294 | 4.241 | 0.0334 |
| 5 | 4.214 | 0.0348 | 4.184 | 0.037 | 4.236 | 0.0381 |
| 6 | 4.208 | 0.0323 | 4.274 | 0.0336 | 4.224 | 0.0322 |
| 7 | 4.154 | 0.0317 | 4.167 | 0.0355 | 4.136 | 0.0336 |
| 8 | 4.289 | 0.0338 | 4.225 | 0.0362 | 4.257 | 0.0367 |
| 9 | 4.242 | 0.0389 | 4.25 | 0.0434 | 4.21 | 0.0338 |
| 10 | 4.301 | 0.0322 | 4.312 | 0.0285 | 4.267 | 0.0369 |
| 11 | 4.177 | 0.0462 | 4.185 | 0.0272 | 4.155 | 0.0341 |
| 12 | 4.189 | 0.0274 | 4.251 | 0.0482 | 4.18 | 0.0305 |

**Table 9.7** Average speed values for each of the treatment combinations in the noise array

| Run | *1* *DiaA* | *2* *DiaB* | *3* *DiaC* | *4* *DiaD* | *5* *SkewA* | *6* *SkewB* | *7* *SkewC* | *8* *SkewD* | *9* *Tension* | *10* *Dirt* | *11* | *Speed* |
|---|---|---|---|---|---|---|---|---|---|---|---|---|
| 1 | .368 | .368 | .368 | .368 | −.007 | −.007 | −.007 | −.007 | 22.5 | Low | 1 | 4.22 |
| 2 | .368 | .368 | .368 | .368 | −.007 | .007 | .007 | .007 | 45 | High | 2 | 4.23 |
| 3 | .368 | .368 | .372 | .372 | .007 | −.007 | −.007 | −.007 | 45 | High | 2 | 4.25 |
| 4 | .368 | .372 | .368 | .372 | .007 | −.007 | .007 | .007 | 22.5 | Low | 2 | 4.24 |
| 5 | .368 | .372 | .372 | .368 | .007 | .007 | −.007 | .007 | 22.5 | High | 1 | 4.21 |
| 6 | .368 | .372 | .372 | .372 | −.007 | .007 | .007 | −.007 | 45 | Low | 1 | 4.24 |
| 7 | .372 | .368 | .372 | .372 | −.007 | −.007 | .007 | .007 | 22.5 | High | 1 | 4.15 |
| 8 | .372 | .368 | .372 | .368 | .007 | .007 | .007 | −.007 | 22.5 | Low | 2 | 4.26 |
| 9 | .372 | .368 | .368 | .372 | .007 | .007 | −.007 | .007 | 45 | Low | 1 | 4.23 |
| 10 | .372 | .372 | .372 | .368 | −.007 | −.007 | −.007 | .007 | 45 | Low | 2 | 4.29 |
| 11 | .372 | .372 | .368 | .372 | −.007 | .007 | −.007 | −.007 | 22.5 | High | 2 | 4.17 |
| 12 | .372 | .372 | .368 | .368 | .007 | −.007 | .007 | −.007 | 45 | High | 1 | 4.21 |

the average of the beginning, middle, and end readings. The resulting array and responses are shown in Table 9.7.

Analysis of means (ANOM) is applied, using WinRobust, to the response, speed, in the factor effects plot shown in Figure 9.11. The figure shows that the noise factors that are the largest contributors to the variability are Tension, Dirt, and DiaD (the diameter of the fourth pair of rollers). These three noise factors can be compounded resulting in the following noise factors:

$$\text{CNF}^{+} = \text{DiaD low} + \text{Tension high} + \text{Dirt low}$$

$$\text{CNF}^{-} = \text{DiaD high} + \text{Tension low} + \text{Dirt high}$$

These are the two combinations that design improvements need to be tested against. These are the two conditions that can be used for the outer array in a parameter design experiment aimed at improving the system's performance by optimizing the control factors.

**Figure 9.11** Factor effects plot for the noise array

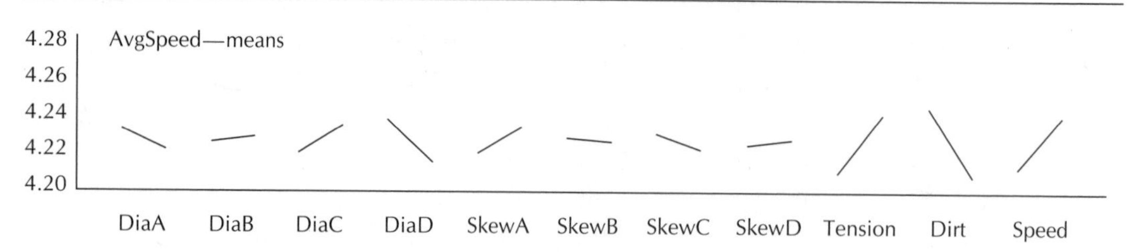

## 9.11 Summary of Chapter 9

The selection, testing, and compounding of noise factors is an essential part of parameter design. It is critical to know the sources of variation to be overcome in order to achieve robustness. The parameter optimization experiment will only make the system robust against those noises that are included in the test. Variations in the energy transformations in a design are the underlying cause of poor quality. Noise factors should be picked for their ability to influence these transformations. It is helpful to categorize the types of noises that can attack a design. Table 9.8 summarizes the four categories described in this chapter.

The noise factor list is challenged in the noise factor test to find the most significant factors. If random variation (experimental error) is large or interactions are large compared to the noise factors, then the noise factors are not well understood. If the noise factors are identified and quantitatively characterized, very efficient parameter optimization experiment tests can be made using compound noises. If the noises are not well understood, then noise factor arrays or surrogate noises (e.g., time, replication, location) will have to be used. These lead to more experimentation and are thus much less efficient. The noise factor experiment can also be used to test and verify the measurement system, which may be important in difficult metrology cases.

The noise factor test should be done using the L12 array, which is included in the enclosed WinRobust Lite software. This array is ideal for a screening-type experiment as well. In the case of the noise factor test, the results of the screening test are applied directly in the parameter optimization experiment. To improve confidence in the compounding, the noise factor experiment should be verified.

Note that these stress tests are substantially different from reliability stress tests in two ways:

1. The level of the noise factors can be kept modest; it is the compounding of several moderate noises that make for an effective stress level. If the distribution of noise factor levels is Gaussian, $\pm 1$ standard deviation is reasonable for the noise factor experiment. For a flat distribution, choose the upper and lower limits.
2. The goal of the stress test is not necessarily to cause the system to fail. Rather, it is to induce a significant level of variation for the parameter optimization experiment to follow.

If the noise factors are severe enough to cause many failures in the parameter optimization experiment, consider reducing the stress level. Do this carefully by having the team perform a reality check on the noises selected. With an on-target approach to engineering, robustness improvements can be achieved while running the system well within failure limits. The key issue here is to be realistic. Such an approach is representative of the actual field performance. Reliance on strictly failure-mode testing is another example of within-specification thinking, which is generally not encouraged here. It is not a bad idea to understand what the failure limits are for the system, but this is done as a separate screening or one-factor experiment.

**Table 9.8** Four categories of noise factors

| | |
|---|---|
| 1. External | Variability from outside of the product or process |
| 2. Unit-to-unit | Variability due to manufacturing processes |
| 3. Deterioration | Variability due to wear, aging, and internal material degradation |
| 4. Surrogate | Variability due to location, position, or time |

**Figure 9.12** The P-diagram ideal function backbone and noise factors

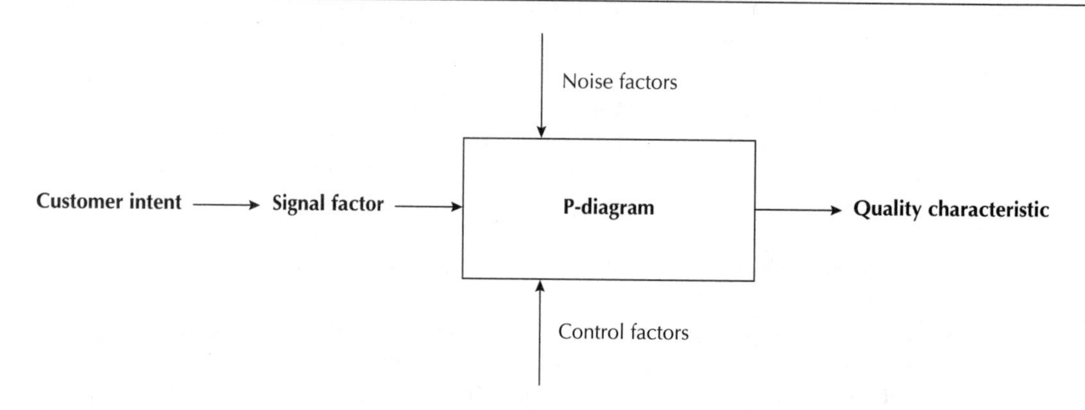


Whatever technique is used to understand the noise factors, it should be clear that these are *preliminary* studies. Where noises that can easily be eliminated are encountered, by all means do so. But many noises are not easily or cheaply eliminated. For these, a quantitative understanding of their effects is needed as a precursor to the main experiment in Robust Design, parameter optimization.

Summarizing Chapter 9, at this point in the process, the most important and difficult parts of the P-diagram have been developed (Figure 9.12) and the lab work can begin.

The process discussed thus far can take considerable time, but it is time well spent. Rushing to the lab frequently results in ineffective experiments that waste time and money. The planning process discussed in this and the previous chapter is an essential part of the robustness process. Do not consider these discussions to be time-wasting meetings. They should be focused, working sessions devoted to planning the critical noise and control-factor experiments to follow. Dr. Taguchi suggests that 80% of the total time spent should be in the up-front planning sessions. This is a remarkable statement, but the authors have found it to be true. By preparing properly, the actual time spent in the lab can be significantly reduced.

One final topic worth discussing is the analysis of noise factor experimental data using analysis of variance (ANOVA). The ANOVA process is useful in decomposing the effect each noise has on the overall response. ANOVA is presented in Chapter 15. The magnitude of the noise factors' variance found using ANOVA can be compared to the experimental error variance. The ANOVA feature in WinRobust provides a ranked order of the strength of each noise factor. The error variance is estimated by replicating several runs from the noise experiment and manually computing the variance of the replicates.

## Exercises for Chapter 9

1. Define what is meant by:
   a. external noise
   b. unit-to-unit noise
   c. deterioration noise
   d. proximity noise
   e. surrogate noise
2. Explain why robust design uses noise to perturb the design during the parameter design experiment.
3. Explain the process details of identifying and deploying noise in a robustness experiment (be specific).

4. What are some consequences of not inducing noise in a designed experiment?
5. Explain the process of compounding noise and why it is done.
6. What is done with the control factor set points during a noise experiment? Explain why this is so.
7. Which orthogonal array is best suited for noise experiments? Why?
8. Why is it recommended to take three to five replicates for one of the noise experiment runs? What can be done with this information? What portion of WinRobust uses this information?
9. When a noise factor experiment is completed, what are three ways noise may then be included into the main parameter design experiment?
10. In a noise factor experiment, we quantify the magnitude and directionality of the noise factors being evaluated. What is meant by the two terms *magnitude* and *directionality?*
11. How is the measurement of experimental error carried out in a noise factor experiment?
12. Explain the difference between experimental error and induced noise.
13. Explain the difference between repetition and replication.
14. Explain the general process of setting up a noise factor experiment for the development of a new lawn mower.
15. Explain the significance of the factor effect for column 11 in the film transport case study (Figure 9.11).

CHAPTER 10

# The Selection of Control Factors

## 10.1 Introduction

Control factors are the design parameters of a concept or technology that need to be optimized. Generally, product performance improves over time as mistakes are eliminated and as components of the system have their performance optimized. The parameter optimization experiment should be used as a major tool to find the optimum set points prior to completing the design. Continual improvement after drawings have been released, tooling has been completed, and manufacturing processes are on-line is good, but difficult and expensive. It is better to find and demonstrate the design optimization prior to reducing the concept to final design. It is in the early stages of product development, when the options are most wide open, that the greatest opportunity for improving robustness presents itself.

The selection of control factors completes the P-diagram and prepares the way for the main robustness experiment. Control factors are the parameters that engineers can specify on production drawings or specification documents. In this step, the multidisciplinary team selects the control factors that govern the performance of the design. Think of control factor levels as the means of minimizing the effect of noise factors on the response. The goal of the main experiment is parameter optimization. The control factors are the parameters that are to be optimized. The levels are explored to find minimal sensitivity to noise. There are two optimization criteria: first, to reduce variability by increasing the S/N ratio; second, to put the quality characteristic on target. The selection of the control factors should be made to maximize the likelihood that the parameter optimization yields a satisfactory result. The control factors and their levels are also chosen to maximize the likelihood of achieving an additive model for the S/N ratio. This makes the optimization as stable as possible over time, when engineering changes and leveraging of the products occur.

Existing systems and leveraged systems usually have few control factors available to change, and the range of their levels is limited by operating constraints. In such cases, pick as many control factors as possible and as wide a set-point range as possible. The guidelines described in this chapter apply best in an unconstrained situation such as "clean sheet" design. Where possible, though, consider the following guidelines for improving tunability and additivity in your results.

## 10.2 Selecting Control Factors to Improve Tunability and Robustness

*Tunability* means that at least one control factor exists that can be used to put the system on target while maintaining optimum stability. It is not necessary or desirable to have many factors that can adjust the mean. Engineering is not a juggling sport, although many try to practice it as one. It is desirable to have many factors contributing to reducing variability. Once robustness is achieved, more than one factor for adjusting the mean on target is redundant and an opportunity for interaction. However, many factors contributing to reducing variability, working together in an additive (noninteractive) manner, greatly improve the overall quality of the system. Think clearly about the purpose of each design element and how it will contribute to improved robustness.

The ability of a control factor to improve robustness is tested in the parameter optimization experiment. This is parameter design's primary purpose. Ideally, the tuning factor affects the mean of the quality characteristic while leaving the S/N ratio relatively unchanged. In the cases of smaller-the-better and larger-the-better, the mean value and variability are automatically co-optimized using the quality loss as the optimization criterion. The tuning considerations discussed here apply to the nominal-the-best (NTB) S/N ratios and dynamic S/N ratios.

The ideal situation is achieved by using the dynamic S/N ratio. In that case, the signal factor, which is chosen based upon the engineering analysis of the ideal function, is the tuning factor. Whenever possible, especially early on in technology development, dynamic analysis should be used, because adjustment of the system over a wide range is imperative for technologies that may be applied to a variety of products. Also, a signal factor in a product may show up as a control knob or feature that gives customers the ability to alter performance to their liking. In some systems, this controllability may be used by a process control system that automatically adjusts the performance onto target. This is often used when the optimized robustness is not good enough to meet the customer requirements without some compensation. In these cases, the use of the dynamic S/N analysis is recommended in order to optimize the performance over the full range of use.

In the evaluation and optimization of existing product concepts and systems, many of the targets are well-defined and the system is capable of running on target, though it may not do so robustly. These situations can be studied using the static nominal-the-best S/N ratios. When doing a static nominal-the-best parameter design experiment, the issue of tuning still exists. Here, however, the signal factor is not chosen a priori. Rather, the results of the experiment are analyzed to find which control factor is best to use for adjustment after variability has been minimized (two-step optimization). For this reason it is critical to choose enough control factors, including some that are potential tuning factors, for the experiment to offer some likelihood of success. The NTB optimization is best done using a fairly large number of parameters. Including potential tuning factors is actually not difficult. Engineers tend to be proficient at putting a system on target. Engineering analysis is required to identify at least a few factors that have the ability to put a system on target without seriously compromising robustness. It is in cases where budget constraints or the desire to study several interactions limits the experiment to only two to four control factors that difficulty is encountered. It is imperative that there be enough control factors to provide a strong likelihood of finding opportunities to improve robustness, as well as to identify a tuning factor. As a guideline, six to eight control factors should be chosen wherever possible.

It is highly desirable to select three levels for each quantitative control factor. This helps span a wide experimental region and provides the ability to study curvature in the response. Using three-level factors allows the analysis to show whether the best level is in fact an optimum level. The criterion for a factor level to be considered an optimum is that the factor average is the maximum for the S/N ratio, suggesting no further optimization is likely.[1] It is very important to select factor levels wide enough that they allow a large effect. Small changes in control factor levels usually result in weak responses; large changes in control factor levels allow the main effects to be significant compared to experimental variation. Choosing wide levels maximizes the probability of including the "sweet spot" in the experimental space. It is always possible to go back and do smaller studies if the optimum lies between two levels that are far apart. Avoid picking levels very conservatively, such as using ±30% around the nominal or expected optimum level for the three levels (i.e., 0.7, 1.0, 1.3). This is not very good, unless there is evidence that suggests a greater spread leads to poorer results. Without such prior evidence, aim for ranges of ±2× (i.e., 0.5, 1.0, 2.0) or more. The team needs to exercise engineering judgment here.

It is not necessary to have uniformly spaced levels. That approach has some benefits for model building (although with a powerful analysis package it is not a requirement). Here, uniformly spaced levels are not required for the analysis of means. Quantitative (continuous) and classification type control factors are all treated the same. The overarching guideline should be to cover as wide a range of experimental space as possible, without running tests that are known to cause the system to fail. Note that deviation from the target, *at this stage,* is acceptable. Failure here refers to treatment combinations (experimental runs) that produce no useful data.

Control-factor levels that are known to be outside of the range of interest should not be included in the experiment. Choose levels from the range you want to *learn* about, and exclude any levels from the range that you *know* will not be useful. When a designed experiment is run, it is very important to let the results speak for themselves. Statisticians tell horror stories of well-meaning engineers and technicians correcting "problems" caused by the treatment combination effects partway through a large designed experiment. Once the experiment starts, it is very important to hold everything constant but the control factors and noise factors in the orthogonal arrays. If those combinations produce bad stuff, then that is useful information. However, it makes no sense to design an experiment that you *already know* produces bad results at some of the levels. If you are ignorant of a control factor combination that produces bad results, then you are discovering something useful—but if you already know that some of the experimental runs produce poor results, even total failure, then spending money to show this can be a waste of effort.

The factors selected should explore new opportunities for improvement, if possible. In the case of brand-new technology, concepts, or designs, this is not a problem. In the case of improving existing designs, considerable testing may have already been done. In any event, the same guideline given for the noise factor study applies here:

- *Don't test what you already know; do test what you don't know for sure.*

## 10.3 Selecting and Grouping Engineering Parameters to Promote Additivity

The control factors should be chosen to help promote additivity. Previously, in the discussion of the S/N ratio in Chapter 5 and the quality characteristics in Chapter 8, a great deal of attention was given to eliminating interactions that are merely artifacts of the mathematics or that are due to the choice of

1. Note that there is always the possibility that the maximum is a local maximum, applicable only within the experimental space studied. This is partly why choosing widely spaced levels is highly desirable.

response. Another source of interactivity is physical interactions, i.e., interactions that are due to one factor affecting another factor's influence on the energy transformations. Physical interactions cannot be eliminated, because they are part of the phenomenon being studied. Their effect on the analysis can, however, be reduced by following the guidelines and techniques discussed in this chapter.

Control factors should be as specific as possible in their effect on the quality characteristic. If there are several control factors that affect the same energy transformation in the system, then it is highly probable that they will have a physical interaction effect on the response. Choose control factors whenever possible so that they affect easily distinguishable physical processes in the system. Group the control factors logically, based on the energy transformations they jointly control. If they are redundant, use one to represent the rest, holding the others constant. Another approach is to compound the control factors. This is done for factors that have similar effects on the energy transformations: Change them together, so that their effects add up (see the discussion of compounding noise factors in Chapter 9).[2]

When choosing control factor levels, use engineering knowledge to improve the efficiency of tests and avoid interactions. Think about the interrelationships (interactions) between the factors due to the known energy transformations in the system. Consider unique factor level combinations that you have reason to believe *cannot* have a beneficial effect on the system being studied. Such a condition is a sure indicator of an interaction. Change the levels, or change the spacing between the levels, as required to maximize the probability that every single run in your experiment will bring you a step closer to your solution. One good technique for minimizing the effects of a known or suspected physical interaction between two control factors is to use *sliding levels* for the control factors.

## 10.4 Sliding Levels for Control Factors

A good example of a known interaction between control factors comes from photography. There are two key energy transformations when photographs are made.

1. Transfer of photons (light energy) from the subject to the film.
2. Chemical reaction of the film due to the photons striking silver halide crystals.

The transfer of photons from the subject to the film can be analyzed using two dynamic relationships. The first, where the photons strike the film, a function of the origin of the photons on the subject. The second, how many photons strike the film, a function of the subject lightness.

The distinguishable control parameters that would control these functions are depth of focus (aberration), which has a large effect on the location of the incident photons, but not on the total number of photons; and the incident energy on the film, which determines the number of photons striking the film, but not the location of the photons. These should be the control factors of interest. However, there are no energy knobs or depth-of-focus knobs on (most) cameras. Manually operated cameras commonly provide two control factors to adjust: aperture and exposure time. A matrix analysis, shown in Table 10.1, can be used between the control factors available and the physical effects of interest to identify potential interactions.

The correlations indicate that aperture and exposure are likely to interact because they both influence the incident energy. Depth of focus (D.O.F.), on the other hand, is controlled solely by aper-

2. The authors acknowledge that this technique produces compound control factors that confound the individual control factor main effects. This is considered poor practice in classical statistical math modeling, but is highly useful in certain cases of engineering optimization. Use of sliding levels is preferred over compounded control factors to avoid the problem of confounding.

**Table 10.1** Control factor matrix for a camera

| | *Physical effects* | |
|---|---|---|
| *Control factors* | *Incident energy* | *Depth of focus* |
| Aperture | X | X |
| Exposure | X | |

ture. Sliding levels can be used to transform the available control factors into the fundamental control factors, as shown in Table 10.2. This table shows the first two columns of an L9 array with three levels of energy and three levels of depth of focus assigned to columns 1 and 2. Column 1′ shows the actual levels of aperture that are used to control the depth of focus. Column 2′ shows the levels of exposure that are required to achieve the energy levels shown in column 2 in a manner that is independent of aperture by *sliding* the exposure level as required.

As another example, consider a chemical reaction where it is well known that the speed of the reaction increases as the temperature is increased (time–temperature effect). In this case it would be inappropriate to independently vary time and temperature in an experiment. Typically, of course, this is exactly what is done in factorial experiments, as shown in Table 10.3.

Here, low time in combination with low temperature produces low yield because the system is underreacted. Also, high time in combination with high temperature produces low yield because the system is overreacted. The interaction plot shown in Figure 10.1 indicates the expected behavior for the control factor levels given in Table 10.3.

**Table 10.2** Orthogonal array showing sliding level for camera example

| | *1* | *2* | *1′* | *2′* |
|---|---|---|---|---|
| *Run* | *D.O.F.* | *Energy* | *Aperture* | *Exposure* |
| 1 | 1 | 1× | f/16 | 1/250 |
| 2 | 1 | 2× | f/16 | 1/125 |
| 3 | 1 | 4× | f/16 | 1/60 |
| 4 | 2 | 1× | f/11 | 1/500 |
| 5 | 2 | 2× | f/11 | 1/250 |
| 6 | 2 | 4× | f/11 | 1/125 |
| 7 | 3 | 1× | f/8 | 1/1000 |
| 8 | 3 | 2× | f/8 | 1/500 |
| 9 | 3 | 4× | f/8 | 1/250 |

**Table 10.3** Experimental design for time–temperature

| *Temp* | *Time1* | *Time2* | *Time3* |
|---|---|---|---|
| Temp1 = 200°C | 10 (hr.) | 17.5 (hr.) | 25 (hr.) |
| Temp2 = 250°C | 10 | 17.5 | 25 |
| Temp3 = 300°C | 10 | 17.5 | 25 |
| Temp4 = 350°C | 10 | 17.5 | 25 |
| Temp5 = 400°C | 10 | 17.5 | 25 |

**Figure 10.1** Anticipated interaction for time–temperature effect

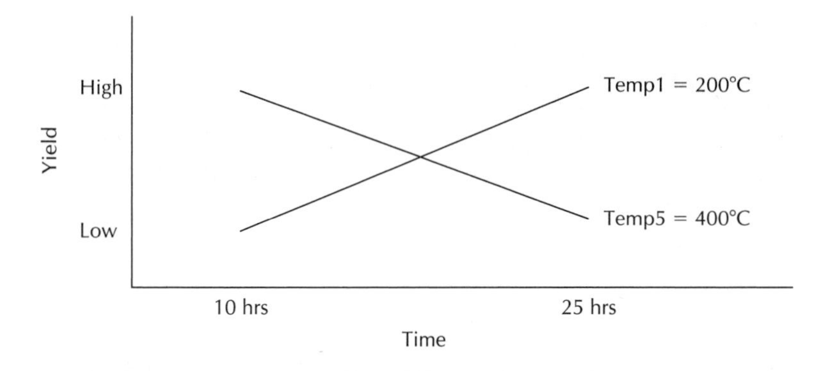


Running the experiment as it is described in Table 10.3, and then "discovering" a strong interaction, is a good example of violating the maxim: *Don't test what you already know; do test what you don't know for sure.* A better design can be achieved if the time factor levels are redefined to take advantage of the known time–temperature effect. For example, the 3×5 full factorial could look like Table 10.4.

Now, in the table, the Time2 level is sliding from 25 hours to 5 hours as the temperature increases. The combinations of time and temperature levels for each row in the new experi-

**Table 10.4** Experimental design for time–temperature including sliding levels

| *Temp* | *Time1* | *Time2* | *Time3* |
|---|---|---|---|
| Temp1 = 200°C | 18.75 (hr.) | 25 (hr.) | 31.25 (hr.) |
| Temp2 = 250°C | 15 | 20 | 25 |
| Temp3 = 300°C | 11.25 | 15 | 18.75 |
| Temp4 = 350°C | 7.5 | 10 | 12.5 |
| Temp5 = 400°C | 3.75 | 5 | 6.25 |

mental design represent equivalent energy transfer levels. Notice that Time1 is everywhere defined as Time1 = 0.75Time2, and Time3 is everywhere defined as Time3 = 1.25Time2. The exact functional relationships in this case are only approximated on the basis of prior knowledge. Nevertheless, the resulting interaction in the analysis should be significantly reduced.

Imagine that the results of the study indicated that the optimum conditions are Temp2 and Time1 (250°C and 15 hours). If the temperature is changed at a later date to 350°C, the optimum time is still Time1. Referring to Table 10.4, Time 1 is the optimum reaction time for this case. Compare that to the case of running the simpler experiment shown in Table 10.3, without including the time–temperature relationship. The optimum time and temperature found using that experiment are Temp2 and Time1 (250°C and 10 hours). But here, if the temperature is changed to 350°C, the best time is Time1 = 10 hours. Clearly this would cause too much reaction, a nonoptimum result. The consequences are confusion, reoptimization (rework), and loss of time and money.

## 10.5 Example: The Catapult

The catapult device, shown in Figure 10.2, has a number of control factors that have a high potential for interaction. Recall that the aim of this device is to throw small objects and hit a target eight feet from the device. The control factors are as follows:

1. Factor A: Release Cup Location—controls the geometric point of projectile release.
2. Factor B: Elastic Band Tie Point—controls the energy input to the system by varying the amount of band deflection.
3. Factor C: Number of Rubber Bands—controls the energy input to the system by varying the spring rate, which controls the amount of energy stored per unit of pull-back deflection.

**Figure 10.2** The catapult

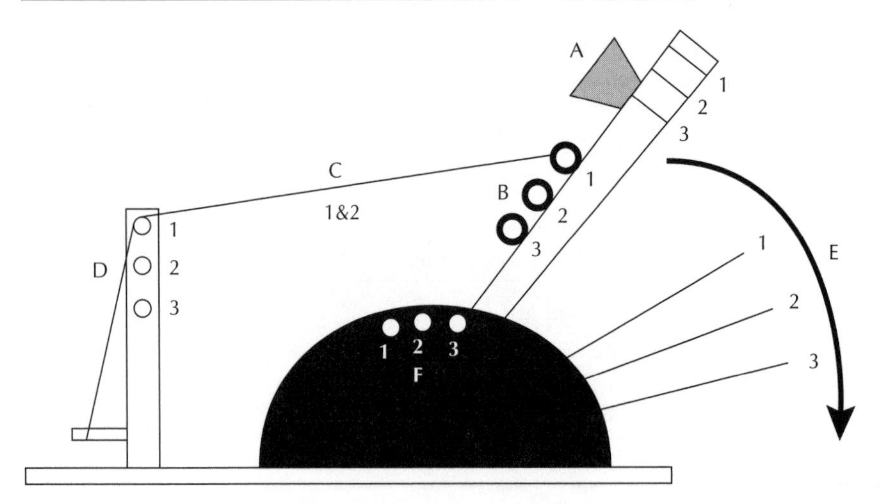

**Table 10.5** Control factor matrix for the catapult

| *Control factors* | *Physical effects* | | |
|---|---|---|---|
| | *Angle* | *Height* | *Speed* |
| *Cup position* | | X | X |
| *Anchor* | | | X |
| *Rubber bands* | | | X |
| *Tensioner* | | | *X* |
| *Stop location* | X | X | X |
| *Pull-back angle* | | | X |

4. Factor D: Elastic Band Stretch Point—controls the energy input to the system by varying the amount of band deflection.
5. Factor E: Pull-back Angle—controls the energy input to the system by varying the amount of band deflection.
6. Factor F: Arm Stop Point—controls geometry of motion by constraining the arm motion and the geometric point of release.

The catapult throws a small ball that then travels through the air with little resistance. Thus, the simple physical equations of projectile motion apply. Their solution for the range, or distance the ball travels before striking the ground, depends upon three initial conditions: the initial speed, the initial angle of flight, and the height of release [B13]. With that understanding of the physics, Table 10.5 is a matrix analysis of the control factors and their correlation with the fundamental parameters of free flight. There is clearly a high potential for interactions, because all of the available control factors influence speed, and a couple of them influence height. However, this analysis can be used to uncouple the interactions by redefining the control factors.

First, since the only factor that controls the angle of release is the stop location, that is a critical control factor assignment. The stop location's influence on height and speed must be minimized. Stop location influences the speed only if the stop location and the pull-back angle are treated independently. In fact, the acceleration and speed of the ball depend upon how far the catapult arm is pulled back *relative* to the stop location. If the pull-back angle is defined with reference to the stop angle, which is determined by the stop location, then, to a first order, the speed is independent of the stop location. To reduce the interactivity, the zero for the pull-back angle should move along with the stop location. In other words, the pull-back angle is now a sliding-level factor referenced to the stop location. It is easier, however, to simply redefine the pull-back angle factor as the *relative pull-back angle.*

The other issue for stop location is the height of release. Since the ball is released from the cup at the end of the catapult arm when the arm strikes the stop, the height of release depends on the stop location as well as on the cup position. Two strategies can be applied here. One is to compensate for the height differences by adjusting the height of the whole catapult to maintain a constant cup elevation as the stop location changes. The other, more interesting approach is to recognize that the variation of height with different stop locations and cup positions is small. If we define a new factor, *catapult*

**Table 10.6** New control factor matrix for the catapult

| | *Angle* | *Height* | *Speed* |
|---|---|---|---|
| *Cup position* | | Weak | X |
| *Anchor* | | | X |
| *Rubber bands* | | | X |
| *Tensioner* | | | X |
| *Stop location* | X | Weak | |
| *Relative pull-back angle* | | | X |
| *Catapult height* | | X | |

*height*, whose levels can be varied widely (e.g., first on the floor, then on a chair, lastly on a table), the stop location and cup position effects become negligible perturbations, thus becoming part of the experimental error.

With these new factors defined, the control factor matrix chart is shown in Table 10.6. There is still a group of factors that controls the same basic energy transformation: potential energy (of the rubber band) $\Rightarrow$ kinetic energy (of the projectile). Here is a good opportunity for compounding control factors. The anchor and tensioner locations determine the length (tension) of the rubber band before pull-back. The number of rubber bands determines, in effect, the net spring constant for the bands. By compounding these factors, a new factor, spring force (SF), can be defined with levels ranging from low to high:

$$\text{SF(low)} = \text{Tensioner(low)} + \text{Anchor(low)} + 1 \text{ rubber band}$$

$$\text{SF(medium)} = \text{Tensioner(med.)} + \text{Anchor(med.)} + 2 \text{ rubber bands}$$

$$\text{SF(high)} = \text{Tensioner(high)} + \text{Anchor(high)} + 3 \text{ rubber bands}$$

This factor can certainly overwhelm the effect of the cup position. The relative pull-back angle must be kept as a control factor because it is the critical tuning factor, useful for putting the balls on target. The pull-back angle is chosen because it is the only factor that can be adjusted continuously. In a dynamic study, it is therefore the obvious candidate for the signal factor. The final control factor matrix (Table 10.7) provides a much higher likelihood that an additive model will be achieved.

**Table 10.7** The final control factor matrix for the catapult

| | *Angle* | *Height* | *Speed* |
|---|---|---|---|
| *Cup position* | | Weak | Weak |
| *Spring force* | | | X |
| *Stop location* | X | Weak | |
| *Relative pull-back angle* | | | X |
| *Catapult height* | | X | |

## 10.6 Example: The Paper Gyrocopter

The gyrocopter, shown in Figure 10.3, also has a number of control factors that have a high potential for interaction. Here, the aim is to produce a paper device that has a low terminal velocity, or high air resistance, so that it floats slowly to the ground. Other requirements are that it be easily built, that it spin in a "pleasing way" as it falls, and that it have good stability in flight. This last requirement means that it is self-righting or that it orients itself properly for stable flight.

For the gyrocopter case, there is nothing simple about the physical analysis. The optimum set points must be found empirically. Nevertheless, the requirements, along with a modicum of knowledge about flight, suggest several fundamental properties that need to be controlled independently. The terminal velocity depends upon the weight of the gyrocopter and the air resistance presented by the wings. The air resistance is assumed to be proportional to the surface area of the wings. The spinning motion is due to the torque provided by the wings. The moment of inertia and the air resistance to spinning determine the angular velocity. Stability depends on the location of the center of gravity. If it is below the point of attachment between the body and wings, then the gyrocopter will be stable in flight. The control factor matrix in Table 10.8 is used as a starting point for the analysis.

It is clear that there is tremendous complexity and potential for interactions. This analysis and the relative lack of analytical understanding suggest that achieving a true additive model will be difficult. Nevertheless, an effort to reduce the level of interaction can greatly simplify the results. It is not necessary to get a perfect additive result in order for the main effects of the control factors to dominate the interaction effects. By a process similar to that used for the catapult, the matrix correlation can be reduced using sliding levels and redefining the factors. (This type of "difficult" problem is representative of complex product design problems. Most engineering students find this sobering by comparison with the "frictionless/massless" tutorial exercises they usually work with.)

The only factor that controls the torque is the wing width. That is the critical control factor assignment. The resistance to falling is proportional to the *wing area.* Thus, wing length must be made a sliding level factor (dependent factor) to allow independent control of the area. Paper weight is

**Figure 10.3** The paper gyrocopter

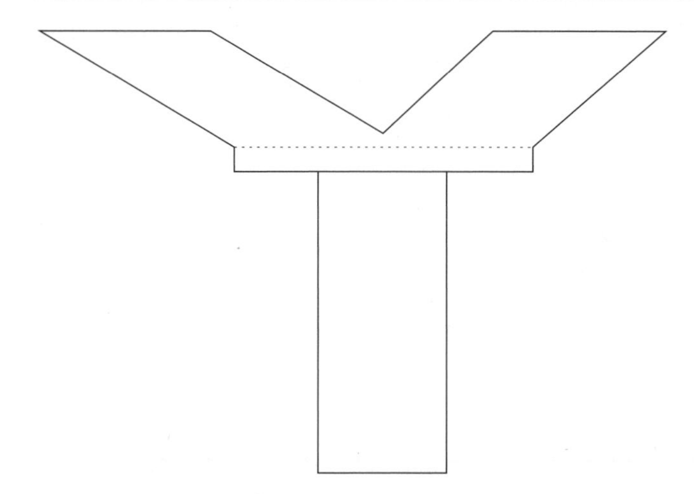

**Table 10.8** Control factor matrix for the gyrocopter

| | *Physical effects* | | | | | |
|---|---|---|---|---|---|---|
| *Control factors* | *Weight* | *Resistance to fall* | *Torque* | *Moment of inertia* | *Resistance to spin* | *Center of gravity* |
| *Paper weight* | X | X | | X | | |
| *Wing length* | X | X | | X | | X |
| *Wing width* | X | X | X | X | | X |
| *Body length* | X | | | | | X |
| *Body width* | X | | | X | X | X |
| *Overall size* | X | | | | X | |
| *Body fold* | | | | X | X | |

important because the stiffer the paper, the less the wings can be deflected by the relative motion of the air and rotation of the gyrocopter. However, a simple addition to the design, *wing gussets,* can stiffen the wing folds independent of paper weight.

Similarly, the location of the center of gravity (COG) is determined by the amount of mass in the wings, the amount of mass in the body, and where these masses are located geometrically. Initially assume that there are no folds or concentrated mass (staples, paper clips, glue) added. A further constraint is that the body width is equal to 2× the wing width. In that case, the COG is determined only by wing length and body length, and the wing length × body length interactive effect on COG can be reduced by choosing the body length as a function of wing length. As a result of these changes, Table 10.9 is obtained.

**Table 10.9** New control factor matrix for the gyrocopter

| | *Weight* | *Resistance to fall* | *Torque* | *Moment of inertia* | *Resistance to spin* | *Center of gravity* |
|---|---|---|---|---|---|---|
| *Paper weight* | X | | | Weak | | |
| *Wing area (WL= WA/WW)* | Weak | X | | X | | |
| *Wing width* | Weak | | X | Weak | | |
| *Body length (BL ∝ WL)* | Weak | | | | | X |
| *Body width* | Weak | | | Weak | Weak | |
| *Overall size* | X | | | | Weak | |
| *Body fold* | | | | Weak | X | |
| *Wing gusset* | | X | | | | |

The new control factor matrix has also been modified to include some prediction of the significance of the remaining correlations. There are still some potential interactions. Nevertheless, this approach provides a much higher likelihood that an additive model will be achieved.

## 10.7 Summary: The P-Diagram

At this point in the Parameter Design process, the P-diagram is complete, as shown in Figure 10.4, and the main experiment can begin.

The work leading up to this point can take considerable time, but it is time well spent. Determining the control factors to be used to try and make the ideal function robust may be the easiest of the steps discussed, but it is also the most satisfying, because it is here that the solution set resides for improving the product or process quality. The effort spent trying to minimize interactions significantly improves one's understanding of these subtle effects. Minimizing the interactions so that the main effects of the control factors stand out offers the most enduring solution. Depending upon interactions between the control factors has several negative consequences. First, using an interaction to improve robustness means that *two* parameters must be specified to get *one* effect. Second, a system that has significant interactive effects is more likely to have uncontrolled interactions with other factors, leading to variation in performance. Third, it takes significantly more lab work to study interactions than to study just main effects. The next few chapters show how to perform and analyze experiments, with and without studying interactions. You will see that the planning effort discussed so far in Part 2 of this book pays rich dividends in reducing the effort required in the remainder of this section.

The time spent in engineering to promote additivity has many other beneficial effects on a design besides simplifying experimental efforts. Additivity of design parameters also provides the following:

- Fewer components (redundant factors are eliminated)
- Higher reliability due to fewer components
- Straightforward dimensioning and tolerancing (GD&T is easier to do and understand)
- Design effort and costs are smaller
- Service is simpler and quicker
- Manufacture and assembly is simplified

**Figure 10.4** The complete P-diagram

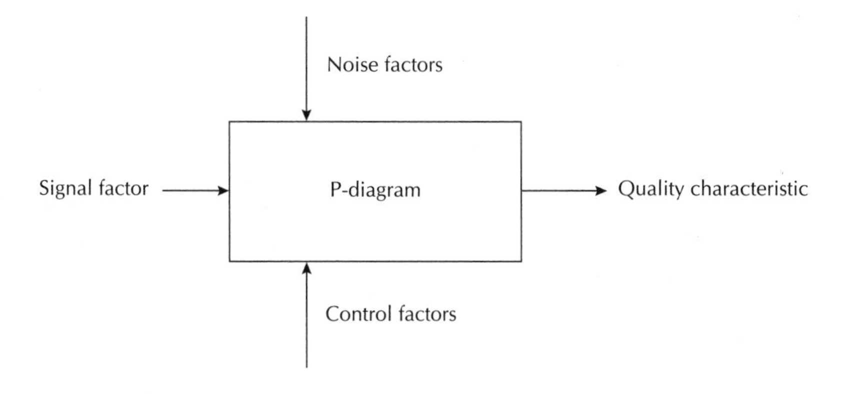

## Exercises for Chapter 10

1. Sometimes engineers who have just learned the Parameter Design process have difficulty deciding which factors should be used as control factors and which should be used as noise factors. What rules can be followed to clarify the differences between control and noise factors?
2. Why is it advisable to select three levels for control factor evaluations?
3. Why is it helpful to select fairly wide differences between the control-factor level set points? What can happen when the control-factor set points are set too close together?
4. What can happen when the control-factor set points are set too far apart?
5. What is meant by the term *control factor additivity grouping?*
6. What is a sliding level for a control factor?
7. Why don't we use sliding levels in noise factor arrays?
8. Discuss the strategy behind studying a broad number of control factors in an experiment, as opposed to studying just two or three control factors (presuming the design has some degree of complexity to it).

CHAPTER 11

# The Parameter Optimization Experiment

## 11.1 Introduction

The control factor optimization experiment has several important goals. The first is to identify the important control factors that can reduce variability caused by the noise factors. The second goal is to identify the control factors that can be used to adjust the mean onto the target (in the nominal-the-better S/N case) or that can be used to adjust the slope onto the target (in the dynamic S/N case). The third goal is to identify which control factors have little influence on the main function so that tolerances on those factors can be relaxed. The parameter optimization experiment should be entirely focused on optimizing the control factors. This is done by identifying the optimal levels at which the control factors should be set. Other experimental goals, such as studying the noise factors or developing a better physical understanding or empirical modeling, should precede this experiment. It is critical here to focus on an engineering solution, i.e., improving the performance in a way that minimizes the cost.

## 11.2 Dr. Taguchi's Parameter Design Approach

Dr. Taguchi's approach to the parameter optimization experiment is featured in this book. There are numerous other approaches [B9, B3, B4, S1]. This experiment attempts to find the optimal levels for the controllable or design parameters. There is no attempt to model the results beyond a simple analysis of means, which gives quantitative information on the control factor effects. In order to avoid the necessity of having to model a large number of CF×NF (Control Factor × Noise Factor) interactions directly, the signal-to-noise ratio is used to summarize the noise factor effects. The S/N ratio also helps reduce unnecessary complications due to interactions between the control factors. It also helps to achieve arithmetic additivity of the control factors while preserving information about the CF×NF interactions. A further simplification that it provides is independence of adjustment. This is necessary for the two-step optimization process, where all of the attention is focused on reducing variability first and reserving one factor for adjusting to target at the end.

The key to running small, saturated control factor experiments (where each orthogonal array column is used, thus keeping the total number of experiments manageable) is to minimize the control factor to control factor (CF×CF) interactions through engineering analysis to promote additivity. This is done in order to help the control factor main effects dominate the results. Dr. Taguchi suggests treating CF×CF interactions as equivalent to experimental error. A traditional approach is to treat CF×CF interactions as model terms to be included in a scientific study. In fact, some CF×CF interactions can be useful, e.g., a synergistic interaction between fuel additives that give better gas mileage than either ingredient by itself (called superadditivity).

Our approach is to treat a CF×CF interaction as one more potential source of robustness, no more or less important than any control factor by itself. The issue, however, is experimental efficiency. As discussed in Chapter 7, an orthogonal array with $n$ runs has associated with it $n - 1$ degrees of freedom (DOF). This means an experimental analysis can be used to quantify, or study, $n - 1$ effects. If all of the DOF (columns) are assigned to control factors, then there is nothing left over for interactions. In such a case, any interactions that do exist would act as nuisance factors, thus, as Dr. Taguchi says, contributing to experimental error. The control factor main effects have to be significant relative to CF×CF interactions as well as to measurement error.

All the control factor interactions[1] can be studied without confounding them with main effects or with each other. This corresponds to a Resolution V design (defined in Chapter 16) in the classical DOE terminology. The DOF required for a Resolution V design can be found by summing the number of main effects and 2-way interactions for the factors being studied. The minimum size of the experiment that is required is one more than the number of DOF.

The number of main effects is given by the number of parameters being studied. The DOF required for each effect is given by $m - 1$, where $m$ is the number of levels for the factor. Thus, one DOF is required for the main effect of a two-level factor, and two DOF are required for the main effect of a three-level factor. The number of two-way interactions is given by the following equation:

$$\text{\# of two-way interactions} = \frac{p!}{(p-2)!2} \qquad \textbf{(11.1)}$$

where $p$ is the number of factors whose interactions are being considered and $p!$ is the factorial of $p$: $p! = (p)(p - 1)(p - 2) \ldots (3)(2)(1)$. The results for several values of $p$ are given in Table 11.1. The DOF required for each two-way interaction of factors with $m$ levels is given by the product $(m - 1)(m - 1)$. Thus, one DOF is required for the interaction of two two-level factors, and four DOF are required for the interaction of two three-level factors. The number of DOF required to study a large number of factors and their interactions quickly becomes so large as to make an experiment impractical, as Table 11.1 shows.

What is happening here is that control factor interactions are replacing additional control factors in the column assignments. From a pragmatic engineering point of view, the question to be asked is: Would you rather spend your precious degrees of freedom (experimental dollars) on CF×CF interactions, or on opportunities available from additional control factors? We do *not* believe there is a single simple answer to this question. Your engineering team must consider the trade-off between additional

1. Chapter 16 is dedicated to the topic of properly evaluating CF×CF interactions. When to study interactions is an engineering decision and is a secondary issue in the robustness process. The primary issue is to promote additivity through appropriate selection and experimental organization of control factors, quality characteristics, and S/N ratios.

**Table 11.1** Table showing the DOF required to study main effects plus interactions

| *# of CF* | *# of 2-way interactions* | *# DOF for 2-level factors* | *# DOF for 3-level factors* |
|---|---|---|---|
| 2 | 1 | 3 | 8 |
| 3 | 3 | 6 | 18 |
| 4 | 6 | 10 | 32 |
| 5 | 10 | 15 | 50 |
| 6 | 15 | 21 | 72 |
| 7 | 21 | 28 | 98 |
| 8 | 28 | 36 | 128 |

runs and less confounding for testing more control factors. Which is more likely to give you useful information (e.g., solve your robustness problem): additional control factors, or few control factors plus their interactions?

The approach that we favor is to use a fully saturated orthogonal array with weakly interacting factors, unless there is a compelling reason to try and model a particular interaction. The model of the factor effects from a fully saturated orthogonal array neglects interactions. Such a model is referred to as an additive model, and interactions are treated as part of the experimental error. The verification experiment checks for interactive effects and other sources of experimental error. Having the main effects of the product's control factors be stronger than the interactions and other error sources is itself a form of robustness. The system is then immune to sudden disturbances in its performance due to one control factor wandering from its optimum value. Such "wandering" often occurs by design; problem-solving activities may result in factors being changed to improve one quality characteristic, to the detriment of other quality characteristics. Thus, control factor interactions lead to performance variation unless they are well understood and vigilantly maintained by all who have access to changing any of the interactive control factors. This includes production, maintenance, and field service technicians, many of whom are distant from the technology centers where the interactions were studied. Having additive control factors eliminates the need to reoptimize control factors because of poorly understood interactions. Only well-understood interactions that are intentionally exploited for their useful effect should be included in the design. In such cases, there is little likelihood of a change being made that would adversely affect an interaction that is necessary for the functionality of the device.

## 11.3 Layout of the Static Experiment

The first step in laying out the parameter optimization experiment is to select approximately six to eight control factors, preferably with three levels each, and assign them to the *inner array.* The best array for the control factors is the L18, because its property of distributing interactions to all the columns treats the interactions as equivalent to noise. The control factor array is referred to as the inner array because control factors, by definition, are factors that are internal to our systems. That is, they are design parameters. However, not all control factors must actually be inside the machine or process. The real test is whether they are controlled. Do not mix control factors and noise factors in the inner array, unless you are using the approach of modeling CF×NF interactions directly. That process will not be discussed here, but, with the help of a statistician or properly featured software package to assist in the analysis, the mixed array procedure can be effective [B4, S2, S3].

The parameter optimization experimental procedure preferred here is the crossed array format. Using a crossed inner and outer (noise) array allows each treatment combination of the inner array to be run at two or more treatment combinations of the noises. The outer array can be a set of compound noises, or it can be an orthogonal array of noise factors. The inner array/outer array format is shown in Figure 11.1, using an L18 for the inner array and an L4 array for the noise factors.

There are several reasons for preferring this experimental design. First, it takes advantage of the engineering distinction between the control and noise factors that is emphasized by the P-diagram. If the goal of the experiment is to study how a number of factors affect the response, without regard for the engineering issues that differentiate between control factors and noise factors, then a mixed array of control and noise factors could, in principle, be more efficient. Here, however, the robustness effect is made clearer by treating the two sets of parameters differently. There is no need to interpret CF×NF interactions to find optimum set points. The control factors are analyzed directly for their effect on

**Figure 11.1** Crossed array layout for the parameter optimization experiment

| | | | | | | | | | | *Noise factor array* | | | | | |
|---|---|---|---|---|---|---|---|---|---|---|---|---|---|---|---|
| | | | | | | | | | G | L | H | H | L | | |
| | | | | | | | | | F | L | H | L | H | | |
| | *Control factor array* | | | | | | | | E | L | L | H | H | | |
| *Run* | 1<br>*A* | 2<br>*B* | 3<br>*C* | 4<br>*D* | 5<br>*E* | 6<br>*F* | 7<br>*G* | 8<br>*H* | | 1 | 2 | 3 | 4 | *S/N* | *Mean* |
| 1 | 1 | 1 | 1 | 1 | 1 | 1 | 1 | 1 | | | | | | S/N1 | $\overline{y1}$ |
| 2 | 1 | 1 | 2 | 2 | 2 | 2 | 2 | 2 | | | | | | ... | ... |
| 3 | 1 | 1 | 3 | 3 | 3 | 3 | 3 | 3 | | | | | | ... | ... |
| 4 | 1 | 2 | 1 | 1 | 2 | 2 | 3 | 3 | | | | | | ... | ... |
| 5 | 1 | 2 | 2 | 2 | 3 | 3 | 1 | 1 | | | | | | ... | ... |
| 6 | 1 | 2 | 3 | 3 | 1 | 1 | 2 | 2 | | | | | | ... | ... |
| 7 | 1 | 3 | 1 | 2 | 1 | 3 | 2 | 3 | | | | | | ... | ... |
| 8 | 1 | 3 | 2 | 3 | 2 | 1 | 3 | 1 | | | | | | ... | ... |
| 9 | 1 | 3 | 3 | 1 | 3 | 2 | 1 | 2 | | | | | | ... | ... |
| 10 | 2 | 1 | 1 | 3 | 3 | 2 | 2 | 1 | | | | | | ... | ... |
| 11 | 2 | 1 | 2 | 1 | 1 | 3 | 3 | 2 | | | | | | ... | ... |
| 12 | 2 | 1 | 3 | 2 | 2 | 1 | 1 | 3 | | | | | | ... | ... |
| 13 | 2 | 2 | 1 | 2 | 3 | 1 | 3 | 2 | | | | | | ... | ... |
| 14 | 2 | 2 | 2 | 3 | 1 | 2 | 1 | 3 | | | | | | ... | ... |
| 15 | 2 | 2 | 3 | 1 | 2 | 3 | 2 | 1 | | | | | | ... | ... |
| 16 | 2 | 3 | 1 | 3 | 2 | 3 | 1 | 2 | | | | | | ... | ... |
| 17 | 2 | 3 | 2 | 1 | 3 | 1 | 2 | 3 | | | | | | ... | ... |
| 18 | 2 | 3 | 3 | 2 | 1 | 2 | 3 | 1 | | | | | | S/N18 | $\overline{y18}$ |

robustness. This is significant when it comes to communicating the results of the parameter optimization experiment to team members or managers who may not be familiar with the subtleties of analyzing designed experiments.

Another reason the crossed array approach works is that very often the noise factors are inexpensive and easy to control and vary in the experiment. This is true because many of the noise factors are external, i.e., they are ways of operating the system that are neither extraordinary nor difficult. In contrast, the control factors are typically design parameters. Different prototypes may have to be built or expensive robustness fixtures used that have plenty of adjustability available in order to set up the control factor treatment combinations. Thus, the additional runs represented by the outer array usually cost much less. Often, an experiment four times as large as the inner array alone only costs 25% more than if there were no noise factor array.

The last reason that crossed arrays are so effective has to do with the experimental design itself. While tuning factors depend on the control factor's effect on the quality characteristic, S/N improvements depend on the CF×NF interactions. Furthermore, as shown in Figure 9.1, even a small CF×NF interaction can be significant in improving robustness. The crossed array design allows the computation of every two-way CF×NF interaction. Smaller mixed arrays, while less expensive, could overlook some useful interactions, thus losing effectiveness. The S/N ratio bypasses the necessity for modeling the CF×NF interactions. The crossed array design makes it possible to use the S/N effects, thus simplifying the analysis and data presentation. Actually running an experiment to plot and study CF×NF interactions is useful for developing a better understanding of the system under study.

## 11.4 Layout of the Dynamic Experiment

The inner array approach for the control factors is not affected by whether the experiment is a static or dynamic type. The layout of the dynamic experiment is shown in Figure 11.2. Usually, the choice of control factors in a dynamic experiment does not need to include tuning factors for the mean quality characteristic level. This is because the signal factor, $M$, included in the outer array is used to control the mean level. In some cases, there is a need to tune the slope of the dynamic relationship. For example, in measurement systems, it is very important to tune the slope to equal 1 for accuracy. This requires analyzing the effect of the control factors on the slope to find a tuning factor. In this case, be sure to include some control factors that might be useful for adjusting the slope. The double-dynamic analysis can also be used in such a case. The double-dynamic layout is shown in Figure 11.3. A process control signal factor, $M^*$, can be included in the outer array and analyses as described in Chapter 6. Thus, for dynamic problems, especially double-dynamic problems, all the control factors are chosen to try to minimize variation due to the noises.

In Figure 11.2, four signal levels and two noises are tested for each of the treatment combinations, resulting in a total of eight data points per row. This is a lot of data (!), but the informational content in this experiment is very high. It is important that the factors in the outer array be easy to control so that the time and cost required to run this experiment do not become excessive. A minimum of three signal factors should be run, but the noises should always be compounded to two levels or at most three.

In Figure 11.3, three process signal factor levels, three functional signal factor levels, and two noises are tested for each of the treatment combinations, resulting in a total of 18 data points per row. Once again, the size of the experiment has grown considerably. This layout and analysis is only recommended where tuning the slope of the dynamic relationship is critical.

**Figure 11.2** Crossed array layout for the dynamic parameter optimization experiment

| | *Inner (control factor) array* | | | | | | | | *Outer (signal & noise factor) array* | | | | | |
|---|---|---|---|---|---|---|---|---|---|---|---|---|---|---|
| | | | | | | | | | M1 | M2 | M3 | M4 | | |
| *Run* | 1 *A* | 2 *B* | 3 *C* | 4 *D* | 5 *E* | 6 *F* | 7 *G* | 8 *H* | N1 N2 | N1 N2 | N1 N2 | N1 N2 | *S/N* | *Slope* |
| 1 | 1 | 1 | 1 | 1 | 1 | 1 | 1 | 1 | | | | | S/N1 | $\overline{\beta 1}$ |
| 2 | 1 | 1 | 2 | 2 | 2 | 2 | 2 | 2 | | | | | ... | ... |
| 3 | 1 | 1 | 3 | 3 | 3 | 3 | 3 | 3 | | | | | ... | ... |
| 4 | 1 | 2 | 1 | 1 | 2 | 2 | 3 | 3 | | | | | ... | ... |
| 5 | 1 | 2 | 2 | 2 | 3 | 3 | 1 | 1 | | | | | ... | ... |
| 6 | 1 | 2 | 3 | 3 | 1 | 1 | 2 | 2 | | | | | ... | ... |
| 7 | 1 | 3 | 1 | 2 | 1 | 3 | 2 | 3 | | | | | ... | ... |
| 8 | 1 | 3 | 2 | 3 | 2 | 1 | 3 | 1 | | | | | ... | ... |
| 9 | 1 | 3 | 3 | 1 | 3 | 2 | 1 | 2 | | | | | ... | ... |
| 10 | 2 | 1 | 1 | 3 | 3 | 2 | 2 | 1 | | | | | ... | ... |
| 11 | 2 | 1 | 2 | 1 | 1 | 3 | 3 | 2 | | | | | ... | ... |
| 12 | 2 | 1 | 3 | 2 | 2 | 1 | 1 | 3 | | | | | ... | ... |
| 13 | 2 | 2 | 1 | 2 | 3 | 1 | 3 | 2 | | | | | ... | ... |
| 14 | 2 | 2 | 2 | 3 | 1 | 2 | 1 | 3 | | | | | ... | ... |
| 15 | 2 | 2 | 3 | 1 | 2 | 3 | 2 | 1 | | | | | ... | ... |
| 16 | 2 | 3 | 1 | 3 | 2 | 3 | 1 | 2 | | | | | ... | ... |
| 17 | 2 | 3 | 2 | 1 | 3 | 1 | 2 | 3 | | | | | ... | ... |
| 18 | 2 | 3 | 3 | 2 | 1 | 2 | 3 | 1 | | | | | S/N18 | $\overline{\beta 18}$ |

**Figure 11.3** Crossed array layout for the double-dynamic parameter optimization experiment

| | *Inner (control factor) array* | | | | | | | | *Outer (signal & noise factor) array* | | | | | | | | | | |
|---|---|---|---|---|---|---|---|---|---|---|---|---|---|---|---|---|---|---|---|
| | | | | | | | | | M1* | | | M2* | | | M3* | | | | |
| | | | | | | | | | M1 | M2 | M3 | M1 | M2 | M3 | M1 | M2 | M3 | | |
| *Run* | 1 *A* | 2 *B* | 3 *C* | 4 *D* | 5 *E* | 6 *F* | 7 *G* | 8 *H* | N1 N2 | N1 N2 | N1 N2 | N1 N2 | N1 N2 | N1 N2 | N1 N2 | N1 N2 | N1 N2 | *S/N* | *Slope* |
| 1 | 1 | 1 | 1 | 1 | 1 | 1 | 1 | 1 | | | | | | | | | | S/N1 | $\overline{\beta 1}$ |
| 2 | 1 | 1 | 2 | 2 | 2 | 2 | 2 | 2 | | | | | | | | | | ... | ... |
| 3 | 1 | 1 | 3 | 3 | 3 | 3 | 3 | 3 | | | | | | | | | | ... | ... |
| 4 | 1 | 2 | 1 | 1 | 2 | 2 | 3 | 3 | | | | | | | | | | ... | ... |
| 5 | 1 | 2 | 2 | 2 | 3 | 3 | 1 | 1 | | | | | | | | | | ... | ... |
| 6 | 1 | 2 | 3 | 3 | 1 | 1 | 2 | 2 | | | | | | | | | | ... | ... |
| 7 | 1 | 3 | 1 | 2 | 1 | 3 | 2 | 3 | | | | | | | | | | ... | ... |
| 8 | 1 | 3 | 2 | 3 | 2 | 1 | 3 | 1 | | | | | | | | | | ... | ... |
| 9 | 1 | 3 | 3 | 1 | 3 | 2 | 1 | 2 | | | | | | | | | | ... | ... |
| 10 | 2 | 1 | 1 | 3 | 3 | 2 | 2 | 1 | | | | | | | | | | ... | ... |
| 11 | 2 | 1 | 2 | 1 | 1 | 3 | 3 | 2 | | | | | | | | | | ... | ... |
| 12 | 2 | 1 | 3 | 2 | 2 | 1 | 1 | 3 | | | | | | | | | | ... | ... |
| 13 | 2 | 2 | 1 | 2 | 3 | 1 | 3 | 2 | | | | | | | | | | ... | ... |
| 14 | 2 | 2 | 2 | 3 | 1 | 2 | 1 | 3 | | | | | | | | | | ... | ... |
| 15 | 2 | 2 | 3 | 1 | 2 | 3 | 2 | 1 | | | | | | | | | | ... | ... |
| 16 | 2 | 3 | 1 | 3 | 2 | 3 | 1 | 2 | | | | | | | | | | ... | ... |
| 17 | 2 | 3 | 2 | 1 | 3 | 1 | 2 | 3 | | | | | | | | | | ... | ... |
| 18 | 2 | 3 | 3 | 2 | 1 | 2 | 3 | 1 | | | | | | | | | | S/N18 | $\overline{\beta 18}$ |

## 11.5 Choosing the Noise Factor Treatment

The noise factor treatment follows from the results of the noise experiment. If a noise experiment has not been performed, then experience must guide the selection of the noises. In either case, compounding is used whenever possible. If several noises are compounded, then the average influence of the noises will have the desired influence, even if one or two noises occasionally reverse their directionality. Strong noise factors overwhelm the effect of any nuisance factors, thus making randomization and blocking [B3, B9, F5] largely unnecessary. Randomization may make the experiment harder to run, and blocking will use precious DOF. However, they are good practices when noises are not thoroughly accounted for in an experiment.[2]

The noise factors go into the outer array. If there are two or three compound noises, then simply repeat each control factor treatment combination using the compound noise levels. It is also possible to use an L4 or a full factorial combination of several noise factors. This may be desirable when there is not enough knowledge for compounding or when the noise factor levels are inexpensive enough to allow running all the combinations. Also, running an orthogonal array for the noise factors allows for the analysis of the CF×NF interactions.

## 11.6 Choosing the S/N Ratio

The S/N ratio is a summary statistic for describing the quality represented by the control factor treatment combination being tested. The S/N ratios were introduced in Part I: Quality Engineering Metrics. There, examples are given showing how they can be used for benchmark comparison purposes. The direct analysis of many noise factor and signal factor effects and their interactions on the quality characteristic is difficult. For a dynamic problem there are signal factors, control factors, noise factors, and CF×NF interactions to analyze. Certainly the techniques exists, but the results are not easily communicated to those who do not have the requisite training. For example, a dynamic problem would require using multivariate analysis, a topic with which few engineers are familiar.

The S/N ratio simplifies the analysis of parameter optimization experiments by breaking down the analysis of the experiment into simpler parts. Too many factors affect the quality characteristic to allow a simple, easily understood analysis otherwise. The S/N ratio found for each row of data includes the effects of the noise factors and signal factors for dynamic problems. The S/N ratio for dynamic problems also contains information about lack of fit to the ideal functional relationship chosen. Once the outer array effects have been reduced to S/N values (and mean or slope values, in some cases), the analysis of the control factor effects can proceed using analysis of means. This focuses one's attention on the critical control factor effects and on finding the optimal levels for the control factors.

---

2. These techniques are tools for the model building process in classical DOE and are not normally recommended for Robust Design. Robust Design engineers the noises into the experimental design to force what can happen to happen. Nuisance factors and random effects are deliberately included in the noise factors (as discussed in Chapter 9).

## 11.7 Summary of Chapter 11

The crossed array format helps clarify what is being done with the S/N analysis. Initially, the experiment produces a large amount of data. Each row of data is analyzed separately to compute the S/N ratio and other statistics, such as the mean or slope, in a dynamic problem. The inner array is then analyzed for the control factor effects on the S/N ratio and mean (or slope). The S/N ratio is always maximized, and only one factor to control the mean (or slope) is required. Thus, the analysis focuses on three simple questions:

1. What factor levels maximize the S/N ratio, and by how much?
2. What factors have little effect on the S/N ratio?
3. Do any of those factors affect the average response?

These questions are easily answered and communicated using simple graphs or tables. As a result, the entire team can see clearly how the controllable factors can be used to improve quality and put the system onto the target. They can further realize which controllable factors are not helping the quality characteristic being analyzed. It is also important that the entire team see clearly how the experiment, using the crossed array layout, leads to the conclusions presented.

## Exercises for Chapter 11

1. Explain the difference between setting up an experiment with a main (inner) orthogonal array coupled with a noise (outer) orthogonal array and a main (inner) orthogonal array coupled with a compound noise array. What are the benefits and deficiencies of each arrangement?
2. Set up a hypothetical experimental layout that would allow one to analyze an automobile tire for a targeted wear rate of 0.100″ per 10,000 miles (a performance tire).
3. What is a combined array, and under what circumstances would one be used?
4. Why is randomizing the experimental run order not a priority in the parameter design experimental process?
5. Why is taking replicate data during a main experiment not recommended?

CHAPTER 12

# The Analysis and Verification of the Parameter Optimization Experiment

## 12.1 Introduction

The parameter optimization experiment, described in the previous chapter, focuses on identifying the effect of the control factors on the quality characteristics and their S/N ratios. The crossed array experiment generates the raw data needed to meet these objectives. This chapter describes the analyses required to process these data to find the control factor effects. The effect of the control factors on the quality characteristic is analyzed using two calculations. First, the response data are used to compute the S/N ratio and other summary statistics such as the mean or slope of a dynamic relationship, as discussed in Chapters 5 and 6. Second, the S/N ratio and the other responses are analyzed using the analysis of means (ANOM) for the control factor effects, as discussed in Chapter 7.

This analysis, like any other statistical analysis, is subject to several caveats that need to be checked. Experimental error or nuisance factors can give false results. Key factors or interactions that are neglected in the analysis could influence the results significantly. Besides a good analysis, the overall objective of parameter design is to demonstrate improved robustness or reduced sensitivity to the noise factors. The analysis is confirmed and the robustness is demonstrated in the verification experiment, where the optimum parameter set points are tested to show repeatable improvement in the expected performance.

The verification experiment is preceded by a prediction step in which the results of the control factor analysis are used to form a simple model. The model predicts the S/N ratio for the optimized design set points. If the prediction is quantitatively verified, then the control factor effects are well understood.

A failure to confirm the predictive model is an indication of difficulty. There may be a general inability to reproduce results, indicating the presence of nuisance factors or a lack of robustness. Results may be reproducible but the optimum performance may miss expectations, indicating that important factors such as interactions are not included in the predictive model. Either of these conditions is cause to consider additional experimentation. Fortunately, even an experiment that fails to verify is a learning opportunity. We have found that groups applying the engineering analysis prior to experimentation as described in Section 12.2 achieve full confirmation in over 85% of the cases. In the remaining cases, virtually all the teams get it right on the second attempt. Of course, not all groups achieve spectacular improvements in performance; their concepts are often already near optimum performance, or the technology may have inherent limitations. But the goal is to understand the system's robustness and to know why the performance achieved is the best available from the current design concept.

## 12.2 The Data Analysis Procedure

Once the experiment is completed, summarize the data in tabular form to make it easy to use when performing calculations. There should be data for each combination of control factor treatments and noise factor levels in the crossed array experiment. Missing data is a significant issue that is discussed later. The data analysis can be done using a spreadsheet program or using a dedicated analysis package such as WinRobust Lite, which is included with your book. The advantage of using a spreadsheet package is that it maximizes flexibility to use custom S/N ratios and orthogonal arrays. WinRobust offers a very simple-to-use solution for the vast majority of analyses. Refer to Appendix B for an introduction to the WinRobust software.

Before analyzing the data, it is necessary to define the control factor array for WinRobust. The outer array is not specified as part of the analysis using WinRobust. The software can be used for a noise study, as described in Chapter 9, but WinRobust does not analyze the noise factors from a crossed array experiment containing an orthogonal outer array. Instead WinRobust is designed to work with compounded noise factors. After defining the inner array, choose the *Response* menu and pick the type of analysis, static or dynamic. The next screen, *Define Response,* is used to name the response, describe how many signal factors (dynamic only), noise conditions (compound noise levels or noise factor treatment combinations), and replicates per noise condition are used. The product of these numbers should equal the total number of runs used for each control factor treatment combination. The type of S/N ratio analysis used is also chosen in the *Define Response* screen. When this information is input and OK is clicked, the next screen is the data input window. The data can be entered manually or using the clipboard function in Windows.

It is not necessary to do a detailed analysis of the noise factor effects in the main experiment, because the purpose of the main experiment is to focus on the control factors and to determine their optimum levels. Instead, for each control-factor treatment combination, the system's average behavior is described by computing the mean (or slope for a dynamic problem). The system's variability is described by computing the S/N ratio. The influence of the control factors on the system's sensitivity to the noise factors is determined by analyzing the S/N values. After entering the data to WinRobust, click on *Crunch* to perform the S/N analysis. The results for each row are displayed on the main window. WinRobust also computes the average factor effects for the mean (or slope in the dynamic case) at each set point level using ANOM. The results are displayed by going to the *Results* menu. Options include displaying tables of the factor effects and displaying factor effect plots. For two-level arrays, where there is no confounding of interactions with main effects or other interactions, interaction plots can also be displayed. These are discussed in Chapter 16.

The factor effects plot makes it possible to easily identify each factor's optimum level, i.e., the set point that gives the maximum S/N value. WinRobust has an optimizer function, found in the *Confirmation and Prediction* menu. The optimizer automatically picks the factor levels that maximize the

S/N ratio and computes the additive model results (discussed later). The user can specify which factors and what levels are to be included in the model. The additive model is used to predict the performance under the optimum combination for comparison to the verification test.

Additional analysis can be made using analysis of variance (ANOVA). ANOVA is used to find each control factor's percent contribution to the response. It is also used to estimate the error in the experiment due to errors in the additive model and experimental error. The estimate of the error provides a criterion for deciding which factors have a significant effect on reducing variability. As a result, a statistically valid predictive model of the optimum S/N ratio can be generated using only the significant control factors. The ANOVA procedure is discussed in Chapter 17.

## 12.3 An Example of the Analysis of the Parameter Optimization Experiment

A study of a two-roller belt-drive mechanism is analyzed. This is a common problem in film manufacture and film use in some products such as copiers. The mechanism is shown in Figure 12.1. The quality characteristic for this problem is the location of the edge of the belt. Ideally the belt does not wander or oscillate laterally as the belt moves around the two rollers. The location can be measured using an edge detecting device.

There are four control factors for this experiment. They are A: drive roller diameter; B: film tension; C: idler roller diameter; and D: idler roller mounting. The last factor is not quantitative. The three mounting schemes tested are D1: fixed, D2: gimbaled, and D3: castered and gimbaled.

The compound noise factors are N1, high belt conicity and misaligned drive shaft, and N2: low conicity and straight drive shaft. The response is film position. The first noise condition has a tendency to cause the film to wander. The second noise condition tends to cause the film not to wander.

While the mean value (the film's neutral position) is not at all critical, it is critical that the edge location not vary. Thus, the nominal-the-best Type II analysis of the edge location data is appropriate for this problem. The S/N ratio is a function of only the variance, as shown in Equation 12.1, which is a measure of the amount of movement of the edge.

$$S/N = -10 \log (S^2) \quad \textbf{(12.1)}$$

Because the film's neutral position is not critical, this static problem only requires an analysis of the S/N ratio. The mean value will not be analyzed, and the two-step optimization process need not be used.

**Figure 12.1** Belt-drive mechanism

**Table 12.1** Data for belt-drive example

| *Run* | *A* | *B* | *C* | *D* | *N1* | *N2* | *S/N* |
|---|---|---|---|---|---|---|---|
| 1 | 1 | 1 | 1 | 1 | .70 | .40 | 13.5 |
| 2 | 1 | 2 | 2 | 2 | .64 | .60 | 31.0 |
| 3 | 1 | 3 | 3 | 3 | .55 | .50 | 29.0 |
| 4 | 2 | 1 | 2 | 3 | .66 | .55 | 22.2 |
| 5 | 2 | 2 | 3 | 1 | .73 | .50 | 15.8 |
| 6 | 2 | 3 | 1 | 2 | .71 | .65 | 27.4 |
| 7 | 3 | 1 | 3 | 2 | .64 | .40 | 15.4 |
| 8 | 3 | 2 | 1 | 3 | .50 | .45 | 29.0 |
| 9 | 3 | 3 | 2 | 1 | .39 | .25 | 20.1 |

**Figure 12.2** Main screen with the data input window for WinRobust

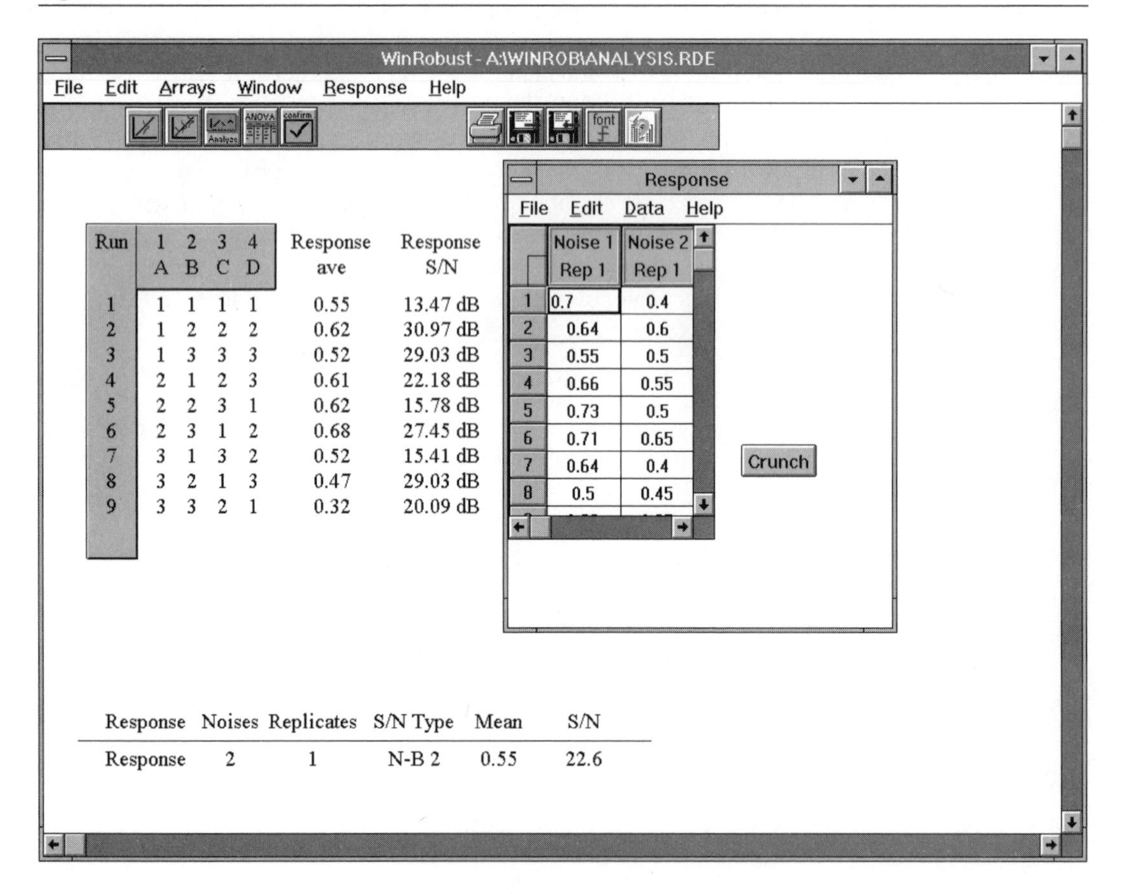

The L9 orthogonal array experiment and the resulting data for the belt transport mechanism are shown in Table 12.1. Figure 12.2 shows the L9 and the input window for WinRobust. The analysis is done when the button marked *Crunch* is clicked. The resulting S/N values for each row are shown on the main screen in Figure 12.2.

## 12.4 Estimating the Effects of Each Factor Using ANOM

The factor effects are found by calculating the average S/N for each control factor level. The first step is to calculate the overall mean of the S/N ratio for the entire experiment, shown in Equation 12.2.

$$\begin{aligned}\overline{S/N}_{exp} &= \frac{1}{9}\left(\sum_{i=1}^{9} S/N_i\right) \\ &= \frac{1}{9}(S/N_1 + S/N_2 + \ldots + S/N_9) \\ &= 22.6 \end{aligned} \quad \textbf{(12.2)}$$

The effect of a factor level is given by the deviation of the response from the overall mean due to the factor level. The S/N values associated with a control factor level from the control factor array are averaged to find the mean S/N for each level of the control factor. For example, the experimental data for runs 4, 5, and 6 are used to evaluate the effect of drive roller diameter at level 2 (factor A at level 2, abbreviated A2):

$$\begin{aligned}\overline{S/N}_{A2} &= \frac{1}{3}(S/N_4 + S/N_5 + S/N_6) \\ &= 21.8 \end{aligned} \quad \textbf{(12.3)}$$

The effect of drive roller diameter at level A2 is:

$$\begin{aligned}\Delta A_2 &= (\overline{S/N}_{A2} - \overline{S/N}_{exp}) \\ &= -0.8 \end{aligned} \quad \textbf{(12.4)}$$

Notice what the other three control factors are doing in experiments 4, 5, and 6:

| | | |
|---|---|---|
| *Film tension* | B | 1,2,3 |
| *Idler roller diameter* | C | 2,3,1 |
| *Idler roller mounting* | D | 3,1,2 |

The averaging process is balanced by the fact that when the mean is calculated for a control factor level, all of the other control factors take on each of their levels an equivalent number of times. *This is the key property of orthogonality.* This cancellation effect is why factor A's effect can be isolated.

The calculation is repeated for the other levels of factor A and for the other factors, which are in columns 2, 3, and 4, to generate a response table. This is found in the *Results* menu from WinRobust, shown in Figure 12.3.

**Figure 12.3** Factor effects table from WinRobust

WinRobust - A:\WINROB\ANALYSIS.RDE - Results

Print Edit Display Help

| Factor | Level | Response (s/n) |
|---|---|---|
| A | 1 | 24.49 |
| | 2 | 21.80 |
| | 3 | 21.51 |
| B | 1 | 17.02 |
| | 2 | 25.26 |
| | 3 | 25.52 |
| C | 1 | 23.32 |
| | 2 | 24.41 |
| | 3 | 20.07 |
| D | 1 | 16.44 |
| | 2 | 24.61 |
| | 3 | 26.75 |

**Figure 12.4** Factor effects plot from WinRobust

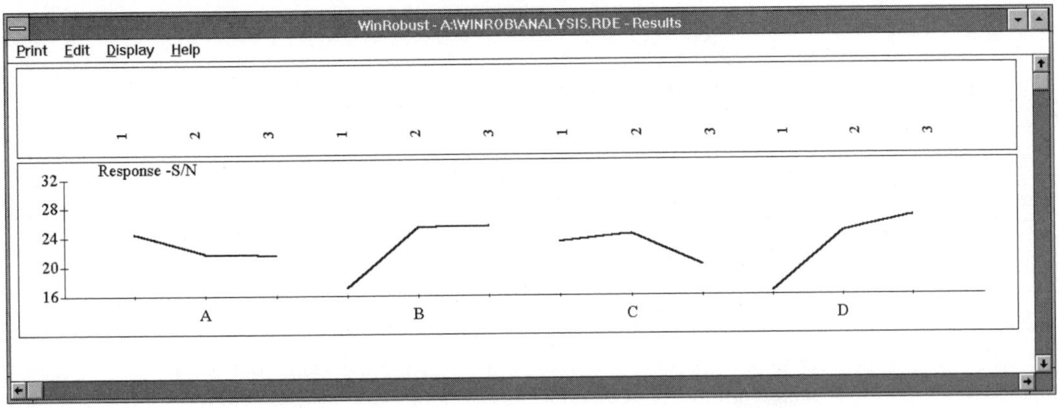

It is even more informative to look at a plot of the factor main effects, shown in Figure 12.4. The plot shown is found in WinRobust in the *Results* menu. This is a good way to visualize the results.

## 12.5 Identifying the Optimum Control Factor Set Points

The control factor plots show which factor levels are best for increasing the S/N ratio. The optimum factor level set points are the set points with the highest S/N ratio. They should give the best response with the smallest effect due to noise. This optimum result is a superior operating point for this experimental arrangement only. The data are generated from an experimental space defined by the factors and the range of their levels. The results may not apply outside that design space.

The optimum factor set points are those control factor set points with the highest average S/N levels:

A1, B3, C2, D3

Notice that in this example the best group of control factors is not identical with any one treatment combination that is run during the experiment. It is the statistical nature of designed experiments that the optimum result is pieced together from the maximum S/N ratios, no matter what experimental run they come from. Also, note that there is some latitude in the choice of optimum levels. For example, the S/N average for B2 ≅ B3, and similarly, C1 ≅ C2. The recommended optimum could thus be changed if a lower film tension or a lower idler roller diameter is desirable for cost or co-optimization reasons.

## 12.6 The Two-Step Optimization Process

For most nominal-the-best and dynamic problems, optimization follows the two-step optimization process. First, minimize the subsystem's sensitivity to noise. Second, get the subsystem's performance

**Table 12.2** Classifications of control factors

| *Factor type* | *Effect on the S/N ratio* | *Effect on the quality characteristic mean* |
|---|---|---|
| S/N factor* | Strong | Strong |
| S/N factor | Strong | Weak |
| Tuning factor | Weak | Strong |
| Cost reduction factor | Weak | Weak |

* Provided that there is an acceptable tuning factor.

on target. This procedure depends upon the S/N ratio's property that it is a quality metric independent of adjustment. This means that control factor set points that give a large S/N ratio are going to give good quality (low quality loss) when the system is put on target, regardless of whether the quality characteristic is on target for the treatment combination. In other words, the S/N ratio detaches minimizing sensitivity to noise from putting the mean on target, thus avoiding a difficult optimization problem. The first step maximizes the S/N ratio; the second step puts the system on target, reducing the S/N ratio as little as possible. There are other benefits to the two-step optimization process: It accommodates evolution of the system requirements. Often during design, adjustments must be made that require a shift in the mean without appreciably increasing the sensitivity to noise. This aspect of robustness can generate large benefits in time savings because the engineering system does not need to be reoptimized every time the target changes. The design can be reused in later products because a means of shifting its optimized performance onto a new desired target has been established.

The two-step optimization requires the classification of the control factors into four groups, as shown in Table 12.2. All the factors that have a strong effect on the S/N ratio should have their level chosen to optimize the S/N ratio. These are referred to as S/N factors in the table. Factors that have a weak effect on the S/N ratio and a strong effect on the mean value of the quality characteristic can be used for tuning. If there is no such factor, it is possible to pick a factor that has a strong effect on the mean *and* the S/N ratio. This compromises the quality, but it may be necessary. Factors that have a weak influence on both the S/N ratio and the quality characteristic as well as excess tuning factors should be set to minimize cost or used to optimize other performance considerations.

Figure 12.5 shows how factor effects plots are used to categorize the control factors. Factors A and B are examples of S/N factors used to optimize the S/N ratio. If necessary, factor A could be used to adjust the mean onto target. Factor C is an example of an ideal tuning factor, and factor D is an example of an ideal cost reduction factor.

## 12.7 The Additive Model

The additive model is used to predict the influence of the control factors on the response. Consider the simple example in Table 12.3, showing the results of a two-factor experiment, where A(Low) $= -1$, A(High) $= +1$, B(Low) $= -1$, and B(High) $= +1$.

**Figure 12.5** Factor effects plot for mean and S/N ratio

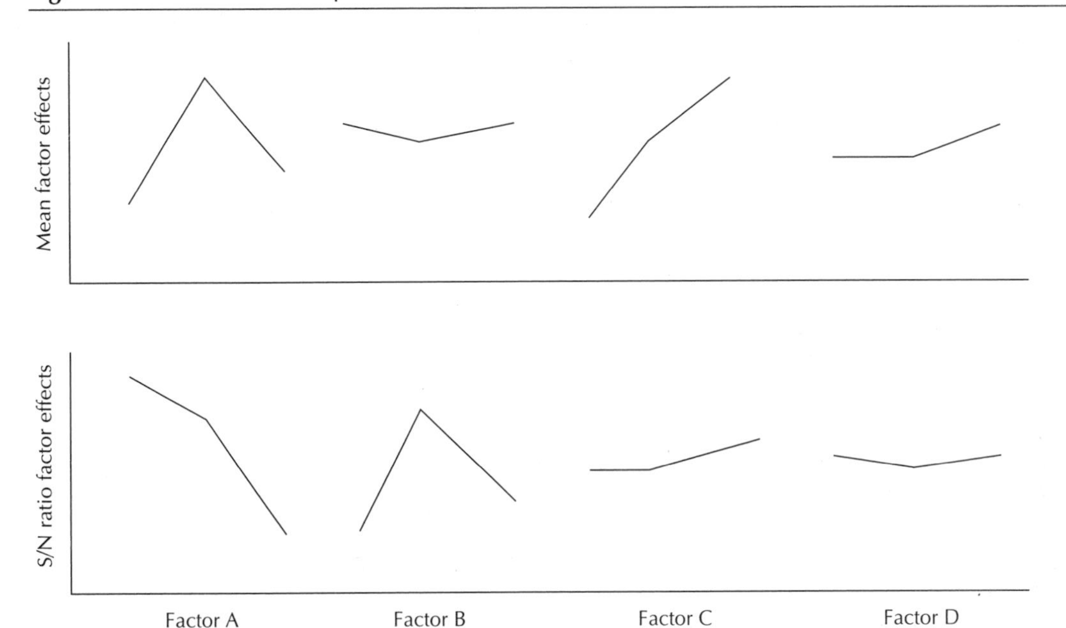


A simple additive model that describes these results is given by Equation 12.5.

$$S/N = 5.5 + 3A + 1.5B \tag{12.5}$$

The first term, 5.5, represents the average S/N for the entire experiment. The coefficient of the second term, 3, represents the effect of factor A. The coefficient of the third term, 1.5, represents the effect of factor B. This equation is the additive model describing the results of the four experiments. The additive model refers to the sum of the individual factor effects, without cross-terms (interactions). Predicted values for the S/N ratio can be found by substituting the appropriate values for the factors A and B in the equation. For example, substituting A = +1 (High) and B = +1 (High) gives the following result for the predicted S/N ratio:

$$S/N = 5.5 + 3 + 1.5 = 10 \text{ dB} \tag{12.6}$$

**Table 12.3** Two-factor full factorial and sample data for additive model example

| *Run* | *A* | *B* | *Response (S/N)* |
|---|---|---|---|
| 1 | Low | Low | 1.00 dB |
| 2 | Low | High | 4.00 dB |
| 3 | High | Low | 7.00 dB |
| 4 | High | High | 10.00 dB |

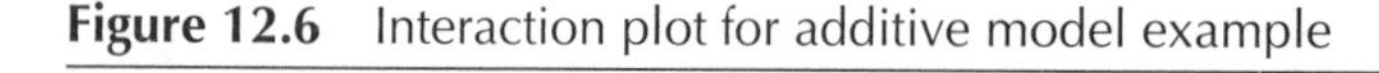

**Figure 12.6** Interaction plot for additive model example

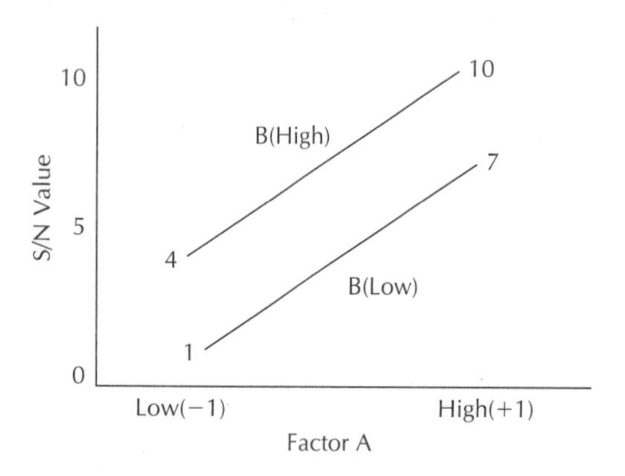


Of course, A(High) and B(High) is actually run #4, which gave S/N = 10 dB. In this case the additive model agrees with the experiment because there is no interaction (or experimental error) in this example. To show that graphically, the data can be displayed in an interaction plot, as shown in Figure 12.6.

The data form parallel lines, showing that there is no interaction between the factors. The additive model works in this case because the effect of changing factor A from low to high is a 6 dB change in the S/N *regardless of the level of factor B.* Also, the effect of changing factor B from low to high is a 3 dB change in the S/N *regardless of the level of factor A.* Thus, Equation 12.5 is accurate for any combination of levels for factors A and B. This would not be true if there was a significant interaction.

The data in Table 12.3 can also be interpreted using analysis of means to find the factor effects. Here, the factor effect of A is (8.5 − 2.5)/2 = 3 dB, and the factor effect of B is (7 − 4)/2 = 1.5 dB. Graphically, the factor effects plot shown in Figure 12.7 is obtained.

The applicability of the additive model cannot be shown simply from the factor effects plot. The analysis of means produces a predictive equation, but cannot prove that additivity applies if a fully saturated orthogonal array is used. However, if the predictive equation is shown to successfully predict the results for various combinations of control factors, then this is evidence that the additive equation applies and interactions are low. This is one major purpose of the verification experiment.

**Figure 12.7** Factor effects plot for additive model example

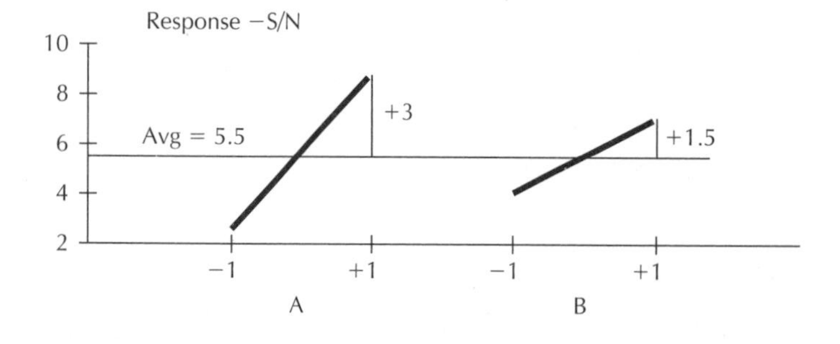

## 12.8 The Predictive Equation

The general form of the predictive equation is as follows:

$$y_{predicted} = \overline{y}_{exp} + (\overline{y}_A - \overline{y}_{exp}) + (\overline{y}_B - \overline{y}_{exp}) + (\overline{y}_C - \overline{y}_{exp}) + \ldots \quad \textbf{(12.7)}$$

where $\overline{y}_{exp}$ is the overall average response (or S/N ratio) for the entire orthogonal array and $\overline{y}_A$, $\overline{y}_B$, $\overline{y}_C$ are the response averages for factors A, B, and C, respectively. The factor effects corresponding to the factor levels being modeled (typically the optimum levels) are used in the predictive equation.

The additive model example given in the previous section can be used to illustrate the predictive equation using the data given in Table 12.3. The ANOM gives the following results:

$$\overline{S/N}_{A(Low)} = (1.00 + 4.00)/2 = 2.5 \text{ dB}$$
$$\overline{S/N}_{A(High)} = (7.00 + 10.00)/2 = 8.5 \text{ dB}$$
$$\overline{S/N}_{B(Low)} = (1.00 + 7.00)/2 = 4 \text{ dB}$$
$$\overline{S/N}_{B(High)} = (4.00 + 10.00)/2 = 7 \text{ dB}$$
$$\overline{S/N}_{exp} = (1.00 + 4.00 + 7.00 + 10.00)/4 = 5.5 \text{ dB}$$

Thus, the predictive equation for A(High), B(High) is given by

$$\overline{S/N}_{predicted} = 5.5 + (8.5 - 5.5) + (7 - 5.5)$$
$$= 5.5 + 3 + 1.5 = 10 \text{ dB}$$

This is the same result obtained before. In the predictive equation, the deviation from the experimental average given for any particular factor effect is the contribution due to that factor at the level chosen. With this simple approach, it is not necessary to fit a model or develop a continuous equation or contour plots to describe the results of the experiment. Those other approaches to modeling results can be done, but require methods and computer tools that are beyond the scope of this book and that do not, in themselves, contribute any additional engineering knowledge.

Returning to the belt drive example begun earlier in this chapter, the creation and use of the predictive equation can be illustrated. The results of the ANOM on the S/N response, shown in Figure 12.4, is displayed in Table 12.4.

The overall average S/N ratio is given by $\overline{S/N}_{exp} = 22.60$ dB. The optimum factor set points are those control factor set points with the highest average S/N levels: A1, B3, C2, D3. The predicted S/N ratio for the optimum configuration is

$$S/N_{opt} = \overline{S/N}_{exp} + (\overline{S/N}_{A1} - \overline{S/N}_{exp}) + (\overline{S/N}_{B3} - \overline{S/N}_{exp}) + (\overline{S/N}_{C2} - \overline{S/N}_{exp}) + (\overline{S/N}_{D3} - \overline{S/N}_{exp})$$
$$= 22.60 + (24.49 - 22.6) + (25.52 - 22.6) + (24.41 - 22.6) + (26.75 - 22.6)$$
$$= 22.60 + 1.89 + 2.92 + 1.81 + 4.15 = 33.37 \text{ dB}$$

**Table 12.4** S/N averages for the belt-drive example

| *Factor* | *Level 1* | *Level 2* | *Level 3* |
|---|---|---|---|
| A (drive roller diameter) | 24.49 | 21.80 | 21.51 |
| B (film tension) | 17.02 | 25.26 | 25.52 |
| C (idler roller diameter) | 23.32 | 24.41 | 20.07 |
| D (mounting) | 16.44 | 24.61 | 26.75 |

This predicted optimum, 33.37 dB, is clearly better performance than any of the runs in the original experiment. This is not unusual because the optimum configuration is not any of the runs used in the original array. Another reason the verification test is needed is to verify that this new configuration will indeed outperform any of the other treatment combinations tested. Note that the optimum configuration is a unique set of parameter combinations within the experimental space—it is not simply one of the trials used to sample the experimental space in this statistically designed experiment.

## 12.9 The Verification Tests

This step in the robustness process is done to verify that the optimum control factor treatment combination found from the data analysis is valid. "Valid" means that the optimum is predictable, verifiable, and reproducible. Verification tests check the results of the parameter optimization experiment, demonstrate the optimum performance, and confirm freedom from interactions. The verification experiment checks that the additive model is predictive, that the system is insensitive to noise, and that the design performance is repeatable under both induced and random noise conditions.

The verification experiment consists of two tests: first, a test using the control factors set to their optimum levels; and second, a test using one of the *nonoptimal* control-factor treatment combinations from the main experiment. Each of these tests is done under the same noise conditions as in the main experiment. This means repeating the outer array exactly as it was run in the crossed array experiment.

The results of each of these tests is checked against the respective predicted results from the additive model. Good agreement by both tests confirms predictability. Good agreement of the predicted best configuration verifies the robustness optimization. The nonoptimum configuration can also be compared to the previously obtained result to confirm reproducibility.

There are four possible outcomes of the verification test:

1. **Strong Confirmation:** Here the optimum configuration gives a result close to the predicted result. The replicated result is also close to the previously obtained result, verifying repeatability. This is the strongest verification and indicates a solid success.
2. **Weak Confirmation:** Here the optimum configuration and the replicate test do not match the expected result, *but both are in error by about the same amount!* This indicates that the factor effects from the experiment are understood. But some other significant parameter has changed and is affecting the results. Thus, the optimum configuration has been correctly identified, but there are other important factors yet to be discovered. More experimentation is required to isolate the responsible factor.
3. **Bad Confirmation Due to an Interaction:** Here the replicated result is close to the previously obtained result, verifying repeatability. On the other hand, the optimum configuration does not agree with the predicted result. Since the experiment is reproducible, there may be a problem with the additive model due to strong interactions that are not present in the model. Thus, the additive model is not accurately representing reality.
4. **Bad Confirmation Due to Experimental Error (Random):** Here, the optimum configuration and replicate tests do not match the expected results, but the errors are random. Generally, this conclusion is drawn after the observation is repeated several times and is attributed to measurement problems or unknown nuisance effects.

The predicted optimum—A1, B3, C2, D3—for the belt-drive system is 33.37 dB, according to the additive model (Equation 12.7). The first treatment combination from the L9, A1, B1, C1, D1, has

**Table 12.5** Examples of verification results

| | *Test 1 Optimum* | *Test 2 Replicated* | *Test 1 − Test 2 Difference* |
|---|---|---|---|
| *Predicted result* | 33.37 | 13.47 | 19.9 |
| *1. Good Confirmation* | 34 | 13 | 21 |
| *2. Weak Confirmation* | 29 | 9 | 20 |
| *3. Interaction* | <23 or >43 | 13 | <10 >30 |
| *4. Random Error* | 38 | 22 | 16 |

an S/N value of 13.47 dB. The additive model prediction is also 13.47 dB. These numbers are used here to illustrate the possible verification outcomes listed. Table 12.5 shows sample results that correspond to the four cases just described. The results shown are made up for illustrative purposes only.

The first set of results listed are very close to the predicted results. This is considered a good confirmation. In practice it is very difficult to state with precision how close the actual numbers must come to the predicted values for the agreement to be considered good. Texts on design of experiments devote many pages to exploring various statistical tests. The ANOVA discussion in Chapter 17 gives one simple analysis to answer this question. Ultimately, however, the statistical analyses give only guidelines. The final determination is up to your engineering judgment.

The second set of results is not in good agreement with the predicted results, but the difference, 20, is very close to the predicted difference of 19.9. This means that there is a uniform shift in the data, corresponding to a change in the mean, $\overline{y}_{exp}$, in Equation 12.7. Such a shift does not necessarily mean that the factor effects are in error. It is more likely that some factor is causing irreproducible results.

The third set of results illustrates a case where interactions are considered to be too significant to be simply nuisance factors. The expected improvement (test 1) vs. the replicated result (test 2) is 20 dB. If only half the expected improvement or more than twice the expected improvement is found, then strong interactive effects are suggested. Thus, anything less than 23 dB or more than 43 dB for the optimum is suspect. Note that the replication result (test 2) is in agreement with the previous result, indicating that there is no overall shift or other reproducibility problems.

Bad confirmation due to experimental error is difficult to prove rigorously and does not have a consistent pattern. The fourth set of results in Table 12.5 looks suspicious. Additional tests, typically several replicates, can help confirm whether there is some form of random error taking place in the verification experiment.

If there is not good agreement between the predicted and the verification values, then the disparity must be understood. Often a failure to confirm is the result of a lack of additivity (significant interactions). Lack of additivity is due to improper interpretation of energy transformation. When this occurs, revisit the physics and engineering concepts and make sure that the S/N selections are appropriate. Consider a different approach for accomplishing the same ideal function, one that does not employ interactive terms. Take a good look at the response metric and the additivity grouping of the control factors. Look for better ways of accounting for the exchange of energy.

Of course, a failure to verify can also be due to experimental and human error. This is especially likely if the nominal condition does not give a result that is close to the S/N value obtained in the original experiment. Error can be caused by meter error, poor experimental technique, mathematical mistakes, broken equipment, misread/misrecorded data, improperly set control factors, or poor communication.

After a course of action is considered, the robustness steps can be redone as often as needed to verify the experiment. *The result is being correct and knowing why!*

## 12.10 Summary: Succeeding at Parameter Design

Chapters 7 through 12 have detailed an approach to planning designed experiments that relies heavily on engineering knowledge. Faced with the prospect of planning, followed by noise experiments, followed by more planning, followed by parameter optimization experiments, and finally followed by analysis and verification, you might consider a failure to verify disastrous. Nothing could be further from the truth. We have never seen a group that did not acquire significant knowledge from these experiments, "bust" or otherwise. The failed experiment always gets turned around to a success later.

Here are some key recommendations to follow to maximize your chances of success on the first try:

1. **Good selection of noise factors:** Be sure to include noise factors for each energy transformation and process in your design. Failure to do so may produce good S/N values, but poor quality when the system is tested in the field. If that is the case, pay special attention to the cause of failure and build those noise factors into your future tests.
2. **Good selection of the quality characteristic:** Be sure you are measuring the physics, not the aftermath of variation in your design. It may not be easy, but picking quality characteristics that are fundamental is critical to your success. These two recommendations are clearly the most important factors contributing to success in improving robustness.
3. **Good selection of control factors:** Be sure to choose a reasonable number of control factors and include factor levels that are set far apart. The size of the experimental space is determined by the number of control factors and the span of their levels. If the solution lies outside this design space, it cannot be found using these techniques. Be bold in your choice of factors and levels to produce breakthrough results.
4. Consider the interactive effects likely to be present in the control factors. This takes practice, but work at it to produce simple additive designs that will promote reusability and, as a result, low cost and short development time.

Remember that there is no substitute for engineering analysis. The planning process described in this book, especially the P-diagram and the two-step optimization process, will significantly improve your productivity, regardless of experimental approach. Rushing into the lab to begin experimentation on a schedule leads to tremendous wasted effort. Identify the ideal function, measure it, and minimize interactions to greatly simplify the analysis and interpretation of results. Engineering is a team effort, and the most elegantly conceived and executed experiment is a true bust if no one on the team understands or believes in the results. In this case, the KISS (Keep It Short and Simple) approach is important. The S/N ratio and the crossed array experiments lead to clear, unambiguous results, especially in cases where the control factor interactions have been minimized. There can be nothing more comprehensible and useful than a simple additive equation describing the quality of your system.

In the event that the experimental effort does not verify, the need to go back and try again should not be viewed as rework or failure. What you are trying to do, produce the best quality products in a highly competitive world, is challenging. Anyone who tries to tell you that their method is easy and foolproof is selling snake oil. Dr. Taguchi's methods require hard work and perseverance, as does any

**Figure 12.8** Flow chart of the Parameter Design process

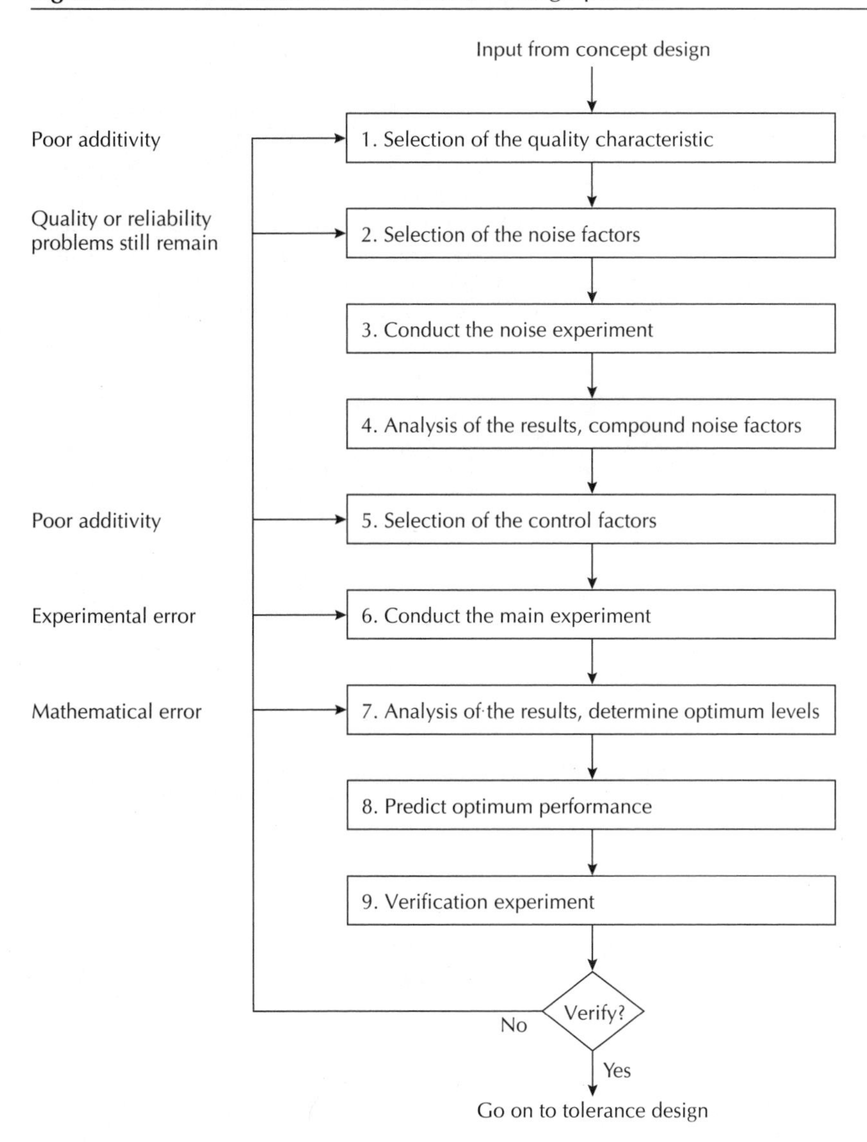


other legitimate approach. In some cases, the parameter design improvements may be inadequate for meeting the quality objective, and it may be necessary to consider tolerance design. In other cases, the current level of engineering technology may not be able to adequately solve the problem. Again, it is important to show this with data in order to convince team members to agree that alternative solutions must be considered.

The chart in Figure 12.8 helps summarize the Parameter Design process and indicates likely re-entry points in the event of a failure to verify. Several illustrations of successful applications of parameter design are given in Chapters 13 and 14 for you to learn from.

## Exercises for Chapter 12

1. Explain what is done in the process of calculating the analysis of the mean (ANOM).
2. Why are the mean and the S/N ratio calculated for the NTB (I) case?
3. Why are the slope and the S/N ratio calculated for the dynamic case?
4. What does the overall experimental average S/N value depend on?
5. Why is the overall experimental average S/N value subtracted from the individual control factor level S/N averages in the additive prediction model?
6. Why are control factors that have a weak effect on the S/N values excluded from the additive prediction model?
7. What is the predicted S/N used for?
8. What is being verified in the verification experiment?
9. What factors and levels should be used in a verification experiment?
10. Why is it recommended that one of the original experiments be repeated and compared to its original S/N value?
11. What noises should be included in the verification experiment?
12. How many times should one repeat a verification experiment, and why?

CHAPTER 13

# Examples of Parameter Design

## 13.1 The Ice Water Experiment: Smaller-the-Better

This first example is a static optimization experiment that the reader can easily repeat at home or work. In this experiment the goal is to cool down one-half cup of water to as low a temperature as possible in the presence of certain noises. The various steps of parameter design that have been introduced in Part II are illustrated here:

1. Selection of the quality characteristic
2. Selection of the noise factors
3. Conducting the noise experiment
4. Analysis and compounding of the noise factors
5. Selection of the control factors
6. Conducting the main experiment
7. Analysis and prediction of the optimum performance
8. Verification

### 13.1.1 Selection of the Quality Characteristic

In this first step, the ideal function is determined. The energy transformation for this problem is clear. It is the transfer of heat energy from the water to the ice. Ideally, that occurs rapidly and efficiently. Constructing an energy flow map helps in visualizing the engineering analysis.

Figure 13.1 indicates that the water temperature is determined by two competing influences. The effect of the ice is to cool the water, while the effect of the environment is to warm it. Ultimately, the ice melts and the water eventually reaches equilibrium with the temperature of the room. This is not the time frame of interest, however. The customer is interested in receiving a drink seconds after ordering and wants that first sip to be cool and refreshing. Thus, one possible response that could be used as the quality characteristic is the temperature of the water after specific periods of time. This is a direct physical measure of the product's quality. However, the quality characteristic should be a response that is

**Figure 13.1** Energy flow map for the ice water experiment

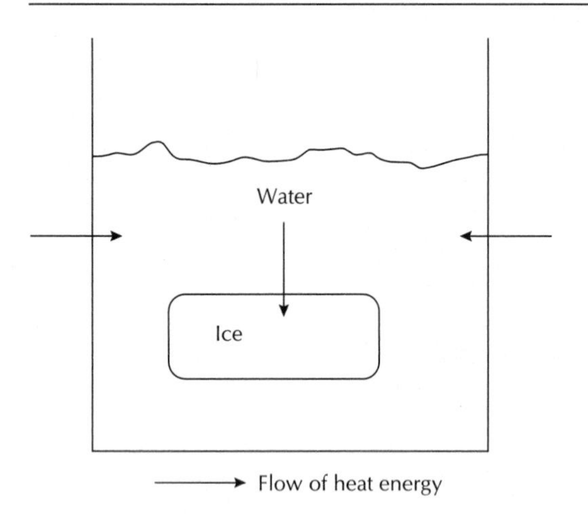


more directly related to the energy transformation of primary interest, i.e., the heat flow from the water to the ice. A more appropriate quality characteristic is the difference between the water temperature and the freezing point of the water:

$$y = \text{water temperature} - \text{freezing point}$$

This quality characteristic has the following advantages:

1. It has an absolute zero, since the water temperature will not fall below the freezing point.[1]
2. The effect of the temperature scale is reduced by subtracting the freezing point from the temperature. The energy transformation is the same whether Fahrenheit or Celsius is used.
3. The smaller-the-better signal-to-noise ratio is appropriate for the parameter optimization experiment because the quality characteristic, $y$, is continuous, has an absolute zero, and the smaller it gets, the better the quality. The STB S/N ratio has the following mathematical form:

$$\text{S/N} = -10 \log\left(\frac{1}{n}\sum_{i=1}^{n} y_i^2\right) \qquad \textbf{(13.1)}$$

## 13.1.2 Selection of the Noise Factors

The following parameters are chosen as the noise factors for this experiment:

1. The volume of water being cooled. This represents variability in filling the cup. The levels chosen for the study are 0 and 6 extra tablespoons added to the nominal one-half cup of water.
2. The physical location of the thermometer in the cup. The levels chosen are bottom and surface measurement locations. These represent the two modes of customer use—sipping from the top of the cup or using a straw to drink from the bottom.

1. Here we are neglecting subtleties such as supercooling or the possible alcohol content of the drink!

**Table 13.1** Noise experiment for ice water problem

| *Run #* | *$H_2O$* | *Thermometer* | *Time* | *Replicate 1* | *Replicate 2* |
|---|---|---|---|---|---|
| 1 | 0 Tbs. | Bottom | 10 s | 65 | 66 |
| 2 | 0 Tbs. | Surface | 60 s | 60 | 61 |
| 3 | 6 Tbs. | Bottom | 60 s | 56 | 57 |
| 4 | 6 Tbs. | Surface | 10 s | 65 | 67 |

3. The time elapsed before measuring the temperature. This represents how quickly the customer takes the first sip. The levels chosen are 10- and 60-second delay times before the temperature measurement is recorded.

### 13.1.3 Conducting the Noise Experiment

The experiment is performed using the three noise factors in an L4 orthogonal array. With only three noise factors for this simple problem, an L12 array would be too large and inefficient. The L4 is used to study the effects of the three noises on a nominal set of conditions of the control factors. The nominal conditions of the control factors are: two solid ice cubes, a paper cup, and unstirred water after the ice is placed in the cup. Each experiment is replicated once to obtain an independent estimate of the experimental error due to human error and other unaccounted-for random effects. The experimental design and the water temperatures are shown in Table 13.1.

Note that the variation between the replicates is small (1 or 2°F) compared to the variation due to the noise factors in the orthogonal array. This is highly desirable. It means that after the noise factors are compounded, replication will not be necessary in the main experiment. This saves considerable time and effort in the bigger experiment.

#### Calculating the Experimental Error

The average experimental error variance is given by

$$S_e^2 = \frac{1}{4}\sum_{i=1}^{4} S_i^2$$

where $S_i^2$ is the variance of the experimental runs in Table 13.1.

$$S_1^2 = [(65 - 65.5)^2 + (66 - 65.5)^2] = 0.25$$
$$S_2^2 = [(60 - 60.5)^2 + (61 - 60.5)^2] = 0.25$$
$$S_3^2 = [(56 - 56.5)^2 + (57 - 56.5)^2] = 0.25$$
$$S_4^2 = [(65 - 66)^2 + (67 - 66)^2] = 2.0$$
$$S_e^2 = 0.875$$

The experimental error is given by $S_e$, the standard deviation due to error:

$$S_e = 0.94$$

**Figure 13.2** Analysis of noise experiment by WinRobust

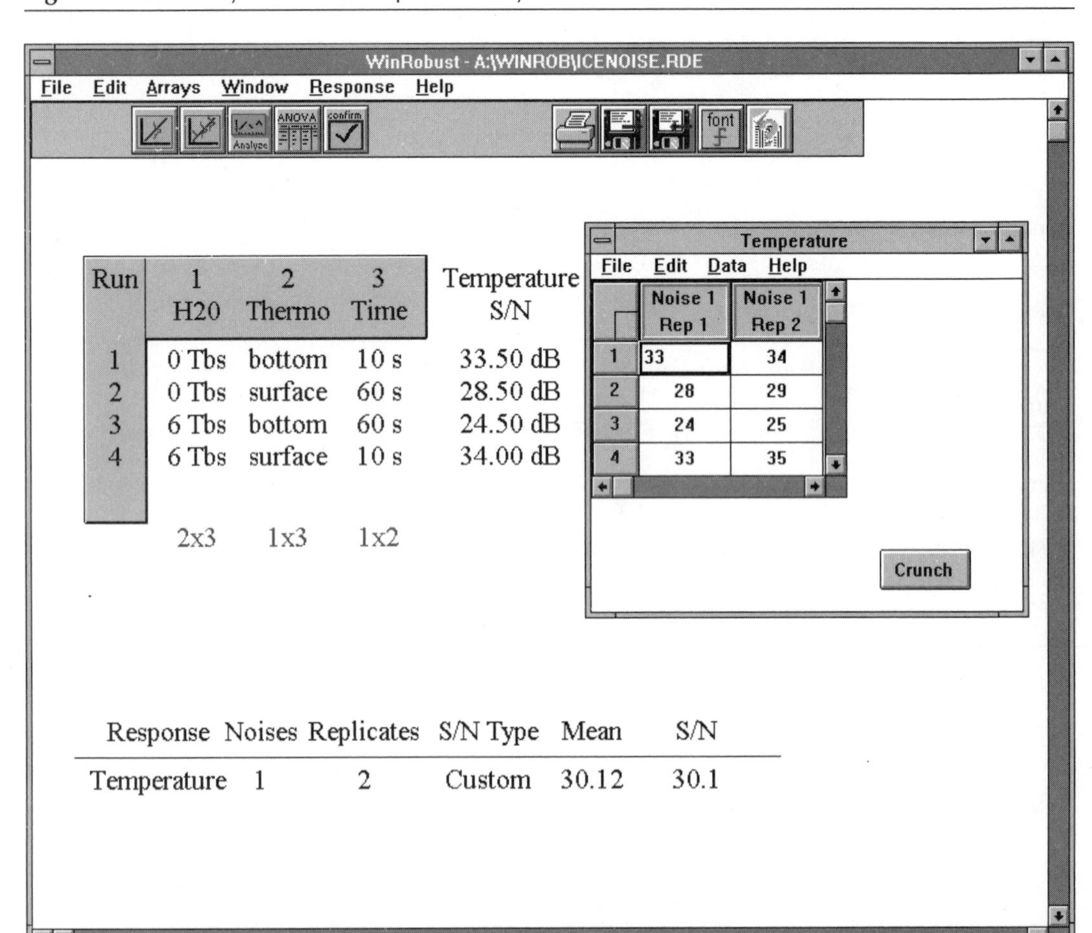


The WinRobust window in Figure 13.2 illustrates the L4 noise experiment with the freezing point of water (32°F) subtracted from the water temperatures. In the noise experiment, the response studied is the mean temperature. The S/N ratio is used for the parameter optimization experiment. To get WinRobust to analyze only mean values, treat the mean values as "custom" S/N values when inputting the response information. (Thus, the WinRobust output for the noise factor experiment says "S/N," but it is actually the mean response being analyzed, and the decibel units should be ignored.)

## 13.1.4 Analysis and Compounding of the Noise Factors

The analysis of the noise experiment is done using analysis of means. The results of the ANOM analysis are shown in Table 13.2 and the factor effects plot is shown in Figure 13.3.

**Table 13.2** Factor response averages table for the noise experiment

| *Factor* | *Level* | *Temperature* |
|---|---|---|
| $H_2O$ | 0 Tbs. | 31.00 |
| | 6 Tbs. | 29.25 |
| Thermo | Bottom | 29.00 |
| | Surface | 31.25 |
| Time | 10 s | 33.75 |
| | 60 s | 26.50 |

**Figure 13.3** Factors effect plot for the noise experiment

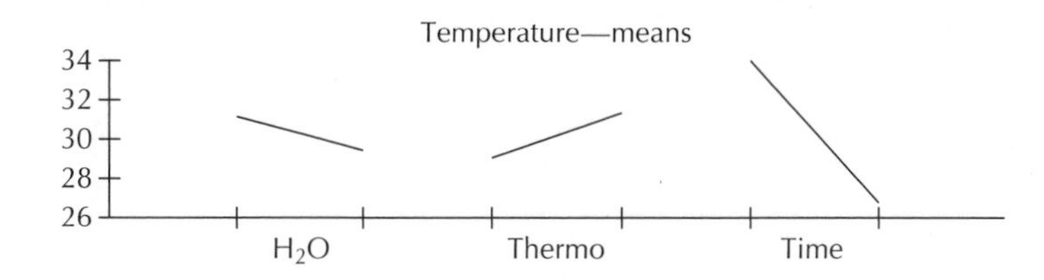


Here is a sample ANOM shown here for the first noise factor, $H_2O$, only:

$$\bar{y}_{(0\ \text{Tbs.})} = (33 + 34 + 28 + 29)/4 = 31°\text{F}$$

$$\bar{y}_{(6\ \text{Tbs.})} = (24 + 25 + 33 + 35)/4 = 29.25°\text{F}$$

Note that the replicate temperatures are each included in the calculation of the average factor effects.

As a result of this analysis, the following compounding is suggested:

- **Compound Noise Factor 1 (High Temp.):** 0 tablespoons additional water added to the cup, thermometer location at the surface, and 10 seconds before recording the temperature.
- **Compound Noise Factor 2 (Low Temp.):** 6 tablespoons additional water added to the cup, thermometer location at the bottom, and 60 seconds before recording the temperature.

The predicted temperature should be verified, because the compounding is based on the results of a saturated orthogonal array (in this case, all columns are used in the L4). The results of the verification test for the ice water noise conditions are given in Table 13.3.

**Table 13.3** Verification results for noise experiment

| | *Predicted* | *Verification result* |
|---|---|---|
| *CNF1* | 35.75°F | 36.0°F |
| *CNF2* | 24.50°F | 24.5°F |

**Figure 13.4** P-diagram for ice water parameter optimization experiment (static case)

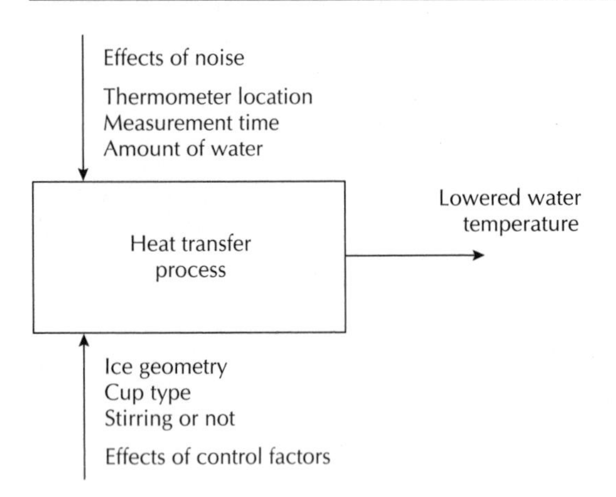


## 13.1.5 Selection of the Control Factors

For this experiment, the following are factors that the engineers (e.g., bartenders) can control:

**A. Ice Geometry** (controls surface area available for heat transfer):
*Level 1* = one solid cube
*Level 2* = one cube crushed into small pieces

**B. Container Material** (controls thermal barrier between water and air):
*Level 1* = thin paper cup
*Level 2* = Styrofoam cup

**C. Mixing Method** (controls conduction versus forced convection heat transfer):
*Level 1* = letting contents stand unstirred
*Level 2* = stirring for 10 seconds

The P-diagram in Figure 13.4 summarizes the information on the parameters chosen for this example.

The control factors are studied using an appropriately sized orthogonal array. Here again, an L4 array is chosen because there are only three control factors. (Remember that this is an example only; typically, more factors and a larger orthogonal array are used.) The L4 array is shown in Table 13.4.

**Table 13.4** Control factor array for the ice water parameter optimization experiment

| *Run #* | *Ice cubes* | *Cup type* | *Mixing method* |
|---|---|---|---|
| 1 | Whole | Paper | Unstirred |
| 2 | Whole | Styrofoam | Stirred |
| 3 | Crushed | Paper | Stirred |
| 4 | Crushed | Styrofoam | Unstirred |

**Figure 13.5** Analysis of parameter optimization experiment by WinRobust

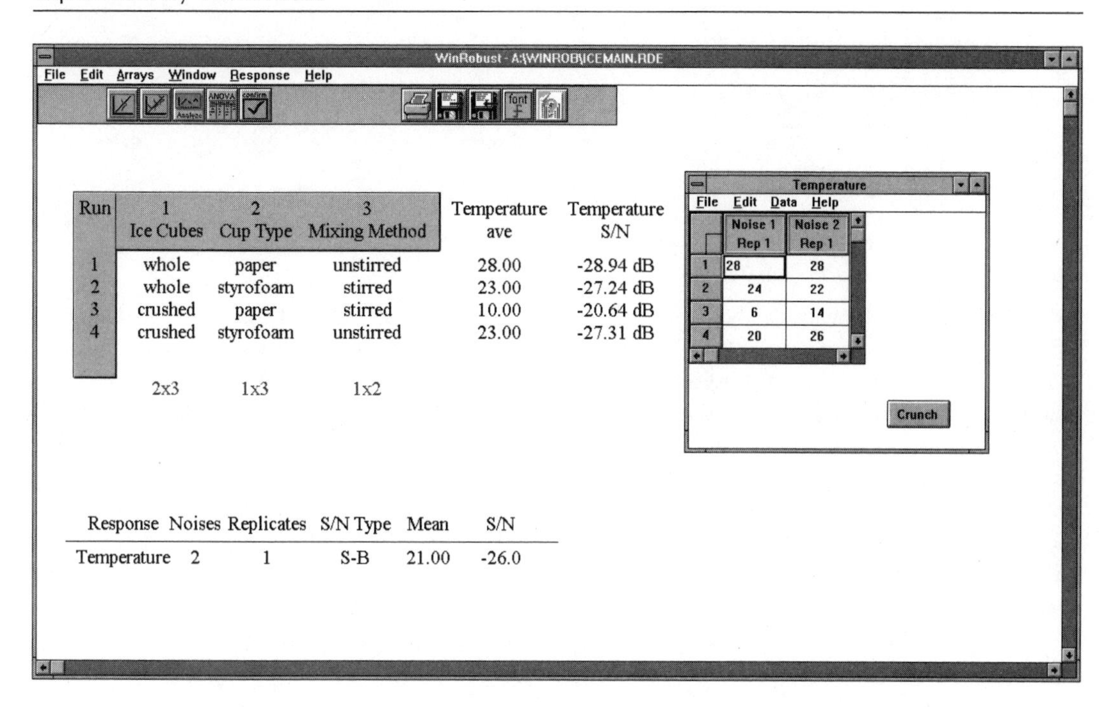


## 13.1.6 Conducting the Main Experiment

The parameter optimization experiment can now be run using the control factor L4 array and the two compound noise factors. The data are shown in Figure 13.5 as entered into WinRobust. Note that the freezing point has already been subtracted from the water temperatures.

## 13.1.7 Analysis and Prediction of the Optimum Performance

The analysis and prediction are done in several stages:

1. Calculation of the S/N ratios from the quality characteristic data.
2. The decomposition of the S/N's into the specific control-factor level contributions.
3. The construction of a predictive model of the optimum S/N ratio based on the optimum set point for each control factor.

WinRobust does all of these steps automatically when *Crunch* is clicked. Some of the results can be seen in Figure 13.5. Some of the calculations are illustrated below; the rest of the results are obtained using the software.

For example, the smaller-the-better S/N ratio for run #1 is found using the following calculation:

$$\text{S/N} = -10 \log\left(\frac{1}{n}\sum_{i=1}^{n} y_i^2\right)$$
$$= -10 \log\left[\frac{1}{2}(28^2 + 28^2)\right]$$
$$= -28.94 \text{ dB}$$

The analysis of means can be performed on both the S/N ratios ($\text{ANOM}_{\text{S/N}}$) and the mean values of the quality characteristic ($\text{ANOM}_{\bar{y}}$). The STB S/N ratio is based on the mean square deviation ($\text{MSD} = S^2 + \bar{y}^2$) because of the noise, which includes both the mean and variance of the quality characteristic. Thus, it is not necessary to analyze the mean values as well as the S/N ratios. It is nevertheless a good idea to do so, for several reasons. The results of the $\text{ANOM}_{\bar{y}}$ are in engineering units; they provide an additional view of the results; and they provide a check on how reasonable the data are. All of these improve comprehension and learning from the experiment. In this particular case, reducing the mean is the dominant effect in the S/N results. The noise factor effects are not as significant. In other cases, variation dominates and the mean and S/N analyses provide valuable insights into different aspects of the results.

Here is the overall mean S/N ratio calculation:

$$\overline{\text{S/N}} = \frac{1}{4}\left[(-28.94) + (-27.24) + (-20.64) + (-27.31)\right]$$
$$= -26.03 \text{ dB}$$

The average S/N ratio for the entire experiment is used in building the predictive model. The overall mean quality characteristic from all four experimental runs is:

$$\overline{\text{S/N}} = \frac{1}{4}(28 + 23 + 10 + 23) = 21.0°\text{F}$$

The ANOM is now used to find out how the control factor effects caused deviations from these overall averages. The goal is to calculate the average S/N ratio and average mean quality characteristic for each level of each control factor. The response values that are averaged for each level of each control factor are determined by the orthogonal array level assignment pattern from each column.

Here is an example of the factor effect analysis for Factor A, Ice Cubes:

$$\overline{\text{S/N}}_{(\text{whole ice})} = (-28.94 + -27.74)/2 = -28.34 \text{ dB}$$

$$\overline{\text{S/N}}_{(\text{crushed ice})} = (-20.64 + -27.31)/2 = -23.98 \text{ dB}$$

In order to get an overall estimate of the significance of factor A, Ice Cubes, the S/N response at A1, is subtracted from the response at A2:

$$\Delta_A = (-23.98) - (-28.34) = 4.36 \text{ dB}$$

This number, when compared to $\Delta_B$ and $\Delta_C$, is an indicator of the relative strength of control factor A. Thus, the same decomposition process can be repeated for the other two control factors. The delta values provide an estimate of the relative factor importance, similar to the information obtained from the ANOVA process (Chapter 17).

Applying $ANOM_{S/N}$ to the S/N response for the other two factors:

$$\overline{S/N}_{(paper)} = (-28.94 + -20.64)/2 = -24.79 \text{ dB}$$
$$\overline{S/N}_{(Styrofoam)} = (-27.74 + -27.31)/2 = -27.52 \text{ dB}$$
$$\Delta = (-27.52) - (-24.79) = -2.73 \text{ dB}$$

$$\overline{S/N}_{(unstirred)} = (-28.94 + -27.31)/2 = -28.21 \text{ dB}$$
$$\overline{S/N}_{(stirred)} = (-27.24 + -20.64)/2 = -23.94 \text{ dB}$$
$$\Delta = (-23.94) - (-28.21) = 4.27 \text{ dB}$$

Applying $ANOM_{\bar{y}}$ for the mean quality characteristic response:

$$\bar{y}_{(whole\ ice)} = (28 + 23)/2 = 25.5° \text{ F}$$
$$\bar{y}_{(crushed\ ice)} = (10 + 23)/2 = 16.5°\text{F}$$
$$\Delta = (16.5) - (25.5) = -9.0°\text{F}$$

$$\bar{y}_{(paper)} = (28 + 10)/2 = 19°\text{F}$$
$$\bar{y}_{(Styrofoam)} = (23 + 23)/2 = 23°\text{F}$$
$$\Delta = (23) - (19) = 4.0°\text{F}$$

$$\bar{y}_{(unstirred)} = (28 + 23)/2 = 25.5°\text{F}$$
$$\bar{y}_{(stirred)} = (23 + 10)/2 = 16.5°\text{F}$$
$$\Delta = (16.5) - (25.5) = -9.0°\text{F}$$

Note how the quality characteristic values are lower in value when the S/N values are higher in value.

The factor response averages can be found in WinRobust in the *Response* menu under *Factor Effect Table.* The output is shown in Table 13.5.

It is evident from the calculations of the factor effect Δ's that Ice Cubes and Mixing Method are about twice as significant as Cup Type. For a statistical analysis of the significance of the factor effects, an ANOVA analysis is required. In most cases, however, an examination of the factor effects plot is sufficient for interpretation of the results. The plots give a visual estimate of the Δ's. For a two-level factor, the factor significance is shown by the slope of the factor effect plot, as shown in Figure 13.6.

The factor effect plots clearly show which level is optimum for each of the control factors. It is always the level at which the S/N ratio is maximized. In this case, the optimum set points are crushed

**Table 13.5** Factor response averages table for the parameter optimization experiment

| *Factor* | *Level* | *Temperature (ave.)* | *Temperature (S/N)* |
|---|---|---|---|
| Ice Cubes | Whole | 25.50 | −28.09 |
| | Crushed | 16.50 | −23.98 |
| Cup Type | Paper | 19.00 | −24.79 |
| | Styrofoam | 23.00 | −27.28 |
| Mixing Method | Unstirred | 25.50 | −28.13 |
| | Stirred | 16.50 | −23.94 |

**Figure 13.6** Factor effects plots for the parameter optimization experiment

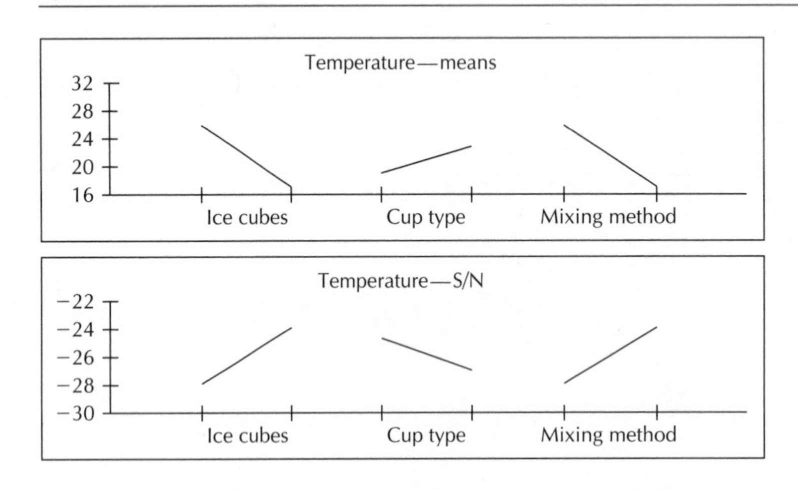


Ice Cubes (factor A, level 2), paper Cup Type (B1), and stirred Mixing Method (C2). As expected for the case of smaller-the-better, the mean response is at a minimum when the S/N is at a maximum.

The factor effect calculations now allow the construction of a predictive model that gives a good estimate of the actual performance expected when the optimum control factor set points just defined (A2, B1, and C2) are tested and ultimately used.

In generating the predictive equation, it can be argued that factor B, Cup Type, should be left out of the estimate for two reasons. First, it has a weak effect on the S/N ratio and the mean temperature as the level changes from paper to Styrofoam. Second, the degrees of freedom contained in B can be used to help estimate experimental error. Both of these arguments are based on the inability to distinguish the factor effect from random variation, a statistical test of significance. In a nonreplicated array, it is very difficult to prove factor significance, and there would be considerable uncertainty about this factor effect.

In this case, however, replicates taken in the noise experiments can be used as an estimate of experimental error. From those data, a mean error (standard deviation) of 0.94°F is found. This sug-

**Figure 13.7** Confirmation analysis from WinRobust

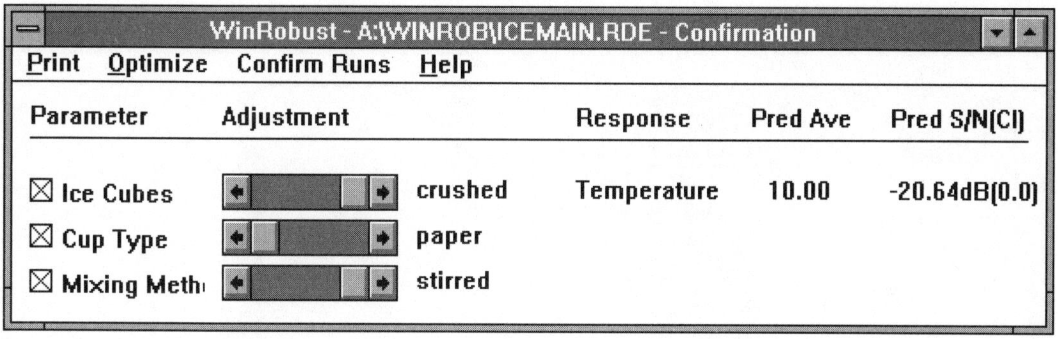

WinRobust - A:\WINROB\ICEMAIN.RDE - Confirmation

Print Optimize Confirm Runs Help

| Parameter | Adjustment | | Response | Pred Ave | Pred S/N(CI) |
|---|---|---|---|---|---|
| ☒ Ice Cubes | | crushed | Temperature | 10.00 | -20.64dB(0.0) |
| ☒ Cup Type | | paper | | | |
| ☒ Mixing Meth | | stirred | | | |

gests that even the 4°F effect due to the cup type may be significant, and so factor B's effects should be included in the predictive model.

The predictive model takes the following form:

$$
\begin{aligned}
S/N_{optimum} &= \overline{S/N}_{exp} + (S/N_{A_2} - \overline{S/N}_{exp}) + (S/N_{B_1} - \overline{S/N}_{exp}) + (S/N_{C_2} - \overline{S/N}_{exp}) \\
&= S/N_{A_2} + S/N_{B_1} + S/N_{C_2} - 2\overline{S/N}_{exp} \\
&= -23.98 + (-24.79) + (-23.94) - 2(-26.03) \\
&= -20.65
\end{aligned}
$$

$S/N_{optimum} = -20.65$ dB . . . *the predicted value*

The *Confirmation* window in WinRobust automatically performs the additive model calculation, as shown in Figure 13.7. Notice that −20.65 dB is better (higher in value) than the majority of the actual experimental runs. In fact, it is the same as experiment #3 from the matrix experiment that was actually run. In large orthogonal array experiments, it is rare that the optimum set points will actually be one of the runs from the original orthogonal array. Therefore, a result such as this is not surprising. Remember, *do not* "pick a winner" from the initial experiment. Go through the ANOM process to fully isolate the main effect of each control factor level on the response. This is why the discipline of an orthogonal array is used in the first place. Use the power it provides.

### 13.1.8 Verification

The verification experiment consists of testing the control factor combination A2, B1, C2. The result is compared to the predicted S/N ratio, $S/N_{optimum} = -20.64$ dB. Figure 13.8 shows the verification results for the ice water example in the *Confirm Runs* window in WinRobust. Data from as many as five runs can be input into WinRobust to check for repeatability.

A comparison of the optimized results shows:

−21.74 dB average actual S/N vs. −20.65 dB predicted S/N . . . very close!

As mentioned in the previous chapter, it is also useful to rerun one of the original experimental runs to see if the results are repeatable. This experimental replication is done to check for drift in the measured values from the initial experiment. Generally, it is not necessary to take replicate data points in robustness experiments, because the noise factors purposefully induce variation well beyond the

**Figure 13.8** Verification analysis from WinRobust

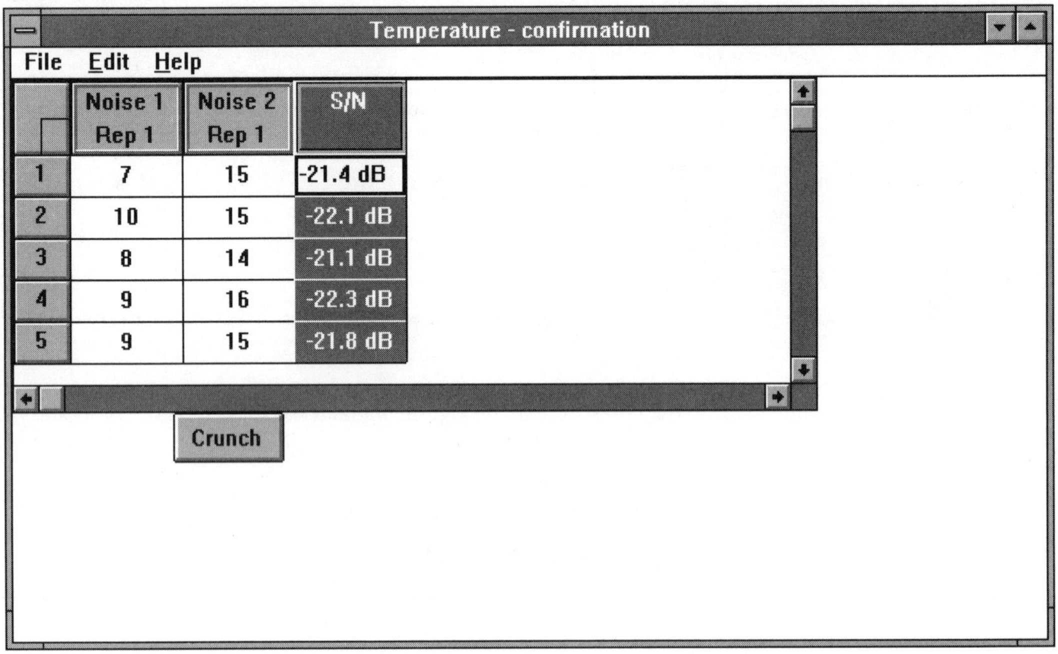

Temperature - confirmation

File Edit Help

| | Noise 1 Rep 1 | Noise 2 Rep 1 | S/N |
|---|---|---|---|
| 1 | 7 | 15 | -21.4 dB |
| 2 | 10 | 15 | -22.1 dB |
| 3 | 8 | 14 | -21.1 dB |
| 4 | 9 | 16 | -22.3 dB |
| 5 | 9 | 15 | -21.8 dB |

Crunch

variation due to conducting the experiment. The variation found in replicates, as seen in the noise experiment, tends to be small errors associated with experimental technique and equipment. Craftsmanship in experimental technique and well-calibrated equipment can keep this to a minimum during the main optimization experiment.

A comparison of the rerun of treatment combination 1 with the original gives the following results:

$$-28.59 \text{ dB rerun S/N vs. } -28.94 \text{ dB original S/N} \ldots \text{ no real drift!}$$

Rarely are such close confirmation values seen, but for such a simple and controlled experiment (i.e., small value for $S_e$), this is reasonable.

## 13.2 The Gyrocopter Experiment: Dynamic Larger-the-Better

This second example is another one that the reader can easily repeat at home or work. In this experiment, the goal is to construct a paper gyrocopter, a device that will spin (autogyrate) and achieve a low terminal velocity as a result of the drag and lift caused by the wings as it spins. This device[2] is intro-

2. The authors wish to acknowledge Shin Taguchi, American Supplier Institute, for introducing us to this device as a teaching aid.

**Figure 13.9** Sketch of the paper gyrocopter

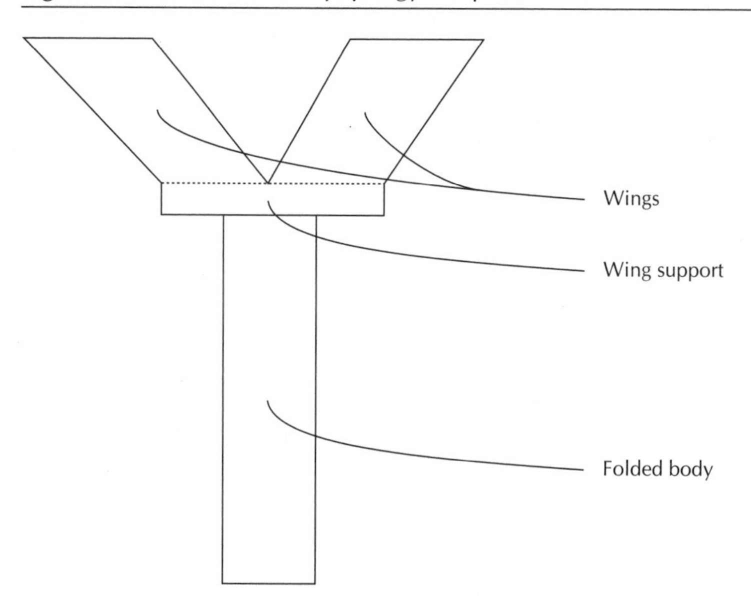


duced in Chapter 10 during the discussion on choosing control factors to minimize interactions. A sketch of the gyrocopter is shown in Figure 13.9.

Imagine the following customer scenario: A gyrocopter design is to be published in a Sunday Comics section as a do-it-yourself project for 6- to 12-year-olds. (This device can also be found in the bookstore of your local science museum.) The customer (the kids) will be happy if they can easily build the device and fly it. Some of the desired flight characteristics include long flight time and fast rotation. There are plenty of noise factors including construction variation (paper type, accuracy of measuring and cutting), launch variation (height of drop, angle of release, angle of wings), and deterioration (loss of rigidity of the paper and creases as they are handled, foreseeable damage to gyrocopter after man[kid!]handling).

Optimizing the gyrocopter is very challenging for several reasons:

1. It is difficult to analyze or get a good intuitive feel for the behavior of the gyrocopter in flight. (If the reader has some experience in aerodynamic engineering, it is a definite plus, but most do not.)
2. The construction materials, paper and glue, are not very well controlled. Experimental error due to variation in the control factor levels has a confounding effect similar to that of interactions.
3. The noise factors are numerous and can have a very strong effect on the gyrocopter flight.

Nevertheless, this device is used for teaching robustness at Kodak and the Rochester Institute of Technology. It is a good class project, and the students usually make progress toward optimization, although many new questions are often raised that require follow-up experiments. The gyrocopter is also used in classical design of experiments courses because it can be empirically modeled, and, without taking countermeasures, has significant control factor interactions.

**Figure 13.10** Ideal function for gyrocopter

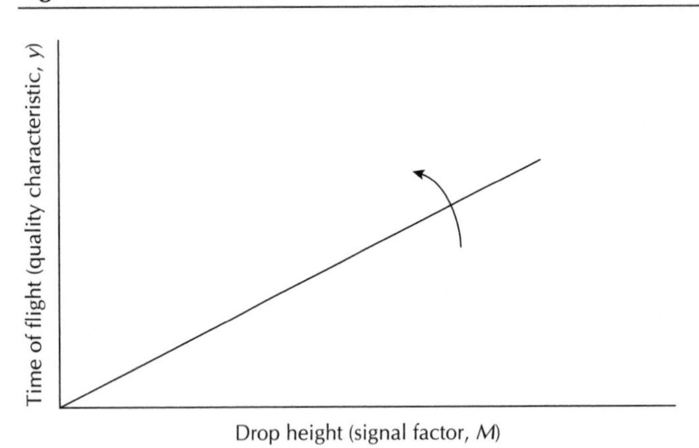


## 13.2.1 Selection of the Quality Characteristic

One obvious quality characteristic for this device is time of flight, which can be analyzed as larger-the-better. But for a thorough robustness parameter design, it is important to determine the dynamic objective function. In fact, most static problems are special cases of an underlying dynamic problem. Identifying the objective function requires finding the dynamic problem. Once the dynamic problem is understood, it is always possible to go ahead and choose a fixed level for the signal factor and do the analysis using a static S/N ratio. For the gyrocopter, the ideal behavior is a constant (low) terminal velocity with *no* free fall period before terminal velocity is achieved. The objective function describing that behavior is the time of flight of a zero-point linear function of drop height, as shown in Figure 13.10. Optimization means maximizing the slope of this function.

To analyze this system for additivity, it helps to visualize the energy flows in this device as it is dropped. The gravitational potential energy causes the gyrocopter to move and fall when dropped. But there is a lot more going on here than free fall. The force of the air on the gyrocopter as it falls provides the pleasing spinning motion. The spinning behavior helps develop lift and maximize drag. The spin and the center of gravity provide flight stability. In order to be robust, the gyrocopter should right itself immediately, regardless of the launch angle. The simple free body diagrams in Figure 13.11 illustrate the engineering analysis and energy flows for the gyrocopter as it falls.

This analysis suggests that there might be multiple quality characteristics to be optimized. However, that would greatly increase the complexity of the optimization process. In order to keep things simple for this illustrative example, the quality characteristic used here is simply the time of flight measured in seconds. For the dynamic experiment, the signal factor levels are three heights: 3 feet, 6 feet, and 9 feet. Although there are many important noises that should be tested in a noise experiment, only one noise, paper weight, is used. The focus here is on using a dynamic analysis for a larger-the-better type problem, and illustrating how to optimize a product with that characteristic. It is left as an exercise for the reader to include the noises that make this a complete robustness study.

**Figure 13.11** Forces acting on the gyrocopter

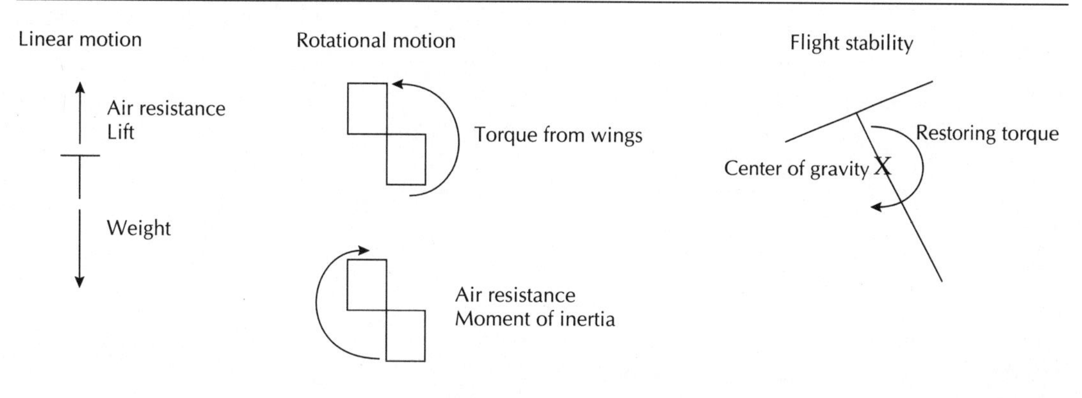


## 13.2.2 Selection of the Control Factors

In Chapter 10, a control factor matrix analysis, Table 10.8, is used to show that there is the potential for many interactions in this system. Because some of these can be anticipated, it is possible to use sliding levels to reduce the interactive effects. The sliding levels help keep the experimental space closer to optimum, thus improving data acquisition (i.e., no "dud" gyrocopters). Using sliding levels also makes the interactions explicit in the definition of the control factors. Table 10.9 shows the expected interactions after these countermeasures are applied. For ease of reference, it is replicated here (Table 13.6).

The torque depends only on wing width, which has weak effects on all of the other anticipated physical processes in the array and is thus treated as an independent factor. The levels chosen for wing width are 0.50 in., 0.75 in., and 1.00 in. Wing length is chosen to achieve three independent levels of wing area, since the area is likely to be important for drag and lift. Wing length is made a dependent

**Table 13.6** Control factor matrix for the gyrocopter (from Table 10.9)

| | *Weight* | *Resistance to fall* | *Torque* | *Moment of inertia* | *Resistance to spin* | *Center of gravity* |
|---|---|---|---|---|---|---|
| *Paper weight* | X | | | Weak | | |
| *Wing area (WL= WA/WW)* | Weak | X | | X | | |
| *Wing width* | Weak | | X | Weak | | |
| *Body length (BL ∝ WL)* | Weak | | | | | X |
| *Body width* | Weak | | | Weak | Weak | |
| *Overall size* | X | | | | Weak | |
| *Body fold* | | | | Weak | X | |
| *Wing gusset* | | X | | | | |

**Table 13.7** Sliding levels for wing width and wing area

| | *WA* = *1.0* | *WA* = *1.5* | *WA* = *2.0* |
|---|---|---|---|
| *WW* = *0.5* | WL = 2.0 | WL = 3.0 | WL = 4.0 |
| *WW* = *0.75* | WL = 1.33 | WL = 2.0 | WL = 2.67 |
| *WW* = *1.0* | WL = 1.0 | WL = 1.5 | WL = 2.0 |

variable by defining WL = WA/WW. The levels chosen for wing area are 1.0 in.$^2$, 1.5 in.$^2$, and 2.0 in.$^2$. The required wing length values are shown in Table 13.7.

In a similar fashion, body length is chosen as a function of wing length in order to keep the center of gravity below the junction of the wings with the body. To keep things simple, the unfolded body width is made equal to the combined width of the two wings. It can be shown, using the parallel axis theorem, that if the body length is equal to the wing length, then the gyrocopter will have poor flight stability, because the center of gravity would be located too close to their junction. The sliding levels chosen for body length are 1.33 × WL, 1.67 × WL, and 2.0 × WL. This results in many possible body lengths, because there are already seven possible wing lengths. Note that body length is still treated as a three-level factor according to the wing length multipliers. All the gyrocopter bodies were folded in thirds and glued along the full length, leaving a 1/4-inch unfolded region to support the wings (see Figure 13.9).

The control factor overall size refers to the physical size reduction (using a photocopier) of the gyrocopter. This is tested in order to get the effect of substantially reducing the gyrocopter weight while maintaining proportional dimensions. Body fold refers to the folding up of the bottom 15% or 30% of the body length to try and achieve greater stability. Gussets refers to an additional design feature intended to keep the wings flat. Note that gussets are a considerable design complication and would only be used if they substantially improve flight performance. There are only two levels for gussets, none or gussets cut at 45°. (The 45° gussets hold the wing at a 90° angle to the body.) This is handled using a dummy level (see Chapter 17) in the last column of the control factor array.

As a result of the experimental planning, it is now possible to construct the P-diagram shown in Figure 13.12.

### 13.2.3 Conducting the Main Experiment

The control factor array chosen is the L18. It provides an additional countermeasure to interactions, as discussed in Chapter 16. Running a few preliminary replicates showed that measurement error using a stopwatch was not significant. It is useful to have only one timekeeper to avoid human variation. The outer array consists of a signal factor at three levels (3 feet, 6 feet, and 9 feet) and the noise factor at two levels. For the noise factor, two gyrocopters are dropped at each height, one made of 20# bond paper (light weight) and one made of 24# laser print paper (heavy weight). The control factor array is shown in Table 13.8.

**Figure 13.12** P-diagram for the paper gyrocopter (dynamic case)

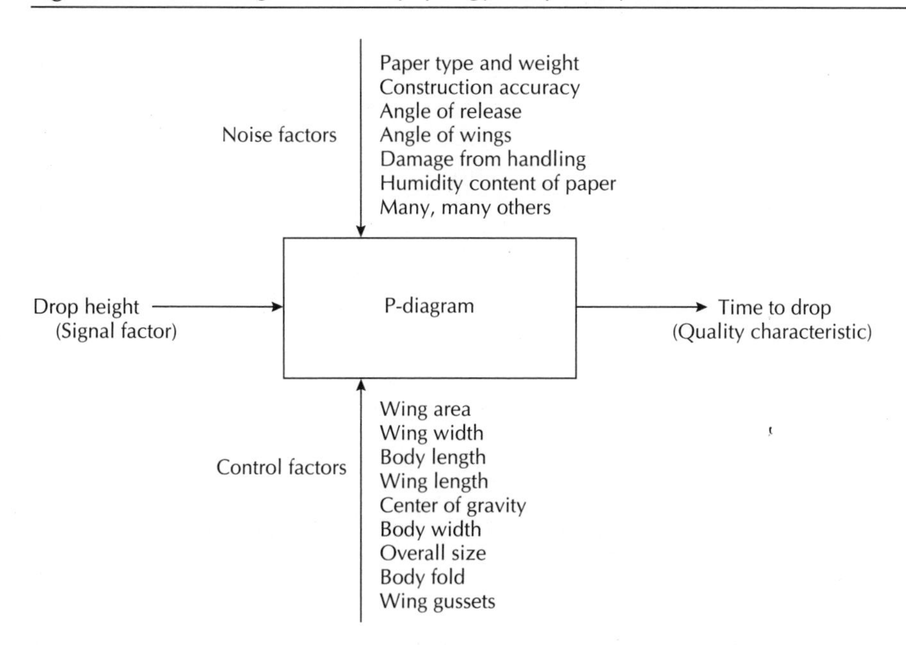


**Table 13.8** Control factor array for the paper gyrocopter parameter optimization experiment

| *Run* | *1* | *2* *WL* | *3* *WW* | *4* *BL* | *5* *Size* | *6* | *7* *B_Fold* | *8* *Gussets* |
|---|---|---|---|---|---|---|---|---|
| 1 | 1 | 1.0/ww | 0.50 | 1.33 × WL | 100% | 1 | 0 | None |
| 2 | 1 | 1.0/ww | 0.75 | 1.67 × WL | 75% | 2 | 15% | 45deg |
| 3 | 1 | 1.0/ww | 1.00 | 2.00 × WL | 50% | 3 | 30% | 45deg |
| 4 | 1 | 1.5/ww | 0.50 | 1.33 × WL | 75% | 2 | 30% | 45deg |
| 5 | 1 | 1.5/ww | 0.75 | 1.67 × WL | 50% | 3 | 0 | None |
| 6 | 1 | 1.5/ww | 1.00 | 2.00 × WL | 100% | 1 | 15% | 45deg |
| 7 | 1 | 2.0/ww | 0.50 | 1.67 × WL | 100% | 3 | 15% | 45deg |
| 8 | 1 | 2.0/ww | 0.75 | 2.00 × WL | 75% | 1 | 30% | None |
| 9 | 1 | 2.0/ww | 1.00 | 1.33 × WL | 50% | 2 | 0 | 45deg |
| 10 | 2 | 1.0/ww | 0.50 | 2.00 × WL | 50% | 2 | 15% | None |
| 11 | 2 | 1.0/ww | 0.75 | 1.33 × WL | 100% | 3 | 30% | 45deg |
| 12 | 2 | 1.0/ww | 1.00 | 1.67 × WL | 75% | 1 | 0 | 45deg |
| 13 | 2 | 1.5/ww | 0.50 | 1.67 × WL | 50% | 1 | 30% | 45 deg |
| 14 | 2 | 1.5/ww | 0.75 | 2.00 × WL | 100% | 2 | 0 | 45deg |
| 15 | 2 | 1.5/ww | 1.00 | 1.33 × WL | 75% | 3 | 15% | None |
| 16 | 2 | 2.0/ww | 0.50 | 2.00 × WL | 75% | 3 | 0 | 45deg |
| 17 | 2 | 2.0/ww | 0.75 | 1.33 × WL | 50% | 1 | 15% | 45deg |
| 18 | 2 | 2.0/ww | 1.00 | 1.67 × WL | 100% | 2 | 30% | None |

**Table 13.9** Data from the parameter optimization experiment

| | *3 feet* | | *6 feet* | | *9 feet* | |
|---|---|---|---|---|---|---|
| | *20# paper* | *24# paper* | *20# paper* | *24# paper* | *20# paper* | *24# paper* |
| 1 | 0.68 s | 0.55 s | 1.48 s | 1.48 s | 2.31 s | 2.38 s |
| 2 | 0.74 | 0.58 | 1.19 | 1.58 | 2.25 | 2.44 |
| 3 | 0.68 | 0.45 | 1.35 | 1.03 | 1.48 | 1.96 |
| 4 | 0.58 | 0.71 | 1.25 | 1.22 | 2.34 | 1.75 |
| 5 | 0.71 | 0.68 | 1.58 | 1.41 | 2.28 | 2.41 |
| 6 | 0.67 | 0.55 | 1.64 | 1.51 | 2.44 | 2.08 |
| 7 | 0.65 | 0.7 | 1.16 | 1.21 | 2.68 | 2.7 |
| 8 | 0.71 | 0.6 | 1.93 | 1.75 | 2.61 | 2.73 |
| 9 | 0.84 | 0.63 | 1.83 | 1.64 | 2.09 | 2.5 |
| 10 | 0.74 | 0.61 | 1.7 | 1.22 | 2.09 | 2.31 |
| 11 | 0.61 | 0.45 | 1.22 | 1.03 | 1.48 | 1.96 |
| 12 | 0.61 | 0.58 | 1.38 | 1.22 | 2.28 | 2.3 |
| 13 | 0.87 | 0.68 | 1.64 | 1.19 | 2.02 | 2.41 |
| 14 | 0.81 | 0.65 | 2.09 | 1.51 | 2.27 | 2.67 |
| 15 | 0.84 | 0.63 | 1.7 | 1.22 | 1.51 | 2.5 |
| 16 | 0.68 | 0.68 | 1.54 | 1.64 | 2.44 | 2.5 |
| 17 | 0.71 | 0.58 | 1.7 | 1.51 | 2.6 | 2.6 |
| 18 | 0.61 | 0.84 | 1.96 | 1.64 | 2.73 | 3.05 |

The way the gyrocopter is held and dropped does have an influence on flight time and stability (additional noise factors!). Different flight characteristics are observed when the body is held prior to release and when the wings are held prior to release. The gyrocopter is held by the body just below the wings, and the wings are fluffed upwards prior to release.[3] This gives the best flight.

The outer array and the data obtained for the 18 gyrocopters are shown in Table 13.9.

### 13.2.4 Analysis of Data—Reducing the Raw Data to Slopes and S/N Ratios

The data analysis for the slope and zero-point S/N ratio values of each treatment combination is done using WinRobust. The *Add Dynamic Characteristic* option is chosen from the *Response* menu. The numbers of signal levels (3) and noise levels (2) are input, and zero-point proportional analysis is chosen. After *OK* is clicked, the data input screen shown in Figure 13.13 appears. The data are entered as

3. In the experiment this could have been treated as a noise factor.

**Figure 13.13** Data input window from WinRobust for a dynamic problem

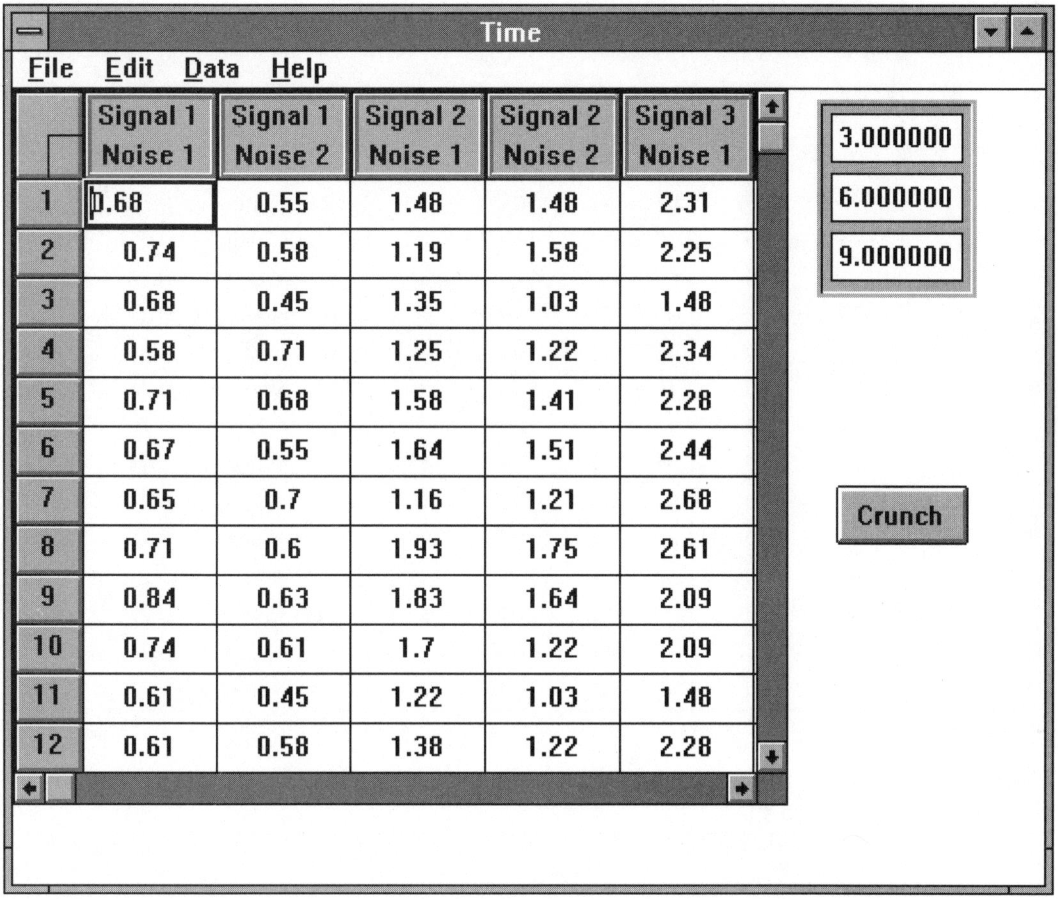

Time

File Edit Data Help

| | Signal 1 Noise 1 | Signal 1 Noise 2 | Signal 2 Noise 1 | Signal 2 Noise 2 | Signal 3 Noise 1 |
|---|---|---|---|---|---|
| 1 | 0.68 | 0.55 | 1.48 | 1.48 | 2.31 |
| 2 | 0.74 | 0.58 | 1.19 | 1.58 | 2.25 |
| 3 | 0.68 | 0.45 | 1.35 | 1.03 | 1.48 |
| 4 | 0.58 | 0.71 | 1.25 | 1.22 | 2.34 |
| 5 | 0.71 | 0.68 | 1.58 | 1.41 | 2.28 |
| 6 | 0.67 | 0.55 | 1.64 | 1.51 | 2.44 |
| 7 | 0.65 | 0.7 | 1.16 | 1.21 | 2.68 |
| 8 | 0.71 | 0.6 | 1.93 | 1.75 | 2.61 |
| 9 | 0.84 | 0.63 | 1.83 | 1.64 | 2.09 |
| 10 | 0.74 | 0.61 | 1.7 | 1.22 | 2.09 |
| 11 | 0.61 | 0.45 | 1.22 | 1.03 | 1.48 |
| 12 | 0.61 | 0.58 | 1.38 | 1.22 | 2.28 |

3.000000

6.000000

9.000000

Crunch

shown. (Note that there are more data not displayed.) The signal factor levels are entered in the boxes to the right.

Clicking on *Crunch* causes each row of data to be analyzed for the slope and S/N ratio using the zero-point proportional equations given in Chapter 6. As a result, each row in the control factor array has a pair of responses associated with it, as shown in Table 13.10.

It is often useful to plot some of the key experimental runs in order to interpret the S/N ratio. The best S/N ratio is found for treatment combination 16. The worst is found for treatment combination 15. The highest slope is found for run 18; the lowest for run 3 (or 11). Figures 13.14 and 13.15, plots of six points per run, show comparisons of these pairs of conditions, giving a visual representation of the range of data variation from this experiment.

**Table 13.10** Results of the parameter optimization experiment

| Run | *1* | *2* *WL* | *3* *WW* | *4* *BL* | *5* *Size* | *6* | *7* *B_Fold* | *8* *Gussets* | *Time (slope)* | *Time (S/N)* |
|---|---|---|---|---|---|---|---|---|---|---|
| 1 | 1 | 1.0/ww | 0.50 | 1.33 × WL | 100% | 1 | 0 | None | 0.25 | 6.94 dB |
| 2 | 1 | 1.0/ww | 0.75 | 1.67 × WL | 75% | 2 | 15% | 45deg | 0.25 | 2.67 dB |
| 3 | 1 | 1.0/ww | 1.00 | 2.00 × WL | 50% | 3 | 30% | 45deg | 0.19 | −0.24 dB |
| 4 | 1 | 1.5/ww | 0.50 | 1.33 × WL | 75% | 2 | 30% | 45deg | 0.22 | 0.69 dB |
| 5 | 1 | 1.5/ww | 0.75 | 1.67 × WL | 50% | 3 | 0 | None | 0.26 | 9.04 dB |
| 6 | 1 | 1.5/ww | 1.00 | 2.00 × WL | 100% | 1 | 15% | 45deg | 0.25 | 3.81 dB |
| 7 | 1 | 2.0/ww | 0.50 | 1.67 × WL | 100% | 3 | 15% | 45deg | 0.26 | −1.95 dB |
| 8 | 1 | 2.0/ww | 0.75 | 2.00 × WL | 75% | 1 | 30% | None | 0.29 | 4.73 dB |
| 9 | 1 | 2.0/ww | 1.00 | 1.33 × WL | 50% | 2 | 0 | 45deg | 0.26 | 2.64 dB |
| 10 | 2 | 1.0/ww | 0.50 | 2.00 × WL | 50% | 2 | 15% | None | 0.24 | 2.81 dB |
| 11 | 2 | 1.0/ww | 0.75 | 1.33 × WL | 100% | 3 | 30% | 45deg | 0.19 | 0.76 dB |
| 12 | 2 | 1.0/ww | 1.00 | 1.67 × WL | 75% | 1 | 0 | 45deg | 0.24 | 3.87 dB |
| 13 | 2 | 1.5/ww | 0.50 | 1.67 × WL | 50% | 1 | 30% | 45 deg | 0.24 | 1.62 dB |
| 14 | 2 | 1.5/ww | 0.75 | 2.00 × WL | 100% | 2 | 0 | 45deg | 0.28 | 0.87 dB |
| 15 | 2 | 1.5/ww | 1.00 | 1.33 × WL | 75% | 3 | 15% | None | 0.23 | −3.96 dB |
| 16 | 2 | 2.0/ww | 0.50 | 2.00 × WL | 75% | 3 | 0 | 45deg | 0.27 | 9.04 dB |
| 17 | 2 | 2.0/ww | 0.75 | 1.33 × WL | 50% | 1 | 15% | 45deg | 0.28 | 4.88 dB |
| 18 | 2 | 2.0/ww | 1.00 | 1.67 × WL | 100% | 2 | 30% | None | 0.31 | 2.99 dB |

**Figure 13.14** Plot of the raw data for the best S/N result (run #16) and worst S/N result (run #15)

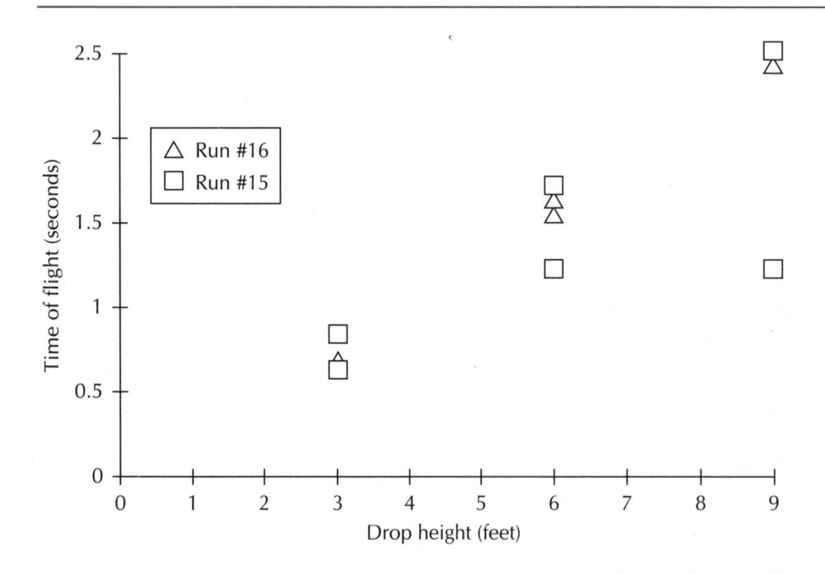

**Figure 13.15** Plot of the raw data for the highest (best) slope result (run #18) and lowest (worst) slope result (run #3)

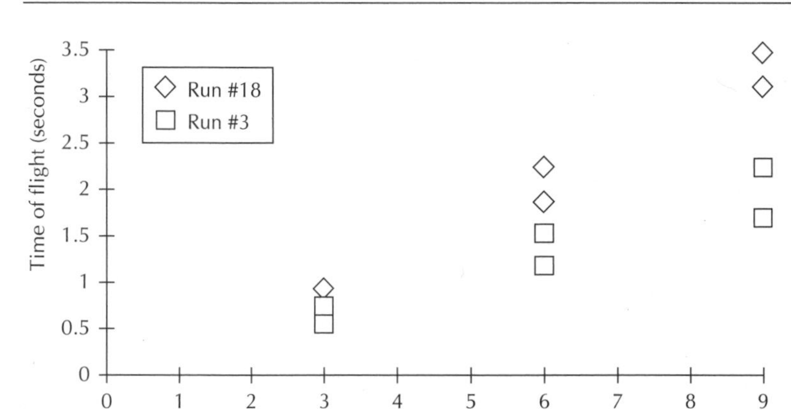


## 13.2.5 Analysis of Data—ANOM

The next step is to analyze the control factor effects that are responsible for the variation observed in the slopes and S/N ratios. Table 13.11 and Figure 13.16 show the factor response averages and factor effects plots obtained from the *Results and Analysis* window in WinRobust.

Examination of the plots in Figure 13.16 reveals the optimum levels for the control factors. Recall the two-step optimization procedure. For this larger-the-better problem, the two-step optimization takes the form of first optimizing the quality by choosing factor levels that increase the S/N ratio, and second, optimizing the slope by choosing factor levels that increase the slope. There is no slope target; the larger the slope is, the longer the flight time is for any height.

Factors are categorized according to how they influence the S/N ratio and the slope. Here it is apparent that body fold has the largest influence on S/N ratio, and wing length has the largest influence on slope. The remaining factors contribute somewhat to both S/N and slope. For this problem, there is little contradiction between the factor levels that optimize S/N and slope. If that is not true, some judgment as to the optimum configuration (balancing or compromising between insensitivity to the noise factors and achieving a high slope) has to be exercised.

## 13.2.6 Data Analysis—Prediction and Verification

The analysis of the L18 orthogonal array is used to construct an additive model. The additive model requires factor effects that are significant enough to produce predictable results without resorting to the constraints of interactions. The *Confirmation and Prediction* option in WinRobust from the *Window* menu uses the additive model.

The optimizer can be used to choose the optimum levels for the control factors to maximize either S/N or slope. Figure 13.17 shows the results of the optimizer when maximizing the S/N ratio is chosen. The factor levels that give the largest S/N ratios from the plot in Figure 13.16 give the same result as the WinRobust optimizer. Applying the optimizer again using maximum slope as the criterion gives the results found in Figure 13.18.

**Table 13.11** Factor response averages table for the parameter optimization experiment

| *Factor* | *Level* | *Time (slope)* | *Time (S/N)* |
|---|---|---|---|
| WL | 1.0/ww | 0.23 | 2.80 |
| | 1.5/ww | 0.25 | 2.01 |
| | 2.0/ww | 0.28 | 3.72 |
| WW | 0.50 | 0.25 | 3.19 |
| | 0.75 | 0.26 | 3.82 |
| | 1.00 | 0.25 | 1.52 |
| BL | 1.33 × WL | 0.24 | 1.99 |
| | 1.67 × WL | 0.26 | 3.04 |
| | 2.00 × WL | 0.25 | 3.50 |
| Size | 100% | 0.26 | 2.23 |
| | 75% | 0.25 | 2.84 |
| | 50% | 0.25 | 3.46 |
| B_Fold | 0 | 0.26 | 5.40 |
| | 15% | 0.25 | 1.38 |
| | 30% | 0.24 | 1.76 |
| Gussets | None | 0.26 | 3.76 |
| | 45deg | 0.25 | 2.39 |

**Figure 13.16** Factor effects plots for the parameter optimization experiment

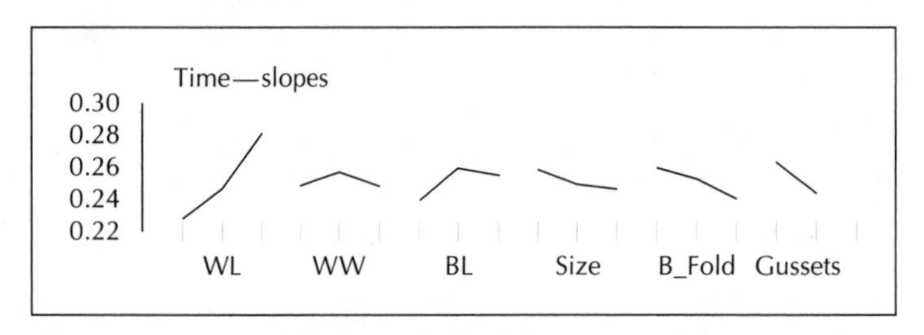


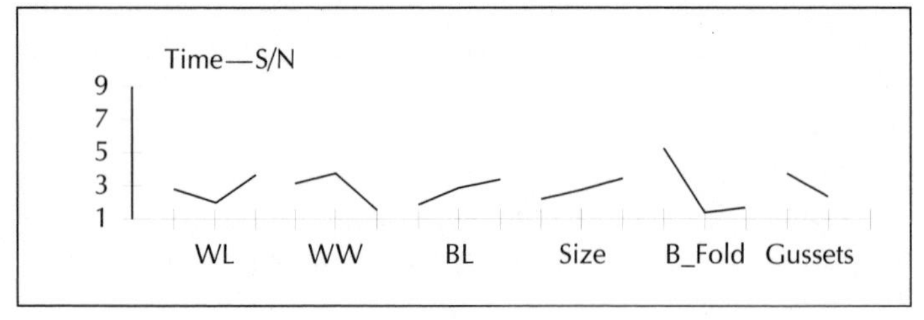

Figure 13.17 Optimum set points for maximizing the S/N ratio

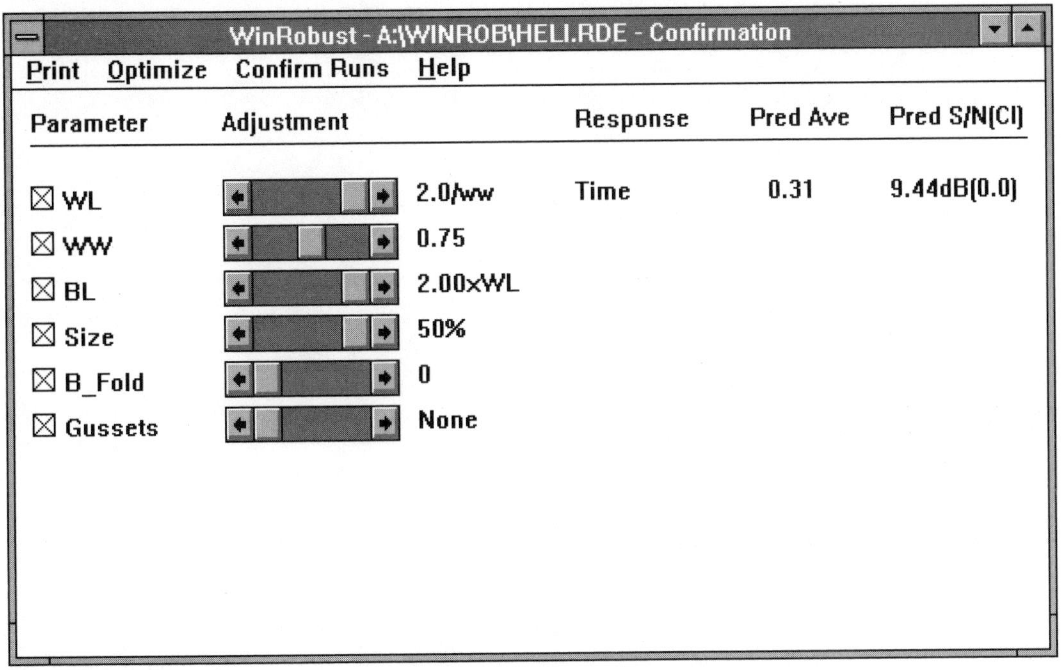

WinRobust - A:\WINROB\HELI.RDE - Confirmation

Print Optimize Confirm Runs Help

| Parameter | Adjustment | Response | Pred Ave | Pred S/N(CI) |
|---|---|---|---|---|
| ☒ WL | 2.0/ww | Time | 0.31 | 9.44dB(0.0) |
| ☒ WW | 0.75 | | | |
| ☒ BL | 2.00xWL | | | |
| ☒ Size | 50% | | | |
| ☒ B_Fold | 0 | | | |
| ☒ Gussets | None | | | |

Figure 13.18 Optimum set points for maximizing the slope

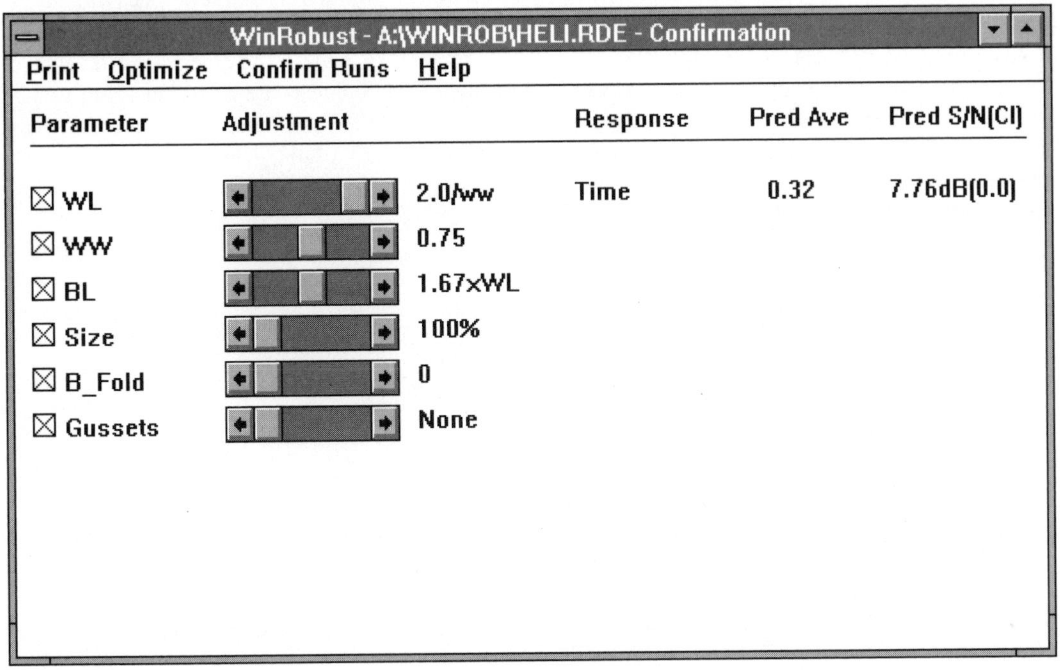

WinRobust - A:\WINROB\HELI.RDE - Confirmation

Print Optimize Confirm Runs Help

| Parameter | Adjustment | Response | Pred Ave | Pred S/N(CI) |
|---|---|---|---|---|
| ☒ WL | 2.0/ww | Time | 0.32 | 7.76dB(0.0) |
| ☒ WW | 0.75 | | | |
| ☒ BL | 1.67xWL | | | |
| ☒ Size | 100% | | | |
| ☒ B_Fold | 0 | | | |
| ☒ Gussets | None | | | |

**Table 13.12** Results of replicates of treatment combination #8

| | *3 feet* | | *6 feet* | | *9 feet* | | | |
|---|---|---|---|---|---|---|---|---|
| | *20#* | *24#* | *20#* | *24#* | *20#* | *24#* | *Slope* | *S/N* |
| 1 | 0.87 s | 0.74 s | 2.02 s | 1.64 s | 2.86 s | 2.57 s | 0.30 s/ft | 4.98 dB |
| 2 | 0.87 | 0.61 | 1.83 | 1.67 | 2.63 | 2.57 | 0.29 | 7.11 |
| 3 | 0.80 | 0.71 | 1.83 | 1.77 | 2.86 | 2.76 | 0.30 | 8.07 |
| 4 | 1.13 | 0.74 | 2.06 | 1.70 | 2.86 | 2.67 | 0.31 | 4.73 |
| 5 | 0.97 | 0.84 | 1.83 | 1.64 | 2.89 | 2.73 | 0.30 | 8.47 |

**Table 13.13** Verification test results checking for reproducibility

| *Run #8* | *Previous* | *Predicted* | *Verification* |
|---|---|---|---|
| *Slope* | 0.29 s/ft | 0.29 | 0.30 |
| *S/N ratio* | 4.73 dB | 5.18 | 6.68 |

Changing body length and size results in a slope increase of 0.01 s/foot, but at a S/N decrease of more than 2.5 dB. (Note that *Pred Ave* refers to the slope in this dynamic case.) Recalling that a 3 dB decrease represents an increase in quality loss of 2×, this is not a very good trade-off. The recommended optimum in this particular case is therefore the factor levels that maximize the S/N ratio.

Two different tests are run for the verification. Run number 8 is replicated five times to test for repeatability. Brand-new gyrocopters are made using a wing width of 0.75 inches, a wing length of 2.0/0.75 = 2.67 inches, and a body length of 2.0 × 2.67 = 5.33 inches. The other factor levels are size = 75%, body fold = 30%, and no gussets. The results of the five tests are shown in Table 13.12. The previous slope and S/N ratio results, the predicted results, and the verification results are shown in Table 13.13. The verification gyrocopter results given in Table 13.15 are the average of the five replicate tests shown in Table 13.14. The results are somewhat larger, but reasonably close to the previously obtained results and the additive model predictions.

**Table 13.14** Results of replicates of optimum configuration

| | *3 feet* | | *6 feet* | | *9 feet* | | | |
|---|---|---|---|---|---|---|---|---|
| | *20# paper* | *24# paper* | *20# paper* | *24# paper* | *20# paper* | *24# paper* | *Slope* | *S/N* |
| 1 | 0.80 s | 0.74 s | 1.96 s | 1.70 s | 2.73 s | 2.89 s | 0.31 s/ft | 6.77 dB |
| 2 | 0.97 | 0.97 | 1.93 | 2.02 | 2.95 | 3.18 | 0.34 | 11.48 |
| 3 | 0.87 | 0.87 | 2.09 | 1.86 | 3.05 | 2.86 | 0.33 | 8.83 |
| 4 | 0.90 | 1.06 | 1.96 | 1.93 | 3.05 | 2.99 | 0.33 | 13.98 |
| 5 | 0.80 | 0.87 | 1.96 | 1.70 | 2.92 | 2.82 | 0.31 | 8.25 |

**Table 13.15** Verification test results for optimum configuration

| *Optimum S/N* | *Predicted* | *Verification* |
|---|---|---|
| *Slope* | 0.31 | 0.32 |
| *S/N ratio* | 9.44 | 9.86 |

Next, the optimum control factor combination for maximizing the S/N ratio is tested by building five gyrocopters using the optimum set points shown in Figure 13.17. The optimum control factor levels are: wing width = 0.75 inches, wing length = 2.0/0.75 = 2.67 inches, and body length = 2.00 × 2.67 = 5.33 inches. The other factor levels are size = 50%, no body fold, and no gussets. The results of the five tests are shown in Table 13.14. The slope and S/N ratio from the additive model and the verification test are shown in Table 13.15. The results for the verification tests indicate good repeatability and good slope predictability.

Very little noise is used in the experiment just described. The P-diagram in Figure 13.12, based on observing many students testing gyrocopters, should give the reader plenty of ideas for planning a noise experiment using the optimum gyrocopter described here as a starting point. After analysis and compounding, consider new control factors for the parameter optimization experiment. The results obtained in this experiment suggest that gussets should be dropped. They don't help and they are expensive (actually a nuisance to build). That offers up to three empty columns with which to study additional control factors.

The fact that we are suggesting another round of experimentation to further optimize the gyrocopter should come as no surprise to experienced engineers. Almost all optimization efforts require iterative solutions. Follow-up experiments, using the best robustness optimization of the previous experiment, allow for higher levels of noise factor stress. Ineffective control factors can be dropped and new factors included. Also, the cost and risk of two or three moderate-sized experiments is much more manageable than trying to solve all the design problems in one experimental effort. When doing iterative parameter optimization experimentation, be sure to carry over significant control factors from one test to the next. If completely different sets of control factors are used in successive experiments (for example, running two L9 tests instead of one L18), then unpleasant surprises from control-factor combinations that were never tested together may result. Stick with the six to eight control-factor guideline and drop/add three to five factors for each iteration.

## 13.3 The Catapult Experiment

This third example requires some special equipment to perform yourself. The experiment employs a toy catapult. This device[4] is used in Chapter 8 to illustrate choosing quality characteristics and in Chapter 10 to illustrate choosing control factors. The device is shown in Figure 10.2, and the control factors are described in Chapter 10. The function of the catapult is to launch a projectile to hit a target that can be located anywhere from 24 to 120 inches away.

A significant noise factor is the type of projectile. Golf balls, small rubber balls, and foam balls (indoor practice golf balls) can be used. Error in the pull-back angle and catapult-to-catapult variation

4. The authors wish to acknowledge Bob Launsby, Launsby Consultants, for introducing us to this device as a teaching aid.

are two additional noise factors. There can also be operator variation. Optimizing the catapult means reducing the variation due to these noise factors. It may not be possible to eliminate the noise factor influence entirely, but reducing the spread in impact location due to the noises is the goal. Correction for the ball type, for example, could then be considered in the tolerance phase.

This device is used for teaching the robustness process at Kodak and Rochester Institute of Technology. It is a good class project, and the students usually achieve substantial improvement. Like the gyrocopter, the catapult is also used in classical design-of-experiment courses, as it, too, can be empirically modeled, and, without taking countermeasures, also has significant control-factor interactions. To make the task more challenging for the robustness students, we restrict them to using four control factors in an L9 array. This constraint makes it necessary to minimize interactions through the application of the countermeasures presented here.

First the countermeasures needed to minimize interactions are discussed. Then the results of the parameter optimization experiment using these countermeasures are presented. Interactions are not completely eliminated in this case study. Nevertheless, ANOM applied to the S/N ratio gives the optimum catapult configuration.

### 13.3.1 Selection of the Quality Characteristic

The selection of the quality characteristic for the catapult is discussed in Chapter 8. There it is shown that the best quality characteristic is the distance, measured in inches, that the projectile travels from the base of the catapult mechanism. This quality characteristic has all the desirable properties to minimize interactions. The catapult quality characteristic is illustrated in Figure 8.17.

The dynamic objective function associated with this quality characteristic is a zero-point linear function of range vs. pull-back angle. There are two reasons for selecting pull-back angle as the signal factor. First, the pull-back angle directly provides the energy to project the balls. If there is no pull-back, the range will be zero. The linearity is an assertion that must be verified by the experiment, but there is no reason a priori, using a linear elastic device such as a rubber band, not to expect a zero-point proportional relationship to apply in this case. Second, the pull-back angle is the only continuous control factor. This is an important property of a signal factor, since it is used to put the projectile onto the target. The ideal function expected is shown in Figure 13.19.

**Figure 13.19** Ideal function for the catapult

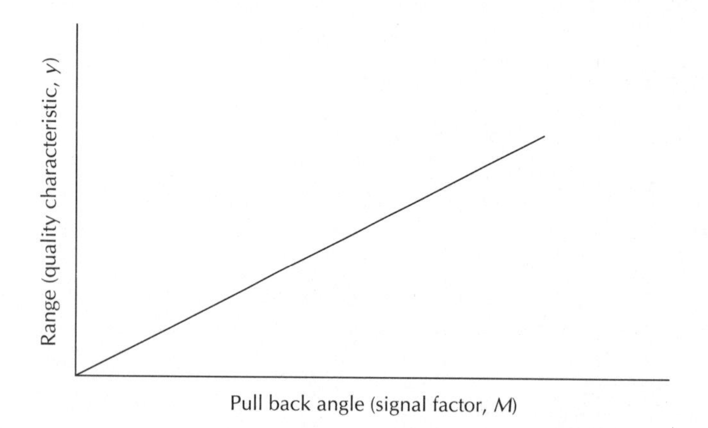

To analyze this system for additivity, it helps to visualize the energy flows as this device is operated. The catapult is fundamentally an energy transformation device. Potential energy due to elastic elongation is transformed into kinetic energy in the form of projectile velocity. The second energy flow is the free-fall projectile motion of the balls. (Here, air resistance is neglected as a minor factor in the analysis.) The effect of gravity causes the upward component of the ball's motion to stop, and then causes the ball to fall. The motion parallel to the ground continues unabated until the projectile strikes the floor. It is possible to solve the equations of motion for the catapult and free fall [S1]. The range is a function of the speed of the projectile and the angle and height of release.

### 13.3.2 Selection of the Control Factors

The six control factors can be grouped into two distinct categories according to how they affect the energy transformation in the catapult. The first group consists of those factors that control the storage and release of the elastic energy (kinetics). The second group consists of those factors that control the geometry of motion of the projectile (kinematics). The factors that control energy storage and release influence the speed of the projectile. The factors that control the geometry of motion influence the angle and height of release. The high degree of functional redundancy in the catapult's design elements is almost certain to induce interactions. The task of the engineer is to promote additivity between the control factors in the catapult. Here, altering and grouping the control factors is used to improve additivity. The control factors are analyzed in Chapter 10 with the results shown in Table 13.16.

The study here is done dynamically using the relative pull-back angle as the signal factor. The relative pull-back angle is referenced to the stop location using sliding levels. Rather than using fixed locations for the pull-back angle levels, the relative pull-back angle is defined to be the distance pulled back from the point where the arm rests on the stop pin. That location slides as the level of the arm stop point is changed. This definition of the pull-back angle minimizes interactivity by reducing the effect of arm stop angle on the catapult dynamics. The levels used for the outer array are 15°, 30°, and 45°. The range available for these levels is limited by the stop location. The smallest stop location used is 45°. Thus, the largest relative pull-back angle used is 45°.

The control factors used and their levels are given in Table 13.17. Stop location (factor C) is the only factor that determines the release angle. Catapult height (factor A) is used as a coarse adjustment of projectile height: The levels chosen are much greater than the variation in height caused by the arm stop point. For the catapult study done here, cup position is left constant in location #1 (fully extended) so that that interaction is avoided.

**Table 13.16** The final control factor matrix for the catapult (from Table 10.7)

| | *Angle* | *Height* | *Speed* |
|---|---|---|---|
| *Cup position* | | Weak | Weak |
| *Spring force* | | | X |
| *Stop location* | X | Weak | |
| *Relative pull-back angle* | | | X |
| *Catapult height* | | X | |

**Table 13.17** Control factors and levels for the catapult parameter optimization experiment

| *Control factors* | *Level 1* | *Level 2* | *Level 3* |
|---|---|---|---|
| A: Catapult height (from floor level) | 0 inches | 15 inches | 30 inches |
| B: Number of rubber bands | 1 | 2 | 3 |
| C: Stop pin angle | 45° | 60° | 75° |
| D: Elastic stretch (compound factor) | ES1 | ES2 | ES3 |

The number of rubber bands (factor B) and the elongation of the rubber bands (factor D) are treated separately here despite the interaction potential. The spring force is given by the product of the elastic constant and the amount of elongation ($F = kx$). The number of rubber bands determines the elastic constant for the system. The amount of elongation, prior to pull-back, is a function of the elastic-band tie point (EBT) and the elastic-band stretch point (EBS); thus, these are redundant. In such a case, the control factors should be grouped. It would be possible to hold one constant and only vary the other, but that reduces the experimental space. It is better to change them *together.* The new, compound control factor (factor D), referred to simply as elastic stretch (ES), has the following levels:

$$\text{ES1} = \text{EBT1} + \text{EBS1 (high tension)}$$

$$\text{ES2} = \text{EBT2} + \text{EBS2 (medium tension)}$$

$$\text{ES3} = \text{EBT3} + \text{EBS2 (low tension)}$$

As a result of the experimental planning, it is now possible to construct the P-diagram shown in Figure 13.20.

**Figure 13.20** P-diagram for the catapult

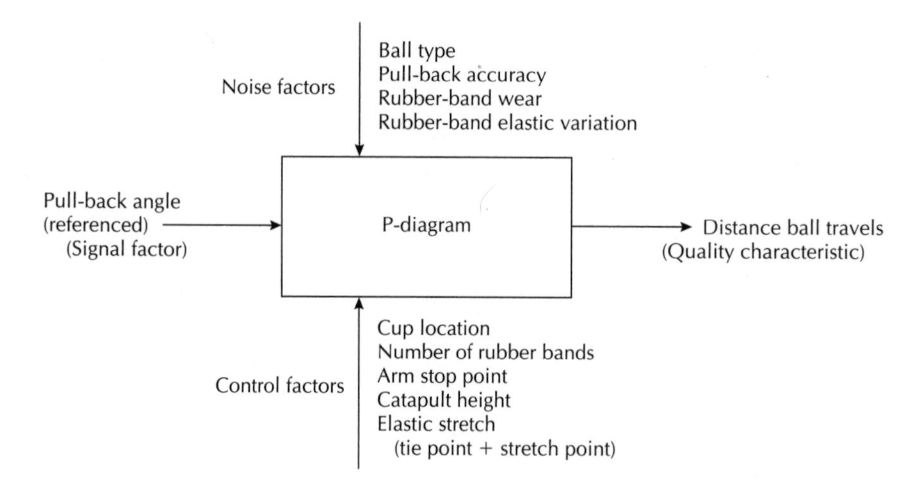

### 13.3.3 The Noise Factor

Measurement error is a significant concern because it is difficult to see exactly where the swiftly moving projectile strikes the floor. Marking techniques such as chalking the balls or carbon-paper landing pads are very useful. Also effective is placing a reference object at various positions on the ground and repeating each run until it strikes the object. Our approach is to run each condition twice. The first throw is used to locate the impact spot, which is marked with a pencil. The impact spot for the second throw is recorded. The pencil acts as a reference, thus minimizing measurement error. No replications or averaging is done in the data collection. The only noise factor chosen is ball type with three levels: (1) Regulation golf ball (white), (2) indoor foam practice golf ball (yellow), (3) small rubber ball (red).

### 13.3.4 Conducting the Main Experiment

Table 13.18 shows the inner array (control factor array) used for this study. The results obtained are shown in Table 13.19.

### 13.3.5 Analysis of Results

The data for each row are used to calculate the zero-point proportional dynamic S/N ratio and the slope, $\beta$. The results, obtained using WinRobust, are shown in Table 13.20.

The control factor effects for the S/N and slope responses are calculated by doing the ANOM. The goal is to reduce the sensitivity and to develop a linear function that can be used to hit the target. Notice that no target is specified. That is because *the machine's function* is being optimized, not its ability to hit a specific target. This concept is what makes the dynamic S/N analysis very powerful. The two-step optimization takes the following form. First, the control-factor set points that optimize the

**Table 13.18** Control factor array for the catapult parameter optimization experiment

| Run | *1* *Height* | *2* *No. bands* | *3* *StopPin* | *4* *Stretch* |
|---|---|---|---|---|
| 1 | 0 | 1 | 45 | ES3 |
| 2 | 0 | 2 | 60 | ES2 |
| 3 | 0 | 3 | 75 | ES1 |
| 4 | 15 | 1 | 60 | ES1 |
| 5 | 15 | 2 | 75 | ES3 |
| 6 | 15 | 3 | 45 | ES2 |
| 7 | 30 | 1 | 75 | ES2 |
| 8 | 30 | 2 | 45 | ES1 |
| 9 | 30 | 3 | 60 | ES3 |

**Table 13.19** Data from the parameter optimization experiment

| | *15 degrees* | | | *30 degrees* | | | *45 degrees* | | |
|---|---|---|---|---|---|---|---|---|---|
| | *white* | *yellow* | *red* | *white* | *yellow* | *red* | *white* | *yellow* | *red* |
| 1 | 18 in. | 26 in. | 22 in. | 40 in. | 58 in. | 52 in. | 58 in. | 83 in. | 78 in. |
| 2 | 50 | 64 | 61 | 90 | 118 | 11 | 136 | 161 | 158 |
| 3 | 58 | 64 | 54 | 91 | 100 | 94 | 154 | 182 | 156 |
| 4 | 64 | 75 | 67 | 118 | 135 | 133 | 163 | 187 | 188 |
| 5 | 18 | 27 | 20 | 33 | 37 | 34 | 42 | 53 | 42 |
| 6 | 74 | 106 | 91 | 160 | 178 | 183 | 228 | 239 | 262 |
| 7 | 34 | 41 | 34 | 67 | 73 | 71 | 96 | 113 | 108 |
| 8 | 88 | 113 | 105 | 183 | 203 | 206 | 252 | 273 | 308 |
| 9 | 40 | 52 | 50 | 85 | 103 | 93 | 152 | 148 | 140 |

**Table 13.20** Results of the S/N calculations from WinRobust

| *Run* | *1 Height* | *2 No. bands* | *3 StopPin* | *4 Stretch* | *Response (slope)* | *Response (S/N)* |
|---|---|---|---|---|---|---|
| 1 | 0 | 1 | 45 | ES3 | 1.62 | −14.35 dB |
| 2 | 0 | 2 | 60 | ES2 | 3.46 | −10.61 dB |
| 3 | 0 | 3 | 75 | ES1 | 3.53 | −10.50 dB |
| 4 | 15 | 1 | 60 | ES1 | 4.11 | −8.54 dB |
| 5 | 15 | 2 | 75 | ES3 | 1.09 | −14.49 dB |
| 6 | 15 | 3 | 45 | ES2 | 5.56 | −8.75 dB |
| 7 | 30 | 1 | 75 | ES2 | 2.35 | −6.71 dB |
| 8 | 30 | 2 | 45 | ES1 | 6.33 | −9.27 dB |
| 9 | 30 | 3 | 60 | ES3 | 3.08 | −8.43 dB |

S/N ratio are specified, resulting in low sensitivity to the noise factors and good linearity with respect to the signal factor. Second, the zero-point function is used to define exactly what pull-back angle to use to hit any target within the dynamic range of the catapult.

The factor effects for the slope can be used to adjust the dynamic range if desired. An increased slope offers the ability to hit more distant targets. A decreased slope reduces sensitivity to error in the pull-back angle. This ability to tune the slope is another feature of the dynamic analysis.

The factor response averages are given in Table 13.21 and plotted in Figure 13.21. It is apparent from the plots that height and elastic stretch have a substantial effect on both the S/N and the slope. Number of bands and stop pin location, on the other hand, have a larger influence on the slope than on the S/N ratio and could be used for tuning if desired.

**Table 13.21** Factor effect table for the parameter optimization experiment

| *Factor* | *Level* | *Response (slope)* | *Response (S/N)* |
|---|---|---|---|
| | 0 | 2.87 | −11.82 |
| Height | 15 | 3.59 | −10.59 |
| | 30 | 3.92 | −8.14 |
| | 1 | 2.70 | −9.87 |
| No. bands | 2 | 3.62 | −11.46 |
| | 3 | 4.06 | −9.22 |
| | 45 | 4.50 | −10.79 |
| StopPin | 60 | 3.55 | −9.19 |
| | 75 | 2.32 | −10.57 |
| | ES3 | 1.93 | −12.42 |
| Stretch | ES2 | 3.79 | −8.69 |
| | ES1 | 4.66 | −9.44 |

**Figure 13.21** Factor effect plots for the parameter optimization experiment

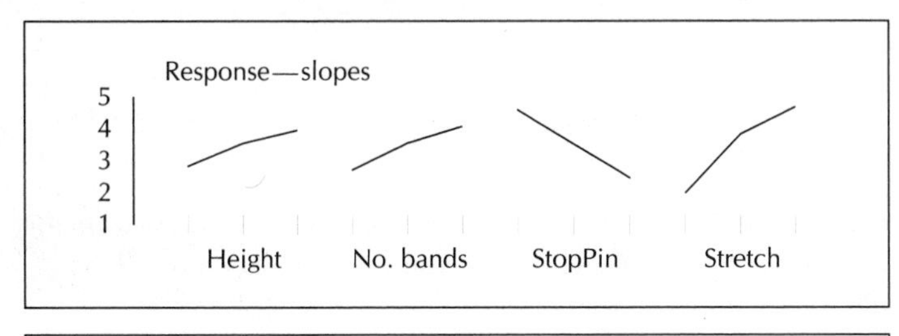


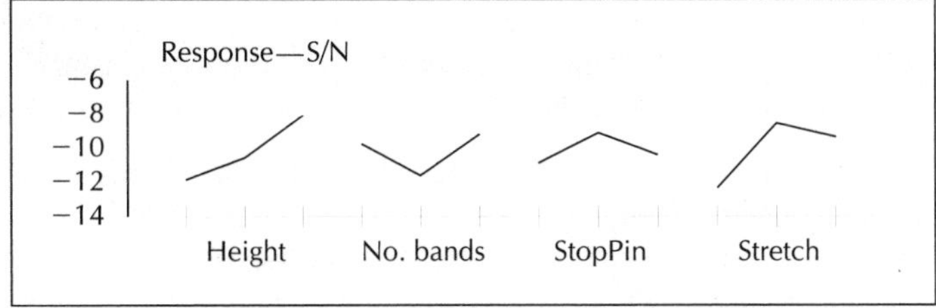

**Figure 13.22** Optimum set points for maximizing the S/N ratio

WinRobust - A:\WINROB\CATAPULT.RDE - Confirmation

Print Optimize Confirm Runs Help

| Parameter | Adjustment | Response | Pred Ave | Pred S/N(CI) |
|---|---|---|---|---|
| ☒ Height | 30 | Response | 4.94 | -4.69dB(0.0) |
| ☒ No.Bands | 3 | | | |
| ☒ StopPin | 60 | | | |
| ☒ Stretch | ES2 | | | |

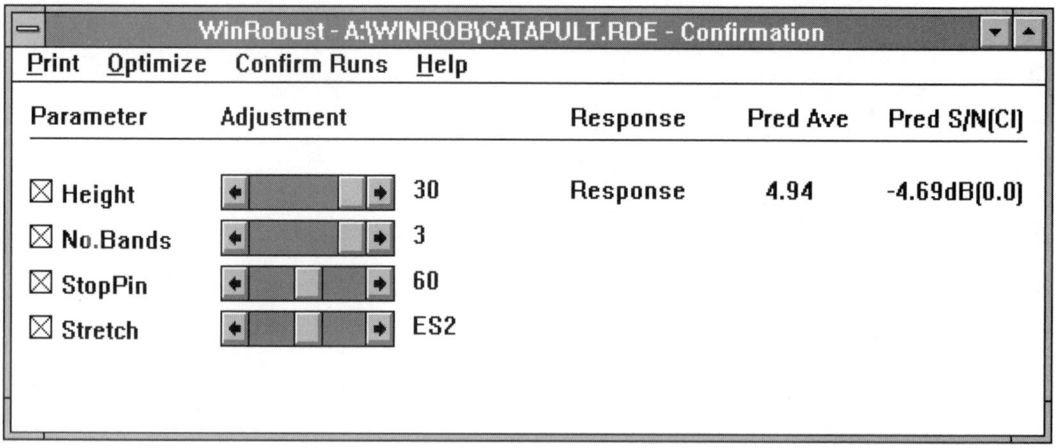

The following observations can be made about the slope behavior:

A. **Catapult height:** The slope increases as height goes up, giving the ball a longer drop time, and thus a greater opportunity to travel horizontally as well.

B. **Number of rubber bands:** The slope increases as the number of rubber bands increases, because the tosses get longer as a result of more energy being put into the throw.

C. **Stop pin angle:** The slope decreases as the stop pin angle increases from 45° to 75°, because the angle of release is forcing the ball into lower trajectories, limiting flight time and thus horizontal travel.

D. **Elastic stretch:** Same as factor B.

## 13.3.6 Prediction and Verification

The WinRobust optimizer is used to choose the optimum levels for the control factors by maximizing the S/N ratio. Figure 13.22 shows the results of the optimizer for maximizing the S/N ratio.

Two different tests are run for the verification experiment:

1. Run number 9, chosen arbitrarily, is replicated to test for repeatability.
2. The optimum configuration—30-inch height, three rubber bands, 60° stop pin, elastic tie point 2, and elastic stretch point 2 (compound factor level ES2)—is run to test for robustness.

The raw data are shown in Table 13.22. The results of the S/N analysis are shown in Table 13.23 for Run #9 and Table 13.24 for the optimum configuration.

Checking for reproducibility, Table 13.23 and the raw data indicate good reproducibility. The experimental error appears to be small. The predicted S/N ratio, however, does not appear to be close to the original or verification result. This is not surprising because the control factor analysis did indicate the presence of some significant interactions. And the fully saturated L9 is not a sufficient countermeasure, especially for the interaction between factors B and D.

**Table 13.22** Raw data from verification tests

| | *15 degrees* | | | *30 degrees* | | | *45 degrees* | | |
|---|---|---|---|---|---|---|---|---|---|
| | *white* | *yellow* | *red* | *white* | *yellow* | *red* | *white* | *yellow* | *red* |
| *Repeat #9* | 37 | 47 | 48 | 86 | 110 | 106 | 124 | 135 | 151 |
| *Optimum* | 82 | 88 | 85 | 145 | 164 | 166 | 214 | 223 | 238 |

**Table 13.23** Verification test results checking for reproducibility

| *Run #9* | *Previous* | *Predicted* | *Verification* |
|---|---|---|---|
| *Slope* | 3.08 inches/degree | 5.81 | 3.12 |
| *S/N ratio* | −8.43 dB | −5.44 | −11.0 |

**Table 13.24** Verification test results for optimum configuration

| *Optimum S/N* | *Predicted* | *Verification* |
|---|---|---|
| *Slope* | 4.94 | 5.13 |
| *S/N ratio* | −4.69 | −6.5 |

Table 13.24 shows the results for the predicted optimum. Here it turns out that the prediction and verification are fairly close, although it is already known that the additive model is not entirely accurate for this case. In spite of the model's quantitative inaccuracy, the verification test for the optimum configuration gives better results (i.e., a larger S/N value) than any of the previous runs in the L9 orthogonal array.

These results are typical for cases where there are some synergistic interactions that are not included in the additive model. The optimum configuration is a true optimum. Additional testing confirms that each factor is at its optimum level, independent of the other factors. However, the additive model is not a quantitatively accurate predictive model. Consider what would be required to develop a better model. Doubling the size of the experiment (using an L18) would slightly improve the accuracy of the additive model. Tripling the size of the experiment (using an L27) would allow modeling of all the interactions. Once again, the choice comes down to a lean experiment that gives the optimum configuration, or a larger experiment that buys the experimenter a mathematical model.

Figure 13.23 shows the plot of the raw data for the optimum configuration. The least-squares best-fit zero-point proportional line, $y = 5.13\ M$, is included. The plot shows that the linearity is excellent and the influence of three different projectiles is minimal. This result also illustrates that the targeting function is given by the zero-point proportional relationship. The correct firing angle to hit a target at any range $y$ is given by Equation 13.3:

$$M[\text{degrees}] = \frac{y[\text{range in inches}]}{5.13} \tag{13.3}$$

**Figure 13.23** Optimum configuration raw data plot showing the zero-point proportional function

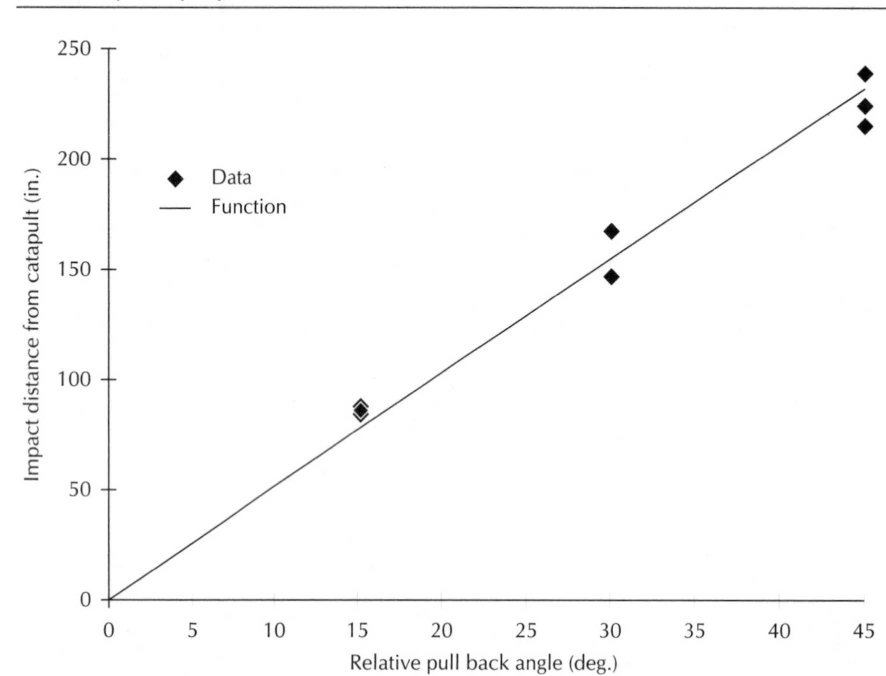


## 13.4 Conclusion

The examples in this chapter are used by the authors to illustrate the Robust Design process in the classroom. They exhibit many of the complexities and difficulties of real-life problems. They are challenging to do well. We have intentionally left control factors and noise factors out of this presentation so that the reader can try these experiments.

The ice water problem is a simple, very additive example. The energy transformation is clear, as shown by Figure 13.1. Of course, the result is entirely predictable but the demonstration using an L4 is instructive. The gyrocopter and catapult are much more difficult to analyze.

The gyrocopter problem defies a closed-form solution. In fact, flying the gyrocopter involves some skill and practice. A simple analysis of the control factors is done using fundamental physical principles such as free body diagrams. It indicates that there are plenty of interaction possibilities. We show how the liberal use of sliding levels and the L18 array, along with a good quality characteristic and the dynamic S/N analysis, produce a nearly perfect additive result. This example was intentionally crafted to demonstrate the "art" of Robust Design.

The catapult can actually be solved analytically [B13, S1]. The analysis is difficult and does not make clear what the optimum levels for robustness are. Instead, use is made of simple equations for projectile motion to predict the likely interactions. One interaction is intentionally left in the L9 experiment to demonstrate the ability of robust design to find the optimum configuration despite the presence of mild to moderate interactions. As a result, the additive model is not quantitatively accurate.

Nevertheless, the optimum configuration is found and verified. Exercise 2, at the end of this chapter, challenges the reader to try to find a strategy for improving the additive model for the catapult. (Hint: Look out for functionally redundant control factors!)

## Exercises for Chapter 13

In Table 13.16, it is shown that spring force and pull back angle both have a strong effect on projectile speed. Thus, one would expect there to be an interaction between them. However, pull-back angle is not included in the inner array. It is the signal factor found in the outer array. The number of rubber bands affects the spring constant. The elastic tie point and elastic stretch point are compounded and referred to as the elastic stretch. They both affect the force applied to the arm as described by Hooke's Law:

$$F = kx$$

1. Does the product of $k$ and $x$ given by Hooke's Law indicate that there is an interaction? Choose two physically reasonable levels for $k$ and $x$ and test Hooke's Law using an L4 with the third column left empty. Compare the effect in the empty column with the other two columns to answer this question. (See Chapter 16.)
2. The factors $k$ and $x$ in Hooke's Law are compounded in the catapult example. Consider the effect of the stop pin location on the elastic stretch prior to pull-back. Suggest a sliding-level approach based on measuring the actual rubber-band length with the arm at rest on the stop pin, and show how one would set up the inner array to reduce the effect of this interaction.

CHAPTER 14

# Parameter Design Case Studies

## 14.1 Introduction

Real-life case studies tend to be unique and varied. The goal of the engineering team is almost always to solve a problem by improving the robustness of a system in the most expedient way possible. For that reason, actual case studies rarely follow the exact steps prescribed for Parameter Design. The best case studies, however, do follow most of the guidelines given in the previous chapters. We conclude Part II by looking at three very different case studies conducted by Kodak engineers.

The first case study was originally done without the calculation of any signal-to-noise ratio. However, the experiment is performed using a one-sided operating window approach. That is, there is only a lower threshold for the experiment. The upper threshold is determined by cost restrictions on the amount of force that could be designed into an Elastomer nip configuration. Since the upper threshold is fixed, the problem is analyzed here using the smaller-the-better threshold.

Some of the conclusions made about the control factors are reached without any analysis of the designed experiment. This should neither surprise nor concern the reader. Engineering is done by problem-solving, and engineers cannot stand still waiting for the statistical analysis to catch up. The robust design approach maximizes one's ability to find solutions to design problems. But never forget that solutions come from the human mind. Sometimes enough evidence is amassed to point to a clear conclusion. The engineering team can be tempted to consider interrupting the experiment to speed up implementation. This is usually not a good thing to do. Going back to finish an experiment after a lapse of time can be problematic from a logistical point of view. It also allows for a lot of human error. It doesn't make for good case studies. However, it does occasionally improve time-to-market.

Although the dynamic approach illustrated in the previous chapter is preferred by Dr. Taguchi, it is not always possible in case studies to use the underlying dynamic relationship. The second case study is a well-executed L18 experiment using a smaller-the-better characteristic. The third case study is an L18 using a nominal-the-best quality characteristic as well as a related smaller-the-better quality characteristic. It is not unusual for there to be multiple responses. Trying to satisfy multiple needs is one of the reasons for choosing numerous control factors. Two or three control factors simply do not offer enough latitude to satisfy conflicting needs.

**Figure 14.1** Sketch of the transfer nip region showing control factors

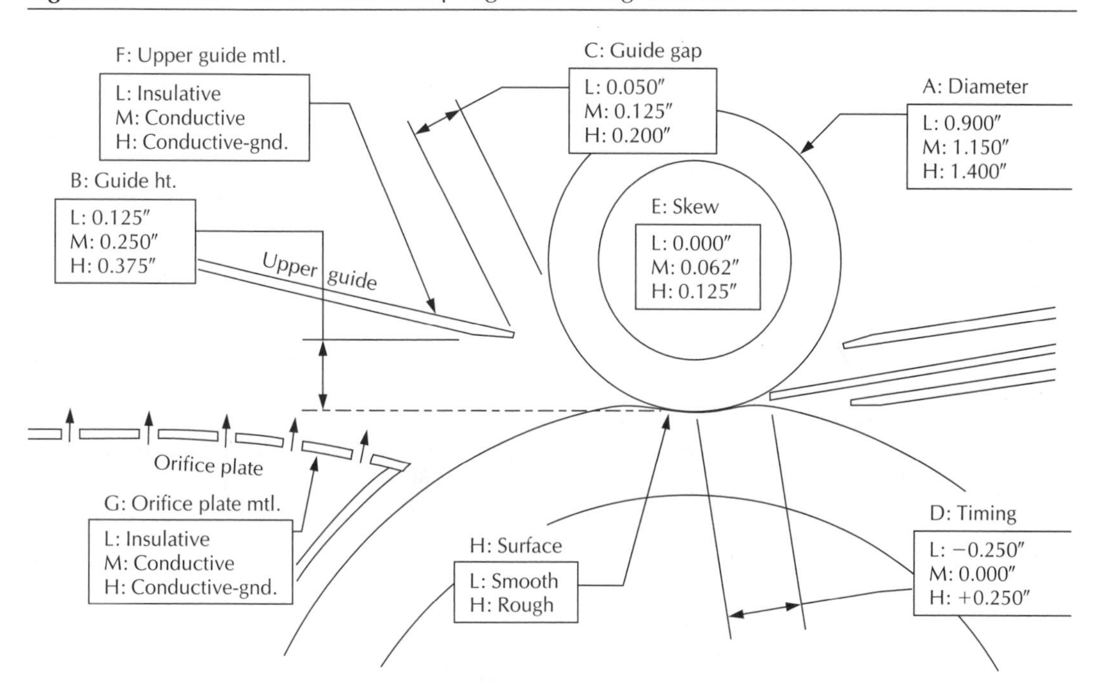


## 14.2 Paper Handling—An Operating Window Example with Two Signal Factors

### 14.2.1 Introduction and Quality Characteristic

This case study[1] optimizes the paper-handling reliability in a laser printer. The study is focused on the transport of paper during the electrostatic transfer of the toner particles from a donor roller to the paper. A sketch of the subsystem of interest is shown in Figure 14.1. Shown there is a 6-inch diameter donor drum, that transports the toner image to the nip, and a smaller backup roller. The backup roller is biased with an electrical potential and the donor drum is grounded. This creates a high electrostatic field that causes the charged toner to transfer from the donor to the paper as the cut sheet is passed through the nip. Unfortunately, the electric current that passes through the nip also charges the paper. This results in paper-handling problems, such as failure of the paper to release, because the charged paper sticks to the donor drum or otherwise fails to pass cleanly through the nip.

The configuration shown gives fairly reliable paper-handling performance as a result of previous development work. Some of the important noise factors considered are paper weight (thickness), paper moisture-content, and environment. The conclusion is that for reliable transfer release under worst-case conditions (thin paper, dry paper, dry environment), the minimum backup roller force required is on the order of 150 lbs. However, the greatest force a low-cost paper roller mechanism can supply is on the order of 60 lbs. Even at that level, flexing of the structure and other undesirable effects are observed.

1. Contributed by Mike Parsons, Steve Russel, and Mark Zaretsky, Eastman Kodak Company.

**Table 14.1** Noise conditions showing compounding

| *Description* | *Worst condition* | *Best condition* |
|---|---|---|
| Environment (°F/%RH) | 80/10 | 70/50 |
| Paper weight (bond) | 16 | 32 |
| Paper condition | Conditioned to environment | Fresh out of ream |
| Paper curl | Down | None |
| Backup roller current ($\mu$A) | 12 | 7 |
| Toner laydown | None | Full page |

The goal of this parameter design study is to find a combination of control factors that minimizes the necessary backup force. The threshold release force is the chosen response. Thus, this is a one-sided operating window problem, with a smaller-the-better threshold. The threshold force is defined as the force at which 10 successful releases (successful transports of the paper through the nip) occur. The threshold release force is determined by starting at a level low enough to produce release failures and then increasing incrementally until 10 successful releases occur consecutively.

## 14.2.2 Noise Factors

Table 14.1 shows the noise factors used for the parameter design study. Previous experience allows grouping of the noise factor levels into worst and best conditions.

## 14.2.3 Initial Experimentation

The control factors are shown in Table 14.2. The orthogonal array used for testing these factors is an L18 array. After the first nine conditions are run with the results shown in Table 14.3, the experiment is halted. Note that for the best-case noise conditions, all nine runs produce equal results. On the other hand, with the exception of runs 7, 8, and 9, the worst-case noise conditions could not achieve the threshold release force even at the highest possible force between the rollers, 30 lbs. The maximum

**Table 14.2** Control factors for the initial experiment

| *Label* | *Description* | *Level 1* | *Level 2* | *Level 3* |
|---|---|---|---|---|
| A | Backup roller smoothness | Smooth | Rough | |
| B | Backup roller diameter (in.) | 0.900 | 1.150 | 0.375 |
| C | Upper guide height (in.) | 0.125 | 0.250 | 0.375 |
| D | Upper guide gap (in.) | 0.050 | 0.125 | 0.250 |
| E | Backup roller lead distance (in.) | −0.25 | 0.00 | +0.25 |
| F | Backup roller skew (in.) | 0.000 | 0.062 | 0.125 |
| G | Upper guide material | Insulative | Conductive | Grounded |
| H | Orifice plate material | Insulative | Conductive | Grounded |

**Table 14.3** Results of the initial experiment

| | Control factors | | | | | | | | Noise condition | |
|---|---|---|---|---|---|---|---|---|---|---|
| *Run* | *A* | *B* | *C* | *D* | *E* | *F* | *G* | *H* | *Best* | *Worst* |
| 1 | Smooth | 0.90 | 0.125 | 0.050 | −0.25 | 0.000 | Ins. | Ins. | 5 | >30 |
| 2 | Smooth | 0.90 | 0.250 | 0.125 | 0.00 | 0.062 | Cond. | Cond. | 5 | >30 |
| 3 | Smooth | 0.90 | 0.375 | 0.250 | 0.25 | 0.125 | Grnd. | Grnd. | 5 | >30 |
| 4 | Smooth | 1.15 | 0.125 | 0.050 | 0.00 | 0.062 | Grnd. | Grnd. | 5 | >30 |
| 5 | Smooth | 1.15 | 0.250 | 0.125 | 0.25 | 0.125 | Ins. | Ins. | 5 | >30 |
| 6 | Smooth | 1.15 | 0.375 | 0.250 | −0.25 | 0.000 | Cond. | Cond. | 5 | >30 |
| 7 | Smooth | 1.40 | 0.125 | 0.125 | −0.25 | 0.125 | Cond. | Grnd. | 5 | 20 |
| 8 | Smooth | 1.40 | 0.250 | 0.250 | 0.00 | 0.000 | Grnd. | Ins. | 5 | 15 |
| 9 | Smooth | 1.40 | 0.375 | 0.050 | 0.25 | 0.062 | Ins. | Cond. | 5 | 15 |

force corresponds to 30 psi of piston air pressure. The experiment is terminated because the data for the best-case compound noise indicate nothing about the control factor effects, and the data for the worst-case compound noise is censored at 30 psi, seriously weakening the predictive ability of any model that would result from the study.

Despite the failed experiment, much is learned about transfer release. This is not unusual in an engineering study. Parameter optimization experiments are not always going to be pretty, but they should always result in learning a lot about the control factors and noise factors tested. Table 14.4 lists some of the conclusions that could be made based on the experiments run, as well as some additional observations made while the lab work was proceeding.

One other critical observation is the fact that there are no paper-handling problems, even with the worst-case noise and low nip force, as long as the transfer current is set to zero. This raises the question: Is the transfer current a control factor or a noise factor? The answer, as with many parameters, depends on your point of view. For the toner transfer engineer, current is a control factor. It is the primary means for controlling the transfer field. For the paper-handling engineer, however, current is a

**Table 14.4** Conclusions based upon the initial experiment

| *Label* | *Description* | *Significance* | *Optimum level* |
|---|---|---|---|
| A | Backup roller smoothness | Unknown | Unknown |
| B | Backup roller diameter (in.) | Very important | 1.400 |
| C | Upper guide height (in.) | Unknown | Unknown |
| D | Upper guide gap (in.) | Unknown | Unknown |
| E | Backup roller lead distance (in.) | Not very important | +0.25 |
| F | Backup roller skew (in.) | Not very important | 0.000 |
| G | Upper guide material | Very important | Grounded |
| H | Orifice plate material | Very important | Plastic |

noise factor, as indicated in Table 14.2. As discussed in Chapter 9, the variation in one subsystem can be a proximity noise to another subsystem. The paper-handling engineer is not in a position to specify the transfer current. Rather, his job is to optimize the paper handling for all transfer current levels that might be used. In this case, the range of transfer current specified is 7 to 12$\mu$A.

The significance of the electrical properties of the guide and orifice plate is another important observation that was not previously appreciated. Their significance is consistent with the understanding that electrostatic effects are critical in this study. The recommended levels in Table 14.4 reverse the configuration previously used. The result of this change is an overall improvement in performance.

### 14.2.4 The Double Operating Window Signal-Factor Experiment

For this operating window experiment, the observed behavior is that the threshold level is a strong function of the transfer current. Therefore, the transfer current is treated as another signal factor that can be used to help quantify the threshold force in a one-sided operating window experiment. In the previous test, the transfer current is treated as a noise factor ranging from 7 to 12 $\mu$A. Now, as an operating window signal factor, the test is run at a range of currents, from 1 $\mu$A to 12 $\mu$A. The minimum transfer release force is determined at each of several current levels.

Transfer current is now no longer treated as a noise factor. Previously, the best noise condition gave good release at 5 psi, regardless of the control factor set points. Therefore, this experiment uses only the worst-case noise condition. An L9 array with control factors A, B, C, and D from Table 14.2 is used. Control factor A has only two levels. The dummy level technique, discussed in Chapter 15, is used for the Smooth level of factor A. Table 14.5 shows the experimental layout and raw data. Note that the experimental equipment is upgraded and, as a result, line pressures of up to 40 psi can be sustained. This helps relieve the problem of censored data at higher current levels.

The data is analyzed using WinRobust. The analysis is done by considering the transfer current levels to be seven noise factor conditions. The S/N treatment is smaller-the-better. Table 14.6 shows the results of the analysis.

**Table 14.5** Raw data for double operating window signal-factor experiment

| | *Control factors* | | | | *Current level ($\mu$A)* | | | | | | |
|---|---|---|---|---|---|---|---|---|---|---|---|
| *Run* | *B* | *C* | *D* | *A* | *1* | *2* | *3* | *4* | *5* | *7* | *12* |
| 1 | 0.90 | 0.125 | 0.050 | Smooth | 20 | 40 | 40 | 40 | 40 | 40 | 40 |
| 2 | 0.90 | 0.250 | 0.125 | Smooth | 20 | 40 | 40 | 40 | 40 | 40 | 40 |
| 3 | 0.90 | 0.375 | 0.250 | Rough | 15 | 40 | 40 | 40 | 40 | 40 | 40 |
| 4 | 1.15 | 0.125 | 0.125 | Rough | 5 | 5 | 5 | 5 | 5 | 5 | 5 |
| 5 | 1.15 | 0.250 | 0.125 | Smooth | 5 | 5 | 5 | 5 | 5 | 20 | 40 |
| 6 | 1.15 | 0.375 | 0.050 | Smooth | 5 | 5 | 5 | 5 | 40 | 40 | 40 |
| 7 | 1.40 | 0.125 | 0.250 | Smooth | 5 | 5 | 5 | 5 | 5 | 5 | 5 |
| 8 | 1.40 | 0.250 | 0.050 | Rough | 5 | 5 | 5 | 5 | 5 | 5 | 5 |
| 9 | 1.40 | 0.375 | 0.125 | Smooth | 5 | 5 | 5 | 5 | 5 | 5 | 5 |

**Table 14.6** WinRobust analysis of the data from Table 14.5

| *Run* | *1*<br>*B* | *2*<br>*C* | *3*<br>*D* | *4*<br>*A* | *Threshold*<br>*(ave.)* | *Threshold*<br>*(S/N)* |
|---|---|---|---|---|---|---|
| 1 | 0.90 | 0.125 | 0.050 | Smooth | 37.14 | −31.55 dB |
| 2 | 0.90 | 0.250 | 0.125 | Smooth | 37.14 | −31.55 dB |
| 3 | 0.90 | 0.375 | 0.250 | Rough | 36.43 | −31.47 dB |
| 4 | 1.15 | 0.125 | 0.125 | Rough | 5.00 | −13.98 dB |
| 5 | 1.15 | 0.250 | 0.250 | Smooth | 12.14 | −24.82 dB |
| 6 | 1.15 | 0.375 | 0.050 | Smooth | 20.00 | −28.45 dB |
| 7 | 1.40 | 0.125 | 0.250 | Smooth | 5.00 | −13.98 dB |
| 8 | 1.40 | 0.250 | 0.050 | Rough | 5.00 | −13.98 dB |
| 9 | 1.40 | 0.375 | 0.125 | Smooth | 5.00 | −13.98 dB |

Figure 14.2 shows the control factor effects. It is clear that factor B, the roller diameter, is the dominant control factor. This conclusion confirms the observations from the preliminary experiment. Although seen twice now, this result is actually very surprising. The reason is that higher nip pressure is the mechanism proposed to explain why the higher backup roller pressure is favorable for release. At the same contact force, a smaller roller diameter produces a higher pressure. In fact, in an earlier configuration, different from that shown in Figure 14.1, smaller-diameter rollers were better for release. Note that this indicates a significant antisynergistic interaction between roller diameter and other factors that make the two configurations different. Despite the evidence of an interaction, the truncated L18, the L9, and intense observation lead the team to discover the much more favorable configuration tested here.

**Figure 14.2** Control factor effects plots

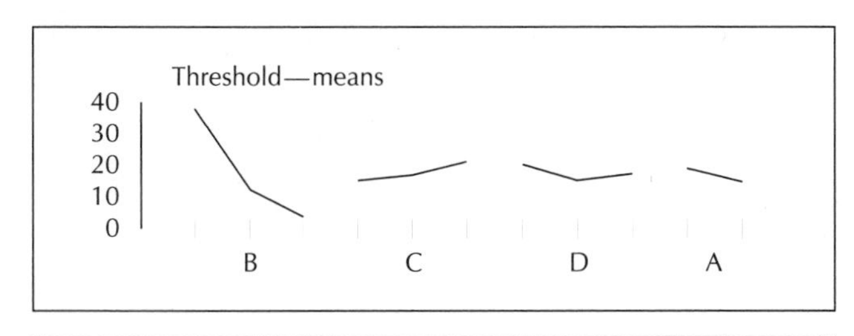


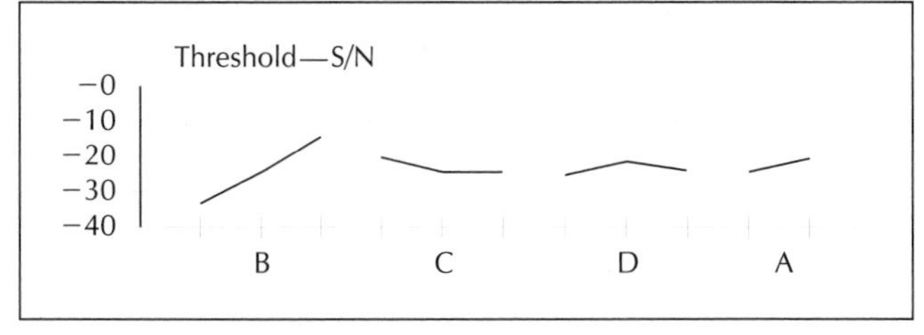

**Table 14.7** Confirmation test conditions

| *Environment (°F/%RH)* | *Media type* | *Condition* |
|---|---|---|
| 80/10 | 16 lb. bond | Conditioned |
| 70/50 | 16 lb. bond | Conditioned |
| 70/75 | 16 lb. bond | Conditioned |
| 65/10 | 13 lb. bond | Conditioned |
| 65/10 | 20 lb. bond, short grain | Conditioned, curled |
| 65/10 | transparency | |

### 14.2.5 Final Experiments and Conclusion

A final experiment using an L4 for only factors B and C (the upper guide height and the upper guide gap) complete the parameter optimization process. In each iteration, strong factors from previous tests are set at their optimum levels and only the remaining factors are tested. This allows the smaller effects to be seen and is an effective way of closing in on an optimum solution.

To confirm the final configuration, a number of test runs are made using new, even more aggressive, noise factor levels. Lighter paper weights, short-grain paper with curl, and transparencies are all tested under a wide range of environments. The noise conditions go beyond the operating specifications. This is entirely appropriate in a robustness experiment and is an important guarantee of overall robustness and reliability. The results of all the tests demonstrate that, under the optimum configuration, the threshold force at 12 $\mu$A is only 5 pounds. The confirmation test conditions are given in Table 14.7, and the final configuration is given in Table 14.8.

The system studied here was not brought to market. However, it did undergo field trials for possible OEM sales. As a result of the experiments done here, and the resulting changes in the configuration, the low-cost, low-force design is completely viable. No more paper-handling problems were observed at this nip after the design changes were implemented.

**Table 14.8** Confirmation design set points (arranged in order of importance)

| *Parameter* | *Final setting* |
|---|---|
| Backup roller diameter | 1.40 in. |
| Upper guide material | Conductive, grounded |
| Orifice plate material | Plastic, insulating |
| Upper guide height (in.) | 0.250 in. |
| Upper guide gap (in.) | 0.125 in. |
| Backup roller smoothness | Smooth |
| Backup roller lead distance (in.) | +0.250 |
| Backup roller skew (in.) | 0.000 |

**Figure 14.3** Sketch of the thermal printer showing the capstan roller system

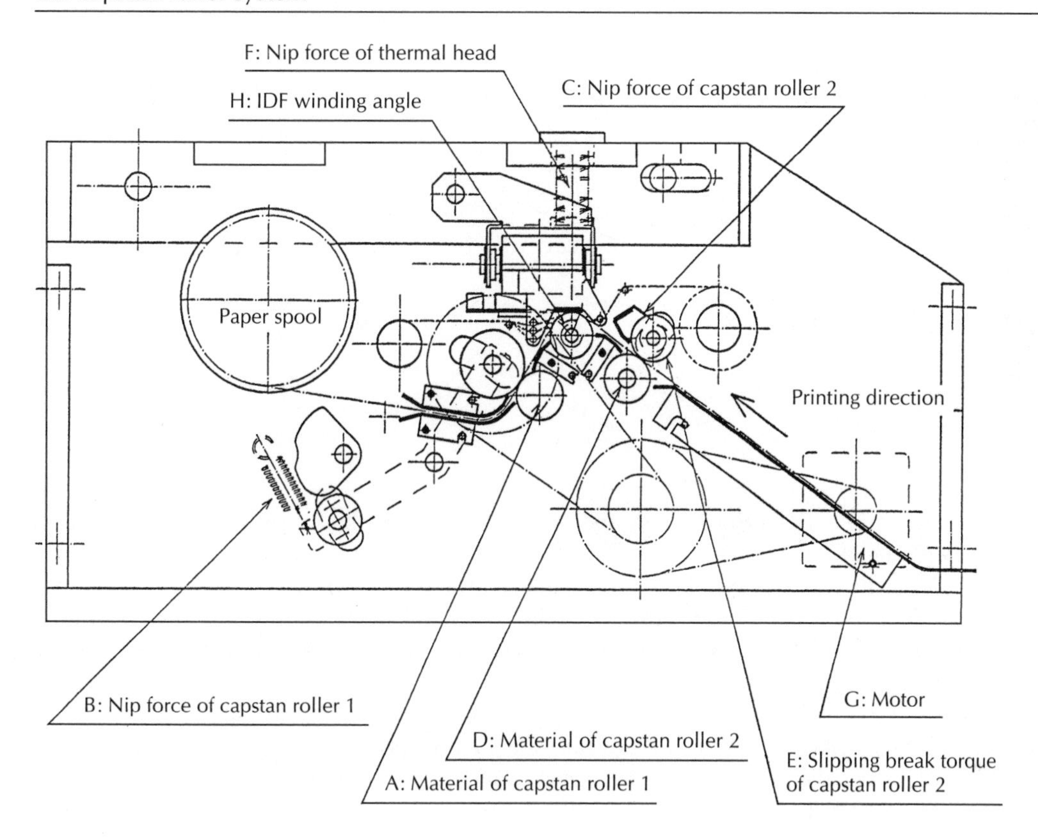


## 14.3 Improvement of a Capstan Roller Printer Registration

### 14.3.1 Introduction and Quality Characteristic

This case study[2] is an optimization of a media drive system in a color dye thermal printer. The printer uses a thermal head that writes information on the medium, typically photographic-grade paper, by the application of heat. The heat causes dye from a donor web to be deposited on the receiver. The process is repeated for three patches of dye, magenta, cyan, and yellow, to produce a photographic-quality color print. The conventional design uses precut sheets of paper and cannot produce prints of different sizes. The new design uses roll paper and a cutting device. A sketch of the system is shown in Figure 14.3. A paper roll and a capstan roller drive mechanism solve the customer's desire to have adjustable size output, but tend to have a larger registration error between the three color images than that of the previous system. In the capstan roller printer, the capstan and pinch roller are thought to be the

2. Contributed by A. Hatakeyama and S. Koshimizu, Eastman Kodak (Japan).

**Figure 14.4** Definition of terms from Equation 14.1

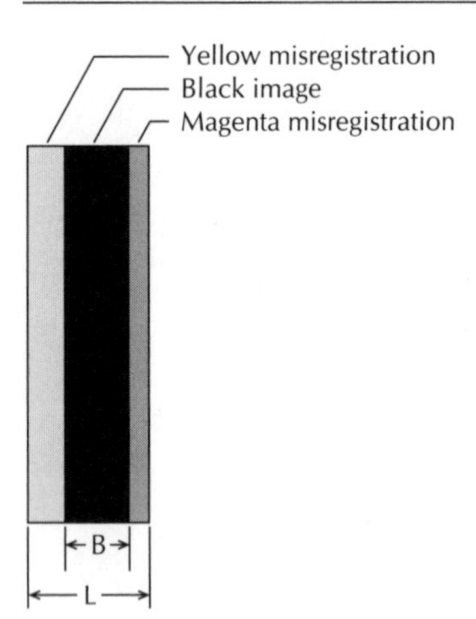


main cause of the registration error because the receiver must move back and forth on these rollers while it is being printed.

The ideal function for this system is that the image location on the receiver be a zero-point proportional linear function of the position information in the data file being printed. Any sources of variation in the location of the image can be considered noise sources for registration errors. However, the functional limit, $\Delta_0$, for registration, at which the average customer is unhappy with the image quality, is approximately 0.004 inches. The image size ranges from 3″ by 4″ to 8″ by 10″ in this product. This means that a dynamic analysis of registration error should have a precision of about 0.01% in order to be useful. This is extremely difficult because of potential measurement errors. For example, the expansion or shrinkage of paper with changes in moisture is comparable to the functional limit, but would not affect registration once the image is laid down on the paper.

A practical quality characteristic is to simply measure the registration error directly and treat it as a smaller-the-better problem. To measure the registration error, a three-color image is printed and the line width is measured. Registration errors cause the apparent line width to be larger than expected. Thus, the quality characteristic is defined as

$$y = L - B, \tag{14.1}$$

where $L$ is the full line width and $B$ is the line width of the black portion only. The quality characteristic is illustrated in Figure 14.4. For this study, the measurement units are millimeters (mm).

## 14.3.2 Noise Factors and Control Factors

Because of the use of the registration error for the quality characteristic, rather than the ideal function, it is not necessary to find noise factors whose levels cause consistent deviations from the zero-point line. Instead, the cyan, magenta, and yellow images are cyclical surrogate noise factors. Their effect is incorporated in the definition of the quality characteristic. In addition, the location of the measured lines on the paper is a positional surrogate noise. For this case study, six locations are measured. The resulting data are analyzed using the STB signal-to-noise ratio, $S/N = -10 \log(\Sigma\, y_i^2/n)$.

The control factors are identified in Figure 14.3, and their levels are shown in Table 14.9.

**Table 14.9** Control factors for the capstan roller experiment

| *Label* | *Description* | *Level 1* | *Level 2* | *Level 3* |
|---|---|---|---|---|
| A | Capstan roller 1 material | Grit | Rubber | |
| B | Capstan roller 1 nip force (kg) | 1.00 | 1.50 | 2.00 |
| C | Capstan roller 2 nip force (kg) | 1.00 | 1.50 | 2.00 |
| D | Capstan roller 2 material | Grit | Rubber | POM |
| E | Capstan roller 2 brake torque (kg-cm) | 0.20 | 0.30 | 0.40 |
| F | Thermal head nip force (kg) | 1.2 | 2.2 | 3.2 |
| G | Motor (horsepower) | 5 | 2 | 4 |
| H | IDF winding angle (degree) | 45 | 55 | 65 |

### 14.3.3 The Parameter Optimization Experiment

The results of the parameter optimization experiment are shown in Table 14.10. The average registration error, the mean square deviation, and the smaller-the-better S/N ratio are all valid responses.

The results of the experiment shown in Table 14.10 are analyzed using ANOM. Figure 14.5 shows the factor effects plot for the S/N ratio. It is apparent that four factors are most significant:

A. Capstan roller 1 material
B. Capstan roller 1 nip force
C. Capstan roller 2 nip force
H. IDF winding angle

The predicted optimum is shown in Figure 14.6. Note that all the factors are included because there is no reason to assert that any of the factors are not significant. Note that the optimum condition is very close to run #4 from the L18 experiment. The resulting S/N ratio, 33.96, can be translated into a predicted registration error assuming that the MSD $= \bar{y}^2$. This, of course, is true only if the registration errors at the six locations on the page are all equal. Nevertheless, such an estimate gives a useful check on the expected magnitude of the error.

Calculation of the expected registration error:

$$33.96 = -10 \log(\bar{y}^2)$$

Therefore,

$$\log(\bar{y}^2) = -3.396$$
$$\bar{y}^2 = 10^{-3.396}$$
$$\bar{y}^2 = 0.0004$$
$$\bar{y} = 0.020 \text{ mm}$$

**Table 14.10** Results of the capstan parameter optimization experiment

| Run | 1<br>A | 2<br>B | 3<br>C | 4<br>D | 5<br>E | 6<br>F | 7<br>G | 8<br>H | $\bar{y}$<br>(mm) | MSD<br>(mm) | S/N ratio<br>(S/N) |
|---|---|---|---|---|---|---|---|---|---|---|---|
| 1 | Grit | 1.00 | 1.00 | Grit | 0.20 | 1.2 | 5 | 45 | 0.35 | 0.13 | 9.01 dB |
| 2 | Grit | 1.00 | 1.50 | Rubber | 0.30 | 2.2 | 2 | 55 | 0.34 | 0.15 | 8.33 dB |
| 3 | Grit | 1.00 | 2.00 | POM | 0.40 | 3.2 | 4 | 65 | 0.23 | 0.08 | 11.10 dB |
| 4 | Grit | 1.50 | 1.00 | Grit | 0.30 | 2.2 | 4 | 65 | 0.02 | 0.00 | 33.73 dB |
| 5 | Grit | 1.50 | 1.50 | Rubber | 0.40 | 3.2 | 5 | 45 | 0.25 | 0.06 | 11.97 dB |
| 6 | Grit | 1.50 | 2.00 | POM | 0.20 | 1.2 | 2 | 55 | 0.26 | 0.07 | 11.54 dB |
| 7 | Grit | 2.00 | 1.00 | Rubber | 0.20 | 3.2 | 2 | 65 | 0.20 | 0.04 | 13.93 dB |
| 8 | Grit | 2.00 | 1.50 | POM | 0.30 | 1.2 | 4 | 45 | 0.25 | 0.06 | 12.15 dB |
| 9 | Grit | 2.00 | 2.00 | Grit | 0.40 | 2.2 | 5 | 55 | 0.90 | 0.82 | 0.86 dB |
| 10 | Rubber | 1.00 | 1.00 | POM | 0.40 | 2.2 | 2 | 45 | 0.22 | 0.05 | 13.25 dB |
| 11 | Rubber | 1.00 | 1.50 | Grit | 0.20 | 3.2 | 4 | 55 | 1.65 | 2.73 | −4.36 dB |
| 12 | Rubber | 1.00 | 2.00 | Rubber | 0.30 | 1.2 | 5 | 65 | 1.22 | 1.48 | −1.70 dB |
| 13 | Rubber | 1.50 | 1.00 | Rubber | 0.40 | 1.2 | 4 | 55 | 0.72 | 0.52 | 2.86 dB |
| 14 | Rubber | 1.50 | 1.50 | POM | 0.20 | 2.2 | 5 | 65 | 0.35 | 0.13 | 9.00 dB |
| 15 | Rubber | 1.50 | 2.00 | Grit | 0.30 | 3.2 | 2 | 45 | 0.83 | 0.70 | 1.57 dB |
| 16 | Rubber | 2.00 | 1.00 | POM | 0.30 | 3.2 | 5 | 55 | 0.46 | 0.21 | 6.75 dB |
| 17 | Rubber | 2.00 | 1.50 | Grit | 0.40 | 1.2 | 2 | 65 | 0.58 | 0.33 | 4.76 dB |
| 18 | Rubber | 2.00 | 2.00 | Rubber | 0.20 | 2.2 | 4 | 45 | 0.00 | 1.14 | −0.57 dB |

**Figure 14.5** Factor effects plot for the capstan roller experiment

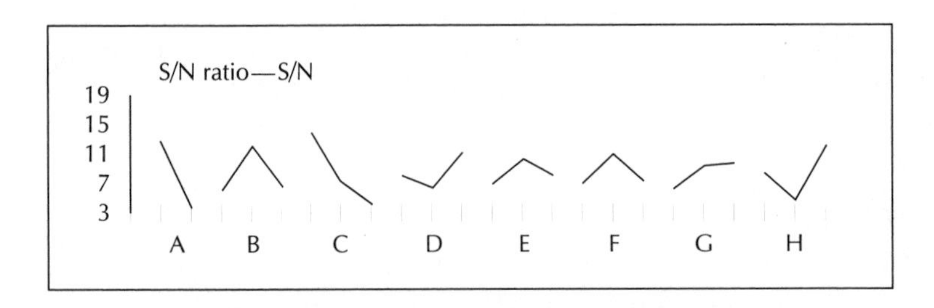


## 14.3.4 The Verification Test and Conclusion

The verification test is done under the optimum conditions, and the registration error is measured at the six locations exactly as before. The results are shown in Table 14.11. The results show that the actual registration error is 0.025 mm, corresponding to an error of 0.001 inch. This is well within the customer requirements.

**Figure 14.6** Predicted optimum for the capstan roller experiment

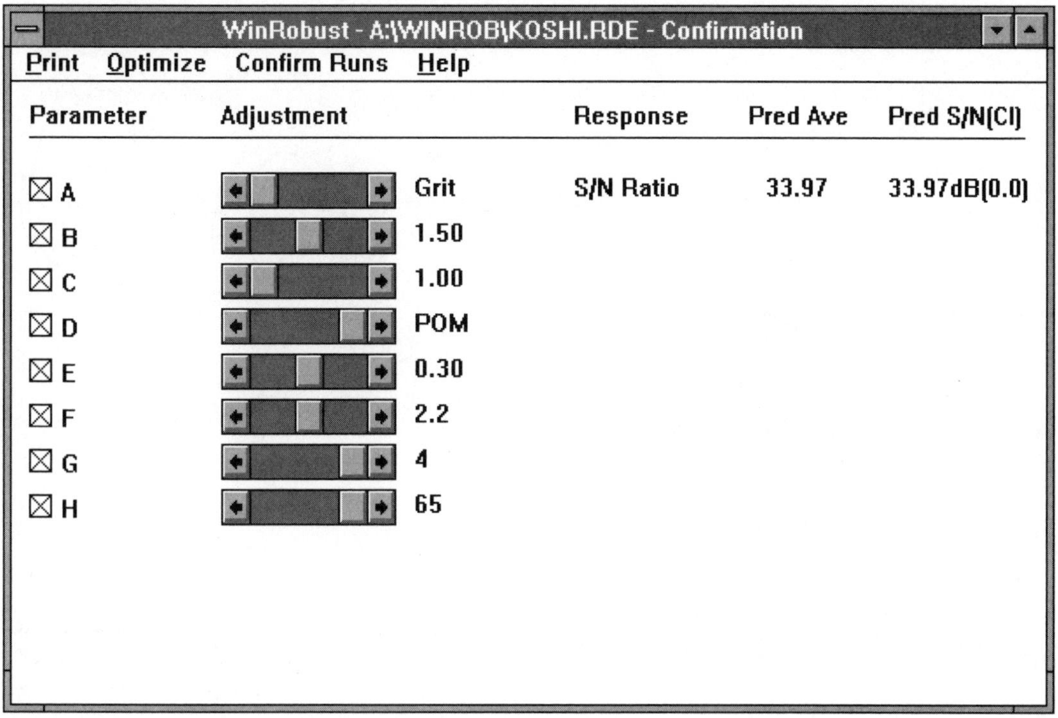

WinRobust - A:\WINROB\KOSHI.RDE - Confirmation

Print Optimize Confirm Runs Help

| Parameter | Adjustment | Response | Pred Ave | Pred S/N(CI) |
|---|---|---|---|---|
| A | Grit | S/N Ratio | 33.97 | 33.97dB(0.0) |
| B | 1.50 | | | |
| C | 1.00 | | | |
| D | POM | | | |
| E | 0.30 | | | |
| F | 2.2 | | | |
| G | 4 | | | |
| H | 65 | | | |

**Table 14.11** Verification results

| | *Predicted* | *Actual* |
|---|---|---|
| S/N ratio | 33.96 | 32.04 |
| Registration error (mm) | 0.020 | 0.025 |

This case study clearly identifies the critical parameters for improving printer registration. They are the capstan roller material, the force in the two capstan nips, and the IDF winding angle. A very satisfactory registration error is achieved when the control factors are all set at their optimum levels.

## 14.4 Enhancement of a Camera Zoom Shutter Design[3]

### 14.4.1 Introduction and Quality Characteristic

The Kodak Cameo Camera aperture/shutter is an iris-style shutter, as shown in Figure 14.7. Three blades are driven through gearing by a 10 mm stepper motor. The 10 mm motor indexes in 18° full step increments. Ten steps of the motor are utilized. The camera specifications call for four levels of expo-

3. Contributed by Chuck Bennett, Eastman Kodak Company.

**Figure 14.7** Iris style shutter

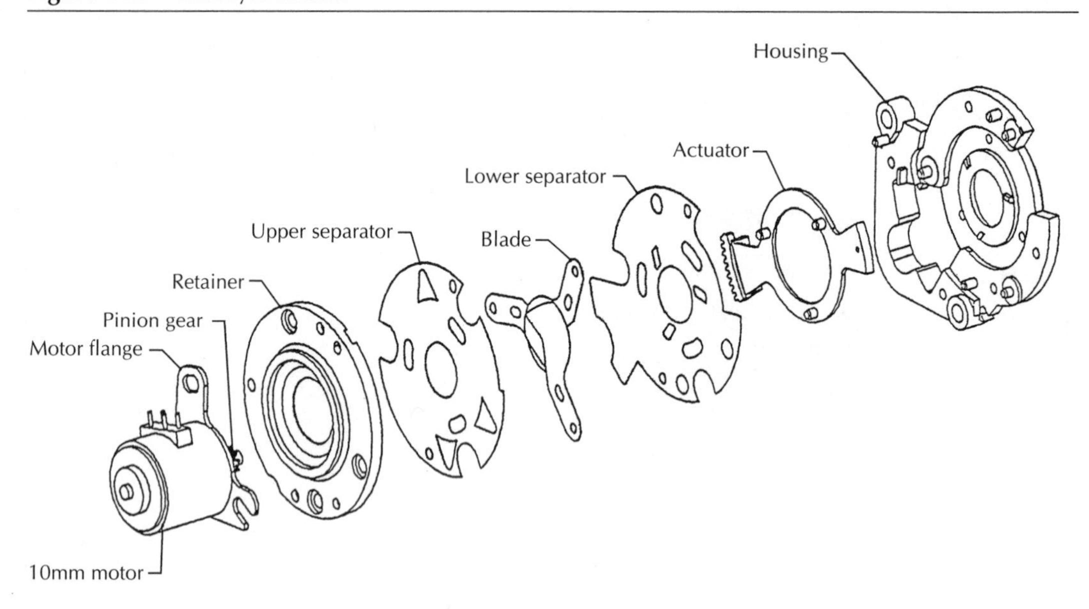


sure, which are controlled by the aperture size and duration of opening. Changing the number of steps yields different-sized aperture openings.

The shutter motor is driven by an open loop system without the use of sophisticated controls. An undesirable property of all stepper motors is a condition known as ringing (oscillation) that occurs when the motor stops at its predetermined step. Improper exposure due to errors in the movement, including ringing, ultimately affects the photographic quality produced by the camera.

Ideally, the motor takes uniform steps and the shutter blade locations are determined only by the motor drive. Because stepper motors do not have hard stops, it is possible to have an error between the intended stop position and the actual stop position. Figure 14.8 shows the ideal relationship between the step number and the area of the aperture. Note that the nominal aperture for step number 10 does not lie on the line formed by the other three aperture settings. This is because aperture 1, fully open, is defined not by the blades, but by a fixed hole in the lower separator. The ideal function is, therefore, a reference point proportional for the aperture area, with steps number 6, 7, and 8 as the signal factor levels.

The engineering team, however, decided to optimize the system only at the step #8 set point for the signal factor, using nominal-the-best to keep the analysis simple. The second quality characteristic used is related to the magnitude of the sinusoidal ringing. To insure good photographic quality, it is desirable to minimize this ringing. Thus, this is a smaller-the-better quality characteristic. The ringing magnitude is given by the amplitude of the ring multiplied by the duration of the ring, Equation 14.3.

$$\text{Ring Area} = 1/2\,(\text{max. light intensity} - \text{min. light intensity}) \times \text{ring duration} \qquad \textbf{(14.3)}$$

Both the shutter area and the ring area are measured using a light source and photodetector which measure the amount of light passing through the shutter. The average amount is proportional to the nominal shutter area. An oscilloscope trace is used to measure the ringing using predetermined thresholds.

**Figure 14.8** Ideal zero-point relationship for the shutter aperture

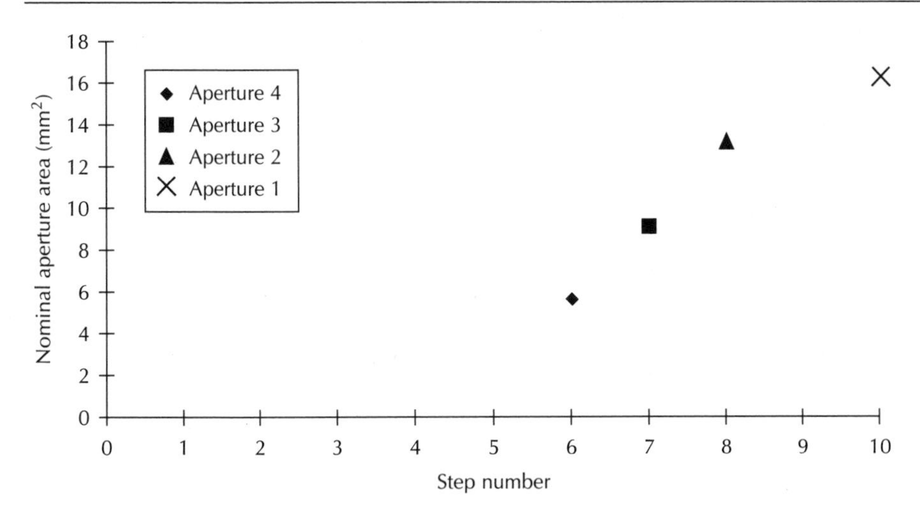


## 14.4.2 Noise Factors

The camera is powered by a 3 volt lithium battery. The shutter mechanism has to operate reliably over a range of voltages produced by the battery from fully charged to a weakened voltage over various environmental conditions. The customer requirements indicate a voltage range of 2.1 volts to 3.1 volts. The simplest method to drive the shutter at low voltage with the old design is to reduce the friction in the shutter mechanism. When this is done, however, the oscillation at high voltage (3.1 volts) is severe. The initial design has no operating latitude between the high- and low-voltage conditions.

For the outer array, two levels of voltage, 2.15 volts and 3.1 volts, are tested. Besides voltage, two other noise factors, environment and life, are included in the outer array. Environment consists of two temperatures, 20°F and 96°F. Life consists of two levels, new and after 70,000 (70K) actuations. The expected life of the shutter is only a fraction of this upper level. The L4 that is used for the outer array of this experiment is shown in Table 14.12. Thus, each of the control factor runs is tested using the four noise factor combinations shown here.

It is interesting to note that the initial noise study shows that 70 K, low voltage, and cold temperature is the worst-case noise condition. This compound noise could not be tested, however, because it results in binding and a failure to actuate under baseline conditions and control factor treatment

**Table 14.12** Noise array for camera shutter

| *Run* | *Life* | *Voltage* | *Temperature* |
|---|---|---|---|
| 1 | 0 K | Low | Cold |
| 2 | 0 K | High | Hot |
| 3 | 70 K | Low | Hot |
| 4 | 70 K | High | Cold |

**Table 14.13** Control factors for the shutter experiment

| *Label* | *Description* | *Level 1* | *Level 2* | *Level 3* |
|---|---|---|---|---|
| A | Hard stop location (in.) | −0.003 | +0.003 | |
| B | Pinion to motor alignment (deg.) | −8 | Nominal | +8 |
| C | Gear mesh (in.) | Nominal | +0.006 | |
| D | Pulse step rate (pulses/s) | −100 | Nominal | +100 |
| E | Blade clearance (in.) | Nominal | +0.003 | 2S |
| F | Hard stop angle (deg.) | Nominal | 30 | 45 |
| G | Closing sequence | Nominal | 4/6 | 6/4 |
| H | Shake sequence | With | Without | |

combinations 8 and 14 (see Table 14.14). For that reason, the levels are assigned to the noise array, Table 14.12, in such a way as to avoid this combination.

### 14.4.3 Control Factors

The control factors are shown in Table 14.13. An explanation of each control factor is given in the following list.

A. **Hard stop location:** The hard stop limits the blades' motion. The two levels tested are ±0.003 inches from the nominal. This is done by a combination of milling and shimming the nominal parts. Nominal = 0.000 inches.

B. **Pinion to motor alignment:** The pinion tooth is aligned to one of the magnetic poles of the motor when the stepper motor is energized. Each setting is controlled to ±1 degree. Nominal = 0.

C. **Gear mesh:** Here only two levels are tested, using the dummy level technique discussed in Chapter 15. Nominal = 0.

D. **Pulse step rate:** Changing the pulse step rate changes the duration of each step and the effective motor torque. Nominal = 0.

E. **Blade clearance:** These are separators between the shutter blades and the housing. Level 1 is one 0.003″ separator. Level 2 is an additional 0.003″ separator. Level 3 is two 0.003″ separators, one on each side of the blades, capturing the blades in an envelope. Nominal = 0.003″.

F. **Hard stop angle:** The nominal location of the hard stop is determined by factor A. Here, this factor refers to the shape of the stop. It is thought that varying the angle might absorb the impact energy better and thus reduce bounce. Nominal = 0.

G. **Extended closing sequence:** This factor is programmed in the software. The numbers refer to additional pulse counts before actuation/after actuation. It is thought that additional counts at the end of the actuation would insure that the actuator stays closed more reliably. Nominal = 4/2 (four counts before actuation and two counts after actuation).

H. **Shake sequence:** This is also handled in software. The shake sequence is hoped to start the blade vibrating so as to minimize any delay due to friction when tripping the shutter. Nominal = with.

**Table 14.14** Results of camera shutter experiment

| Run | 1 A | 2 B | 3 C | 4 D | 5 E | 6 F | 7 G | 8 H | Apt. mean | Apt S/N (S/N) | Ring S/N (S/N) |
|---|---|---|---|---|---|---|---|---|---|---|---|
| 1 | −0.003 | −8 | 0 | −100 | 0 | 0 | 4/2 | W/o | 13.06 | 44.79 dB | −41.10 dB |
| 2 | −0.003 | −8 | 0 | 0 | +0.003 | 30 | 4/6 | W | 10.39 | 30.20 dB | −41.07 dB |
| 3 | −0.003 | −8 | +0.006 | +100 | 2S | 45 | 6/4 | W | 12.86 | 40.72 dB | −7.43 dB |
| 4 | −0.003 | 0 | 0 | −100 | +0.003 | 30 | 6/4 | W | 13.90 | 42.53 dB | −41.15 dB |
| 5 | −0.003 | 0 | 0 | 0 | 2S | 45 | 4/2 | W/o | 13.55 | 36.39 dB | −14.06 dB |
| 6 | −0.003 | 0 | +0.006 | +100 | 0 | 0 | 4/6 | W | 14.33 | 44.24 dB | −41.13 dB |
| 7 | −0.003 | +8 | 0 | 0 | 0 | 45 | 4/6 | W | 15.62 | 43.37 dB | −32.08 dB |
| 8 | −0.003 | +8 | 0 | +100 | +0.003 | 0 | 6/4 | W/o | 11.71 | 6.17 dB | −60.42 dB |
| 9 | −0.003 | +8 | +0.006 | −100 | 2S | 30 | 4/2 | W | 15.15 | 40.72 dB | −10.07 dB |
| 10 | +0.003 | −8 | 0 | +100 | 2S | 30 | 4/6 | W/o | 11.77 | 30.28 dB | −18.53 dB |
| 11 | +0.003 | −8 | 0 | −100 | 0 | 45 | 6/4 | W | 12.16 | 37.79 dB | −41.88 dB |
| 12 | +0.003 | −8 | +0.006 | 0 | +0.003 | 0 | 4/2 | W | 12.41 | 44.92 dB | −39.63 dB |
| 13 | +0.003 | 0 | 0 | 0 | 2S | 0 | 6/4 | W | 13.28 | 28.59 dB | −25.91 dB |
| 14 | +0.003 | 0 | 0 | +100 | 0 | 30 | 4/2 | W | 10.11 | 5.23 dB | −40.86 dB |
| 15 | +0.003 | 0 | +0.006 | −100 | +0.003 | 45 | 4/6 | W/o | 13.33 | 23.16 dB | −29.27 dB |
| 16 | +0.003 | +8 | 0 | +100 | +0.003 | 45 | 4/2 | W | 14.81 | 34.26 dB | −35.21 dB |
| 17 | +0.003 | +8 | 0 | −100 | 2S | 0 | 4/6 | W | 14.69 | 33.81 dB | −17.61 dB |
| 18 | +0.003 | +8 | +0.006 | 0 | 0 | 30 | 6/4 | W/o | 14.61 | 34.20 dB | −36.13 dB |

### 14.4.4 Experimental Results and Confirmation

The control factor array and results are shown in Table 14.14. The results of the experiment are analyzed using ANOM. Figures 14.9–14.11 show the factor effect plots for this experiment.

It is apparent from Figure 14.10 and engineering knowledge (A and H were deemed less important) that the most important factors for the aperture area S/N are these:

- B. Pinion to motor alignment, baseline = 0, best = −8 deg.
- C. Gear mesh, baseline = 0, best = +0.006 in.
- D. Pulse step rate, baseline = 0, best = 0 or −100.

It is apparent from Figure 14.11 that the most important factor for the ring area S/N is this:

- E. Blade clearance, baseline = one 0.003 separator, best = two 0.003 in. separators, one on each side of the shutter blade.

The optimum configuration chosen for the verification experiment is shown in Table 14.15. It is based in part on the results of the experiment. Other considerations such as cost and ease of implementation are also taken into account.

Figure 14.9 Factor effects plot for the aperture area means

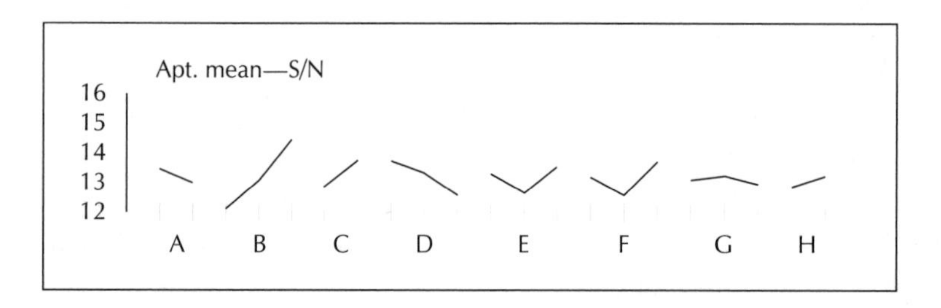


Figure 14.10 Factor effects plot for the aperture area S/N ratios

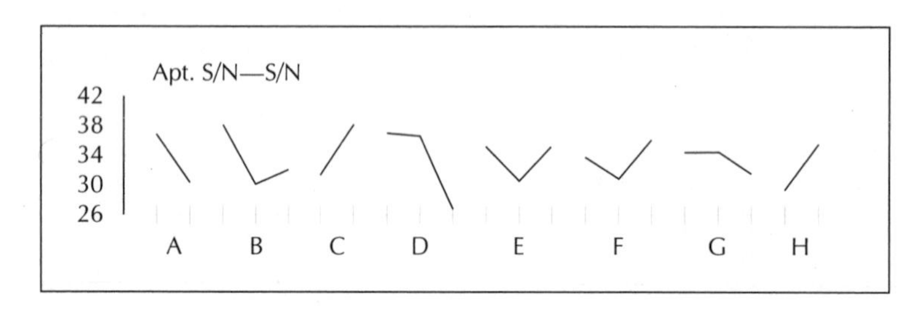


Figure 14.11 Factor effects plot for the ring area S/N ratios

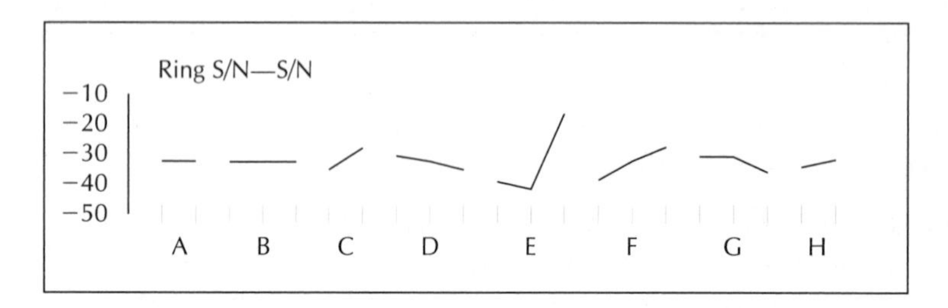


The predicted S/N ratios for the baseline and confirmation configuration are also given in Table 14.15. Five shutter assemblies for each configuration were built and tested. The results are given in Table 14.16. As expected, there is a very substantial improvement in ring elimination. The actual improvement is greater than 20 dB, a 100× improvement in quality or 10× reduction in ringing. The aperture area S/N had only a slight improvement, 2 dB vs. the expected 10 dB. This is not a problem, since the results for aperture area are well within the customer tolerance. The 2 dB improvement in on-target performance is beneficial, but the most significant result is that the optimum configuration performed well even at the extreme noise condition of 70 K life, cold temperature, and low voltage. This condition was previously excluded as being too harsh!

Table 14.15 Control factors for the shutter experiment

| *Label* | *Description* | *Baseline* | *Optimum* |
|---|---|---|---|
| A | Hard stop location (in.) | 0 | 0 |
| B | Pinion to motor alignment (deg.) | 0 | −8 |
| C | Gear mesh (in.) | 0 | 0 |
| D | Pulse step rate (pulses/sec) | 0 | 0 |
| E | Blade clearance (in.) | 0 | 2S |
| F | Hard stop angle (deg.) | 0 | 45 |
| G | Closing sequence | 4/2 | 4/6 |
| H | Shake sequence | With | With |
| | Predicted aperture area S/N | 35.57 | 45.80 |
| | Predicted ring S/N | −44.34 | −9.44 |

Table 14.16 Control factors for the shutter experiment

| | *Baseline* | *Optimum* |
|---|---|---|
| Predicted Aperture Area S/N | 28.8 | 30.1 |
| Predicted Ring S/N | −40.0 | −19.2 |

### 14.4.5 Conclusion

As a result of these experiments, ringing is reduced 10-fold over the baseline configuration at a cost of $0.06 per assembly (one additional separator). These experiments were instrumental to the success of this project. They provided the team with critical information in a timely, cost-effective manner.

## 14.5 Summary

The three case studies presented in this chapter are representative of hundreds of case studies that have been performed at the Eastman Kodak Company. Not all case studies result in breakthrough performance improvements, because very often the system has been optimized previously through tedious one-factor-at-a-time experimentation. But where the experiments have provided new levels of performance (about half of the cases), the results have been surprising, and at times breathtaking.

# PART III

# Advanced Topics

The Robust Design methodology combined with Design of Experiments is a powerful process for design optimization. In Part II: Parameter Design, we have explored and demonstrated that process using several examples. Methods that are useful for achieving additive designs have been discussed. In Part III, we discuss three advanced topics and conclude the book with an examination of Robust Design in the context of other quality optimization processes.

Ideally, Parameter Design can be accomplished using only the standard orthogonal arrays. However, in some cases it is necessary to modify orthogonal arrays to allow for unusual factor requirements. Three techniques: column downgrading, column upgrading, and the analysis of compound factors, are discussed in Chapter 15. Column downgrading is typically used for the study of a two-level factor in a three-level column. This increases the usefulness of three-level arrays such as the L18. Column upgrading is used to allow the study of a four-level factor in a two-level array. Analysis of compound factors allows two two-level factors to be assigned to a single three-level column. These techniques greatly broaden the flexibility of experimental design when trying to match an array with the factors that are to be studied.

The techniques described in Part II are suitable for optimizing the performance and additivity of a design. But there are plenty of circumstances where an interaction needs to be studied. The power of classical designed experiments, where the study of interactions is explicitly included, is useful in research and technology studies. There may also be superadditive cases where a synergistic interaction is exploited to achieve performance benefits. Chapter 16 shows how to study interactive effects using the standard orthogonal arrays.

Going beyond parameter optimization to do tolerance design requires quantitative information on the relative significance of factor effects, especially sources of variation. The standard technique for getting such information from a designed experiment is called analysis of variance. Chapter 17 discusses this useful analysis technique.

Lastly, Robust Design does not exist in a vacuum. Engineers and management need to use Robust Design in the context of product development, which increasingly is managed using Quality Function Deployment. Classical DOE and Six Sigma are two very powerful techniques that are also used to improve product quality. Unfortunately, there is a common misconception that these methods are in conflict with Robust Design. We do not believe that that needs to be the case. Chapter 18 explores these issues to conclude the book.

With the exception of column downgrading, the techniques discussed in Part III are not available in WinRobust Lite. If the need arises, the fully featured version of WinRobust can be obtained. Information on how to do so is given on the disk and in Appendix B.

CHAPTER 15

# Modifying Orthogonal Arrays

## 15.1 Introduction

It is very common in doing parameter design to find some factors with two levels, some with three levels, and even some factors with more than three levels. For *continuous factors* (e.g., time, temperature, size, force), two levels give information about the slope of the factor effect only; three levels give curvature information as well, as shown in Figure 15.1. This is very useful for finding the optimum conditions. Normally, more than three levels are not very useful. For *classification factors,* however, any number of levels may be required to test the various factor classes that may exist.

Examples of classification factors include the following:

- Type of solvent, in chemical engineering
- Production machine, in manufacturing
- Paper type, in printing
- Transistor type, in electrical engineering

One array that can handle two-level and three-level factors, the L18, has already been discussed. The L36 (see Appendix C) is another well-known example. It is also possible to take some other arrays

**Figure 15.1** Two-level and three-level factor effects

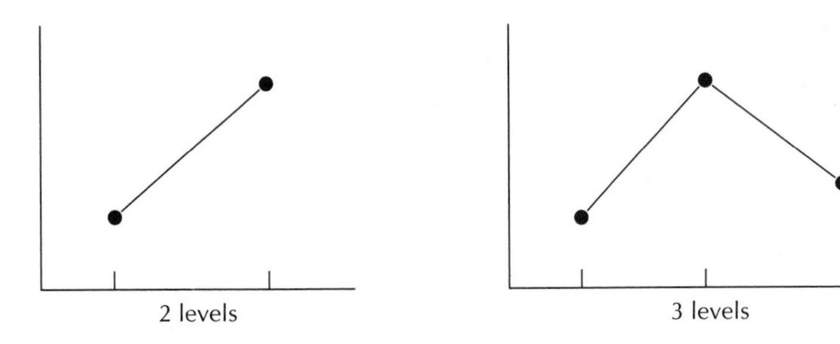

and alter the number of levels in their columns. There are three special techniques that can be used to alter the number of levels in orthogonal arrays:

1. Downgrading a column (dummy level)
2. Upgrading a column (column merging)
3. Compounding factors

## 15.2 Downgrading a Column

Downgrading is a modification to an orthogonal array's column having three or more levels in order to allow the use of a parameter having fewer levels. For example, an L9 array has four three-level columns, as shown in Table 15.1. The downgrading modification allows for the use of two-level and three-level factors in this array.

Suppose factor A has two levels, A1 and A2. Factor A can be assigned to a three-level column by creating a *dummy level* A3, which has the same value as either level A1 or level A2. The resulting array is shown in Table 15.2. The following assignments have been made in column 1: A1 at level 1 in column 1, A2 at level 2 in column 1, and A1 at level 3 in column 1. Notice that the resulting array is still orthogonal, although it is no longer exactly balanced. The degrees of freedom needed to estimate the factor effects (the Total $DOF_f$) are:

$$\text{Total DOF}_f = 1(2-1) + 3(3-1) = 7$$

which is one less than the usual Total $DOF_f$ of eight for an L9. The L9 still generates eight $DOF_{exp}$, so there is one DOF left over! The reason for the additional degree of freedom is that there are now six runs with A1. As a result, the estimate of the effect of A1 is twice as precise as that of A2.

There are some key issues to consider when choosing between making the dummy level (level 3) A1 or A2.

1. Choose the level that requires more precise information. If there is already a lot known about A1, then choose A2 for the dummy level.
2. Consider the availability of experimental resources and ease of experimentation. If A1 and

**Table 15.1** An unmodified L9 orthogonal array

| Run | 1<br>A | 2<br>B | 3<br>C | 4<br>D |
|---|---|---|---|---|
| 1 | 1 | 1 | 1 | 1 |
| 2 | 1 | 2 | 2 | 2 |
| 3 | 1 | 3 | 3 | 3 |
| 4 | 2 | 1 | 2 | 3 |
| 5 | 2 | 2 | 3 | 1 |
| 6 | 2 | 3 | 1 | 2 |
| 7 | 3 | 1 | 3 | 2 |
| 8 | 3 | 2 | 1 | 3 |
| 9 | 3 | 3 | 2 | 1 |

**Table 15.2** An L9 orthogonal array with dummy level

| Run | 1<br>A | 2<br>B | 3<br>C | 4<br>D |
|---|---|---|---|---|
| 1 | 1 | 1 | 1 | 1 |
| 2 | 1 | 2 | 2 | 2 |
| 3 | 1 | 3 | 3 | 3 |
| 4 | 2 | 1 | 2 | 3 |
| 5 | 2 | 2 | 3 | 1 |
| 6 | 2 | 3 | 1 | 2 |
| 7 | 1′ | 1 | 3 | 2 |
| 8 | 1′ | 2 | 1 | 3 |
| 9 | 1′ | 3 | 2 | 1 |

**Table 15.3** An unmodified L8 orthogonal array

| *Run* | *1* | *2* | *3* | *4* | *5* | *6* | *7* |
|---|---|---|---|---|---|---|---|
| 1 | 1 | 1 | 1 | 1 | 1 | 1 | 1 |
| 2 | 1 | 1 | 1 | 2 | 2 | 2 | 2 |
| 3 | 1 | 2 | 2 | 1 | 1 | 2 | 2 |
| 4 | 1 | 2 | 2 | 2 | 2 | 1 | 1 |
| 5 | 2 | 1 | 2 | 1 | 2 | 1 | 2 |
| 6 | 2 | 1 | 2 | 2 | 1 | 2 | 1 |
| 7 | 2 | 2 | 1 | 1 | 2 | 2 | 1 |
| 8 | 2 | 2 | 1 | 2 | 1 | 1 | 2 |
| | 2×3 | 1×3 | 1×2 | 1×5 | 1×4 | 1×7 | 1×6 |
| | 4×5 | 4×6 | 4×7 | 2×6 | 2×7 | 2×4 | 2×5 |
| | 6×7 | 5×7 | 5×6 | 3×7 | 3×6 | 3×5 | 3×4 |

A2 are raw materials and A1 is very scarce or expensive, make A2 the dummy level so that the experiment can be completed on time and within budget.

## 15.3 Upgrading a Column

It is also possible to go the other way and modify an array to handle a factor with more levels than exist in any column. Upgrading is a modification to a two-level orthogonal array that is done by combining several columns. The result is a column with multiple levels, allowing the use of a parameter that has more than two levels. For example, an L8 array has seven two-level columns, as shown in Table 15.3. The upgrading modification allows for the use of one four-level or three-level factor in this array.

The L8 has one $DOF_f$ per column. To include a four-level factor requires $DOF_f = 4 - 1 = 3$. This can be accomplished by merging three columns, each of which provides one of the required DOF. The choice of which columns to merge is made using the interaction information that is given at the bottom of each column. Three columns that represent an interacting set can be merged. Columns 1, 2, and 3 have this special relationship in the L8 (Table 15.3). That is, Column 3 = 1×2 interaction, Column 2 = 1×3 interaction, and Column 1 = 2×3 interaction. With this relationship in hand, columns 1, 2, and 3 can be merged as shown in Table 15.4.

**Table 15.4** Column merging table

| *Column 1 level* | *Column 2 level* | *Column 3 level* | *Merged column level* |
|---|---|---|---|
| 1 | 1 | 1 | 1 |
| 1 | 2 | 2 | 2 |
| 2 | 1 | 2 | 3 |
| 2 | 2 | 1 | 4 |

**Table 15.5** L8 orthogonal array after column upgrading

| *Run* | *(1, 2, 3)* | *4* | *5* | *6* | *7* |
|---|---|---|---|---|---|
| 1 | 1 | 1 | 1 | 1 | 1 |
| 2 | 1 | 2 | 2 | 2 | 2 |
| 3 | 2 | 1 | 1 | 2 | 2 |
| 4 | 2 | 2 | 2 | 1 | 1 |
| 5 | 3 | 1 | 2 | 1 | 2 |
| 6 | 3 | 2 | 1 | 2 | 1 |
| 7 | 4 | 1 | 2 | 2 | 1 |
| 8 | 4 | 2 | 1 | 1 | 2 |

The L8 orthogonal array, after columns 1, 2, and 3 are merged, is shown in Table 15.5. Once a four-level column is generated, downgrading can be applied to use a three-level factor.

There are a number of four-level column arrays that can be constructed from the L16. These can be found in the WinRobust upgrade package. They are listed here in Tables 15.6–15.10 and in Appendix C for your reference.

**Table 15.6** L16-1 ($4^1 2^{12}$)

| *Run* | *1* | *2* | *3* | *4* | *5* | *6* | *7* | *8* | *9* | *10* | *11* | *12* | *13* |
|---|---|---|---|---|---|---|---|---|---|---|---|---|---|
| 1 | 1 | 1 | 1 | 1 | 1 | 1 | 1 | 1 | 1 | 1 | 1 | 1 | 1 |
| 2 | 1 | 1 | 1 | 1 | 1 | 2 | 2 | 2 | 2 | 2 | 2 | 2 | 2 |
| 3 | 1 | 2 | 2 | 2 | 2 | 1 | 1 | 1 | 1 | 2 | 2 | 2 | 2 |
| 4 | 1 | 2 | 2 | 2 | 2 | 2 | 2 | 2 | 2 | 1 | 1 | 1 | 1 |
| 5 | 2 | 1 | 1 | 2 | 2 | 1 | 1 | 2 | 2 | 1 | 1 | 2 | 2 |
| 6 | 2 | 1 | 1 | 2 | 2 | 2 | 2 | 1 | 1 | 2 | 2 | 1 | 1 |
| 7 | 2 | 2 | 2 | 1 | 1 | 1 | 1 | 2 | 2 | 2 | 2 | 1 | 1 |
| 8 | 2 | 2 | 2 | 1 | 1 | 2 | 2 | 1 | 1 | 1 | 1 | 2 | 2 |
| 9 | 3 | 1 | 2 | 1 | 2 | 1 | 2 | 1 | 2 | 1 | 2 | 1 | 2 |
| 10 | 3 | 1 | 2 | 1 | 2 | 2 | 1 | 2 | 1 | 2 | 1 | 2 | 1 |
| 11 | 3 | 2 | 1 | 2 | 1 | 1 | 2 | 1 | 2 | 2 | 1 | 2 | 1 |
| 12 | 3 | 2 | 1 | 2 | 1 | 2 | 1 | 2 | 1 | 1 | 2 | 1 | 2 |
| 13 | 4 | 1 | 2 | 2 | 1 | 1 | 2 | 2 | 1 | 1 | 2 | 2 | 1 |
| 14 | 4 | 1 | 2 | 2 | 1 | 2 | 1 | 1 | 2 | 2 | 1 | 1 | 2 |
| 15 | 4 | 2 | 1 | 1 | 2 | 1 | 2 | 2 | 1 | 2 | 1 | 1 | 2 |
| 16 | 4 | 2 | 1 | 1 | 2 | 2 | 1 | 1 | 2 | 1 | 2 | 2 | 1 |

**Table 15.7** L16-2 ($4^2 2^9$)

| Run | 1 | 2 | 3 | 4 | 5 | 6 | 7 | 8 | 9 | 10 | 11 |
|---|---|---|---|---|---|---|---|---|---|---|---|
| 1 | 1 | 1 | 1 | 1 | 1 | 1 | 1 | 1 | 1 | 1 | 1 |
| 2 | 1 | 2 | 1 | 1 | 1 | 2 | 2 | 2 | 2 | 2 | 2 |
| 3 | 1 | 3 | 2 | 2 | 2 | 1 | 1 | 1 | 2 | 2 | 2 |
| 4 | 1 | 4 | 2 | 2 | 2 | 2 | 2 | 2 | 1 | 1 | 1 |
| 5 | 2 | 1 | 1 | 2 | 2 | 1 | 2 | 2 | 1 | 2 | 2 |
| 6 | 2 | 2 | 1 | 2 | 2 | 2 | 1 | 1 | 2 | 1 | 1 |
| 7 | 2 | 3 | 2 | 1 | 1 | 1 | 2 | 2 | 2 | 1 | 1 |
| 8 | 2 | 4 | 2 | 1 | 1 | 2 | 1 | 1 | 1 | 2 | 2 |
| 9 | 3 | 1 | 2 | 1 | 2 | 2 | 1 | 2 | 2 | 1 | 2 |
| 10 | 3 | 2 | 2 | 1 | 2 | 1 | 2 | 1 | 1 | 2 | 1 |
| 11 | 3 | 3 | 1 | 2 | 1 | 2 | 1 | 2 | 1 | 2 | 1 |
| 12 | 3 | 4 | 1 | 2 | 1 | 1 | 2 | 1 | 2 | 1 | 2 |
| 13 | 4 | 1 | 2 | 2 | 1 | 2 | 2 | 1 | 2 | 2 | 1 |
| 14 | 4 | 2 | 2 | 2 | 1 | 1 | 1 | 2 | 1 | 1 | 2 |
| 15 | 4 | 3 | 1 | 1 | 2 | 2 | 2 | 1 | 1 | 1 | 2 |
| 16 | 4 | 4 | 1 | 1 | 2 | 1 | 1 | 2 | 2 | 2 | 1 |

**Table 15.8** L16-3 ($4^3 2^6$)

| Run | 1 | 2 | 3 | 4 | 5 | 6 | 7 | 8 | 9 |
|---|---|---|---|---|---|---|---|---|---|
| 1 | 1 | 1 | 1 | 1 | 1 | 1 | 1 | 1 | 1 |
| 2 | 1 | 2 | 2 | 1 | 1 | 2 | 2 | 2 | 2 |
| 3 | 1 | 3 | 3 | 2 | 2 | 1 | 1 | 2 | 2 |
| 4 | 1 | 4 | 4 | 2 | 2 | 2 | 2 | 1 | 1 |
| 5 | 2 | 1 | 2 | 2 | 2 | 1 | 2 | 1 | 2 |
| 6 | 2 | 2 | 1 | 2 | 2 | 2 | 1 | 2 | 1 |
| 7 | 2 | 3 | 4 | 1 | 1 | 1 | 2 | 2 | 1 |
| 8 | 2 | 4 | 3 | 1 | 1 | 2 | 1 | 1 | 2 |
| 9 | 3 | 1 | 3 | 1 | 2 | 2 | 2 | 2 | 1 |
| 10 | 3 | 2 | 4 | 1 | 2 | 1 | 1 | 1 | 2 |
| 11 | 3 | 3 | 1 | 2 | 1 | 2 | 2 | 1 | 2 |
| 12 | 3 | 4 | 2 | 2 | 1 | 1 | 1 | 2 | 1 |
| 13 | 4 | 1 | 4 | 2 | 1 | 2 | 1 | 2 | 2 |
| 14 | 4 | 2 | 3 | 2 | 1 | 1 | 2 | 1 | 1 |
| 15 | 4 | 3 | 2 | 1 | 2 | 2 | 1 | 1 | 1 |
| 16 | 4 | 4 | 1 | 1 | 2 | 1 | 2 | 2 | 2 |

**Table 15.9** L16-4 ($4^4 2^3$)

| *Run* | *1* | *2* | *3* | *4* | *5* | *6* | *7* |
|---|---|---|---|---|---|---|---|
| 1 | 1 | 1 | 1 | 1 | 1 | 1 | 1 |
| 2 | 1 | 2 | 2 | 1 | 2 | 2 | 2 |
| 3 | 1 | 3 | 3 | 2 | 3 | 1 | 2 |
| 4 | 1 | 4 | 4 | 2 | 4 | 2 | 1 |
| 5 | 2 | 1 | 2 | 2 | 1 | 2 | 1 |
| 6 | 2 | 2 | 1 | 2 | 2 | 1 | 2 |
| 7 | 2 | 3 | 4 | 1 | 4 | 2 | 2 |
| 8 | 2 | 4 | 3 | 1 | 3 | 1 | 1 |
| 9 | 3 | 1 | 3 | 1 | 4 | 2 | 2 |
| 10 | 3 | 2 | 4 | 1 | 3 | 1 | 1 |
| 11 | 3 | 3 | 1 | 2 | 2 | 2 | 1 |
| 12 | 3 | 4 | 2 | 2 | 1 | 1 | 2 |
| 13 | 4 | 1 | 4 | 2 | 2 | 1 | 2 |
| 14 | 4 | 2 | 3 | 2 | 1 | 2 | 1 |
| 15 | 4 | 3 | 2 | 1 | 4 | 1 | 1 |
| 16 | 4 | 4 | 1 | 1 | 3 | 2 | 2 |

**Table 15.10** L16-5 ($4^5$)

| *Run* | *1* | *2* | *3* | *4* | *5* |
|---|---|---|---|---|---|
| 1 | 1 | 1 | 1 | 1 | 1 |
| 2 | 1 | 2 | 2 | 2 | 2 |
| 3 | 1 | 3 | 3 | 3 | 3 |
| 4 | 1 | 4 | 4 | 4 | 4 |
| 5 | 2 | 1 | 2 | 1 | 4 |
| 6 | 2 | 2 | 1 | 2 | 3 |
| 7 | 2 | 3 | 4 | 4 | 2 |
| 8 | 2 | 4 | 3 | 3 | 1 |
| 9 | 3 | 1 | 3 | 4 | 2 |
| 10 | 3 | 2 | 4 | 3 | 1 |
| 11 | 3 | 3 | 1 | 2 | 4 |
| 12 | 3 | 4 | 2 | 1 | 3 |
| 13 | 4 | 1 | 4 | 2 | 3 |
| 14 | 4 | 2 | 3 | 1 | 4 |
| 15 | 4 | 3 | 2 | 4 | 1 |
| 16 | 4 | 4 | 1 | 3 | 2 |

**Table 15.11** L18-1 ($1^6 3^6$)

| *Run* | *(1, 2)* | *3* | *4* | *5* | *6* | *7* | *8* |
|---|---|---|---|---|---|---|---|
| 1 | 1 | 1 | 1 | 1 | 1 | 1 | 1 |
| 2 | 1 | 2 | 2 | 2 | 2 | 2 | 2 |
| 3 | 1 | 3 | 3 | 3 | 3 | 3 | 3 |
| 4 | 2 | 1 | 1 | 2 | 2 | 3 | 3 |
| 5 | 2 | 2 | 2 | 3 | 3 | 1 | 1 |
| 6 | 2 | 3 | 3 | 1 | 1 | 2 | 2 |
| 7 | 3 | 1 | 2 | 1 | 3 | 2 | 3 |
| 8 | 3 | 2 | 3 | 2 | 1 | 3 | 1 |
| 9 | 3 | 3 | 1 | 3 | 2 | 1 | 2 |
| 10 | 4 | 1 | 3 | 3 | 2 | 2 | 1 |
| 11 | 4 | 2 | 1 | 1 | 3 | 3 | 2 |
| 12 | 4 | 3 | 2 | 2 | 1 | 1 | 3 |
| 13 | 5 | 1 | 2 | 3 | 1 | 3 | 2 |
| 14 | 5 | 2 | 3 | 1 | 2 | 1 | 3 |
| 15 | 5 | 3 | 1 | 2 | 3 | 2 | 1 |
| 16 | 6 | 1 | 3 | 2 | 3 | 1 | 2 |
| 17 | 6 | 2 | 1 | 3 | 1 | 2 | 3 |
| 18 | 6 | 3 | 2 | 1 | 2 | 3 | 1 |

A six-level column can be made using the first two columns of the L18. The six-level column requires $DOF_f = 5$. The required $DOF_f$ come from column 1 ($DOF_f = 1$) and column 2 ($DOF_f = 2$). The two additional required degrees of freedom are available from the column 1×2 interaction. The L18 has two additional degrees of freedom: The L18 $DOF_{exp} = 17$, but the Total $DOF_f = 15$. The "missing" DOF are the 1×2 interaction. Thus, those columns can be merged without sacrificing any other columns. The result is shown in Table 15.11.

Upgrading can be done using the L32 to give eight-level columns. Three-level columns such as in the L27 can also be upgraded to give a nine-level column. In that case, four columns, all forming an interaction set, are merged. The L12 cannot be upgraded because of its special nature with respect to interactions. The topic of interactions and their analysis is fully discussed in the following chapter. It should be understood that the upgrading process complicates the interaction analysis, and that these arrays can only be used in additive cases.

## 15.4 Compound Factors

In the case of noise factors (the outer array), compounding factors to greatly reduce the experimental effort has already been discussed. However, compounding usually results in the loss of information about individual factor effects. Thus, compounding is not usually applied to the inner (control factor) array. Generally, it is better to go to a larger array to accommodate all the control factors being analyzed. In situations where this is not economically feasible, it may be preferable to attempt some analysis of compound control factors. There is some risk in this technique because orthogonality is sacrificed to reduce the size of the experiment by reducing the number of runs for each factor level. With those caveats in mind, here is another tool to consider.

The technique of compounding factors allows the analysis of two 2-level factors in a simple 3-level column. [Note that the DOF required for two 2-level factors is $2(2 - 1) = 2$, which are available in a 3-level column.] Let there be two 2-level factors A and B, and three 3-level factors C, D, and E. Normally an L18 would be recommended, with one 3-level column downgraded. However, the total number of factor-effect DOF required is only

$$\text{Total DOF}_f = 2(2 - 1) + 3(3 - 1) = 8$$

The L18 provides 17 DOF, 9 more than are needed. This translates to a 47% efficiency. If there are budgetary constraints, the required DOF could be obtained using the more economical L9 design.

There are four possible combinations of A and B: A1B1, A1B2, A2B1, A2B2. Three of them are arbitrarily chosen and assigned as follows:

A1B1= level 1 in column 1

A1B2= level 2 in column 1

A2B1= level 3 in column 1

This would result in the array shown in Table 15.12.

The individual factor effects for A and B are found by first applying the analysis of means to the three-level column 1—that is, the average response for A1B1, for A1B2, and for A2B1. Pairwise differences are used to get the individual effects. Thus, the factor effect for factor A is given by

$$A1 - A2 = A1B1 - A2B1 \tag{15.1}$$

The factor effect for factor B is given by

$$B1 - B2 = A1B1 - A1B2 \tag{15.2}$$

**Table 15.12** L9 with compound factors in the first column

| Run | *1* *A,B* | *2* *C* | *3* *D* | *4* *E* |
|---|---|---|---|---|
| 1 | 1,1 | 1 | 1 | 1 |
| 2 | 1,1 | 2 | 2 | 2 |
| 3 | 1,1 | 3 | 3 | 3 |
| 4 | 1,2 | 1 | 2 | 3 |
| 5 | 1,2 | 2 | 3 | 1 |
| 6 | 1,2 | 3 | 1 | 2 |
| 7 | 2,1 | 1 | 3 | 2 |
| 8 | 2,1 | 2 | 1 | 3 |
| 9 | 2,1 | 3 | 2 | 1 |

## 15.5 Summary of Chapter 15

Orthogonal arrays are powerful tools for studying several factors together in a balanced manner that allows for independent factor effect analysis from a single experimental effort. However, it may be difficult to find an array that matches precisely the number of factors and levels per factor that one wants to study. Going to larger arrays is inefficient and usually impractical. Here, three techniques have been shown that can be used to introduce flexibility into the orthogonal array. WinRobust includes the ability to analyze dummy levels and certain upgraded arrays.The techniques shown conserve the DOF requirements in analyzing factor effects. In most cases, the ability to analyze interactions in the array is completely disrupted. With dummy levels, balance is partially lost. With compound factors, orthogonality is seriously compromised. Is this a significant problem? In some cases, yes. Whenever possible, try to stick with an established orthogonal array. But if this technique is needed to run an experiment, then use it with the understanding that the results could be slightly compromised. Applying the techniques described in Part II to achieve additivity is very helpful. However, it is always critical to verify the results of the analysis to test for difficulties.

## Exercises for Chapter 15

1. Under what circumstances is it useful to study more than three levels for a control factor?
2. Explain the process of downgrading a column.
3. Explain the process of upgrading a column.
4. Explain the process of compounding control factors.
5. Why is the compounding method controversial?
6. Set up a dummy level column in an L9 experiment studying the performance of an ink pen.

# CHAPTER 16

# Working with Interactions

*It is more desirable to treat interactions by including it [sic] in noise. Only a main effect which exceeds the magnitude of the interactions can safely be used in [Robust Design]*—G. TAGUCHI *[T6, p. 149]*

## 16.1 The Nature of Interactions in Robust Design

The issue of interactions between parameters is essential in any discussion of designed experiments. What is the role of interactions and additivity between control factors in parameter design? Remember, an additive model is one that describes the response without including any interaction terms. An interaction between control factors exists when the effect of one control factor is dependent on the level of another control factor. There are two approaches to dealing with such interactions:

1. Study control factor interactions to quantify their effects.
2. Engineer the design to minimize the likelihood of significant interactions and thus avoid having to estimate them.

Significant interactions cause the response to deviate from the value predicted by the simple additive model. When verification experiments fail, this is one of the probable causes. When the S/N ratio behavior follows the additive model, including only the main effects and no interaction terms, then the design is free of major interactions. The recommended strategy is to spend a good deal of time engineering for additivity and then verifying that the additive model works for that system.

This approach has been vehemently opposed in statistical literature. This is primarily because classical DOE is not structured or oriented toward minimizing interactivity in a design—it assumes interactions exist and seeks to examine them. Most statisticians do not have the background to enable them to assess a design for interactivity at the conceptualization stage. Let us see what that means from the engineer's point of view. An additive approach means focusing attention on *average control factor main effects.* Robust Design assumes that the main effects are strong enough to stand out from the random noise (experimental error) and that they are stronger than interactive effects between the control factors. As shown in Table 11.1, if a design has eight critical-to-function (CTF) factors, then there are 28 two-way interactions to consider. From the engineering point of view, each of these interactions represents a potential constraint dictating what the levels of the interacting control factors must be. Consider the design complexity that an additional 28 constraints imply. In a highly interactive paradigm, each of the eight CTFs would require considering seven possible interactions before specifying the parameter. (Of course, it is all too common to ignore these possible *codependencies* when making an engineering change. Interactions that are ignored quickly surface when one fix causes two new problems, which then become the subject of their own corrective actions . . . and so on, and so on, until the money runs out or the project is canceled.)

When experimental data are analyzed to rank-order each control factor for the magnitude of its effect on the response, it is always wise to compare all the values to the experimental error, $S_e$. This determines whether the main effects for each factor are stronger than the experimental error. Remember, the experimental error is the measure of random experimental variation, including, for an L12 or L18 orthogonal array, interactive effects.

It is not the intention here to ignore the fact that interactions between control factors exist. Rather, the goal is to optimize functional performance without depending upon interactions. This is possible if the interactions are weak or can be made weak. In that case, the additive model approximates the complex relationships between the S/N ratio and the control factors. The benefits of achieving an additive model are many, including using far fewer degrees of experimental freedom (i.e., shorter experiments). To rigorously study the math model relating the S/N ratio to the design parameters would be a lengthy process. The additive model allows one to proceed down a very efficient path using orthogonal array–based experiments.

In order to assure additivity, the interactivity of factors must first be recognized and then minimized. However, there are circumstances in which studying the interactions between control factors is necessary. For example, in the early stages of research on a new technology, the control factor effects are not yet well understood. This includes a lack of knowledge of the interactive effects. Also, in some systems, there are levels of performance that can only be achieved by the use of a synergistic interaction (some authors refer to this case as superadditivity). For these cases, the engineering team must have the tools with which to carry out a proper interactivity evaluation. The remainder of this chapter is devoted to building skills in this important area of experimental design and analysis.

## 16.2 Interactions Defined

An interaction between two control factors means that the effect of one factor on the response depends upon the level of another factor. Interactions are defined in Chapter 7. An example of an interaction is shown in Figure 16.1. When Factor B is Low, changing factor A from Low to High causes the response to get smaller. When Factor B is High, changing factor A from Low to High causes the response to get bigger. Thus, the effect of Factor A on the response depends upon the level of Factor B. This dependency is reciprocal—that is, the effect of Factor B on the response depends upon the level of Factor A.

An interaction such as that shown in Figure 16.1 is referred to as *antisynergistic,* because the slope of the factor effect plots changes sign depending upon Factor B's level. If the response is the S/N ratio and the antisynergistic interaction is between control factors, then the optimum level for factor A depends on B and vice versa. If one factor changes, then the effect of both factors on the response changes. The antisynergistic aspect means that if one of the factors is changed, the other one can end up at its nonoptimum level, *even though it is not changed!* As an engineering issue, this leads to reoptimization of the rest of the system after a change is made in just one factor. This is a nonrobust condition, because it implies that the system is very sensitive to changes in one of the control factors. In an additive situation, the effect of a change in one factor is limited to that factor's main effect only.

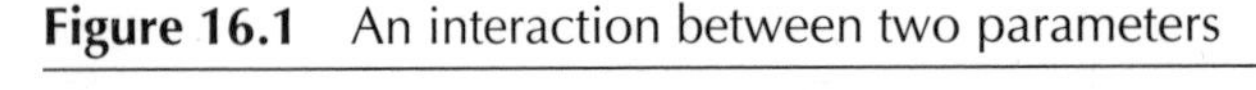

**Figure 16.1** An interaction between two parameters

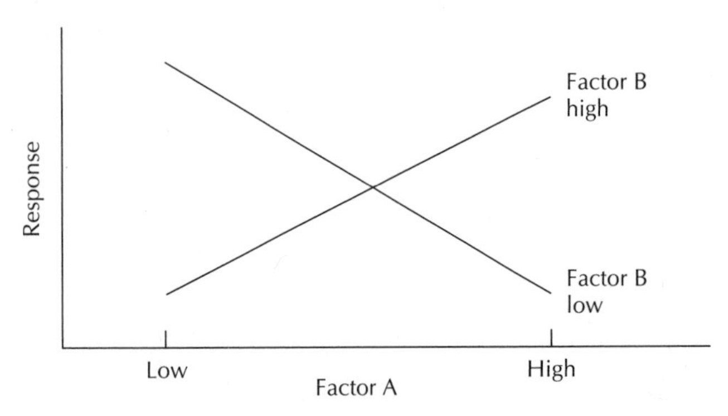


Reoptimization of the other factors is not needed because their effect is unchanged in the absence of an interaction.

A mild interaction between control factors, as shown in Figure 16.2, is a tolerable situation. Here, although there is an interaction, it is referred to as *synergistic* because the optimum factor levels do not depend on each other. The optimum level for control factor A is High, regardless of the level of factor B. Similarly, the optimum level for control factor B is High, regardless of the level of factor A. The *magnitude* of the effect of changing factor A depends on the level of factor B. Similarly, the *magnitude* of the effect of changing factor B depends upon the effect of changing factor A. This situation is described as *monotonic*. Since the optimal levels stay the same, reoptimization is not needed. Nevertheless, the situation is not ideal. A relatively small loss in robustness, which might result from a factor level change, can be amplified by the interactions.

Effectiveness at engineering a Robust Design depends upon the ability to find and exploit even small interactions between noise factors and control factors as well as the ability to avoid strong anti-synergistic interactions between control factors.

**Figure 16.2** A mild interaction between control factors

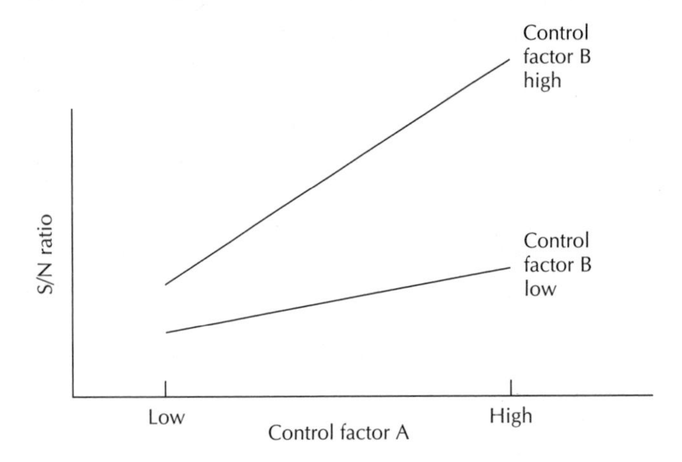

**Table 16.1** The L4 array

| Run | 1<br>*A* | 2<br>*B* | 3<br>*C = A×B* |
|---|---|---|---|
| 1 | 1 | 1 | 1 |
| 2 | 1 | 2 | 2 |
| 3 | 2 | 1 | 2 |
| 4 | 2 | 2 | 1 |

## 16.3 How Interactions Are Measured

Interactions can be estimated using certain types of orthogonal arrays—for example, the $L_4$ orthogonal array shown in Table 16.1. This array has the property that Column 3 can be used to represent the Column 1 and Column 2 interaction (A×B). Thus, if factors A and B are assigned to columns 1 and 2, respectively, *and* column 3 is left empty, then the calculation of the effect of column 3 can be used to study the A×B interaction. Column 3 has one degree of freedom available to account for either the A×B interaction or the effect of a third, new (additive) control factor C.

Let us see how this works using a quantitative example. Start by assuming that there is a response, *y*, that is a function of factors A and B, shown in Equation 16.1. Note that no interaction is indicated between A and B.

$$y = 5 + 3A - B \tag{16.1}$$

If this response is studied using the L4 array, the data shown in Table 16.2 are generated. The plot of the data, shown in Figure 16.3, shows no interaction. (The lines are parallel.)

**Table 16.2** Data for Figure 16.3

| *Run* | *A* | *B* | *C* | *y* |
|---|---|---|---|---|
| 1 | 1 | 1 | 1 | 7 |
| 2 | 1 | 2 | 2 | 6 |
| 3 | 2 | 1 | 2 | 10 |
| 4 | 2 | 2 | 1 | 9 |

**Figure 16.3** Interaction plot showing no interaction

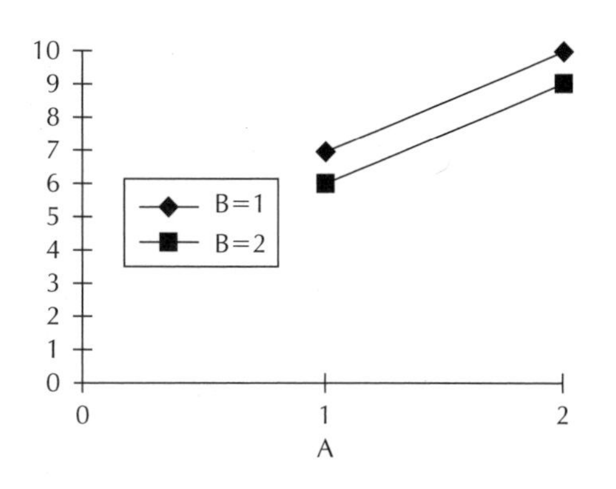

Here is the analysis of means for the data shown in Table 16.2:

Overall average = (7 + 6 + 10 + 9)/4 = 8

$$\bar{y}_{A2} = (10 + 9)/2 = 9.5$$
$$\bar{y}_{A1} = (7 + 6)/2 = \underline{6.5}$$
$$\Delta_A = 3.0$$

$$\bar{y}_{B2} = (6 + 9)/2 = 7.5$$
$$\bar{y}_{B1} = (7 + 10)/2 = \underline{8.5}$$
$$\Delta_B = -1.0$$

$$\bar{y}_{C2} = (6 + 10)/2 = 8$$
$$\bar{y}_{C1} = (7 + 9)/2 = \underline{8}$$
$$\Delta_C = 0 \Rightarrow \text{Confirming that there is no interaction!}$$

Now let us modify the function so that the response is given by Equation 16.2:

$$y = 5 + 3A - B + 2C \tag{16.2}$$

Factor C represents the A×B interaction. If this response is studied using the L4 array, the data shown in Table 16.3 are generated. The plot of the data is shown in Figure 16.4. Now the lines are no longer parallel. In fact, they cross, clearly indicating some sort of interaction.

The data given for the four runs in the L4 array can be analyzed, using analysis of means (ANOM), to quantify the strength of the effects of factors A and B, as well as the A×B interaction.

Here is the analysis of means for the data shown in Table 16.3:

Overall average = (9 + 10 + 14 + 11)/4 = 11

$$\bar{y}_{A2} = (14 + 11)/2 = 12.5$$
$$\bar{y}_{A1} = (9 + 10)/2 = \underline{9.5}$$
$$\Delta_A = 3.0$$

$$\bar{y}_{B2} = (10 + 11)/2 = 10.5$$
$$\bar{y}_{B1} = (9 + 14)/2 = \underline{11.5}$$
$$\Delta_B = -1.0$$

$$\bar{y}_{C2} = (10 + 14)/2 = 12$$
$$\bar{y}_{C1} = (9 + 11)/2 = \underline{10}$$
$$\Delta_C = 2$$

The addition of factor +2C to the response equation and the L4 array causes a change in the ANOM for factors A and B. However, the differences $\Delta_A$ and $\Delta_B$, which are the factor effects for factors A and B, stay the same. This is because the main effect is determined only by the dependence of the response on the factor. Only the factor effect, $\Delta_C$, has changed from the analysis of Table 16.2 to the

**Table 16.3** Data for Figure 16.4

| *Run* | *A* | *B* | *C* | *y* |
|---|---|---|---|---|
| 1 | 1 | 1 | 1 | 9 |
| 2 | 1 | 2 | 2 | 10 |
| 3 | 2 | 1 | 2 | 14 |
| 4 | 2 | 2 | 1 | 11 |

**Figure 16.4** Interaction plot showing an interaction

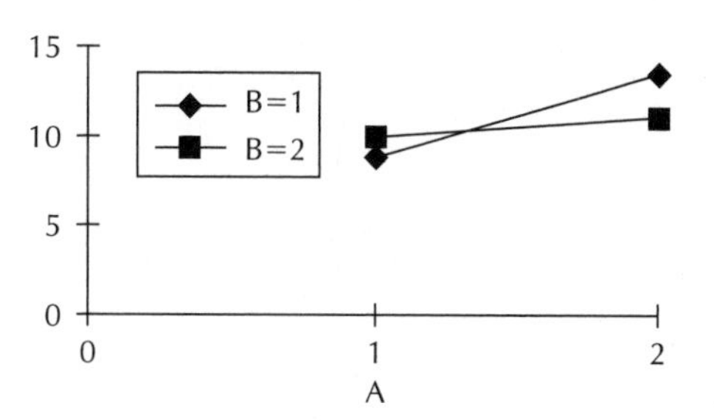


analysis of Table 16.3. This addition does cause the interaction plot, Figure 16.4, to change, *even though* the factor effects are the same.

The difference between Figures 16.3 and 16.4 is due to the data. If there is *no* factor assigned to Column 3, and the experimental data produces a calculated effect such as seen in Figure 16.4, then the effect would be attributed to an A×B interaction. There are three points to learn from this discussion:

1. The factor effect measured in column 3 is due to the A×B interaction, if there is no physical factor assigned to column 3 (i.e., no factor C).
2. The interaction plot can be used to visualize the A×B interaction, again only if there is no factor assigned to column 3.
3. The net result of assigning a factor to Column 3, as shown in Table 16.3, is that the calculated effect is due to factor C, or to the A×B interaction, or to a confounded blend of the two.

Thus, things can get confusing when a control factor is assigned to a column that is capable of quantifying both an interactive effect and the effect of a control factor. The way to avoid this problem is to refrain from using interaction columns that are related to factors for which strong interactions may exist. Thus, an approach for choosing the proper column assignments to study interactions is needed. If all the columns of an array are used, then there is some uncertainty as to how the results are interpreted. Is the analysis of means due to the factor assigned to that column (the main effect) or due to the interactions also attributed to that column?

## 16.4 Degrees of Freedom for Interactions

One other issue to explore is how many degrees of freedom are required to study (model) an interaction. The number of degrees of freedom required to study main effects and interactions goes up quickly with the number of factors. Here, this is shown by looking at the column assignments and the number of factors each orthogonal array can accept without confounding main effects with two-way interactions (i.e., interactions between two factors). Remember, for each unknown quantity that is calculated in a matrix experiment, one treatment combination must be run. To describe the interaction between two 2-level factors, it is sufficient to find the difference in slope between the two main effects, as shown in Figure 16.5.

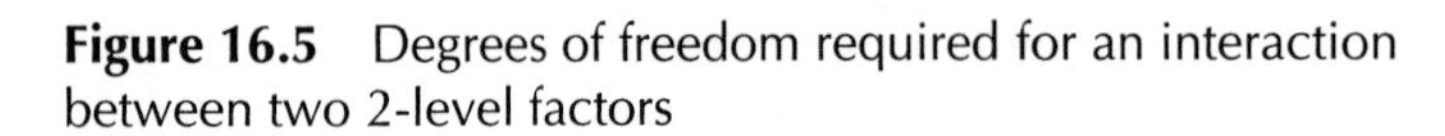

**Figure 16.5** Degrees of freedom required for an interaction between two 2-level factors

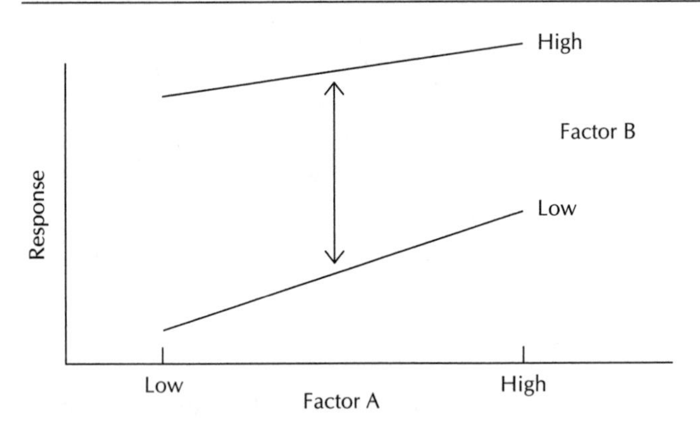


In the L4 example, one 2-level column (Column 3), with $DOF_f = (2 - 1) = 1$, is adequate to estimate the interaction between factors A and B. To describe the interaction between two 3-level factors, four differences in slope between the two main effects are required, as shown in Figure 16.6.

Thus, for the L9, two 3-level columns (Columns 3 and 4), each with $DOF_f = (3 - 1) = 2$, are required to estimate the interaction between factors assigned to Columns 1 and 2. The general equation for the $DOF_{int}$ required to estimate an interaction between an $n$-level factor and an $m$-level factor is

$$DOF_{int} = (m - 1)(n - 1) \quad \textbf{(16.3)}$$

As is shown in Table 11.1, studying a small number of 2-level factors, including interactions, is feasible in terms of total DOF. When 3-level factors are studied, inclusion of interactions is very expensive in terms of total DOF.

**Figure 16.6** Degrees of freedom required for an interaction between two 3-level factors

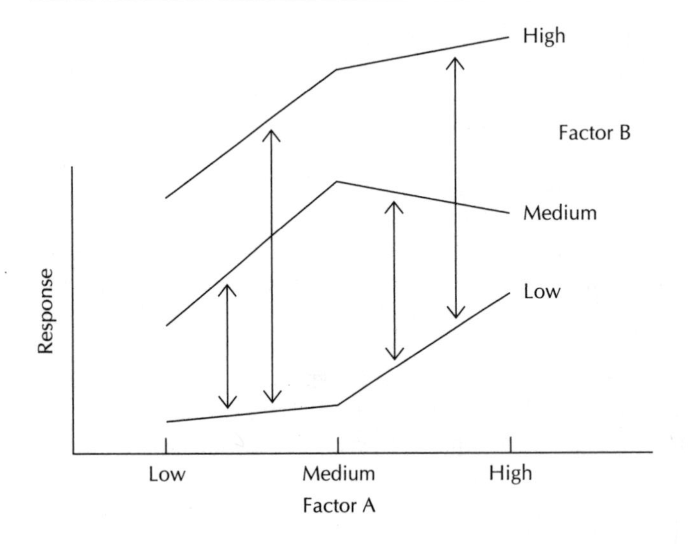

## 16.5 Setting up the Experiment When Interactions Are Included

The standard approach for studying interactions is to use fractional factorial orthogonal arrays. Depending upon the degree of fractionation, two-way interactions can be studied free of confounding with main effects and other two-way interactions. When all the two-way interactions are free of confounding with other two-way interactions and main effects, this is referred to as a *Resolution V* array. When two-way interactions confound with two-way interactions but not main effects, this is referred to as *Resolution IV.* When two-way interactions confound with main effects, this is referred to as *Resolution III.* This simple classification allows some general statements to be made about the amount of confounding in an experiment.

### *16.5.1 Two-Level Factor Arrays*

There are rigorous mathematical techniques for building fractional factorial arrays with the desired resolution. A simpler approach is to use the orthogonal arrays, which are equivalent to fractional factorials with appropriate column assignments. For example, consider the L8 shown in Table 16.4. The column assignments shown for factors A, B, and C allow the study of the three two-way interactions with *no* confounding. This Resolution V design is equivalent to a $2^3$ full factorial design. The three main effects and three interactions require a total of 6 DOF. Thus, six columns are used in this design.

If a fourth factor is added to the experiment, it is assigned to column 7 in order to avoid confounding with the two-way interactions as shown in Table 16.5. Now there are a total of 4 DOF for main effects and 6 DOF for the interactions. Thus, some confounding is inevitable. This is because the $DOF_{exp}$ available from an L8 is $8 - 1 = 7$. However, if the columns are chosen as shown, the confounding is only between the interactions, *not* with the main effects. This Resolution IV design is equivalent to a $2^{4-1}$ fractional factorial. It is highly suitable for a Parameter Design study, where the primary interest is in quantifying the main effects. For a full math model including interactions, a Resolution V experiment is more suitable. If it is known that the factor D has a negligible interaction with the other three factors, then the confounding can be ignored and the empty columns (3, 5, and 6)

**Table 16.4** L8 ($2^3$), Resolution V

| *Run* | *1* *A* | *2* *B* | *3* | *4* *C* | *5* | *6* | *7* |
|---|---|---|---|---|---|---|---|
| 1 | 1 | 1 | 1 | 1 | 1 | 1 | 1 |
| 2 | 1 | 1 | 1 | 2 | 2 | 2 | 2 |
| 3 | 1 | 2 | 2 | 1 | 1 | 2 | 2 |
| 4 | 1 | 2 | 2 | 2 | 2 | 1 | 1 |
| 5 | 2 | 1 | 2 | 1 | 2 | 1 | 2 |
| 6 | 2 | 1 | 2 | 2 | 1 | 2 | 1 |
| 7 | 2 | 2 | 1 | 1 | 2 | 2 | 1 |
| 8 | 2 | 2 | 1 | 2 | 1 | 1 | 2 |
| | | | 1×2 | | 1×4 | 2×4 | |

**Table 16.5** L8 ($2^{4-1}$), Resolution IV

| *Run* | *1* *A* | *2* *B* | *3* | *4* *C* | *5* | *6* | *7* *D* |
|---|---|---|---|---|---|---|---|
| 1 | 1 | 1 | 1 | 1 | 1 | 1 | 1 |
| 2 | 1 | 1 | 1 | 2 | 2 | 2 | 2 |
| 3 | 1 | 2 | 2 | 1 | 1 | 2 | 2 |
| 4 | 1 | 2 | 2 | 2 | 2 | 1 | 1 |
| 5 | 2 | 1 | 2 | 1 | 2 | 1 | 2 |
| 6 | 2 | 1 | 2 | 2 | 1 | 2 | 1 |
| 7 | 2 | 2 | 1 | 1 | 2 | 2 | 1 |
| 8 | 2 | 2 | 1 | 2 | 1 | 1 | 2 |
| | | | 1×2<br>4×7 | | 1×4<br>2×7 | 1×7<br>2×4 | |

**Table 16.6** L16 ($2^{5-1}$) Resolution V

| Run | *1* *A* | *2* *B* | *3* | *4* *C* | *5* | *6* | *7* | *8* *D* | *9* | *10* | *11* | *12* | *13* | *14* | *15* *E* |
|---|---|---|---|---|---|---|---|---|---|---|---|---|---|---|---|
| 1 | 1 | 1 | 1 | 1 | 1 | 1 | 1 | 1 | 1 | 1 | 1 | 1 | 1 | 1 | 1 |
| 2 | 1 | 1 | 1 | 1 | 1 | 1 | 1 | 2 | 2 | 2 | 2 | 2 | 2 | 2 | 2 |
| 3 | 1 | 1 | 1 | 2 | 2 | 2 | 2 | 1 | 1 | 1 | 1 | 2 | 2 | 2 | 2 |
| 4 | 1 | 1 | 1 | 2 | 2 | 2 | 2 | 2 | 2 | 2 | 2 | 1 | 1 | 1 | 1 |
| 5 | 1 | 2 | 2 | 1 | 1 | 2 | 2 | 1 | 1 | 2 | 2 | 1 | 1 | 2 | 2 |
| 6 | 1 | 2 | 2 | 1 | 1 | 2 | 2 | 2 | 2 | 1 | 1 | 2 | 2 | 1 | 1 |
| 7 | 1 | 2 | 2 | 2 | 2 | 1 | 1 | 1 | 1 | 2 | 2 | 2 | 2 | 1 | 1 |
| 8 | 1 | 2 | 2 | 2 | 2 | 1 | 1 | 2 | 2 | 1 | 1 | 1 | 1 | 2 | 2 |
| 9 | 2 | 1 | 2 | 1 | 2 | 1 | 2 | 1 | 2 | 1 | 2 | 1 | 2 | 1 | 2 |
| 10 | 2 | 1 | 2 | 1 | 2 | 1 | 2 | 2 | 1 | 2 | 1 | 2 | 1 | 2 | 1 |
| 11 | 2 | 1 | 2 | 2 | 1 | 2 | 1 | 1 | 2 | 1 | 2 | 2 | 1 | 2 | 1 |
| 12 | 2 | 1 | 2 | 2 | 1 | 2 | 1 | 2 | 1 | 2 | 1 | 1 | 2 | 1 | 2 |
| 13 | 2 | 2 | 1 | 1 | 2 | 2 | 1 | 1 | 2 | 2 | 1 | 1 | 2 | 2 | 1 |
| 14 | 2 | 2 | 1 | 1 | 2 | 2 | 1 | 2 | 1 | 1 | 2 | 2 | 1 | 1 | 2 |
| 15 | 2 | 2 | 1 | 2 | 1 | 1 | 2 | 1 | 2 | 2 | 1 | 2 | 1 | 1 | 2 |
| 16 | 2 | 2 | 1 | 2 | 1 | 1 | 2 | 2 | 1 | 1 | 2 | 1 | 2 | 2 | 1 |
| | | | 1×2 | | 1×4 | 2×4 | 8×15 | | 1×8 | 2×8 | 4×15 | 4×8 | 2×15 | 1×15 | |

can be analyzed for the A×B, A×C, and B×C interactions. Any additional factors assigned to the L8 result in a high degree of confounding. In such cases, the L12 is preferred because it distributes the confounding between any pair of factors over all the other columns. This makes the interaction appear as experimental error, thus minimizing its confounding with any one main effect.

The next example is the L16 shown in Table 16.6. The greatest number of factors that can be assigned to an L16 for a Resolution V experiment is five. The Resolution V design has five main effects and 10 two-way interactions for a total of 15 DOF, the maximum that can be assigned to an L16 without confounding. If the five factors are assigned to the columns labeled A, B, C, D, and E, then the resulting design is a Resolution V design, equivalent to a $2^{5-1}$ fractional factorial.

Up to three more factors can be added, still without confounding main effects with two-way interactions. Table 16.7 shows the column assignments that preserve freedom of confounding for the main effects. Note that the first column is now left empty because many two-way interactions are confounded there. This Resolution IV design is equivalent to a $2^{8-4}$ fractional factorial.

If it is acceptable to confound main effects with interactions, the appropriate column assignment for the next factor would be factor I in column 1, resulting in a high degree of confounding for that factor and low confounding for the remaining main effects. Any additional factors would result in a high degree of confounding. Again, the L12 is preferable for that circumstance.

The largest two-level array looked at is the L32, shown in Table 16.8. The greatest number of factors that can be assigned to the L32 in a Resolution V design is six. This is true despite the fact that seven factors have 21 two-way interactions for a total of 28 DOF, which is three less than the total

**Table 16.7** L16 ($2^{8-4}$) Resolution IV

| Run | 1 | 2<br>A | 3<br>F | 4<br>B | 5<br>G | 6 | 7 | 8<br>C | 9<br>H | 10 | 11 | 12 | 13 | 14<br>E | 15<br>D |
|---|---|---|---|---|---|---|---|---|---|---|---|---|---|---|---|
| 1 | 1 | 1 | 1 | 1 | 1 | 1 | 1 | 1 | 1 | 1 | 1 | 1 | 1 | 1 | 1 |
| 2 | 1 | 1 | 1 | 1 | 1 | 1 | 1 | 2 | 2 | 2 | 2 | 2 | 2 | 2 | 2 |
| 3 | 1 | 1 | 1 | 2 | 2 | 2 | 2 | 1 | 1 | 1 | 1 | 2 | 2 | 2 | 2 |
| 4 | 1 | 1 | 1 | 2 | 2 | 2 | 2 | 2 | 2 | 2 | 2 | 1 | 1 | 1 | 1 |
| 5 | 1 | 2 | 2 | 1 | 1 | 2 | 2 | 1 | 1 | 2 | 2 | 1 | 1 | 2 | 2 |
| 6 | 1 | 2 | 2 | 1 | 1 | 2 | 2 | 2 | 2 | 1 | 1 | 2 | 2 | 1 | 1 |
| 7 | 1 | 2 | 2 | 2 | 2 | 1 | 1 | 1 | 1 | 2 | 2 | 2 | 2 | 1 | 1 |
| 8 | 1 | 2 | 2 | 2 | 2 | 1 | 1 | 2 | 2 | 1 | 1 | 1 | 1 | 2 | 2 |
| 9 | 2 | 1 | 2 | 1 | 2 | 1 | 2 | 1 | 2 | 1 | 2 | 1 | 2 | 1 | 2 |
| 10 | 2 | 1 | 2 | 1 | 2 | 1 | 2 | 2 | 1 | 2 | 1 | 2 | 1 | 2 | 1 |
| 11 | 2 | 1 | 2 | 2 | 1 | 2 | 1 | 1 | 2 | 1 | 2 | 2 | 1 | 2 | 1 |
| 12 | 2 | 1 | 2 | 2 | 1 | 2 | 1 | 2 | 1 | 2 | 1 | 1 | 2 | 1 | 2 |
| 13 | 2 | 2 | 1 | 1 | 2 | 2 | 1 | 1 | 2 | 2 | 1 | 1 | 2 | 2 | 1 |
| 14 | 2 | 2 | 1 | 1 | 2 | 2 | 1 | 2 | 1 | 1 | 2 | 2 | 1 | 1 | 2 |
| 15 | 2 | 2 | 1 | 2 | 1 | 1 | 2 | 1 | 2 | 2 | 1 | 2 | 1 | 1 | 2 |
| 16 | 2 | 2 | 1 | 2 | 1 | 1 | 2 | 2 | 1 | 1 | 2 | 1 | 2 | 2 | 1 |
| | 2×3<br>4×5<br>8×9<br>14×15 | | | | | 2×4<br>3×5<br>8×14<br>9×15 | 2×5<br>3×4<br>8×15<br>9×14 | | | 2×8<br>3×9<br>4×14<br>5×15 | 2×9<br>3×8<br>4×15<br>5×14 | 2×14<br>3×15<br>4×8<br>5×9 | 2×15<br>3×14<br>4×9<br>5×8 | | |

available. It is simply not possible to assign seven factors without some two-way interaction confounding. Table 16.8 shows the column assignments for this near Resolution V experiment (seven factors). By omitting factor D, a true Resolution V experiment, with six main effects and 15 two-way interactions for a total of 21 DOF, can be achieved.

For eight, nine, or ten factors, there are progressively more two-way interactions that become confounded. You can explore the confounding patterns using WinRobust, which was used to generate these figures. Table 16.9 shows the best column assignments for 10 factors. Again, the L12 can be used for up to 11 factors, resulting in fewer runs and a better (more uniformly distributed) type of confounding.

It *is* possible to get up to 16 factors into the L32 without confounding main effects with two-way interactions, using a technique introduced by Jerry Roslund, GMC.[1] This is based on the observation that interactions between odd-numbered columns are always in even-numbered columns, and vice versa. Thus, up to 16 factors can be assigned to the odd-numbered columns, resulting in the interaction pattern shown in Table 16.10.[2] However, this approach quickly leads to two-way interaction confounding and cannot be used for Resolution V studies.

---

1. NTU Telecourse Notes.
2. Note that additional odd-column interactions are given in Appendix C.

**Table 16.8** L32 ($2^{7-2}$) Resolution IV

| Run | 1<br>A | 2<br>B | 3 | 4<br>C | 5 | 6 | 7<br>D | 8<br>E | 9 | 10 | 11 | 12 | 13 | 14 | 15 | 16<br>F |
|---|---|---|---|---|---|---|---|---|---|---|---|---|---|---|---|---|
| 1 | 1 | 1 | 1 | 1 | 1 | 1 | 1 | 1 | 1 | 1 | 1 | 1 | 1 | 1 | 1 | 1 |
| 2 | 1 | 1 | 1 | 1 | 1 | 1 | 1 | 1 | 1 | 1 | 1 | 1 | 1 | 1 | 1 | 2 |
| 3 | 1 | 1 | 1 | 1 | 1 | 1 | 1 | 2 | 2 | 2 | 2 | 2 | 2 | 2 | 2 | 1 |
| 4 | 1 | 1 | 1 | 1 | 1 | 1 | 1 | 2 | 2 | 2 | 2 | 2 | 2 | 2 | 2 | 2 |
| 5 | 1 | 1 | 1 | 2 | 2 | 2 | 2 | 1 | 1 | 1 | 1 | 2 | 2 | 2 | 2 | 1 |
| 6 | 1 | 1 | 1 | 2 | 2 | 2 | 2 | 1 | 1 | 1 | 1 | 2 | 2 | 2 | 2 | 2 |
| 7 | 1 | 1 | 1 | 2 | 2 | 2 | 2 | 2 | 2 | 2 | 2 | 1 | 1 | 1 | 1 | 1 |
| 8 | 1 | 1 | 1 | 2 | 2 | 2 | 2 | 2 | 2 | 2 | 2 | 1 | 1 | 1 | 1 | 2 |
| 9 | 1 | 2 | 2 | 1 | 1 | 2 | 2 | 1 | 1 | 2 | 2 | 1 | 1 | 2 | 2 | 1 |
| 10 | 1 | 2 | 2 | 1 | 1 | 2 | 2 | 1 | 1 | 2 | 2 | 1 | 1 | 2 | 2 | 2 |
| 11 | 1 | 2 | 2 | 1 | 1 | 2 | 2 | 2 | 2 | 1 | 1 | 2 | 2 | 1 | 1 | 1 |
| 12 | 1 | 2 | 2 | 1 | 1 | 2 | 2 | 2 | 2 | 1 | 1 | 2 | 2 | 1 | 1 | 2 |
| 13 | 1 | 2 | 2 | 2 | 2 | 1 | 1 | 1 | 1 | 2 | 2 | 2 | 2 | 1 | 1 | 1 |
| 14 | 1 | 2 | 2 | 2 | 2 | 1 | 1 | 1 | 1 | 2 | 2 | 2 | 2 | 1 | 1 | 2 |
| 15 | 1 | 2 | 2 | 2 | 2 | 1 | 1 | 2 | 2 | 1 | 1 | 1 | 1 | 2 | 2 | 1 |
| 16 | 1 | 2 | 2 | 2 | 2 | 1 | 1 | 2 | 2 | 1 | 1 | 1 | 1 | 2 | 2 | 2 |
| 17 | 2 | 1 | 2 | 1 | 2 | 1 | 2 | 1 | 2 | 1 | 2 | 1 | 2 | 1 | 2 | 1 |
| 18 | 2 | 1 | 2 | 1 | 2 | 1 | 2 | 1 | 2 | 1 | 2 | 1 | 2 | 1 | 2 | 2 |
| 19 | 2 | 1 | 2 | 1 | 2 | 1 | 2 | 2 | 1 | 2 | 1 | 2 | 1 | 2 | 1 | 1 |
| 20 | 2 | 1 | 2 | 1 | 2 | 1 | 2 | 2 | 1 | 2 | 1 | 2 | 1 | 2 | 1 | 1 |
| 21 | 2 | 1 | 2 | 2 | 1 | 2 | 1 | 1 | 2 | 1 | 2 | 2 | 1 | 2 | 1 | 1 |
| 22 | 2 | 1 | 2 | 2 | 1 | 2 | 1 | 1 | 2 | 1 | 2 | 2 | 1 | 2 | 1 | 2 |
| 23 | 2 | 1 | 2 | 2 | 1 | 2 | 1 | 2 | 1 | 2 | 1 | 1 | 2 | 1 | 2 | 1 |
| 24 | 2 | 1 | 2 | 2 | 1 | 2 | 1 | 2 | 1 | 2 | 1 | 1 | 2 | 1 | 2 | 2 |
| 25 | 2 | 2 | 1 | 1 | 2 | 2 | 1 | 1 | 2 | 2 | 1 | 1 | 2 | 2 | 1 | 1 |
| 26 | 2 | 2 | 1 | 1 | 2 | 2 | 1 | 1 | 2 | 2 | 1 | 1 | 2 | 2 | 1 | 2 |
| 27 | 2 | 2 | 1 | 1 | 2 | 2 | 1 | 2 | 1 | 1 | 2 | 2 | 1 | 1 | 2 | 1 |
| 28 | 2 | 2 | 1 | 1 | 2 | 2 | 1 | 2 | 1 | 1 | 2 | 2 | 1 | 1 | 2 | 2 |
| 29 | 2 | 2 | 1 | 2 | 1 | 1 | 2 | 1 | 2 | 2 | 1 | 2 | 1 | 1 | 2 | 1 |
| 30 | 2 | 2 | 1 | 2 | 1 | 1 | 2 | 1 | 2 | 2 | 1 | 2 | 1 | 1 | 2 | 2 |
| 31 | 2 | 2 | 1 | 2 | 1 | 1 | 2 | 2 | 1 | 1 | 2 | 1 | 2 | 2 | 1 | 1 |
| 32 | 2 | 2 | 1 | 2 | 1 | 1 | 2 | 2 | 1 | 1 | 2 | 1 | 2 | 2 | 1 | 2 |
| | | | 1×2<br>4×7 | | 1×4<br>2×7 | 1×7<br>2×4 | | | 1×8 | 2×8 | | 4×8 | 16×29 | | 7×8 | |

| 17 | 18 | 19 | 20 | 21 | 22 | 23 | 24 | 25 | 26 | 27 | 28 | 29<br>G | 30 | 31 |
|---|---|---|---|---|---|---|---|---|---|---|---|---|---|---|
| 1 | 1 | 1 | 1 | 1 | 1 | 1 | 1 | 1 | 1 | 1 | 1 | 1 | 1 | 1 |
| 2 | 2 | 2 | 2 | 2 | 2 | 2 | 2 | 2 | 2 | 2 | 2 | 2 | 2 | 2 |
| 1 | 1 | 1 | 1 | 1 | 1 | 1 | 2 | 2 | 2 | 2 | 2 | 2 | 2 | 2 |
| 2 | 2 | 2 | 2 | 2 | 2 | 2 | 1 | 1 | 1 | 1 | 1 | 1 | 1 | 1 |
| 1 | 1 | 1 | 2 | 2 | 2 | 2 | 1 | 1 | 1 | 1 | 2 | 2 | 2 | 2 |
| 2 | 2 | 2 | 1 | 1 | 1 | 1 | 2 | 2 | 2 | 2 | 1 | 1 | 1 | 1 |
| 1 | 1 | 1 | 2 | 2 | 2 | 2 | 2 | 2 | 2 | 2 | 1 | 1 | 1 | 1 |
| 2 | 2 | 2 | 1 | 1 | 1 | 1 | 1 | 1 | 1 | 1 | 2 | 2 | 2 | 2 |
| 1 | 2 | 2 | 1 | 1 | 2 | 2 | 1 | 1 | 2 | 2 | 1 | 1 | 2 | 2 |
| 2 | 1 | 1 | 2 | 2 | 1 | 1 | 2 | 2 | 1 | 1 | 2 | 2 | 1 | 1 |
| 1 | 2 | 2 | 1 | 1 | 2 | 2 | 2 | 2 | 1 | 1 | 2 | 2 | 1 | 1 |
| 2 | 1 | 1 | 2 | 2 | 1 | 1 | 1 | 1 | 2 | 2 | 1 | 1 | 2 | 2 |
| 1 | 2 | 2 | 2 | 2 | 1 | 1 | 1 | 1 | 2 | 2 | 2 | 2 | 1 | 1 |
| 2 | 1 | 1 | 1 | 1 | 2 | 2 | 2 | 2 | 1 | 1 | 1 | 1 | 2 | 2 |
| 1 | 2 | 2 | 2 | 2 | 1 | 1 | 2 | 2 | 1 | 1 | 1 | 1 | 2 | 2 |
| 2 | 1 | 1 | 1 | 1 | 2 | 2 | 1 | 1 | 2 | 2 | 2 | 2 | 1 | 1 |
| 2 | 1 | 2 | 1 | 2 | 1 | 2 | 1 | 2 | 1 | 2 | 1 | 2 | 1 | 2 |
| 1 | 2 | 1 | 2 | 1 | 2 | 1 | 2 | 1 | 2 | 1 | 2 | 1 | 2 | 1 |
| 2 | 1 | 2 | 1 | 2 | 1 | 2 | 2 | 1 | 2 | 1 | 2 | 1 | 2 | 1 |
| 1 | 2 | 1 | 2 | 1 | 2 | 1 | 1 | 2 | 1 | 2 | 1 | 2 | 1 | 2 |
| 2 | 1 | 2 | 2 | 1 | 2 | 1 | 1 | 2 | 1 | 2 | 2 | 1 | 2 | 1 |
| 1 | 2 | 1 | 1 | 2 | 1 | 2 | 2 | 1 | 2 | 1 | 1 | 2 | 1 | 2 |
| 2 | 1 | 2 | 2 | 1 | 2 | 1 | 2 | 1 | 2 | 1 | 1 | 2 | 1 | 2 |
| 1 | 2 | 1 | 1 | 2 | 1 | 2 | 1 | 2 | 1 | 2 | 2 | 1 | 2 | 1 |
| 2 | 2 | 1 | 1 | 2 | 2 | 1 | 1 | 2 | 2 | 1 | 1 | 2 | 2 | 1 |
| 1 | 1 | 2 | 2 | 1 | 1 | 2 | 2 | 1 | 1 | 2 | 2 | 1 | 1 | 2 |
| 2 | 2 | 1 | 1 | 2 | 2 | 1 | 2 | 1 | 1 | 2 | 2 | 1 | 1 | 2 |
| 1 | 1 | 2 | 2 | 1 | 1 | 2 | 1 | 2 | 2 | 1 | 1 | 2 | 2 | 1 |
| 2 | 2 | 1 | 2 | 1 | 1 | 2 | 1 | 2 | 2 | 1 | 2 | 1 | 1 | 2 |
| 1 | 1 | 2 | 1 | 2 | 2 | 1 | 2 | 1 | 1 | 2 | 1 | 2 | 2 | 1 |
| 2 | 2 | 1 | 2 | 1 | 1 | 2 | 2 | 1 | 1 | 2 | 1 | 2 | 2 | 1 |
| 1 | 1 | 2 | 1 | 2 | 2 | 1 | 1 | 2 | 2 | 1 | 2 | 1 | 1 | 2 |
| 1×16 | 2×16 | | 4×16 | 8×29 | | 7×16 | 8×16 | 4×29 | 7×29 | | 1×29 | | | 2×29 |

**Table 16.9** L32 ($2^{10-5}$) Resolution IV design

| | *1* | *2* | *3* | *4* | *5* | *6* | *7* | *8* | *9* | *10* | *11* | *12* | *13* | *14* | *15* |
|---|---|---|---|---|---|---|---|---|---|---|---|---|---|---|---|
| *Run* | *A* | *B* | | *C* | | | | *D* | | | | | | | *E* |
| 1 | 1 | 1 | 1 | 1 | 1 | 1 | 1 | 1 | 1 | 1 | 1 | 1 | 1 | 1 | 1 |
| 2 | 1 | 1 | 1 | 1 | 1 | 1 | 1 | 1 | 1 | 1 | 1 | 1 | 1 | 1 | 1 |
| 3 | 1 | 1 | 1 | 1 | 1 | 1 | 1 | 2 | 2 | 2 | 2 | 2 | 2 | 2 | 2 |
| 4 | 1 | 1 | 1 | 1 | 1 | 1 | 1 | 2 | 2 | 2 | 2 | 2 | 2 | 2 | 2 |
| 5 | 1 | 1 | 1 | 2 | 2 | 2 | 2 | 1 | 1 | 1 | 1 | 2 | 2 | 2 | 2 |
| 6 | 1 | 1 | 1 | 2 | 2 | 2 | 2 | 1 | 1 | 1 | 1 | 2 | 2 | 2 | 2 |
| 7 | 1 | 1 | 1 | 2 | 2 | 2 | 2 | 2 | 2 | 2 | 2 | 1 | 1 | 1 | 1 |
| 8 | 1 | 1 | 1 | 2 | 2 | 2 | 2 | 2 | 2 | 2 | 2 | 1 | 1 | 1 | 1 |
| 9 | 1 | 2 | 2 | 1 | 1 | 2 | 2 | 1 | 1 | 2 | 2 | 1 | 1 | 2 | 2 |
| 10 | 1 | 2 | 2 | 1 | 1 | 2 | 2 | 1 | 1 | 2 | 2 | 1 | 1 | 2 | 2 |
| 11 | 1 | 2 | 2 | 1 | 1 | 2 | 2 | 2 | 2 | 1 | 1 | 2 | 2 | 1 | 1 |
| 12 | 1 | 2 | 2 | 1 | 1 | 2 | 2 | 2 | 2 | 1 | 1 | 2 | 2 | 1 | 1 |
| 13 | 1 | 2 | 2 | 2 | 2 | 1 | 1 | 1 | 1 | 2 | 2 | 2 | 2 | 1 | 1 |
| 14 | 1 | 2 | 2 | 2 | 2 | 1 | 1 | 1 | 1 | 2 | 2 | 2 | 2 | 1 | 1 |
| 15 | 1 | 2 | 2 | 2 | 2 | 1 | 1 | 2 | 2 | 1 | 1 | 1 | 1 | 2 | 2 |
| 16 | 1 | 2 | 2 | 2 | 2 | 1 | 1 | 2 | 2 | 1 | 1 | 1 | 1 | 2 | 2 |
| 17 | 2 | 1 | 2 | 1 | 2 | 1 | 2 | 1 | 2 | 1 | 2 | 1 | 2 | 1 | 2 |
| 18 | 2 | 1 | 2 | 1 | 2 | 1 | 2 | 1 | 2 | 1 | 2 | 1 | 2 | 1 | 2 |
| 19 | 2 | 1 | 2 | 1 | 2 | 1 | 2 | 2 | 1 | 2 | 1 | 2 | 1 | 2 | 1 |
| 20 | 2 | 1 | 2 | 1 | 2 | 1 | 2 | 2 | 1 | 2 | 1 | 2 | 1 | 2 | 1 |
| 21 | 2 | 1 | 2 | 2 | 1 | 2 | 1 | 1 | 2 | 1 | 2 | 2 | 1 | 2 | 1 |
| 22 | 2 | 1 | 2 | 2 | 1 | 2 | 1 | 1 | 2 | 1 | 2 | 2 | 1 | 2 | 1 |
| 23 | 2 | 1 | 2 | 2 | 1 | 2 | 1 | 2 | 1 | 2 | 1 | 1 | 2 | 1 | 2 |
| 24 | 2 | 1 | 2 | 2 | 1 | 2 | 1 | 2 | 1 | 2 | 1 | 1 | 2 | 1 | 2 |
| 25 | 2 | 2 | 1 | 1 | 2 | 2 | 1 | 1 | 2 | 2 | 1 | 1 | 2 | 2 | 1 |
| 26 | 2 | 2 | 1 | 1 | 2 | 2 | 1 | 1 | 2 | 2 | 1 | 1 | 2 | 2 | 1 |
| 27 | 2 | 2 | 1 | 1 | 2 | 2 | 1 | 2 | 1 | 1 | 2 | 2 | 1 | 1 | 2 |
| 28 | 2 | 2 | 1 | 1 | 2 | 2 | 1 | 2 | 1 | 1 | 2 | 2 | 1 | 1 | 2 |
| 29 | 2 | 2 | 1 | 2 | 1 | 1 | 2 | 1 | 2 | 2 | 1 | 2 | 1 | 1 | 2 |
| 30 | 2 | 2 | 1 | 2 | 1 | 1 | 2 | 1 | 2 | 2 | 1 | 2 | 1 | 1 | 2 |
| 31 | 2 | 2 | 1 | 2 | 1 | 1 | 2 | 2 | 1 | 1 | 2 | 1 | 2 | 2 | 1 |
| 32 | 2 | 2 | 1 | 2 | 1 | 1 | 2 | 2 | 1 | 1 | 2 | 1 | 2 | 2 | 1 |
| | | | 1×2<br>29×30 | | 1×4<br>27×30 | 2×4<br>27×29 | 8×15<br>16×23 | | 1×8<br>23×30 | 2×8<br>23×29 | 4×15<br>16×27 | 4×8<br>23×27 | 2×15<br>16×29 | 1×15<br>16×30 | |

| *16* *F* | *17* | *18* | *19* | *20* | *21* | *22* | *23* *G* | *24* | *25* | *26* | *27* *H* | *28* | *29* *I* | *30* *J* | *31* |
|---|---|---|---|---|---|---|---|---|---|---|---|---|---|---|---|
| 1 | 1 | 1 | 1 | 1 | 1 | 1 | 1 | 1 | 1 | 1 | 1 | 1 | 1 | 1 | 1 |
| 2 | 2 | 2 | 2 | 2 | 2 | 2 | 2 | 2 | 2 | 2 | 2 | 2 | 2 | 2 | 2 |
| 1 | 1 | 1 | 1 | 1 | 1 | 1 | 1 | 2 | 2 | 2 | 2 | 2 | 2 | 2 | 2 |
| 2 | 2 | 2 | 2 | 2 | 2 | 2 | 2 | 1 | 1 | 1 | 1 | 1 | 1 | 1 | 1 |
| 1 | 1 | 1 | 1 | 2 | 2 | 2 | 2 | 1 | 1 | 1 | 1 | 2 | 2 | 2 | 2 |
| 2 | 2 | 2 | 2 | 1 | 1 | 1 | 1 | 2 | 2 | 2 | 2 | 1 | 1 | 1 | 1 |
| 1 | 1 | 1 | 1 | 2 | 2 | 2 | 2 | 2 | 2 | 2 | 2 | 1 | 1 | 1 | 1 |
| 2 | 2 | 2 | 2 | 1 | 1 | 1 | 1 | 1 | 1 | 1 | 1 | 2 | 2 | 2 | 2 |
| 1 | 1 | 2 | 2 | 1 | 1 | 2 | 2 | 1 | 1 | 2 | 2 | 1 | 1 | 2 | 2 |
| 2 | 2 | 1 | 1 | 2 | 2 | 1 | 1 | 2 | 2 | 1 | 1 | 2 | 2 | 1 | 1 |
| 1 | 1 | 2 | 2 | 1 | 1 | 2 | 2 | 2 | 2 | 1 | 1 | 2 | 2 | 1 | 1 |
| 2 | 2 | 1 | 1 | 2 | 2 | 1 | 1 | 1 | 1 | 2 | 2 | 1 | 1 | 2 | 2 |
| 1 | 1 | 2 | 2 | 2 | 2 | 1 | 1 | 1 | 1 | 2 | 2 | 2 | 2 | 1 | 1 |
| 2 | 2 | 1 | 1 | 1 | 1 | 2 | 2 | 2 | 2 | 1 | 1 | 1 | 1 | 2 | 2 |
| 1 | 1 | 2 | 2 | 2 | 2 | 1 | 1 | 2 | 2 | 1 | 1 | 1 | 1 | 2 | 2 |
| 2 | 2 | 1 | 1 | 1 | 1 | 2 | 2 | 1 | 1 | 2 | 2 | 2 | 2 | 1 | 1 |
| 1 | 2 | 1 | 2 | 1 | 2 | 1 | 2 | 1 | 2 | 1 | 2 | 1 | 2 | 1 | 2 |
| 2 | 1 | 2 | 1 | 2 | 1 | 2 | 1 | 2 | 1 | 2 | 1 | 2 | 1 | 2 | 1 |
| 1 | 2 | 1 | 2 | 1 | 2 | 1 | 2 | 2 | 1 | 2 | 1 | 2 | 1 | 2 | 1 |
| 2 | 1 | 2 | 1 | 2 | 1 | 2 | 1 | 1 | 2 | 1 | 2 | 1 | 2 | 1 | 2 |
| 1 | 2 | 1 | 2 | 2 | 1 | 2 | 1 | 1 | 2 | 1 | 2 | 2 | 1 | 2 | 1 |
| 2 | 1 | 2 | 1 | 1 | 2 | 1 | 2 | 2 | 1 | 2 | 1 | 1 | 2 | 1 | 2 |
| 1 | 2 | 1 | 2 | 2 | 1 | 2 | 1 | 2 | 1 | 2 | 1 | 1 | 2 | 1 | 2 |
| 2 | 1 | 2 | 1 | 1 | 2 | 1 | 2 | 1 | 2 | 1 | 2 | 2 | 1 | 2 | 1 |
| 1 | 2 | 2 | 1 | 1 | 2 | 2 | 1 | 1 | 2 | 2 | 1 | 1 | 2 | 2 | 1 |
| 2 | 1 | 1 | 2 | 2 | 1 | 1 | 2 | 2 | 1 | 1 | 2 | 2 | 1 | 1 | 2 |
| 1 | 2 | 2 | 1 | 1 | 2 | 2 | 1 | 2 | 1 | 1 | 2 | 2 | 1 | 1 | 2 |
| 2 | 1 | 1 | 2 | 2 | 1 | 1 | 2 | 1 | 2 | 2 | 1 | 1 | 2 | 2 | 1 |
| 1 | 2 | 2 | 1 | 2 | 1 | 1 | 2 | 1 | 2 | 2 | 1 | 2 | 1 | 1 | 2 |
| 2 | 1 | 1 | 2 | 1 | 2 | 2 | 1 | 2 | 1 | 1 | 2 | 1 | 2 | 2 | 1 |
| 1 | 2 | 2 | 1 | 2 | 1 | 1 | 2 | 2 | 1 | 1 | 2 | 1 | 2 | 2 | 1 |
| 2 | 1 | 1 | 2 | 1 | 2 | 2 | 1 | 1 | 2 | 2 | 1 | 2 | 1 | 1 | 2 |
| | 1×16<br>15×30 | 2×16<br>15×29 | 4×23<br>8×27 | 4×16<br>15×27 | 2×23<br>8×29 | 1×23<br>8×30 | | 8×16<br>15×23 | 2×27<br>4×29 | 1×27<br>4×30 | | 1×29<br>2×30 | | | 1×30<br>2×29<br>4×27<br>8×23<br>15×16 |

**Table 16.10** L32 Resolution IV design

| Run | *1* *A* | *2* | *3* *B* | *4* | *5* *C* | *6* | *7* *D* | *8* | *9* *E* | *10* | *11* *F* | *12* | *13* *G* | *14* | *15* *H* | *16* |
|---|---|---|---|---|---|---|---|---|---|---|---|---|---|---|---|---|
| 1 | 1 | 1 | 1 | 1 | 1 | 1 | 1 | 1 | 1 | 1 | 1 | 1 | 1 | 1 | 1 | 1 |
| 2 | 1 | 1 | 1 | 1 | 1 | 1 | 1 | 1 | 1 | 1 | 1 | 1 | 1 | 1 | 1 | 2 |
| 3 | 1 | 1 | 1 | 1 | 1 | 1 | 1 | 2 | 2 | 2 | 2 | 2 | 2 | 2 | 2 | 1 |
| 4 | 1 | 1 | 1 | 1 | 1 | 1 | 1 | 2 | 2 | 2 | 2 | 2 | 2 | 2 | 2 | 2 |
| 5 | 1 | 1 | 1 | 2 | 2 | 2 | 2 | 1 | 1 | 1 | 1 | 2 | 2 | 2 | 2 | 1 |
| 6 | 1 | 1 | 1 | 2 | 2 | 2 | 2 | 1 | 1 | 1 | 1 | 2 | 2 | 2 | 2 | 2 |
| 7 | 1 | 1 | 1 | 2 | 2 | 2 | 2 | 2 | 2 | 2 | 2 | 1 | 1 | 1 | 1 | 1 |
| 8 | 1 | 1 | 1 | 2 | 2 | 2 | 2 | 2 | 2 | 2 | 2 | 1 | 1 | 1 | 1 | 2 |
| 9 | 1 | 2 | 2 | 1 | 1 | 2 | 2 | 1 | 1 | 2 | 2 | 1 | 1 | 2 | 2 | 1 |
| 10 | 1 | 2 | 2 | 1 | 1 | 2 | 2 | 1 | 1 | 2 | 2 | 1 | 1 | 2 | 2 | 2 |
| 11 | 1 | 2 | 2 | 1 | 1 | 2 | 2 | 2 | 2 | 1 | 1 | 2 | 2 | 1 | 1 | 1 |
| 12 | 1 | 2 | 2 | 1 | 1 | 2 | 2 | 2 | 2 | 1 | 1 | 2 | 2 | 1 | 1 | 2 |
| 13 | 1 | 2 | 2 | 2 | 2 | 1 | 1 | 1 | 1 | 2 | 2 | 2 | 2 | 1 | 1 | 1 |
| 14 | 1 | 2 | 2 | 2 | 2 | 1 | 1 | 1 | 1 | 2 | 2 | 2 | 2 | 1 | 1 | 2 |
| 15 | 1 | 2 | 2 | 2 | 2 | 1 | 1 | 2 | 2 | 1 | 1 | 1 | 1 | 2 | 2 | 1 |
| 16 | 1 | 2 | 2 | 2 | 2 | 1 | 1 | 2 | 2 | 1 | 1 | 1 | 1 | 2 | 2 | 2 |
| 17 | 2 | 1 | 2 | 1 | 2 | 1 | 2 | 1 | 2 | 1 | 2 | 1 | 2 | 1 | 2 | 1 |
| 18 | 2 | 1 | 2 | 1 | 2 | 1 | 2 | 1 | 2 | 1 | 2 | 1 | 2 | 1 | 2 | 2 |
| 19 | 2 | 1 | 2 | 1 | 2 | 1 | 2 | 2 | 1 | 2 | 1 | 2 | 1 | 2 | 1 | 1 |
| 20 | 2 | 1 | 2 | 1 | 2 | 1 | 2 | 2 | 1 | 2 | 1 | 2 | 1 | 2 | 1 | 2 |
| 21 | 2 | 1 | 2 | 2 | 1 | 2 | 1 | 1 | 2 | 1 | 2 | 2 | 1 | 2 | 1 | 1 |
| 22 | 2 | 1 | 2 | 2 | 1 | 2 | 1 | 1 | 2 | 1 | 2 | 2 | 1 | 2 | 1 | 2 |
| 23 | 2 | 1 | 2 | 2 | 1 | 2 | 1 | 2 | 1 | 2 | 1 | 1 | 2 | 1 | 2 | 1 |
| 24 | 2 | 1 | 2 | 2 | 1 | 2 | 1 | 2 | 1 | 2 | 1 | 1 | 2 | 1 | 2 | 2 |
| 25 | 2 | 2 | 1 | 1 | 2 | 2 | 1 | 1 | 2 | 2 | 1 | 1 | 2 | 2 | 1 | 1 |
| 26 | 2 | 2 | 1 | 1 | 2 | 2 | 1 | 1 | 2 | 2 | 1 | 1 | 2 | 2 | 1 | 2 |
| 27 | 2 | 2 | 1 | 1 | 2 | 2 | 1 | 2 | 1 | 1 | 2 | 2 | 1 | 1 | 2 | 1 |
| 28 | 2 | 2 | 1 | 1 | 2 | 2 | 1 | 2 | 1 | 1 | 2 | 2 | 1 | 1 | 2 | 2 |
| 29 | 2 | 2 | 1 | 2 | 1 | 1 | 2 | 1 | 2 | 2 | 1 | 2 | 1 | 1 | 2 | 1 |
| 30 | 2 | 2 | 1 | 2 | 1 | 1 | 2 | 1 | 2 | 2 | 1 | 2 | 1 | 1 | 2 | 2 |
| 31 | 2 | 2 | 1 | 2 | 1 | 1 | 2 | 2 | 1 | 1 | 2 | 1 | 2 | 2 | 1 | 1 |
| 32 | 2 | 2 | 1 | 2 | 1 | 1 | 2 | 2 | 1 | 1 | 2 | 1 | 2 | 2 | 1 | 2 |
| | | 1×3 | | 1×5 | | 1×7 | | 1×9 | | 1×11 | | 1×13 | | 1×15 | | 1×17 |
| | | 5×7 | | 3×7 | | 3×5 | | 3×11 | | 3×9 | | 3×15 | | 3×13 | | 3×19 |
| | | 9×11 | | 9×13 | | 9×15 | | 5×13 | | 5×15 | | 5×9 | | 5×11 | | 5×21 |
| | | 13×15 | | 11×15 | | 11×13 | | 7×15 | | 7×13 | | 7×11 | | 7×9 | | 7×23 |

| *17*<br>*I* | *18* | *19*<br>*J* | *20* | *21*<br>*K* | *22* | *23*<br>*L* | *24* | *25*<br>*M* | *26* | *27*<br>*N* | *28* | *29*<br>*O* | *30* | *31*<br>*P* |
|---|---|---|---|---|---|---|---|---|---|---|---|---|---|---|
| 1 | 1 | 1 | 1 | 1 | 1 | 1 | 1 | 1 | 1 | 1 | 1 | 1 | 1 | 1 |
| 2 | 2 | 2 | 2 | 2 | 2 | 2 | 2 | 2 | 2 | 2 | 2 | 2 | 2 | 2 |
| 1 | 1 | 1 | 1 | 1 | 1 | 1 | 2 | 2 | 2 | 2 | 2 | 2 | 2 | 2 |
| 2 | 2 | 2 | 2 | 2 | 2 | 2 | 1 | 1 | 1 | 1 | 1 | 1 | 1 | 1 |
| 1 | 1 | 1 | 2 | 2 | 2 | 2 | 1 | 1 | 1 | 1 | 2 | 2 | 2 | 2 |
| 2 | 2 | 2 | 1 | 1 | 1 | 1 | 2 | 2 | 2 | 2 | 1 | 1 | 1 | 1 |
| 1 | 1 | 1 | 2 | 2 | 2 | 2 | 2 | 2 | 2 | 2 | 1 | 1 | 1 | 1 |
| 2 | 2 | 2 | 1 | 1 | 1 | 1 | 1 | 1 | 1 | 1 | 2 | 2 | 2 | 2 |
| 1 | 2 | 2 | 1 | 1 | 2 | 2 | 1 | 1 | 2 | 2 | 1 | 1 | 2 | 2 |
| 2 | 1 | 1 | 2 | 2 | 1 | 1 | 2 | 2 | 1 | 1 | 2 | 2 | 1 | 1 |
| 1 | 2 | 2 | 1 | 1 | 2 | 2 | 2 | 2 | 1 | 1 | 2 | 2 | 1 | 1 |
| 2 | 1 | 1 | 2 | 2 | 1 | 1 | 1 | 1 | 2 | 2 | 1 | 1 | 2 | 2 |
| 1 | 2 | 2 | 2 | 2 | 1 | 1 | 1 | 1 | 2 | 2 | 2 | 2 | 1 | 1 |
| 2 | 1 | 1 | 1 | 1 | 2 | 2 | 2 | 2 | 1 | 1 | 1 | 1 | 2 | 2 |
| 1 | 2 | 2 | 2 | 2 | 1 | 1 | 2 | 2 | 1 | 1 | 1 | 1 | 2 | 2 |
| 2 | 1 | 1 | 1 | 1 | 2 | 2 | 1 | 1 | 2 | 2 | 2 | 2 | 1 | 1 |
| 2 | 1 | 2 | 1 | 2 | 1 | 2 | 1 | 2 | 1 | 2 | 1 | 2 | 1 | 2 |
| 1 | 2 | 1 | 2 | 1 | 2 | 1 | 2 | 1 | 2 | 1 | 2 | 1 | 2 | 1 |
| 2 | 1 | 2 | 1 | 2 | 1 | 2 | 2 | 1 | 2 | 1 | 2 | 1 | 2 | 1 |
| 1 | 2 | 1 | 2 | 1 | 2 | 1 | 1 | 2 | 1 | 2 | 1 | 2 | 1 | 2 |
| 2 | 1 | 2 | 2 | 1 | 2 | 1 | 1 | 2 | 1 | 2 | 2 | 1 | 2 | 1 |
| 1 | 2 | 1 | 1 | 2 | 1 | 2 | 2 | 1 | 2 | 1 | 1 | 2 | 1 | 2 |
| 2 | 1 | 2 | 2 | 1 | 2 | 1 | 2 | 1 | 2 | 1 | 1 | 2 | 1 | 2 |
| 1 | 2 | 1 | 1 | 2 | 1 | 2 | 1 | 2 | 1 | 2 | 2 | 1 | 2 | 1 |
| 2 | 2 | 1 | 1 | 2 | 2 | 1 | 1 | 2 | 2 | 1 | 1 | 2 | 2 | 1 |
| 1 | 1 | 2 | 2 | 1 | 1 | 2 | 2 | 1 | 1 | 2 | 2 | 1 | 1 | 2 |
| 2 | 2 | 1 | 1 | 2 | 2 | 1 | 2 | 1 | 1 | 2 | 2 | 1 | 1 | 2 |
| 1 | 1 | 2 | 2 | 1 | 1 | 2 | 1 | 2 | 2 | 1 | 1 | 2 | 2 | 1 |
| 2 | 2 | 1 | 2 | 1 | 1 | 2 | 1 | 2 | 2 | 1 | 2 | 1 | 1 | 2 |
| 1 | 1 | 2 | 1 | 2 | 2 | 1 | 2 | 1 | 1 | 2 | 1 | 2 | 2 | 1 |
| 2 | 2 | 1 | 2 | 1 | 1 | 2 | 2 | 1 | 1 | 2 | 1 | 2 | 2 | 1 |
| 1 | 1 | 2 | 1 | 2 | 2 | 1 | 1 | 2 | 2 | 1 | 2 | 1 | 1 | 2 |
|  | 1×19<br>3×17<br>5×23<br>7×21 |  | 1×21<br>3×23<br>5×17<br>7×19 |  | 1×23<br>3×21<br>5×19<br>7×17 |  | 1×25<br>3×27<br>5×29<br>7×31 |  | 1×27<br>3×25<br>5×31<br>7×29 |  | 1×29<br>3×31<br>5×25<br>7×27 |  | 1×31<br>3×29<br>5×27<br>7×25 |  |

**Table 16.11** L9 ($3^2$) Resolution V

| Run | 1<br>A | 2<br>B | 3 | 4 |
|---|---|---|---|---|
| 1 | 1 | 1 | 1 | 1 |
| 2 | 1 | 2 | 2 | 2 |
| 3 | 1 | 3 | 3 | 3 |
| 4 | 2 | 1 | 2 | 3 |
| 5 | 2 | 2 | 3 | 1 |
| 6 | 2 | 3 | 1 | 2 |
| 7 | 3 | 1 | 3 | 2 |
| 8 | 3 | 2 | 1 | 3 |
| 9 | 3 | 3 | 2 | 1 |
| | | | 1×2 | 1×2 |

### *16.5.2 Three-Level Factor Arrays*

Three arrays are practical for studying three-level factors: the L9, the L18, and the L27. The L9 can be used to study two 3-level factors in a full factorial, as shown in Table 16.11. It has two factors and one interaction. The main effects require two $DOF_f$ each, and the interaction requires four DOF for a total of eight DOF. That means any additional factors, which can be assigned to columns 3 and 4, result in Resolution III arrays, with confounding between main factors and interactions.

Note that two columns are required to study the interaction. This can be understood by considering the degrees of freedom. Each column, with three levels, has two $DOF_f$ available. Quantifying the two-way interaction by applying Equation 16.1 requires four DOF. Hence, two columns are required. This makes studying interactions between three-level factors very inefficient. Basically, there are three choices when considering interactions:

1. Study several 2-level factors and their interactions, but no curvature effects.
2. Study very few 3-level factors and their interactions.
3. Study the main effects of several 3-level factors, but not their interactions.

The last is the additive engineering approach, and the L18 array is recommended for this purpose. This is because the L18, like the L12, distributes the effect of an interaction between any pair of factors over all the other columns, thus minimizing its confounding with any one main effect.

The next largest three-level array is the L27, shown in Table 16.12. Here, three factors can be assigned to result in a full factorial ($3^3$) with all interactions available without confounding. This is done by choosing the columns labeled A, B, and C. An additional factor D is shown, which results in a $3^{4-1}$ fractional factorial. It has four main effects (eight DOF) and six interactions (24 DOF) for a total of 32 DOF. As a result, there is some confounding of interactions.

Obviously it is not easy to perform a Resolution V study on more than three 3-level factors. Also note the basic inefficiency of this array. More than two-thirds of the array is devoted to interactions. Is it likely that interactions are more important than main effects in improving a design? Engineering for additivity is clearly needed to break the cycle of endless studies using small numbers of factors searching for a solution.

**Table 16.12** L27 $3^{4-1}$ Resolution IV design

| Run | 1<br>A | 2<br>B | 3 | 4 | 5<br>C | 6 | 7 | 8 | 9<br>D | 10 | 11 | 12 | 13 |
|---|---|---|---|---|---|---|---|---|---|---|---|---|---|
| 1 | 1 | 1 | 1 | 1 | 1 | 1 | 1 | 1 | 1 | 1 | 1 | 1 | 1 |
| 2 | 1 | 1 | 1 | 1 | 2 | 2 | 2 | 2 | 2 | 2 | 2 | 2 | 2 |
| 3 | 1 | 1 | 1 | 1 | 3 | 3 | 3 | 3 | 3 | 3 | 3 | 3 | 3 |
| 4 | 1 | 2 | 2 | 2 | 1 | 1 | 1 | 2 | 2 | 2 | 3 | 3 | 3 |
| 5 | 1 | 2 | 2 | 2 | 2 | 2 | 2 | 3 | 3 | 3 | 1 | 1 | 1 |
| 6 | 1 | 2 | 2 | 2 | 3 | 3 | 3 | 1 | 1 | 1 | 2 | 2 | 2 |
| 7 | 1 | 3 | 3 | 3 | 1 | 1 | 1 | 3 | 3 | 3 | 2 | 2 | 2 |
| 8 | 1 | 3 | 3 | 3 | 2 | 2 | 2 | 1 | 1 | 1 | 3 | 3 | 3 |
| 9 | 1 | 3 | 3 | 3 | 3 | 3 | 3 | 2 | 2 | 2 | 1 | 1 | 1 |
| 10 | 2 | 1 | 2 | 3 | 1 | 2 | 3 | 1 | 2 | 3 | 1 | 2 | 3 |
| 11 | 2 | 1 | 2 | 3 | 2 | 3 | 1 | 2 | 3 | 1 | 2 | 3 | 1 |
| 12 | 2 | 1 | 2 | 3 | 3 | 1 | 2 | 3 | 1 | 2 | 3 | 1 | 2 |
| 13 | 2 | 2 | 3 | 1 | 1 | 2 | 3 | 2 | 3 | 1 | 3 | 1 | 2 |
| 14 | 2 | 2 | 3 | 1 | 2 | 3 | 1 | 3 | 1 | 2 | 1 | 2 | 3 |
| 15 | 2 | 2 | 3 | 1 | 3 | 1 | 2 | 1 | 2 | 3 | 2 | 3 | 1 |
| 16 | 2 | 3 | 1 | 2 | 1 | 2 | 3 | 3 | 1 | 2 | 2 | 3 | 1 |
| 17 | 2 | 3 | 1 | 2 | 2 | 3 | 1 | 1 | 2 | 3 | 3 | 1 | 2 |
| 18 | 2 | 3 | 1 | 2 | 3 | 1 | 2 | 2 | 3 | 1 | 1 | 2 | 3 |
| 19 | 3 | 1 | 3 | 2 | 1 | 3 | 2 | 1 | 3 | 2 | 1 | 3 | 2 |
| 20 | 3 | 1 | 3 | 2 | 2 | 1 | 3 | 2 | 1 | 3 | 2 | 1 | 3 |
| 21 | 3 | 1 | 3 | 2 | 3 | 2 | 1 | 3 | 2 | 1 | 3 | 2 | 1 |
| 22 | 3 | 2 | 1 | 3 | 1 | 3 | 2 | 2 | 1 | 3 | 3 | 2 | 1 |
| 23 | 3 | 2 | 1 | 3 | 2 | 1 | 3 | 3 | 2 | 1 | 1 | 3 | 2 |
| 24 | 3 | 2 | 1 | 3 | 3 | 2 | 1 | 1 | 3 | 2 | 2 | 1 | 3 |
| 25 | 3 | 3 | 2 | 1 | 1 | 3 | 2 | 3 | 2 | 1 | 2 | 1 | 3 |
| 26 | 3 | 3 | 2 | 1 | 2 | 1 | 3 | 1 | 3 | 2 | 3 | 2 | 1 |
| 27 | 3 | 3 | 2 | 1 | 3 | 2 | 1 | 2 | 1 | 3 | 1 | 3 | 2 |
| | | | 1×2<br>5×9 | 1×2 | | 1×5<br>2×9 | 1×5 | 1×9<br>2×5 | | 1×9 | 2×5 | 2×9 | 5×9 |

## 16.6 Summary of Chapter 16

The orthogonal arrays shown in this chapter are fractional factorials that can be used in designed experiments for the study of main effects and interactions. The Resolution V examples allow such studies to be done without confounding. There are also other designs, such as central composite designs [B9], that should be used for empirical modeling of classical DOE experiments.

If there is some knowledge of the expected interactions, then Resolution IV arrays can also be used. This is not easy, and it requires good engineering knowledge of the interactions *prior* to running the experiment. A useful tool for planning such experiments is linear graphs, which are discussed elsewhere [P1, P2, T6]. Resolution IV designs are useful in Robust Design as one of the countermeasures to interactions. Assuming that the design for additivity guidelines discussed in Part II have been used, control factor interactions should be mild. For that reason, the L18 is an ideal array. However, if circumstances dictate that an array of another size is to be used, it is much safer to have the main effects distinct from the interactions. The Resolution IV array, like many of the other techniques taught in this book, focuses one's attention on the control-factor main effects and away from interactive effects. Thus, we recommend, when the L12 or L18 arrays are not used, that the interaction assignments indicated in this chapter be used whenever possible.

WinRobust is capable of analyzing interactions in the manner shown in the L4 example of this chapter. The ANOM applied to empty columns indicates whether or not an interaction is active. However, it is usually necessary to create an interaction plot to visualize the interaction itself. WinRobust creates interaction plots automatically for Resolution V designs at two levels only. For Resolution IV designs, the software can create the plots, provided that the user identifies the significant interaction from the confounded interactions for a given column.

## Exercises for Chapter 16

1. Define what an interaction between two control factors is.
2. Define and sketch the three types of interactions that can occur between control factors.
3. Which type of interaction is used in Dr. Taguchi's justification to treat interactions as part of the overall experimental noise (error)?
4. How does the parameter design process detect the presence of a strong interaction between control factors? How does this contrast with the methods of classical DOE?
5. What does it mean to saturate an orthogonal array with control factors?
6. What is a benefit of saturating an array with control factors?
7. What harm can come from saturating an array?
8. Using the classic dynamic equation from Sir Isaac Newton's Second Law of Motion, $F = ma$ (force = mass times acceleration), show whether there is an interaction between the mass and acceleration of two differing masses as acted upon by two differing amounts of acceleration. You should provide the actual numbers for the mass and acceleration as part of this exercise (or just use the ones and twos in Table 16.13, if you prefer). Plot the interaction between mass and acceleration.
9. Why is it difficult to study a lot of control factors at three levels when you are also planning on studying all the interactions between the var-

**Table 16.13**

| *L4 experiment* | *Mass (column 1)* | *Acceleration (column 2)* | *Empty third column (m × a)* | *Calculated force* |
|---|---|---|---|---|
| 1 | 1 | 1 | 1 | $m1 \times a1$ |
| 2 | 1 | 2 | 2 | $m1 \times a2$ |
| 3 | 2 | 1 | 2 | $m2 \times a1$ |
| 4 | 2 | 2 | 1 | $m2 \times a2$ |

ious control factors? Does it help to study just a few interactions between some of the control factors?

10. Define Resolution III, IV, and V designs.
11. When is it wise to study interactions in a matrix experiment?
12. What are some strategies available to engineers to minimize the effect of interactions within an experiment? How would you know if the interactivity is moderate in the experiment?
13. Using WinRobust, try assigning factors to the columns in the order shown by the letter sequence used in the tables in this chapter. Note where the interactions go. Now do the same thing using the odd column approach shown for the L32, Table 16.10. Note where the interactions go in this case. Explain why the odd column assignment technique cannot be used if you want to study two-way interactions.

# CHAPTER 17

# Analysis of Variance (ANOVA)

## 17.1 Introduction

Analysis of variance is a computational technique that enables the engineer to quantitatively estimate the relative contribution each control factor makes to the overall measured response and express it as a percentage. For example: Four control factors are governing a product's response. Factor A is contributing 36% of the response, Factor B is contributing 12% of the response, Factor C is contributing 40% of the response, Factor D is contributing 7% of the response, and the remaining 5% of the response is attributed to random experimental effects (error). Thus, 95% of the measured response is due to the main effects attributable to the four control factors.

ANOVA uses a mathematical technique known as the *sum of squares* to quantitatively examine the deviation of the control factor effect response averages from the overall experimental mean response. This is referred to as the variation *between* the control factors. The significance of the individual control factors is quantified by comparing the variance between the control factor effects against the variance in the experimental data due to random experimental error and the effects of unrepresented interactions. The variation of the data due to random experimental variability and interactions is referred to as the variation *within* the control factors that make up the experiment. (In classical DOE, the interactions are treated as additional effects and are kept strictly separated from the random experimental variability.) Thus, the ratio in Equation 17.1, referred to as the *F-ratio,* can be formed between the control factor effect variance (the mean square due to a control factor) in the numerator and the experimental error variance (the mean square due to experimental error) in the denominator. These terms are fully defined in Section 17.4.

$$F = \frac{MS}{S_e^2} = \frac{\text{mean square due to a control factor}}{\text{mean square due to experimental error}} \tag{17.1}$$

The ANOVA process permits the engineer to gain insight into which control factors are the most important and which are less significant. In fact, if a comparison between the effect of Control Factor D (7%) and the random experimental effects (5%) is made, it is not at all clear that Factor D is truly contributing to the response in any significant way beyond that which is attributable to random experimental effects. This kind of information can be very helpful when decisions have to be made concerning which control factors are worth further expenditure of resources to enhance product robustness.

Typically, ANOVA is used to help make decisions during the Tolerance Design phase. The control factors that contribute high percentages to the measured response are the focus of the cost vs. performance balance that takes place during the Tolerance Design process.

Since ANOVA can be viewed as a procedure that *decomposes the variance* for each factor relative to the overall mean response, it is important to remember that the variance is a statistic that measures the width of a distribution of data about the mean value. In the Taguchi approach, ANOVA can be applied to two forms of data. First, it can be applied to the data as measured in *engineering units* (used in the Tolerance Design approach). Second, it can be applied to the data after it has been transformed into S/N ratios (done in the Parameter Design approach). ANOVA is used in Parameter Design to aid in identifying the strongest contributing control factors, so that an accurate predictive model can be constructed and utilized during the verification process. ANOVA is used in Tolerance Design to identify those control factors that are worth considering for tolerance tightening, improving material grades, or some other means of improving quality at an increased cost.

Tolerance Design is not covered in this text. The discussion of the ANOVA process given here focuses on the Parameter Design process, where it is used to aid in the construction of a predictive S/N model. In this procedure, the response values are the S/N ratios obtained from the results of orthogonal array matrix experiments. The overall mean S/N for the entire matrix experiment must always be calculated. This is the reference point from which the variance of each control factor's S/N averages is calculated by summing the squared deviations from the overall mean. Most of the variation around the overall mean S/N ratio is due to the main effects of the control factors. A small part of it is typically due to random experimental error. Yet another part (ideally a small amount) of it is due to interactive effects between the control factors. The goal of this chapter is to explain how to separate the percentage of parameter contributions to the overall measured response and properly attribute these percentages to the appropriate sources.

## 17.2 An Example of the ANOVA Process

In Chapter 12, a belt-drive system is presented as an example of a parameter optimization experiment. Here the analysis is completed by applying ANOVA to the data. Table 17.1 shows the L9 matrix experiment and the S/N responses obtained previously.

WinRobust uses fairly simple calculations to process the S/N data from the experiments. The following example demonstrates the ANOVA technique known as the *sum of squares* on the S/N values from the experimental runs.

The overall mean from which all the variation is calculated is given by

$$\overline{S/N} = \frac{1}{9}(13.47 + 31.97 + \ldots + 20.09) = 22.60 \tag{17.2}$$

**Table 17.1** Data for the belt-drive system (from Table 12.1)

| A | B | C | D | S/N |
|---|---|---|---|---|
| 1 | 1 | 1 | 1 | 13.5 |
| 1 | 2 | 2 | 2 | 31.0 |
| 1 | 3 | 3 | 3 | 29.0 |
| 2 | 1 | 2 | 3 | 22.2 |
| 2 | 2 | 3 | 1 | 15.8 |
| 2 | 3 | 1 | 2 | 27.4 |
| 3 | 1 | 3 | 2 | 15.4 |
| 3 | 2 | 1 | 3 | 29.0 |
| 3 | 3 | 2 | 1 | 20.1 |

The ANOVA process begins by determining the *grand total sum of squares* (GTSS):

$$\begin{aligned} \text{GTSS} &= \sum_{i=1}^{9} (\text{S/N})_i^2 \\ &= (13.47)^2 + (31.97)^2 + \ldots + (20.09)^2 \\ &= 4{,}961.6 \end{aligned} \quad \textbf{(17.3)}$$

The GTSS can be decomposed into two parts:

1. The sum of squares due to the *overall experimental mean:*

$$\begin{aligned} \text{SS due to the mean} &= (\text{\# of experiments}) \times \overline{\text{S/N}}^2 \\ &= 9 \times (22.60)^2 = 4{,}596.8 \end{aligned} \quad \textbf{(17.4)}$$

2. The sum of the squares due to variation *about the mean,* referred to as the total sum of the squares:

$$\begin{aligned} \text{Total SS} &= \sum_{i=1}^{9} (\text{S/N}_i - \overline{\text{S/N}})^2 \\ &= (13.47 - 22.60)^2 + (31.97 - 22.60)^2 + \ldots + (20.09 - 22.60)^2 \\ &= 364.3 \end{aligned} \quad \textbf{(17.5)}$$

Note that the grand total SS = total SS + SS due to mean.

The technique of summing squares can be used to define the contribution of each individual control factor within the total sum of squares. The sum of squares method is based on numerically quantifying the *variation* that is induced by the factor effects around the overall experimental mean response. This is why it is called the analysis of *variance.* The process uses variances to quantify the strength of the control factor main effects. The ANOVA relationships can be summarized by Figure 17.1.

The experiment is structured such that the orthogonal (balanced) nature of the design produces data that must be gathered in the following format, for the L9:

- Three runs for A at level 1 (runs 1, 2, and 3)
- Three runs for A at level 2 (runs 4, 5, and 6)
- Three runs for A at level 3 (runs 7, 8, and 9)

**Figure 17.1** Decomposition of the sums of squares

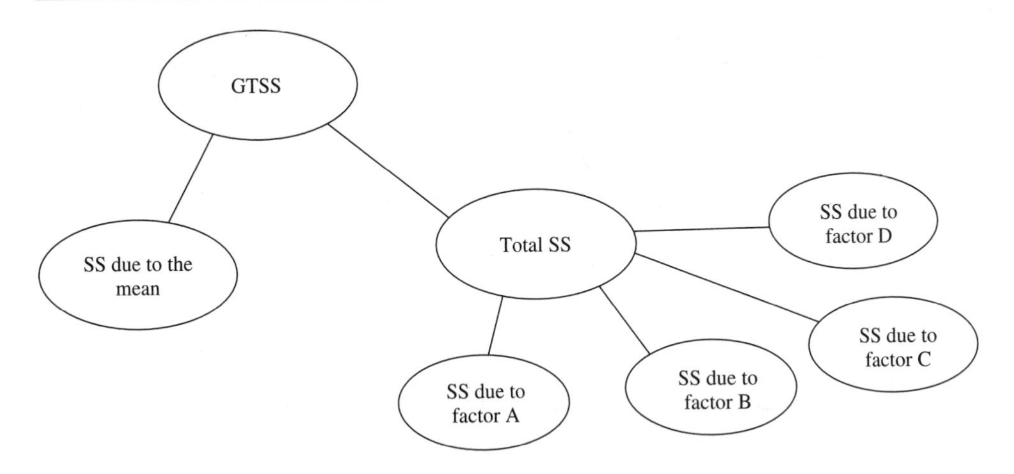


Therefore, for factor A, the sum of the squares due to variation about the mean is

$$\begin{aligned}SS_A &= (\text{\# of exp. at A1})(\overline{S/N}_{A_1} - \overline{S/N})^2 + (\text{\# of exp. at A2})(\overline{S/N}_{A_2} - \overline{S/N})^2 \\ &\quad + (\text{\# of exp. at A3})(\overline{S/N}_{A_3} - \overline{S/N})^2 \\ &= 3(\overline{S/N}_{A_1} - \overline{S/N})^2 + 3(\overline{S/N}_{A_2} - \overline{S/N})^2 + 3(\overline{S/N}_{A_3} - \overline{S/N})^2 \\ &= 3(24.49 - 22.60)^2 + 3(21.80 - 22.60)^2 + 3(21.51 - 22.60)^2 \\ &= 16.2\ (\text{dB})^2\end{aligned} \quad \textbf{(17.6)}$$

where $\overline{S/N}_{A_1}$ is the average of the 3 S/N samples for each level ($i = 1, 2,$ or 3).

If this procedure is repeated for factors B, C, and D, the sums of squares are as follows:

$$SS_B = 140.3\ (\text{dB})^2$$

$$SS_C = 30.6\ (\text{dB})^2$$

$$SS_D = 177.4\ (\text{dB})^2$$

These values represent a measure of the relative importance of each control factor in controlling the measured response. If the components are added (16.2 + 140.3 + 30.6 + 177.4 = 364.5), the total SS as found in Equation 17.5 is obtained (ignoring rounding errors).

The percentage contribution of the relative effect each factor has on the measured response is found using the following formula:

$$\text{Percentage contribution} = (SS_{\text{factor}}/\text{total SS}) \times 100 \quad \textbf{(17.7)}$$

Factor A: $(16.2/364.5) \times 100 = 4.4\%$

Factor B: $(140.3/364.5) \times 100 = 38.5\%$

Factor C: $(30.6/364.5) \times 100 = 8.4\%$

Factor D: $(177.4/364.5) \times 100 = 48.7\%$

## 17.3 Degrees of Freedom

Note that no percentage has been assigned to account for experimental error contributions. Does this mean that there is no error in this experiment? No! The reason there is no sum of squares assigned to account for experimental error can be understood by looking at the degrees of freedom available from the experiment. Remember that the degrees of freedom follow a relationship similar to the one for ANOM. Compare Figure 17.2 to Figure 17.1.

A degree of freedom is an independent parameter associated with one of the following:

- A matrix experiment
- A factor
- A sum of squares computation

For example, an L9 orthogonal array has nine experiments, so it has nine degrees of freedom, and its grand total sum of squares has nine degrees of freedom. Calculating the overall mean for any orthogonal array takes up one degree of freedom, as does the sum of squares due to the overall experimental mean. As a result:

$$\text{Total SS} = (\text{GTSS} - \text{SS due to the mean})$$

$$\text{Total DOF} = (\text{\# of experiments} - 1)$$

The total sum of squares can only have eight degrees of freedom.

Similarly, for factors in an orthogonal array, one degree of freedom is allocated for each level of a factor. For the overall experimental mean of each factor, subtract one degree of freedom. Therefore, for each three-level factor, there are $3 - 1$ or two degrees of freedom. In general, the number of degrees of freedom associated with a factor is one less than the total number of levels for that factor.

In this L9 example, there are eight DOF available to estimate control factor effects. All eight DOF are employed in evaluating the four control factors. No DOF is available to be allocated to quantifying experimental error effects. No experimental "energy" is spent on accounting for experimental error. To do so would incur the cost of running larger experiments, consuming time, effort, and money.

**Figure 17.2** Decomposition of the degrees of freedom

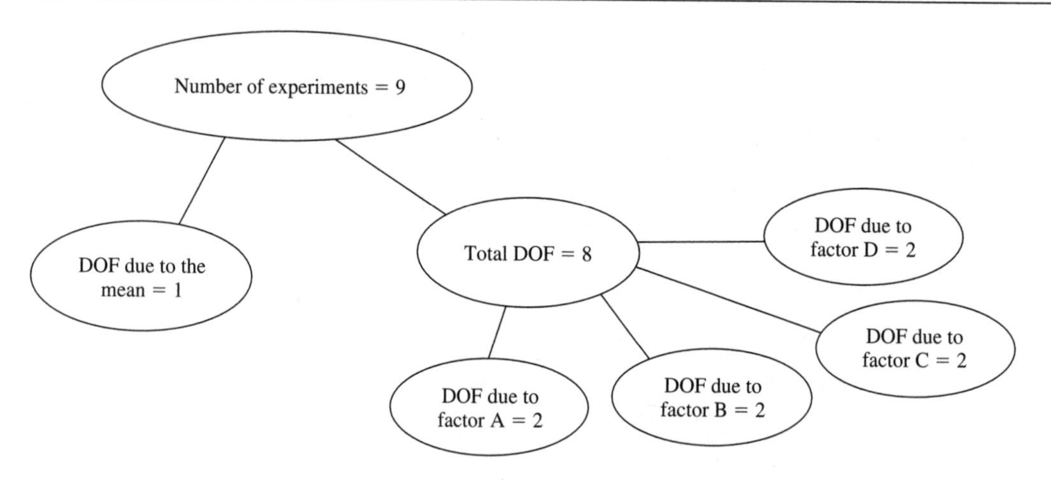

## 17.4 Error Variance and Pooling

In order to apply Equation 17.1, experimental error needs to be estimated. Here's one way to do it. The mean square for a control factor is defined by

$$\text{MS} = \frac{\text{factor effect sum of the squares}}{\text{factor degrees of freedom}} \tag{17.8}$$

The error mean square (error variance) for the control factors is defined by

$$S_e^2 = \frac{\text{error sum of the squares}}{\text{error degrees of freedom}} \tag{17.9}$$

In the interest of seeking the most information possible from an orthogonal array matrix experiment, all the available degrees of freedom are used to focus on the study of control factors. That's why the mean square of the error in the belt-drive example is not estimated—at least, not directly. A method of estimating the error variance does, however, exist. The method is called *pooling* the sum of the squares of the control factors that have a small contribution to the overall S/N ratio.

A reasonable guideline is to pool the sum of squares of up to half of the control factors (the half defined by their percent contribution to be least significant). That makes about half the degrees of freedom available to estimate the error variance. In this instance, the weak control factors stand in to estimate the experimental error. Note that this is only a guideline. In some experiments, all the factors may be highly significant; in others, few factors may be significant. Therefore, the method of pooling typically provides an exaggerated estimate of experimental error and is viewed as a conservative approach. An example of pooling follows shortly.

## 17.5 Error Variance and Replication

A more direct way of estimating the error variance in the experiment is to repeat the measurement of the response three to five times at the same factor set points. This means setting up the same control factor levels and noise factor levels to replicate the S/N measurement three to five times. Remember, a true replicate requires physically changing the setup and then returning to the same conditions. Simply repeating measurements on a given setup is called *repetition* and does not give the same effects as replication.

One common strategy for the replicate case is to do a repeat setup at the beginning, middle, and end of the overall experiment. Since the control parameters are the same for each repeated measurement, all the variability can be attributed to the experimental error variance. This technique, however, misses any error due to interactions.

The sum of squares due to the experimental error determined by replication is calculated as follows:

$$S_e^2 = \frac{1}{(r-1)} \sum_{j=1}^{r} (\text{S/N}_j - \overline{\text{S/N}}_\text{R})^2 \tag{17.10}$$

where $r$ is the number of replicate measurements, $\text{S/N}_j$ are the individual S/N values (or any other response being considered) for the replicates, and $\overline{\text{S/N}}_\text{R}$ is the average of the replicate S/N values.

This error estimate can now be used in the same way as the one obtained by pooling. In WinRobust, this value can be manually entered in the space provided as an external prediction or estimate of error.

## 17.6 Error Variance and Utilizing Empty Columns

Another common technique for estimating experimental error is to use any empty columns in the matrix experiment to gather data on experimental error, including interactive effects. One or more empty columns can be processed just as if there is a control factor being studied. The numerical values produced in the analysis of variance for the empty columns express the random experimental variability and any interaction effects that show up in the empty columns. This technique then automatically includes interactions as a source of experimental error. WinRobust's ANOVA table displays the numerical values for the sum of squares for any empty columns not assigned to a control factor. Be sure any columns not being occupied by a control factor are left unassigned when setting up the experimental array in WinRobust. This technique is most appropriate for the L12 and L18 arrays for which the interactions are uniformly distributed among all columns.

## 17.7 The *F*-Test

In the ANOVA table there is a term called the *F*-ratio, also referred to as the variance ratio. This ratio is used to test for the significance of factor effects. The *F*-ratio is given by Equation 17.1. When $F$ is much greater than 1, the effect of the control factor is large compared to the variance due to experimental error and interaction effects.

Here are some suggested rankings (generalized values) for *F*-ratios:

- $F < 1$: The experimental error outweighs the control factor effect; the control factor is insignificant and indistinguishable from the experimental error.
- $F \approx 2$: The control factor has only a moderate effect compared to experimental error.
- $F > 4$: The control factor is strong compared to experimental error and is clearly significant.

Use these general guidelines when deciding which factors to include in the additive prediction model, i.e., which factors are critical to the design's robustness. When it is necessary to be exact in your knowledge of significance of the *F*-ratio, the use of an *F*-ratio table is recommended. The top row of an *F*-ratio table contains the degrees of freedom used in the numerator, and the left column of the table contains the degrees of freedom used to estimate the experimental error term. Consult a basic statistics book for a detailed explanation of this formal statistical ANOVA process [F5,C7,R2].

In summary, the *F*-ratio is the statistical analogue to Dr. Taguchi's signal-to-noise ratio[1] for the control factor effect vs. the experimental error. The *F*-ratio uses information based on sample variances (mean squares) to define the relationship between the power of the control factor effects (a type of signal) and the power of the experimental error (a type of noise). Taguchi S/N ratios are dependent on the physical case at hand and come in many forms, depending upon the type of response being measured (STB, NTB, etc.).

1. Do not confuse the signal-to-noise ratio used in the Taguchi Parameter Design approach to analyzing experimental data with the ratio used in defining the $F$ statistic in the ANOVA process. While they are generally describing the same things, they are mathematically different.

**Figure 17.3** WinRobust main screen showing belt drive array

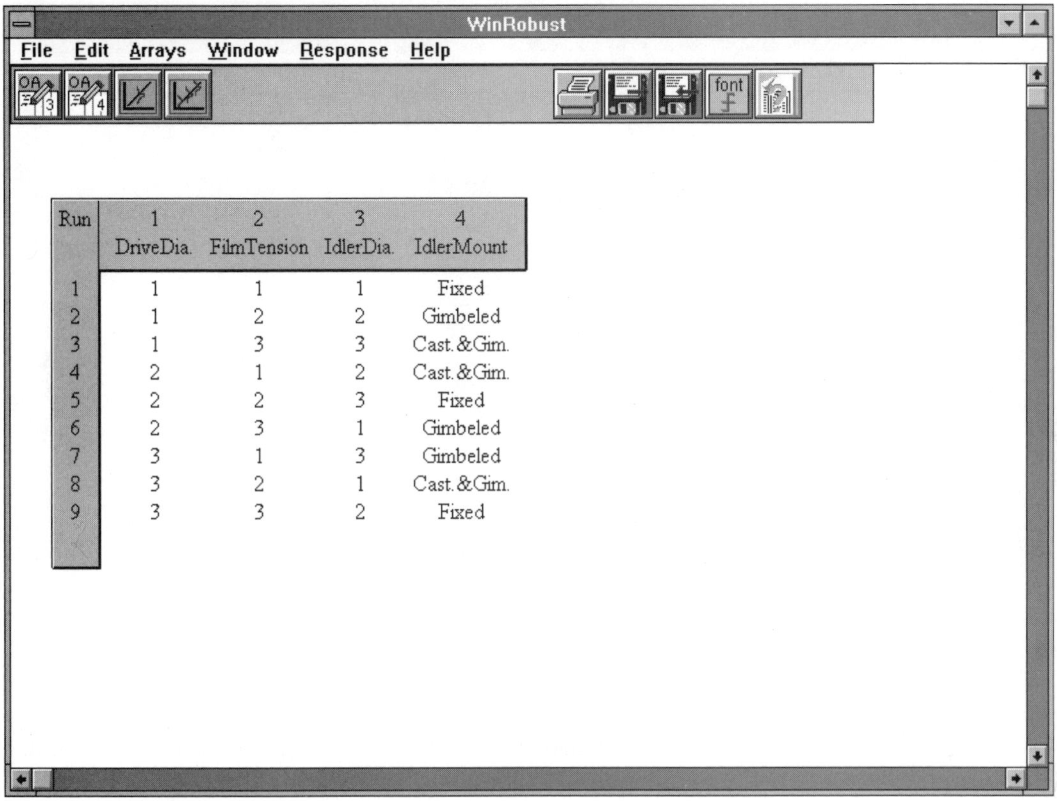

WinRobust

File Edit Arrays Window Response Help

| Run | 1 DriveDia. | 2 FilmTension | 3 IdlerDia. | 4 IdlerMount |
|---|---|---|---|---|
| 1 | 1 | 1 | 1 | Fixed |
| 2 | 1 | 2 | 2 | Gimbeled |
| 3 | 1 | 3 | 3 | Cast.&Gim. |
| 4 | 2 | 1 | 2 | Cast.&Gim. |
| 5 | 2 | 2 | 3 | Fixed |
| 6 | 2 | 3 | 1 | Gimbeled |
| 7 | 3 | 1 | 3 | Gimbeled |
| 8 | 3 | 2 | 1 | Cast.&Gim. |
| 9 | 3 | 3 | 2 | Fixed |

## 17.8 WinRobust Examples

### 17.8.1 Belt-Drive Example (Continued from Section 12.3)

Performing an ANOVA is simplified by using WinRobust. The completed L9 array, loaded and ready for data entry, is shown in Figure 17.3. The data is entered from Table 12.1 using the response window. Recall that the response is film position. The type of signal-to-noise ratio for this case is nominal-the-best (II): S/N $= -10\log(S^2)$.

Figure 17.4 displays the results of the ANOVA calculations performed by WinRobust. The ANOVA data seen here are just for the main control factor effects. In order to compute $F$-ratios, one option for estimating the error must be chosen. The factors are rank-ordered according to the magnitude of their mean squares. Factors D and B contribute far more than half the total variation. Therefore, the pooling guideline certainly applies here. Factors C and A should have their sums of squares (SS) pooled together and attributed to experimental error. When this is done, the effects of drive roller diameter and idle roller diameter are considered indistinguishable from noise (random variation). That is why they should be left out of the additive model when doing prediction and verification.

The window, shown in Figure 17.5, has the three options for estimating experimental error. This window is opened by selecting the *Options* menu from the ANOVA window. Figure 17.5 shows the

**Figure 17.4** Results of the ANOVA

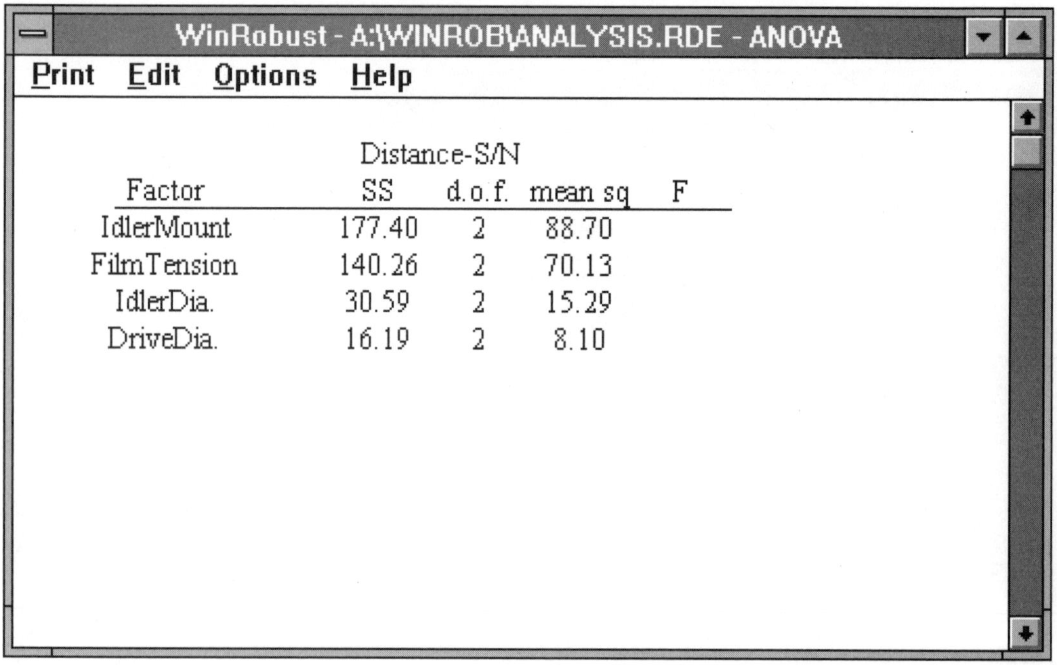

WinRobust - A:\WINROB\ANALYSIS.RDE - ANOVA

Print Edit Options Help

Distance-S/N

| Factor | SS | d.o.f. | mean sq | F |
|---|---|---|---|---|
| IdlerMount | 177.40 | 2 | 88.70 | |
| FilmTension | 140.26 | 2 | 70.13 | |
| IdlerDia. | 30.59 | 2 | 15.29 | |
| DriveDia. | 16.19 | 2 | 8.10 | |

pooling option chosen. The value 4 has been entered to tell WinRobust to use the four degrees of freedom associated with the two control factors that have relatively weak responses compared with the others.

After the pooling operation is performed, the ANOVA table is modified as shown in Figure 17.6. The $F$-ratios appear, indicating the level of significance for each of the control factors. Consulting an $F$-ratio table, for two DOF for the numerator and four DOF for the denominator at the 90% confidence level, shows that 4.32 is the critical $F$ value above which the control factors can be considered statistically significant in their contribution to the S/N performance of the design. At the 95% confidence level, the critical $F$ value is 6.9. By this standard, the two control factors (B and D) are marginally significant with respect to the estimated experimental error in the experiment. This is one reason that pooling for error is somewhat imprecise in comparison to direct measures of experimental error through repetitions.

## 17.8.2 Catapult (Continued from Chapter 13)

The catapult case provides interesting data for a further illustration of how the ANOVA process can be employed. The catapult case contains four control factors. The goal is to identify the significance of each factor on the overall response, for inclusion in the predictive model. The catapult experimental array is shown in Table 13.18.

Using WinRobust's ANOVA menu, the four control factors are ranked in order of numerical importance. The ANOVA table displays the data in a format that does not account for any experimental error (Table 17.2).

**Figure 17.5** WinRobust screen showing options for experimental error

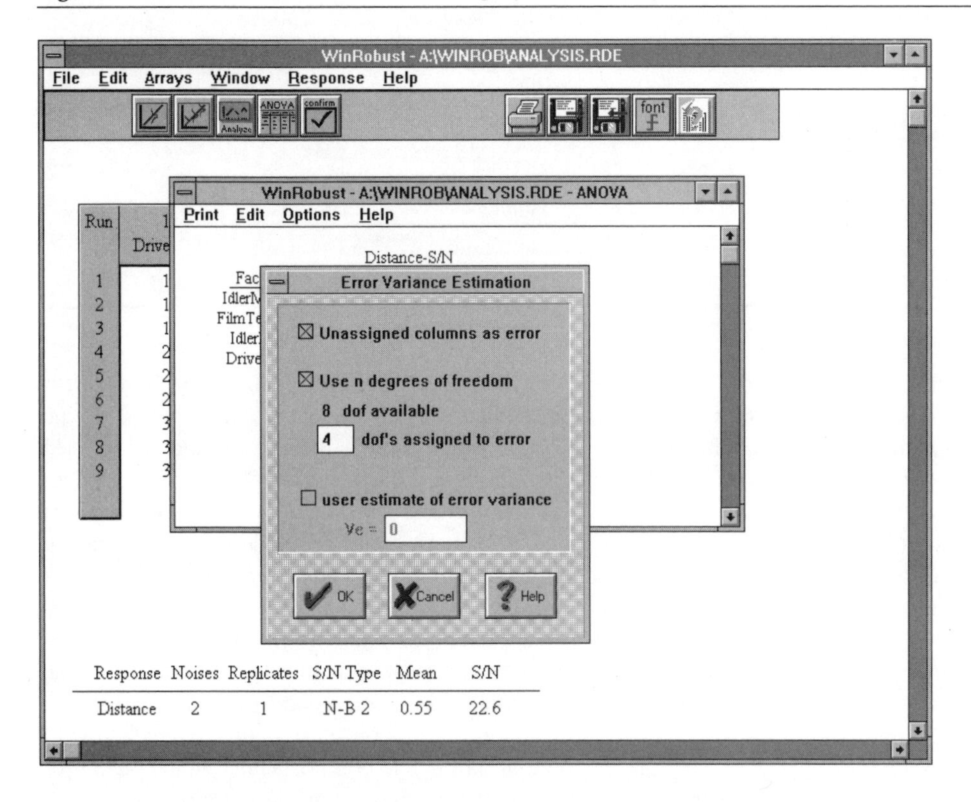


**Figure 17.6** Results of the ANOVA after choosing pooling option

WinRobust - A:\WINROB\ANALYSIS.RDE - ANOVA

Print Edit Options Help

Distance-S/N

| Factor | SS | d.o.f. | mean sq | F |
|---|---|---|---|---|
| IdlerMount | 177.40 | 2 | 88.70 | 7.6 |
| FilmTension | 140.26 | 2 | 70.13 | 6.0 |
| error | 46.78 | 4 | 11.70 | |

**Table 17.2** The catapult ANOVA table before accounting for experimental error

| | Response—S/N | | | |
|---|---|---|---|---|
| *Factor* | *SS* | *d.o.f.* | *mean sq* | *F* |
| Stretch | 23.44 | 2 | 11.72 | |
| Height | 21.10 | 2 | 10.55 | |
| No. bands | 7.93 | 2 | 3.97 | |
| StopPin | 4.50 | 2 | 2.25 | |

**Table 17.3** The catapult ANOVA table after pooling

| | Response—S/N | | | |
|---|---|---|---|---|
| *Factor* | *SS* | *d.o.f.* | *mean sq* | *F* |
| Stretch | 23.44 | 2 | 11.72 | 5.2 |
| Height | 21.10 | 2 | 10.55 | 4.7 |
| No. bands | 7.93 | 2 | 3.97 | — |
| Error | 4.50 | 2 | 2.25 | |

In this case, a single control factor with the lowest value is pooled. Using WinRobust's error pooling feature, two DOF are specified to be used in estimating experimental error. In Table 17.3, the data for the stop pin now represent the experimental error. The $F$ ratios are displayed and can be used to aid in interpreting the significance of the remaining three control factors. The stretch and height factors are clearly significant, while the number of bands is indistinguishable from the estimated amount of experimental error. From these data, the engineering team is left to conclude that the predictive model should be constructed from the stretch and height control factors. The confirmation experiment is run with all four factors at their optimum set points. The pooled factors and any factors with $F$ ratios less than 2 (shown as a dash in WinRobust) should be left out of the predictive model.

The number of bands is not pooled for error because, had it been included in the error estimate, no factors would have been significant ($F < 2$). This would make no sense and would indicate that the true experimental error had probably been overestimated. Use common sense in how many factors to include in the pooling process. Here, pooling the bottom half of the control factors would have produced very disturbing results. Using just one factor in the pooling process makes more sense.

It should be clear from the preceding discussion that pooling involves some arbitrary decision-making about the significance of factors. It is entirely possible that even the lowest-ranking factor in the ANOVA table has an effect, albeit small, on the response that should not be equated to error. It is also possible that even the highest-ranking factor effect is an artifact of experimental error. For these reasons, the approaches for estimating error discussed in Sections 17.5 and 17.6 are considered more reliable. These are illustrated next.

### 17.8.3 Gyrocopter (Continued from Chapter 13)

The gyrocopter case contains some unique applications of the sliding level technique to help minimize interactivity between the control factors. The ANOVA can determine the effectiveness of this countermeasure, applied to limit control factor interactivity. The empty columns in the L18 array can be used to quantify both interactive and experimental error effects. The gyrocopter example L18 array is shown in Table 13.10. Table 13.10 shows that column 1 and column 6 have been left empty and are free to express experimental error effects as discussed in Section 17.6. Recall that the L18 has the property that interactions between any factors in columns 2–8 are distributed uniformly among the remaining columns. Thus, column 6 also contains interactive effects.

Table 17.4 shows the ANOVA table for the gyrocopter experiment. Columns 6 and 1 are shown in the table. Column 6, while empty, is showing a strong response because of its capacity to quantify

**Table 17.4** The gyrocopter ANOVA table before accounting for experimental error

| | Time—S/N | | | |
|---|---|---|---|---|
| *Factor* | *SS* | *d.o.f.* | *mean sq* | *F* |
| B_Fold | 59.21 | 2 | 29.60 | |
| 6 | 19.26 | 2 | 9.63 | |
| WW | 17.03 | 2 | 8.51 | |
| Gussets | 7.53 | 1 | 7.53 | |
| WL | 8.79 | 2 | 4.39 | |
| BL | 7.18 | 2 | 3.59 | |
| Size | 4.50 | 2 | 2.25 | |
| 1 | 1.66 | 1 | 1.66 | |

**Table 17.5** The gyrocopter ANOVA table using unassigned columns for error

| | Time—S/N | | | |
|---|---|---|---|---|
| *Factor* | *SS* | *d.o.f.* | *mean sq* | *F* |
| B_Fold | 59.21 | 2 | 29.60 | 4.2 |
| WW | 17.03 | 2 | 8.51 | — |
| Gussets | 7.53 | 1 | 7.53 | — |
| WL | 8.79 | 2 | 4.39 | — |
| BL | 7.18 | 2 | 3.59 | — |
| Size | 4.50 | 2 | 2.25 | |
| Error | 20.91 | 3 | 6.97 | |

the interactivity present between the other control factors in the experiment. While the ANOVA does not indicate anything about which factors are interacting in this case, it does provide a measure of the strength of whatever interactivity is present. There appears to be a measurable amount of control factor interactivity taking place along with the random experimental error in this experiment. This could not be estimated by pooling, because it is obviously too large an effect to accurately represent random experimental error. When there is an empty column with such a strong response, be sure to run a rigorous confirmation experiment to be sure that the interactive effects are of the benign monotonic type that are not detrimental to the optimum predicted configuration. If the confirmation experiments do not confirm, then further action may be required to minimize the interaction effects or to study which factors are interacting. Ideally, one would like to design a gyrocopter (or any other product) that behaves in an additive fashion, thus making it easy to confirm robust performance.

Table 17.5 shows the gyrocopter ANOVA table after choosing the *Unassigned Columns as Error* option to use columns 1 and 6 as an overall estimate of the experimental error, including interactive effects.

It is also possible in this case to form an estimate of the error from the confirmation experiment replicate data given in Chapter 13. This approach is considered by most to be the purest estimate of experimental error possible. Tables 13.12 and 13.14 give the results of five replicates of run #8 and the optimum configuration, respectively. Table 17.6 summarizes the analysis of the replicate data using Equation 17.10.

The average error variance is found by taking the mean of the two $S_e^2$ values from Table 17.6, $\overline{S_e^2} = 5.6$. This value is reasonably close to the error estimate obtained by pooling the two empty columns, shown in Table 17.5. When an independent estimate of the experimental error is obtained in this manner, it can be entered into WinRobust manually using the *user estimate of error variance* option. The resulting ANOVA table is shown in Table 17.7.

In Table 17.7, only the $F$-ratio for body fold is given. This is because the default in WinRobust is to not display any $F$-ratio values that are less than 2. Thus, these tables indicate that only body fold is statistically significant in its effect on the S/N ratio. Does this mean that body fold is the only critical-to-function control factor for the gyrocopter? Remember that quality is given by on-target perfor-

**Table 17.6** Error variance estimates from replicate data

| *Data* | *Average S/N* | *Variance,* $S_e^2$ |
|---|---|---|
| Table 13.12 | 6.68 | 3.0 |
| Table 13.14 | 9.86 | 8.2 |

**Table 17.7** The gyrocopter ANOVA table using mean replicate variance for error

| | *Time—S/N* | | | |
|---|---|---|---|---|
| *Factor* | *SS* | *d.o.f.* | *mean sq* | *F* |
| B_Fold | 59.21 | 2 | 29.60 | 5.3 |
| 6 | 19.26 | 2 | 9.63 | — |
| WW | 17.03 | 2 | 8.51 | — |
| Gussets | 7.53 | 1 | 7.53 | — |
| WL | 8.79 | 2 | 4.39 | — |
| BL | 7.18 | 2 | 3.59 | — |
| Size | 4.50 | 2 | 2.25 | — |
| 1 | 1.66 | 1 | 1.66 | — |
| Error | 5.60 | 1 | 5.60 | |

mance, not just low variability. In this larger-the-better dynamic problem maximizing time vs. height the slope is important too. Therefore the question cannot be answered until the ANOVA is done on the slope response values.

The overall mean slope from which the variation of the slope data is calculated is given by $\bar{y}_{\text{slope}} = 0.25$. The *grand total sum of squares* (GTSS) is given by

$$\text{GTSS} = \sum_{i=1}^{18} (y_{\text{slope}})_i^2$$

$$= (0.25)^2 + (0.25)^2 + \ldots + (0.31)^2 = 1.1465$$

The sum of squares due to the *overall experimental mean* is given by

$$\text{SS}_{\bar{y}} = (\text{\# of experiments}) \times \bar{y}_{\text{slope}}^2 = 18 \times (0.25)^2 = 1.13$$

Thus, the total sum of the squares is given by

$$\text{Total SS} = \sum_{i=1}^{9} (\bar{y}_i - \bar{y}_{\text{slope}})^2 = \text{GTSS} - \text{SS}_{\bar{y}} = 0.0156$$

The individual factor sums of squares are calculated using Equation 17.6, remembering that there are six experiments for each of the factor levels (12 in the case of the gussets dummy level). The percent contribution and the mean square are given by Equations 17.7 and 17.8, respectively. The results are summarized in Table 17.8.

Now it is clear that wing length and probably body length are critical control factors in addition to body fold. But are they statistically significant? Once again, the error variance from the verification tests can be used for the denominator of the $F$-ratio. For the slopes, the error variance values determined from the data in Tables 13.12 and 13.14 are $S_e^2 = 0.00005$ and $S_e^2 = 0.00018$, respectively, for an average $S_e^2 = 0.00012$. Using this value, the $F$-ratios are computed and shown in Table 17.9.

**Table 17.8** ANOVA table for gyrocopter slopes

| *Factor* | *SS* | *MS* | *% Contribution* |
|---|---|---|---|
| Wing length | 0.00780 | 0.0039 | 50% |
| Column 6 | 0.0036 | 0.0018 | 23% |
| Body length | 0.0012 | 0.0006 | 7.7% |
| Body fold | 0.0012 | 0.0006 | 7.7% |
| Wing width | 0.0006 | 0.0003 | 3.8% |
| Size | 0.0006 | 0.0003 | 3.8% |
| Gussets | 0.0006 | 0.0003 | 3.8% |
| Column 1 | 0 | 0 | 0% |

**Table 17.9** ANOVA table for gyrocopter slopes

| *Factor* | *SS* | *MS* | *F-Ratio* |
|---|---|---|---|
| Wing length | 0.00780 | 0.0039 | 32.5 |
| Column 6 | 0.0036 | 0.0018 | 15.0 |
| Body length | 0.0012 | 0.0006 | 5.0 |
| Body fold | 0.0012 | 0.0006 | 5.0 |
| Wing width | 0.0006 | 0.0003 | 2.5 |
| Size | 0.0006 | 0.0003 | 2.5 |
| Gussets | 0.0006 | 0.0003 | 2.5 |
| Column 1 | 0 | 0 | 0 |

The results in Table 17.9 indicate that wing length, body length, and body fold are all clearly significant ($F > 4$) and are indeed critical to function. Wing width, size, and gussets are marginally significant. This example illustrates the importance of paying attention to the ideal function of the design, not just the S/N ratio.

## 17.9 Summary

Analysis of variance is central to the statistical analysis of designed experiments. ANOVA allows one to decompose the variation in data into the factor effects, including interactions, and the remaining variation, assignable to experimental error. ANOVA is based on the principle that the sums of squares are additive. Geometrically, ANOVA has been compared to Pythagoras's Theorem for the sum of squares of a right triangle. ANOVA is a mathematical treatment of the data that gives quantitative information about the significance of the control factors. Such information can be used to formulate and answer, with statistical rigor, hypotheses concerning the probability that a factor judged significant is indeed active—and conversely, that a factor judged not significant is indeed negligible. Such information is also useful for tolerance design.

For engineering purposes, however, the result of the analysis of means is usually adequate to answer these questions. The optimum level, which factors are most important, and which factors are most likely irrelevant can be judged from a factor effects plot. When you are doing nonreplicated, fully saturated arrays, it is difficult to properly apply ANOVA because the estimate of the experimental error is suspect. As we have recommended throughout this book, it is helpful to use ANOVA and ANOM and the other designed experiment techniques as tools to aid in making engineering decisions. Engineering decisions are not, and never have been, made on the basis of pure statistics. Rather, good engineering decisions require good engineering judgment. Statistical analysis, regardless of which approach you use to support your decision-making, cannot take the place of experience and judgment.

## Exercises for Chapter 17

1. Explain how ANOVA is used in both parameter design and tolerance design.
2. Define experimental error and explain what is included in quantifying its effects during a matrix experiment.
3. How is the mean square due to a control factor calculated?
4. How is the mean square due to experimental error calculated?
5. What is the difference between the sum of the squares due to the overall experimental mean and the total sum of squares?
6. What type of sum of squares holds the information about the individual control factor contributions to the response?
7. How many degrees of freedom are allocated to a three-level control factor? An L18 array? A five-level control factor?
8. What is an *F*-ratio and how does it differ from an S/N ratio?
9. Discuss the three methods for estimating experimental error.

# CHAPTER 18

# The Relationship of Robust Design to Other Quality Processes

This chapter is intended to step back and take a look at the bigger picture of quality processes that are available to industry at the end of the 20th century. There are many books, journals, seminars, educators, and consultants actively promoting a diverse array of quality processes. The authors have read widely and networked extensively with professional industrial consultants, industrial practitioners, educators, academicians, and gurus of all sorts. It is our intent to add a distinct perspective on practicing quality processes by discussing how some of the more successful quality processes relate to the engineering process of Robust Design.

The following section will briefly look into the process best known for keeping product development, including quality processes, focused on the Voice of the Customer—Quality Function Deployment. The focus will then shift to the renowned debate over Classical Design of Experiments in relation to the methods of Robust Design. Once the pages cool down from the DOE debate, the final process discussed is Motorola's intriguing process of attaining Six Sigma product performance. It is our intent to help the reader come to grips with how all of these processes can, in fact, be blended and successfully practiced within any corporate culture. The key is common sense, an open mind, and, of course, a willingness to work *hard* to balance the interests of the customer with those of the business in which the reader participates.

## 18.1 Quality Function Deployment (QFD) and Robust Design

QFD is a large subject that cannot be fully discussed within the scope of this book. There are a number of good references on this topic that should be consulted [A2, C2]. The essence of the process is the use of matrix analysis as a planning and recording tool that displays information about how the customer requirements are to be satisfied by a product under development. Figure 18.1 illustrates the basic structure of QFD.

**Figure 18.1** Quality function deployment

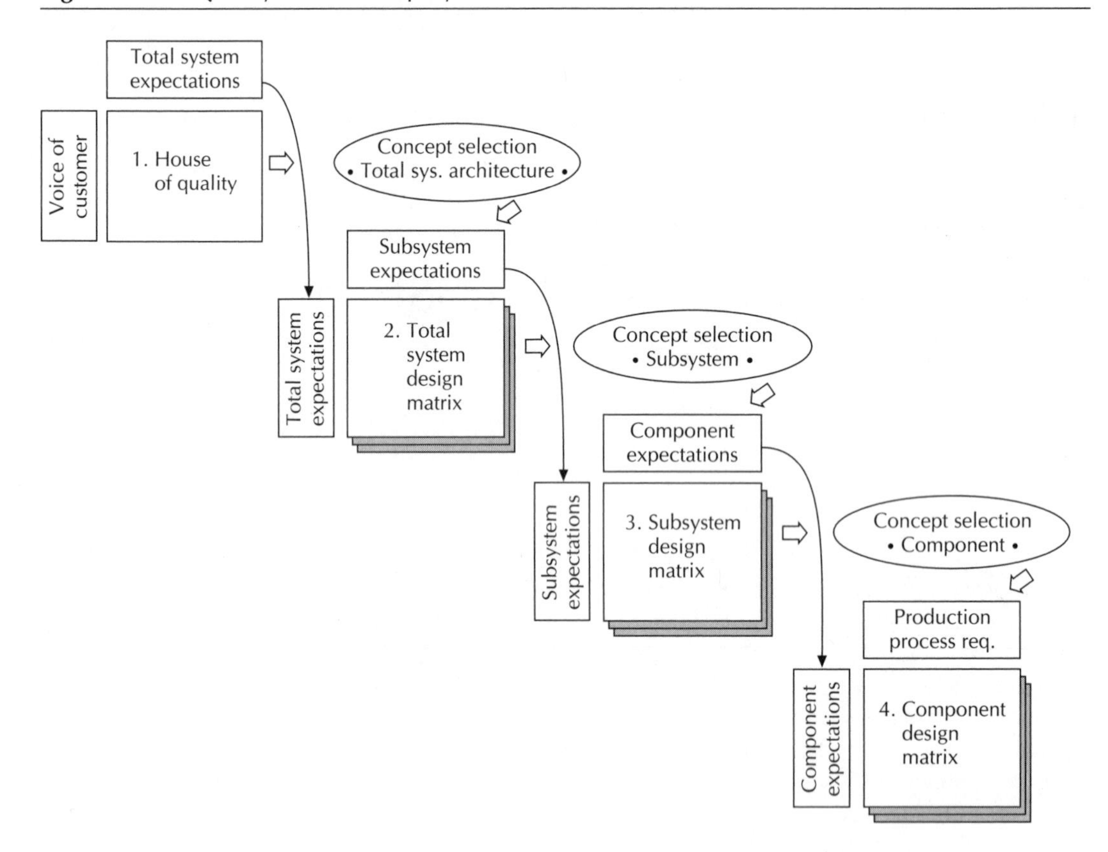


Robust Design is effectively practiced in coordination with the application of QFD to project management. The following descriptions identify the key phases of QFD and indicate which Robust Design activities should be practiced in each phase.

1. The House of Quality (Voice of the Customer vs. Total System Expectations). Here the key design issue is the measurement of customer-observable attributes such as product quality, environmental emissions, and reliability metrics. Products are benchmarked as a system using noise factors to stress the subsystem of interest. The customer functional limits should be determined at this stage.
2. The Total System Design Matrix (System Expectations vs. Subsystem Expectations). Here the key design issue is the measurement of the ideal engineering function using subsystem attributes as quality characteristics. The subsystems are tested using noise factors to induce variability. In addition, the ability of the subsystem of interest to affect the other subsystems is noted in the roof of this matrix. Proximity noise factors, intended to capture the subsystem interactions, are used.
3. The Subsystem Design Matrix (Subsystem Expectations vs. Component Expectations). The results of the parameter optimization experiments are used to specify the component parameters. In addition, measurement of the capability indices ($C_p$, $C_{pk}$) can be applied to critical-to-function component parameters. Tolerance design (introduced in Chapters 1 and 3) is

done using the loss function and the measured variability of the component parameters determined using noise factors.

4. The Component Design Matrix (Component Expectations vs. Production Process Requirements). Here Robust Design allows low-cost manufacturing processes wherever possible. Where necessary, Robust Design is applied to optimize the manufacturing process. More expensive manufacturing processes are applied selectively as justified from a quality loss vs. cost trade-off. Measures of manufacturing and production quality such as first-time yield, as-received quality, and freedom from early-life failure are applied.

## 18.2 Design of Experiments and Robust Design

The issue of design of experiments vs. Dr. Taguchi's approach to Robust Design has been a very contentious one. This is too bad, because the goal should be to achieve a robust design by whatever method works. However, there are substantive issues of experimental efficiency and design for additivity that do represent real departures in approach between the two camps. In this book, we present Dr. Taguchi's approach because we have found that his approach is effective, uses engineering knowledge in the experimental planning, and focuses on the on-target solution we seek. The mathematical DOE approach is more concerned with the statistical purity of the experimental design and the correctness of the resulting model than with the engineering solution. The idea behind this approach is that with a scientific understanding, including a rigorous understanding of the statistical validity of conclusions drawn from DOE studies, the engineering solution will become apparent. This difference in approach is thoroughly discussed in T. Mori's book *The New Experimental Design* [M7], which we recommend for those interested in further understanding this subject.

Here we will review the essential difference between the two practices, including our own opinions based on our experience in practicing both approaches.

### 18.2.1 The Objective of the Experimental Study

The traditional DOE point of view is that the experimental objective is to better understand the engineering system, achieved by an iterative step-by-step learning procedure.

> ***George Box***: *Modeling and response surface methodology is applied to understand specific control factor–noise factor interactions (and non-linearities).*[1] *Engineering know-how is then used to suggest ways of compensating for them, eliminating them, or reducing them.*[2]

Dr. Taguchi's view is this:

> *Our primary objective is to make design and production decisions that draw fully on prior engineering knowledge so that quality and cost are simultaneously and quickly improved, enabling the product to go into production while the market window is still open. . . . Box's approach . . . demonstrates a research philosophy; therefore it is too time consuming for product development and production decision making.*[3]

1. G. Box, *Harvard Business Review,* March–April, 228, 1990.
2. G. Box, *Technometrics* **34,** 131, 1992.
3. G. Taguchi and D. Clausing, *Harvard Business Review,* March–April, 229, 1990.

Research and engineering require different tools. The scientific method, including Western DOE, is useful to analyze, reason, and determine the rules phenomena follow. This is an important precursor to using Taguchi's methods for Robust Design. Taguchi's techniques, however, are more efficient for the optimization and characterization of a design's quality. Dr. Taguchi's approach is a commercialization process. The classical DOE approach is a knowledge-building process.

One very important objective is that engineers actually complete the optimization experiments, including verification of the optimum design set-points. Taguchi's approach is noted for its appeal to practicing engineers. It has been our experience that the probability of an engineer actually using a matrix experiment, ultimately completing it and applying the results, is much higher with the Taguchi approach than with classical DOE.

### 18.2.2 The Response

Traditional DOE practitioners usually offer no particular guidelines for choosing the response. Traditional DOE is done by fitting models to the mean and standard deviation of the response. Sometimes the data are transformed (say, by taking the log) before means and standard deviations are taken in order to fit a simpler model, adjust for varying variances, etc. The optimum configuration is chosen by co-optimizing the mean to be on-target and the standard deviation to be minimized. Most DOE textbooks emphasize the universality of the analytical method of data decomposition to any type of response, instead of the relevance of the response to fundamental causes.

There are, however, Western practitioners of DOE who are beginning to follow Dr. Taguchi's recommendations to place more emphasis on picking good quality characteristics for the response.

> ***Leon, Shoemaker, and Tsui:*** *[Dr. Taguchi's view is] to identify a good response start with the engineering or economic goal. Next identify the fundamental mechanisms and physical laws affecting this goal. Finally, choose the response to increase understanding of these mechanisms and physical laws. . . . Good responses are continuous, complete, practical, and fundamental. . . . A response is complete if it provides all the information needed to understand the basic physical mechanisms that reflect on the engineering and economic goals. . . . A response is practical if it is easy to measure with reasonable frequency. . . . A response is fundamental if it is unaggregated, elemental, and independent of imposed values.*[4]

Shin Taguchi is even more emphatic:

> *In parameter design, the most important job of the engineer is to select an effective characteristic to measure as data. . . . We should measure data that relate to the function itself and not the symptom of variability. . . . Quality problems take place because of variability in the energy transformation. Considering the energy transformation helps to recognize the function of the system.*[5]

The guidelines for choosing the quality characteristic given in this book are absolutely essential for good experimental design and efficiency. How the response relates to the fundamental function of the system being studied should affect how the experiment is designed and analyzed. This is one of the

4. R. Leon, A. Shoemaker, and K. Tsui, *Technometrics* **35,** 21, 1993.
5. S. Taguchi, *Technometrics* **34,** 138, 1992.

reasons that Dr. Taguchi's approach is considered to be more engineering-based than statistically-based. Engineers focus on the consequences of the physics and, therefore, prefer measures of the physics. The S/N ratio acts as an excellent summary statistic for the interaction of the noise factors and the control factors on the response. It is an intuitively satisfying parameter for practicing engineers and is consistent with the P-diagram analysis. The customization needed in selecting the appropriate S/N ratio is not difficult once the basic guidelines are understood. The subsequent analysis of the control factors is easy and leads naturally to the two-step optimization process. The multidimensional visualization needed to interpret the response surface methodology is not required in this approach.

### 18.2.3 Interactions between Control Factors

The traditional DOE point of view is this:

> ***M. Phadke:*** *Interactions are allowed to be present; they are appropriately included in the model; and experiments are planned so that they can be estimated.*[6]

> ***T. Lorenzen:*** *Although different response variables will most assuredly influence the complexity of the required model, neither measurable energy output nor fundamental physics need be additive. Sorry about that, but interactions may be needed. . . . Control [by] noise interactions [is] necessary to improve robustness. The difference between a control and a noise factor is definitional. . . .*[7]

> ***R. Kacker:*** *It is claimed that when the effects of control factors are additive . . . the optimal setting for the laboratory environment [is] likely to remain optimal during manufacturing and customer conditions. . . . I think . . . prior experience rather than ad hoc philosophy concerning the additivity of control-factor effects is the key to extrapolation of laboratory results.*[8]

Madhav Phadke's view is this:

> *Achieving additivity is very critical in Robust Design, because presence of large interactions is viewed as an indication that the optimum conditions obtained through a matrix experiment may prove to be nonoptimum when levels of other control factors (other than those included in the matrix experiment at hand) are changed in subsequent Robust Design experiments. Additivity is considered to be a property that a given quality characteristic and S/N ratio possess or do not possess.*[9]

Dr. Taguchi adds:

> *In the author's opinion, when monotonicity or additivity of factorial effects does not hold, in other words when there is an interaction, it is because insufficient research has been done on the characteristic values; and he predicts that as long as one selects logical characteristic values it will be possible to contrive that there will be monotonicity to the factorial effects.*[10] *The efficiency of the research will drop if it is not possible to find charac-*

6. M. Phadke, *Quality Engineering Using Robust Design,* 177, 1989.
7. T. Lorenzen, *Technometrics* **34,** 138, 1992.
8. R. Kacker, *Technometrics* **34,** 139, 1992.
9. M. Phadke, *Quality Engineering Using Robust Design,* 177, 1989.
10. G. Taguchi, *System of Experimental Design,* 124, 1987.

*teristics that reflect the effects of the individual factors regardless of the influence of other factors.*[11] *It is more desirable to treat interaction by including it in noise. Essentially, only a main effect which exceeds the magnitude of the interaction can safely be used in large scale production.*[12]

Those interactions that are artifacts of the analysis need to be distinguished from those that are fundamental to the physics of the factor's effect on the quality characteristic. In addition, the interactivity of the mean response is different than that of the S/N ratio. Many of Dr. Taguchi's guidelines only apply to the latter. The S/N ratio and monotonicity help remove some of the analysis interactions. The sliding level approach helps minimize the apparent level of the physical interactions.

It is desirable to avoid antisynergistic interactions when possible, and to be sure that the remaining interactions are very well understood so as to avoid constant reoptimization of a design as it passes from research to design to manufacturing. The experimenter needs to decide whether to use a Resolution III or IV design. The number of runs in an experiment can drop substantially by the use of Resolution III designs. That facilitates using three-level and mixed factorial arrays. A good test to use is the question: "Is modeling this interaction likely to give more useful information than the inclusion of an additional control factor?" A key issue is the intent or objective of the study. If modeling is the goal, then low-resolution studies are not appropriate. However, if finding the best set points is the primary goal, then any "screening test" has the potential of succeeding, and the approach of testing for additivity should be routinely adopted before going to higher-resolution experiments. The occasional "failed" low-resolution experiment will be more than compensated for by the frequently successful results obtained. When a "failed" experiment does occur, the Taguchi approach is to define the root cause of the interaction so that it is really understood, not simply as a term in a model, but as an engineering phenomenon.

The issue of interactions has been a particular source of conflict in this debate. A recent article in *Technometrics* (Vol. 34, 1992) was presented in the form of a round-table discussion with proponents from both sides. Here are some additional quotes from this stimulating discussion:

***Jeff Wu:*** *Generally speaking [finding a characteristic possessing monotonicity] is very original, [using sliding levels when the factors are inter-related] is a useful reminder of what has been known but not emphasized.*

***Raghu Kacker:*** *I agree that a study of the underlying mechanism may be more effective than the direct focus on final characteristics. But interactions may still be present in the underlying mechanism.*

***George Box:*** *It seems to me logically indefensible to say that we need an experiment to find out which factors have main effects (first-order effects) and at the same time claim to know which factors have interactions (second-order effects).*

Remember that engineers do not think of control factors as main effects in a model. These factors are put into an experiment in order to find their robust set points. Constraints or dependencies between engineering parameters, which is what an interaction imposes, are well understood by any diligent, practicing engineer.

11. G. Taguchi, *System of Experimental Design,* 61, 1987.
12. G. Taguchi, *System of Experimental Design,* 149, 1987.

### 18.2.4 Experimental Bias—Randomization and Blocking

The traditional DOE point of view on running an orthogonal array is that randomization and blocking, where appropriate, are necessary to protect against nuisance variables confounding with factor effects. The concern here is that during the running of the experiment, some factor outside of the array could influence the response in a way that could be confounded with one of the parameters in the array. Randomization refers to the order in which the treatment combinations are run. In some situations, restricted randomization is forced by physical constraints. However, it is always useful to get as much randomization as possible. Blocking refers to the assignment of a column to a parameter that is not being studied but is suspected of influencing the response.

These approaches derive from the basic assumption that an orthogonal array experiment is sampling from the entire experimental space (all possible combinations of factor levels). Sampling is a statistical concept with specific rules to prevent bias (the statistical term for systematic errors in the data analysis). Nuisance factors are very common in agriculture, one major area of application of DOE. As a result of these experiences, countermeasures to prevent experimental bias are routinely practiced.

Dr. Taguchi has in the past recommended randomization. For example:

> *It has become possible to evaluate . . . the degree of . . . influence of various uncontrollable . . . causes on experimental data . . . by randomizing of experimental sequence.*[13] *. . . The order is to be decided by drawing lots.*[14]

He has also recognized the usefulness of blocking to remove experimental bias:

> *Block factor . . . is a factor that has been selected even though its effect cannot be used even for adjustment, in order to prevent its effect from straying in among the control sources.*[15] *[From the Ina tile case study] factors H and I are block factors. . . . Their purpose is to raise the precision of the experiments for the other factors.*[16]

More recently, however, Dr. Taguchi's primary concern—which we share—has been experimental efficiency. Strategies for ensuring against nuisance factors must be balanced against the likelihood of such problems occurring and the cost of protecting against them. Ease of use and running experiments in less time with less effort is very important. The use of explicit noise factors in the experiment helps mitigate the effect of nuisance factors. In fact, most nuisance factors can be included as noise if their effects are important. Robust Design forces what can happen to happen by design, using explicit noise factors. These then override random effects.

The arrays used have increasing randomization in the columns as one moves from left to right. Thus, partial randomization is achieved by choosing the columns to the right in the orthogonal array for factor assignment. This is similar to restricted randomization. When fully saturated arrays are used (Resolution III), complete randomization is not possible anyway. Randomization is least needed in cases (e.g., mechanical hardware) where the control factors and noise factors are known to be much stronger than nuisance factors. Randomization is recommended in cases where replication is a significant surrogate noise factor. Good experimental techniques, precise transducers, etc., are prerequisites for good results in all cases, randomization notwithstanding.

---

13. G. Taguchi, *System of Experimental Design,* 125, 1987.
14. G. Taguchi, *System of Experimental Design,* 4, 1987.
15. G. Taguchi, *System of Experimental Design,* 148, 1987.
16. G. Taguchi, *System of Experimental Design,* 404, 1987.

### 18.2.5 Modeling

The traditional DOE point of view is that design of experiments allows for fitting models of varying complexity—from screening designs, where only main effects are estimated, to optimizing designs, where quadratic polynomials (nonlinear factor effects) and interactions are modeled. Central composite designs are the usual way of estimating nonlinear effects, since three-level fractional factorials require more runs and have much worse confounding.

Dr. Taguchi's view is that an accurate model of the mean response, as a function of several control factors, is not as important as finding the control factor levels that optimize for robustness. His methods emphasize great simplicity in the calculations. After the variance has been reduced, the mean response can easily be adjusted with the help of one adjustment factor. By limiting the process of putting the system on the target to one parameter, Dr. Taguchi avoids difficult co-optimization issues. The S/N ratio is chosen to facilitate this process.

Response surface models are not necessary for Robust Design. This book makes no use of contour plots. However, the computer tools needed to construct models are readily available to many practitioners and make it possible to interpolate between the levels actually tested for the optimum levels of quantitative control factors. In contrast, the method of ANOM is restricted to choosing optimum factor levels from the discrete levels actually tested.

The modeling of interactions should be done on an exception basis only. When it is needed, however, it should be done—and done correctly. Interactions cannot be ignored in all situations. Modeling the response with respect to the noise factors is primarily useful for understanding purposes only. Noise factors and control factors should not be analyzed in the same array, except for research purposes.

### 18.2.6 Significance Tests

The traditional DOE point of view is that significance tests such as the $F$-test should be used to determine if a particular factor should be included in the model. Such an analysis requires an error estimate best obtained by replicates, which are usually taken only at a center point. This implies a constant-variance assumption, illustrated in Figure 9.3.

Madhav Phadke takes the following view:

> *In Robust Design,* F*-ratios are calculated to determine the relative importance of the various control factors in relation to the error variance. Statistical significance tests are not used because a level must be chosen for every control factor regardless of whether that factor is significant or not.*[17]

Dr. Taguchi adds:

> *To carry out at least six more runs just for information on error variance is highly questionable from the standpoint of efficiency of the experiment. This is the reason why we pool small sources and substitute them for the error variance as a convenient method. . . .*[18]

17. Phadke *Quality Engineering Using Robust Design*, 180, 1989.
18. G. Taguchi, *System of Experimental Design*, 293, 1987.

Dr. Taguchi is correct in acknowledging that the error estimate may be biased significantly (p. 294, *Systems of Experimental Design*), but of what significance is that error? The relative importance of the control factors can be obtained from pooling (see Chapter 17). While not strictly accurate, pooling is easy and useful for factor ranking. It would be more accurate to use half-normal plots of the factor effects. However, this technique must first be generalized to three-level and mixed-level arrays. It must then be taught to engineers before it can be considered useful. Engineers are likely to use a technique that they understand, and with which they are self-reliant. Dr. Taguchi's approach reduces the dependence on statisticians for analysis and interpretation of experimental results.

### 18.2.7 Experimental Layout

The traditional DOE point of view:

> ***G. Box:*** *The cross product designs [inner × outer] can result in rather large experimental runs so that the total amount of work may be excessive.*[19]

> ***A. Shoemaker and K. Tsui:*** *A combined array [noise and control factors together] lets the experimenter choose the interactions to be estimated. This provides more flexibility. . . . Control × noise factor interactions provide special insights. . . . The wealth of techniques for empirical modeling can be more easily applied.*[20]

Dr. Taguchi's view is that the noise factors and control factors are treated differently in parameter design. The noise factors are used to test the treatment combinations in the control factor array. The noise factors can be arrayed in their own matrix, or combined into two or three compound factors. In either case, they provide a means for systematically stress-testing the control-factor combinations. The S/N ratio then becomes a metric for the quality of the control-factor treatment combinations.

The use of a noise array to test the control-factor combinations makes good engineering sense. The goal of the experiment is, after all, to find the most robust set points for the control factors. The crossed array approach addresses this directly. The literature is full of control factor by noise factor interaction discussions. Although this provides the fundamental mechanism for achieving robustness, modeling the control factor by noise factor interactions is an unnecessary complication that is eliminated by Dr. Taguchi's approach. The total number of experimental observations increases, but in most cases the noise factors are inexpensive to run. Thus, the incremental cost is low.

## 18.3 Six Sigma Quality Process and Robust Design

The Six Sigma process was developed at the Motorola Company in the eighties. The dramatic improvements achieved in product quality since its introduction have earned Motorola both commercial success as well as the Malcolm Baldrige award in 1988. Since then, Motorola engineers have published many books and articles on Six Sigma. Motorola also runs courses on Six Sigma. The company has, as a matter of policy, freely shared the process with industry.

---

19. G. Box, *Technometrics* **34,** 147, 1992.
20. A. Shoemaker and K. Tsui, *Technometrics* **34,** 147, 1992.

### 18.3.1 Capability Indices and the Loss Function

The basic Six Sigma metric is the Capability Index for Bilateral Limits (e.g., nominal-the-best):

$$C_p = \frac{|USL - LSL|}{6\sigma} \quad \textbf{(18.1)}$$

where USL refers to the Upper Specification Limit, LSL refers to the Lower Specification Limit, and $\sigma$ is the (population) standard deviation of the process being studied. The numerator is the customer functional limit tolerance range for a design parameter in a product or process. Note that there may be factory tolerances or statistical process control (SPC) limits that are considerably smaller than the functional limits. The denominator is a measure of the manufacturing variability of that parameter. From the capability index, we can see that the fundamental relationships between Quality Engineering and Six Sigma are given by the following observations:

1. The tolerance range can be increased by applying Quality Engineering to increase the design latitude. The customer expectation is not in our control. Robust Design increases the amount of variation that a particular critical dimension or parameter can have without causing the system performance to exceed the customer limit.
2. The manufacturing variability can be decreased by applying Quality Engineering to the manufacturing process.

Three-sigma quality is achieved when $C_p = 1$. Figure 18.2 shows a Gaussian distribution in relationship to the tolerance limits for $C_p = 1$.

This looks pretty good but, as we will show, it is inadequate for most products. Six Sigma quality is achieved when $C_p = 2$. Figure 18.3 shows a Gaussian distribution in relationship to the tolerance limits for $C_p = 2$.

**Figure 18.2** Distribution for $C_p = 1$

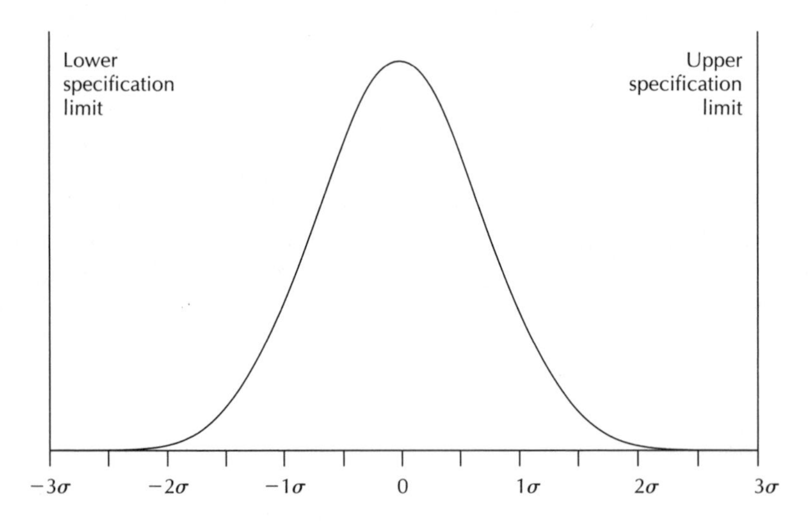

**Figure 18.3** Distribution for $C_p = 2$

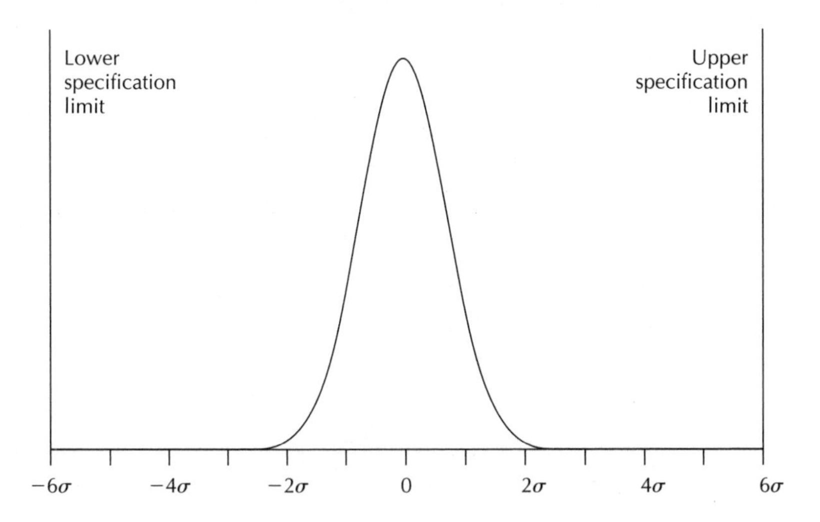


The capability index, $C_p$, is not an adequate description of quality by itself, because the process must be on target as well as capable. Thus, another capability index has been introduced: $C_{pk}$, defined by Equation 18.2.

$$C_{pk} = C_p(1 - k), \qquad k = \frac{|\mu - T|}{(\text{USL} - \text{LSL})/2} \tag{18.2}$$

where $\mu$ is the (population) mean for the distribution and $T$ is the target. Typically, the target is taken to be at the center of the tolerance range; thus:

$$T = \frac{\text{USL} + \text{LSL}}{2} \tag{18.3}$$

Figure 18.4 illustrates the application of $C_{pk}$.

An equivalent form of Equation 18.2, which applies when the target is centered between the specification limits, is given by Equation 18.4:

$$C_{pk} = \min(C_{pu}, C_{pl}), \qquad \text{where}$$
$$C_{pu} = \frac{|\text{USL} - \mu|}{3\sigma} \qquad \text{and} \qquad C_{pl} = \frac{|\mu - \text{LSL}|}{3\sigma} \tag{18.4}$$

Equation 18.4 states that the $C_{pk}$ value is given by the smaller of the two capability indices $C_{pu}$, the capability with respect to the upper limit, and $C_{pl}$, the capability with respect to the lower limit. Thus, we see that the $C_{pk}$ index is in fact measuring how close to the tolerance limit a process is running. This is not consistent with on-target engineering!

The $C_{pk}$ metric does, however, share some of the same properties as the average quadratic loss function. Equation 18.5 shows the quadratic loss function using notation consistent with the capability indices:

$$L = \frac{A_0}{[(\text{USL} - \text{LSL})/2]^2}[(\mu - T)^2 + \sigma^2] \tag{18.5}$$

**Figure 18.4** Distribution off target

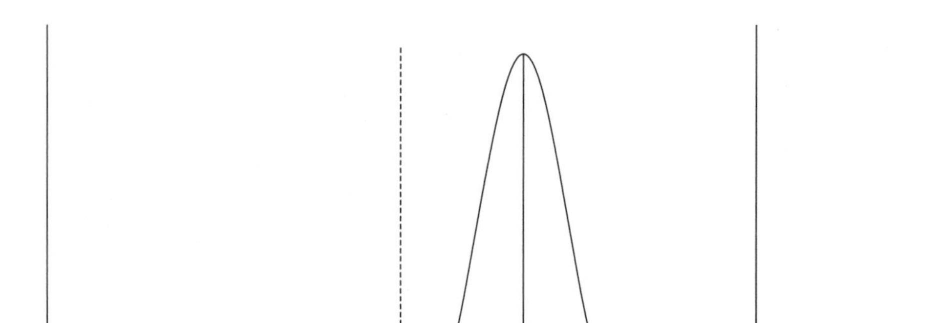


The key distinguishing characteristic of the loss function is the explicit inclusion of $A_0$, the loss incurred at the functional limits (USL and LSL in this notation). This allows financial loss to be the driving force for improving quality in Quality Engineering. As we will see, the power of Six Sigma derives from a series of metrics that relate to manufacturing yield (freedom from defects).

Before closing this section, it is very instructive to compare the quadratic loss function to the $C_{pk}$ over a range of distribution widths and average values. To do so, we introduce a dimensionless metric normalized to the tolerance width, referred to as the unit quality deviate. The unit quality deviate is defined in Equation 18.6:

$$y' = \frac{y - T}{(\text{USL} - \text{LSL})/2} \tag{18.6}$$

Equation 18.6 can be rearranged to solve for $y$, giving

$$y = \left(\frac{\text{USL} - \text{LSL}}{2}\right) y' + T \tag{18.7}$$

Similarly, the mean, variance, and standard deviation of the unit quality deviate are, respectively, given by

$$\mu_y = \left(\frac{\text{USL} - \text{LSL}}{2}\right) \mu_{y'} + T$$

$$\sigma_y^2 = \left(\frac{\text{USL} - \text{LSL}}{2}\right)^2 \sigma_{y'}^2$$

$$\sigma_y = \left(\frac{\text{USL} - \text{LSL}}{2}\right) \sigma_{y'} \tag{18.8}$$

where the subscript indicates the basis of the statistical quantities.

Equations 18.8 and 18.3 can be used to find some values of the unit quality deviate's mean and standard deviation shown here:

$$\begin{array}{lll} \mu_{y'} = 0 & \text{if} & \mu_y = T \\ \mu_{y'} = 1 & \text{if} & \mu_y = \text{USL} \\ \sigma_{y'} = 1 & \text{if} & \sigma_y = \text{USL} \\ \sigma_{y'} = 0.33 & \text{if} & \sigma_y = \text{USL}/3 \\ \sigma_{y'} = 0.167 & \text{if} & \sigma_y = \text{USL}/6 \end{array}$$

Equations 18.8 and 18.3 can be substituted into Equation 18.2 to give $C_{pk}$ as a function of the unit quality deviate:

$$C_{pk} = \frac{1 - \mu_{y'}}{3\sigma_{y'}} \tag{18.9}$$

A contour plot made using Equation 18.9 is shown in Figure 18.5.

Equations 18.8 and 18.3 can be substituted into Equation 18.5 to find the quality loss, $L$, as a function of the unit quality deviate:

$$L = A_0(\sigma_{y'}^2 + \mu_{y'}^2) \tag{18.10}$$

A contour plot made using Equation 18.10, with $A_0 = 1$, is shown in Figure 18.6.

Here we see that both metrics tend toward their optimum value as the mean of $y'$ approaches 0 ($y \rightarrow$ target) and the standard deviation approaches 0 (low variability). However, the loss function treats both quality improvements equally, while the contours of the $C_{pk}$ metric show that it increases more rapidly with reduction in variability than it does with reduction of the mean of $y' \propto (y - T)$.

**Figure 18.5** $C_{pk}$ contours using normalized units

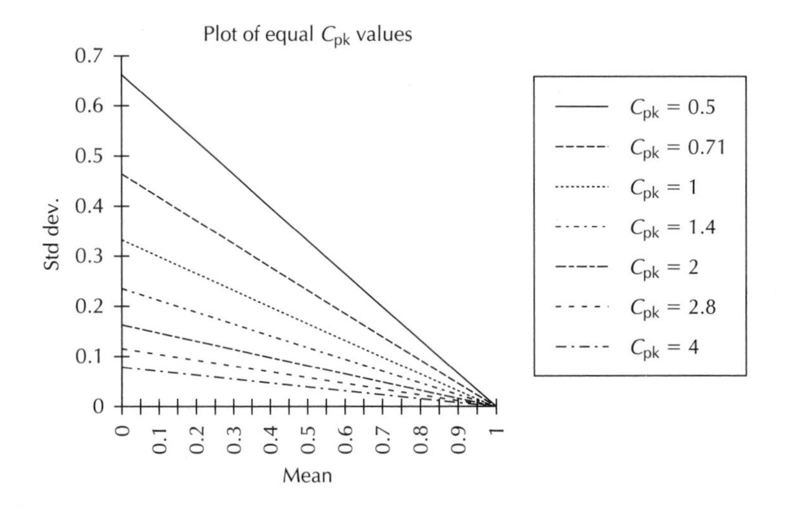

**Figure 18.6** Quadratic loss function contours using normalized units

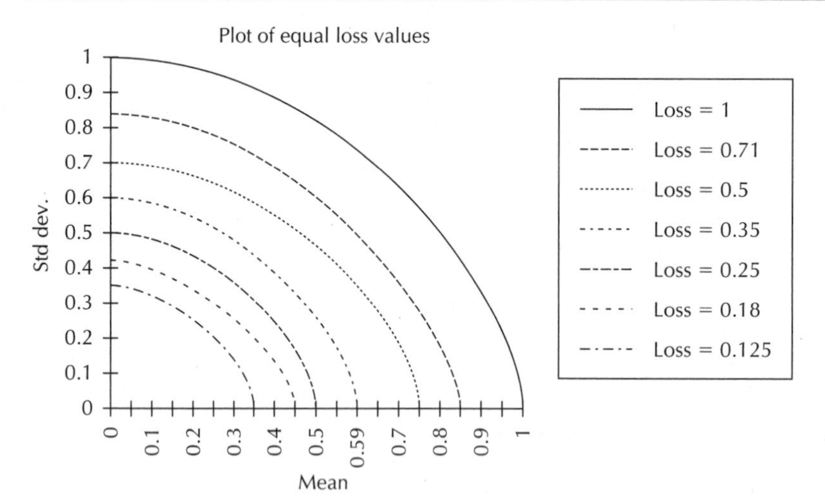


## 18.3.2 Measures of Manufacturing Yield (Freedom from Defects)

There are several metrics in Six Sigma that are closely related to $C_{pk}$. The first one we consider is DPMO, defects per million opportunities. This is found by calculating the probability, $p$, that a process will exceed the tolerance limits. The probability is given by the area under that portion of the normalized probability distribution curve that lies outside the tolerance limits, as shown in Figure 18.7.

Assuming a Gaussian distribution, DPMO is a simple function of the capability index. The probability, $p$, and DPMO are related by Equation 18.11:

$$p = \text{DPMO}/10^6 \tag{18.11}$$

If the process is on-target, then $C_p = C_{pk}$. Typically, however, the process mean wanders about somewhat over time. This is because of long-term drift, special causes, and the common causes within the control limits of statistical process control. Experience has shown that if the short-term process variability is $\sigma$, then the long-term variation of the mean is typically $1.5\sigma$ off target. The resulting $C_{pk}$ is given by Equation 18.12:

$$C_{pk} = \left(1 - \frac{1.5}{3C_p}\right)C_p. \tag{18.12}$$

This results in a higher probability of defects than would be expected if we did not account for this drift. Table 18.1 shows the expected defects per million opportunities at different quality or distribution widths ($\sigma$ levels) as a function of $C_p$, assuming on-target performance, and $C_{pk}$, assuming $1.5\sigma$ off-target behavior.

If we assume that there are $n$ items in a complex system, that the probability of failure, $p$, is the same for each of them, and that there is no correlation between the individual component failures, then the average number of defects per unit, DPU, is given by Equation 18.13:

$$\text{DPU} = np \tag{18.13}$$

**Figure 18.7** Distribution showing some parts exceeding specification limits

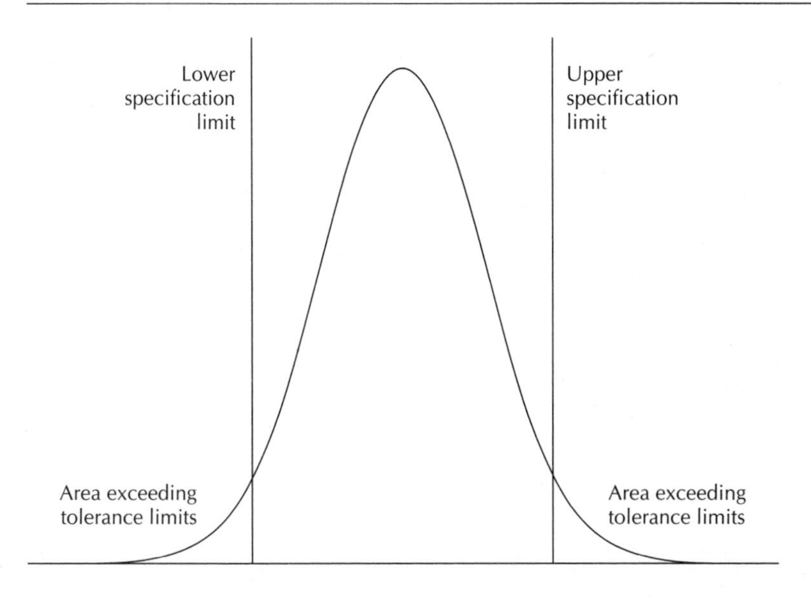


The probability $P\{q\}$ that a complex system consisting of $n$ components has $q$ defects is given by the Poisson distribution, Equation 18.14:

$$P\{q\} = \frac{(\text{DPU})^q e^{-\text{DPU}}}{q!} \tag{18.14}$$

The probability of a unit being completely free of defects as manufactured (requiring no rework) is related to the average defects per unit. It is found by taking the limiting case of $q = 0$ in Equation 18.14. This probability of zero defects, referred to as the first-time yield, FTY, is given in Equation 18.15:

$$\text{FTY} = e^{-\text{DPU}} \tag{18.15}$$

The preceding equations then provide links among the number of defects, yield, probability of defects, and capability indices. Each of these quality metrics is related to the others. Figure 18.8 shows how the first-time yield depends upon the quality level and system complexity.

**Table 18.1** DPMO vs. $C_p$ with and without the mean shift

| | *On target* | | *1.5σ off target* | |
|---|---|---|---|---|
| *Quality* | $C_p$ | *DPMO* | $C_{pk}$ | *DPMO* |
| 3σ | 1 | 2700 | 0.5 | 66,800 |
| 4σ | 1.33 | 63 | 0.83 | 6210 |
| 5σ | 1.67 | 0.57 | 1.17 | 233 |
| 6σ | 2.0 | 0.002 | 1.5 | 3.4 |

**Figure 18.8** Overall yield vs. sigma level

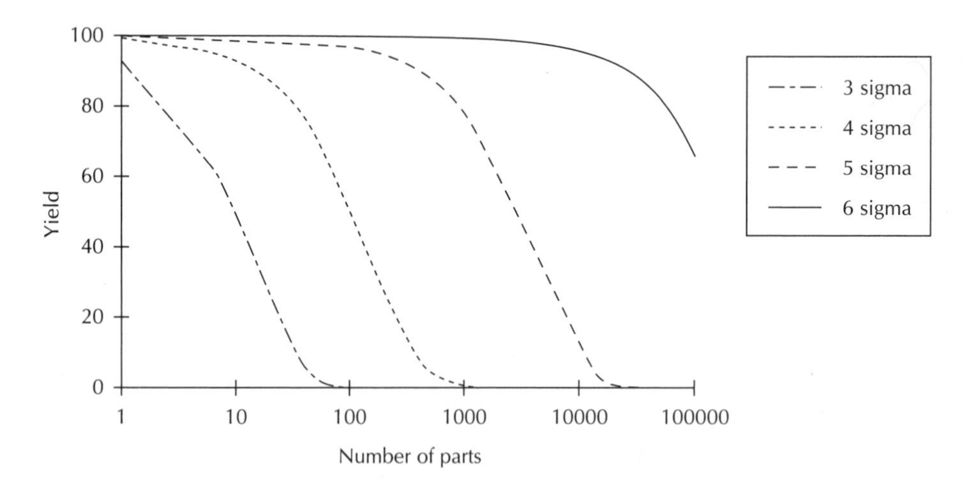


## 18.3.3 Comparison of Six Sigma and Quality Engineering

We can now compare Six Sigma and Robust Design. The following list focuses on the strong and weak points of the Six Sigma process:

1. The capability indices are referenced to the tolerance limits, not to the target. The definition of $C_{pk}$ includes the deviation of the mean from the target, and therefore encourages on-target engineering. However, the $C_{pk}$ value is determined by how many standard deviations the center of the distribution is from the tolerance limit. As shown in Figure 18.5, the $C_{pk}$ value is more sensitive to reducing the distribution width than to putting the process onto the target. It is possible to get very good $C_{pk}$ values ($C_{pk} > 1.5$) while being well off-target. This can lead to quality loss despite good manufacturing yield.
2. The capability indices quantify quality in dimensionless units. This is very useful for calculating probabilities and comparing processes.
3. Six Sigma is consistent with the in-spec paradigm. Thus, we find that the major application of Six Sigma is to reduce DPU. This can improve quality only up to a point. To move beyond reducing waste in the factory and to achieve ultimate quality goals on-target engineering must be used.
4. The metrics are highly sensitive to distribution shape (Gaussian distribution for $C_{pk}$ and DPMO, Poisson distribution for FTY). It is possible to have meaningful means and standard deviations for non-Gaussian distributions. DPMO in particular depends on the shape of the distribution at the tails. Such metrics are notoriously unreliable if the Gaussian distribution assumption is not accurate. Some authors [H5] have stated categorically that the Gaussian distribution is not at all valid in the factory. In fact, for processes that are under statistical control, the Gaussian distribution is fairly good. But special causes or mistakes, which generate parts that are not under control, cause the true DPMO to always be higher than the expected value based on the assumed distribution. This is part of the reason that the $1.5\sigma$ shift is applied to the data before DPMO is calculated.

5. Six Sigma marries the design and manufacturing contributions to product quality. This may be one of the most important contributions of Six Sigma. Equation 18.1 gives a recipe for improving product quality that has equal contributions from the design engineers and manufacturing engineers. Use of this metric makes both communities responsible for manufacturing yield, as they should be. Robust Design is an important tool that both groups can apply to improve their contribution to the capability.
6. Six Sigma is easy to "sell" to design and manufacturing engineers. The use of familiar concepts such as tolerances, defects, and yield helps engineers grasp the applicability of Six Sigma to their task. At the same time, the capability indices challenge them to measure their variability and to reduce variation and (to some extent) put their process onto the target.
7. Six Sigma can lead to an "oscillation" between quality improvement and cost reduction. The true *cost of quality* is waste, quality losses, and service and warranty costs. However, at some point in the pursuit of reduced variation and increased $C_{pk}$ values, unit manufacturing cost will go up. Without a balancing mechanism such as the quality loss function, which seeks to co-optimize cost and quality, it is possible to find companies alternately improving quality and reducing cost. Robust Design inherently avoids this by seeking low-cost solutions to variation reduction. Ideally, Six Sigma requires that the design community have a knowledge of manufacturing variation and then design the product to have good $C_{pk}$. But Six Sigma practitioners who attempt to increase $C_{pk}$ during production, using techniques such as noise reduction and manufacturing process changes, can add cost while improving quality.

The following list gives some of the strong and weak points of Robust Design:

1. The key Robust Design metric, the quadratic loss function, is referenced to the target. The tolerance range is in the function as part of the economic constant. But the actual tolerance limits are not included in the metric, nor are they included in the S/N ratio. Thus, Robust Design focuses on improving the absolute quality, not stopping once some limit is passed.
2. Robust Design measures quality in monetary units. The advantage here is that this leads directly to the ability to co-optimize all facets of costs, including the engineering effort, manufacturing costs, the cost of quality, and life-cycle costs (see Figure 1.4). The problem is in getting good data to calculate the economic constant in the quality loss function.
3. Robust Design partitions quality loss into variance and deviation from target. This leads directly to the derivation of the S/N ratios, as shown in Chapter 4. Such metrics are extremely powerful in quality improvement. The two-step optimization process represents a major improvement in the quality field. The direct application of the loss function, with the squared deviation from target and the variance treated equally, has even been granted a U.S. patent. Also, the use of the quality loss function for tolerancing is included in a Japanese Industrial Standard.
4. Robust Design can be used to develop on-line and manufacturing strategies for optimizing control limits in a cost- and quality-effective manner. However, tolerancing and other quality loss applications require confidence in quantitative loss function calculations.

### 18.3.4 Six Sigma Product Development Steps

The following is the Six Sigma process described by Motorola. We have taken the liberty, for this chapter, of indicating where Robust Design (RD), Quality Function Deployment (QFD), and Design of Experiments (DOE) can be applied within this process.

**Step 1:** Identify the customer requirements. This can be done using the Quality Function Deployment process to determine the House of Quality (Product Planning Matrix). It is also important to determine the functional limits for the customer requirements.

**Step 2:** Identify the product characteristics that contribute to achieving the customer requirements. This can be done using design analysis, DOE, or RD. The results are recorded in the Design Matrix when using QFD.

**Step 3:** Identify the process characteristics that contribute to achieving the customer requirements. This can be done using design analysis, DOE, or RD. The results are recorded in the manufacturing process matrix (Process Planning Matrix) when using QFD.

**Step 4:** Establish customer tolerance limits for the product and process characteristics identified in Steps 2 and 3. The best method for doing this is to use the loss function during tolerance design. This should follow parameter design to ensure the widest possible tolerances.

**Step 5:** Determine the actual process capability for the product and process characteristics identified in Steps 2 and 3. This is a critical metric that should be displayed as part of the design and manufacturing process QFD matrices.

**Step 6:** Assure $C_p \geq 2$ and $C_{pk} \geq 1.5$. The Six Sigma practitioners use all methods available to achieve these quality goals. Tools used include Robust Design, Design of Experiments, noise factor elimination, and Statistical Process Control (SPC).

### 18.3.5 Summary

There are practitioners of the Six Sigma process who view Robust Design as a process in opposition to Six Sigma. Although there are significant differences in emphasis, in fact Robust Design is a key tool for the Six Sigma practitioners [H2]. Six Sigma is a useful process, particularly for getting the attention and cooperation of the design and manufacturing community. Six Sigma is presented as a set of key metrics that define the results of good quality practices. The means for achieving high $C_{pk}$ values and low DPU values are not usually specified in the Six Sigma literature. Rather, engineers are encouraged to use any means available to them, including Robust Design.

## 18.4 Summary

There are many quality engineering processes besides the three discussed here. They all contribute, in their unique ways, to achieving Juran's definition of quality—features and conformance to features—given in Chapter 1. In our experience there is no better approach to achieving the latter than Robust Design. Dr. Taguchi's Robust Design method enjoys a long and well-documented history as an effective engineering process for achieving on-target performance. We recommend it wholeheartedly and hope you will find this book and WinRobust to be useful tools as you endeavor to implement this process.

For university students, this is the first generation of newly educated engineers that will enter the work force with Quality Engineering already in their engineering "tool kit." The authors have seen the positive reaction of many students when they experience, for the first time, the power of understanding and employing an engineering *process* in which they can deploy their many technical courses in engineering and physics. We wish you well as you enter the adventure and challenge of the world of product and process engineering.

APPENDIX A

# Glossary

## A

**Absolute Zero** A property of a quality characteristic that allows all measurements to begin from the number zero and go up the rational number line from that point.

**Additivity** The independence of each individual control factor's effect on the measured response.

**Additive Model** A mathematical function that expresses a dependent variable, *y*, in terms of the sum (simple addition) of functions of *relatively* independent variables represented by the control factors in an optimization experiment. The additive model assumes mild to moderate interactivity between the control factors and assigns the numerical contribution to the response for this interactivity to the error term in the model: $y = f(A) + g(B) + h(C) + \ldots +$ error. This model is approximate in nature and is useful for engineering optimization, but not for rigorous statistical math modeling of fundamental functional relationships.

**Adjustment Factor** Any control factor that has a strong effect on the mean performance of the response. There are two special types of adjustment factors: scaling factors and tuning factors. Adjustment factors are used in static and dynamic applications of the two-step optimization process.

**Aliasing** See **Confounding.**

**ANOM** (ANalysis Of the Mean) An analytical process that quantifies the mean response for each individual control factor level. ANOM can be performed on data in regular engineering units or data transformed into a signal-to-noise ratio. This is also referred to as level average analysis or marginal analysis.

**ANOVA** (ANalysis Of the VAriance) An analytical process that decomposes the contribution of each individual control factor to the overall experimental response. The ANOVA process is also capable of accounting for the contribution of interactive effects between control factors and experimental error in the response, provided enough degrees of freedom are available in the experimental array.

**Aperture** A physical control factor in a camera, usually circular, that adjusts the amount of light that can enter the camera and expose the film.

**Array** An arithmetically derived matrix or

table of rows and columns that is used to impose an order for efficient experimentation. The rows contain the individual experiments. The columns contain the experimental factors and their individual levels or set points.

**Asymmetric Loss Function** A version of the nominal-the-best loss function that accounts for the possibility of different economic consequences as quantified by the economic coefficients, $k_{\text{lower}}$ and $k_{\text{upper}}$, on either side of the target value.

**Average Quality Loss** The application of the quadratic loss function to numerous samples of a product or process resulting in the calculation of a term called the mean square deviation (MSD). The average quality loss approximates the nature of loss over a variety of customer economic scenarios, thus providing a generalized evaluation of the loss.

## B

**Balanced Experiment** An experiment in which each experimental factor is provided with an equal opportunity to express its effect on the measured response. The term *balance* is synonymous with the orthogonality of a matrix experiment.

**Benchmarking** The process of comparative analysis between two or more concepts, components, subsystems, products, or processes. The goal of benchmarking is to quantitatively identify a superior subject within the competing choices. Often the benchmark is used as a standard to meet or surpass.

**Beta** ($\beta$) The Greek letter $\beta$ is used to represent the slope of a line. It indicates the linear relationship between the signal factors and the measured response in a dynamic experiment.

**Blocking** A technique used in classical DOE to remove the effects of unwanted noise or variability from the experimental response, so that only the effects from the control factors are present in the response data. Blocking is a data purification process used to help assure the integrity of the experimental data used in constructing a math model.

**Build–Test–Fix Methods** A traditional engineering process used to iterate product development and design activities until a "successful" product is ultimately produced. These processes can be lengthy and costly to employ. The main difficulty with this approach is inefficiency and production of suboptimal results, requiring even more redesign effort in the production phase of the product life cycle.

## C

**Calibration** Setting a meter or a transducer (measurement instrumentation) to a known standard reference that assures the quantification of reliable measures of physical phenomena. Calibration makes the accuracy and repeatability of measurements as good as possible, given the inherent capability of the instrument.

**Central Composite Design (CCD)** An experimental design technique used in classical DOE to build math models of interactive and nonlinear phenomena.

**Classical Design of Experiments (DOE)** Experimental methods employed to construct math models relating a dependent variable (the response) to the set points of any number of independent variables (the experimental factors). Used to build knowledge of fundamental functional relationships between various factors and a response variable.

**Classification Factor** A parameter that uses integers, simple binary states, or nonnumerical indicators to define the levels that are evaluated in the experimental process. Classes can be stages, groups, or conditions that define the level of the classification factor. This type of factor is used in cases where continuous variables cannot adequately define the state of the factor set point.

**Coefficient of Variation (COV)** The ratio obtained when the standard deviation is divided by the mean. The COV is often expressed as a percentage by multiplying the ratio by 100. It is an indication of how much variability is occurring around the mean response. Often it is of interest for detecting a quantitative relationship between changes in the mean response and changes in the standard deviation.

**Column** The part of an experimental array that contains experimental factors and their various levels or set points.

**Commercialization** A business process that harnesses the resources of a company in the endeavor of conceiving, developing, designing, producing, selling, distributing, and servicing a product.

**Compensation** The use of feedforward or feedback control mechanisms to intervene when certain noise effects are present in a product or process. Compensation should only be used when insensitivity to noise cannot be attained through robustness optimization.

**Complete** The property of a single quality characteristic that promotes the quantification of all the dimensions of the ideal function of the design.

**Compound Control Factors** Two or more control factors whose individual effects on the response are understood and which are grouped together so that they contribute in the same way to the response to improve additivity. Control factors can also be grouped when necessary to allow the use of a smaller orthogonal array.

**Compound Noise Factors** Noise factors that have been strategically grouped by the nature of their strength and effect on the directionality of the measured response. They are used to increase the efficiency of experimentation.

**Concept Design** The initial phase of the Taguchi approach to Quality Engineering. This part of the commercialization process is used to link the Voice of the Customer with the inventive and technical capabilities of the business so as to design superior product concepts. It is then followed by the parameter and tolerance design processes.

**Concurrent Engineering** The process of simultaneously carrying out various product development, design, and manufacturing activities in parallel rather than in series. The outcome is reduced cycle time, higher quality, and lower product commercialization costs. Concurrent engineering is noted for its focus on multifunctional teams that work closely to improve communication of seemingly conflicting requirements and priorities. Their goal is to formulate effective synergy around the customers' needs so as to bring the project to a speedy and optimal conclusion.

**Confounding** The situation created when one control factor main effect is numerically blended with another control factor main effect and/or interactive effects by intentionally or unintentionally allocating the same degrees of freedom in a matrix experiment to multiple factors. Often the term *aliasing* is used interchangeably with confounding. Confounding can be benign or harmful, depending on the amount of interactivity between the confounded factors.

**Continuous Factor** A parameter that can take on any value within a range of rational numbers to define the levels that are to be evaluated in the experimental process.

**Contour Mapping** A graphical representation of the response as a function of two parameters. It is computer generated based on the math model created from any number of classically designed experiments. It is used to predict points or areas of optimum performance based on empirical data. It does not typically account for the effects of noise on the predicted response.

**Control Factor** The factors or parameters in a design or process that the engineer can control and specify to define the optimum combination of set points for satisfying the Voice of the Customer.

**Crossed Array Experiment** The combination of inner and outer orthogonal arrays to introduce noise into the experiment in a disciplined and strategic fashion. This is done to produce data that can be transformed into S/N ratios.

## D

**Degrees of Freedom (DOF)** The capacity of an experimental factor or an entire matrix experiment to produce information. The $DOF_f$ for an experimental factor is always one less than the number of levels assigned to the factor. The $DOF_{exp}$ for an entire matrix experiment is always one less than the number of experimental runs. Whether accounting for experimental factor DOF or matrix experiment DOF, one DOF is always employed to calculate the average response. This is why one DOF is always subtracted from the total.

**Design of Experiments (DOE)** A process for generating data that utilizes a mathematically derived matrix to methodically gather and evaluate the effect of numerous parameters on a response variable. Designed experiments, when properly used, efficiently produce useful data for model building or engineering optimization activities.

**Deterioration Noise Factor** A source of variability that results in some form of physical deterioration or degradation of a product or process. This is also called an *inner noise* because it refers to variation in the control factor levels (inside the design).

**Double-Dynamic Signal-to-Noise Case** A variation of the dynamic S/N case that employs two signal factors. The two signal factors are the functional signal factor, which is a factor that directly and actively controls some form of input energy possessing a linear relationship to the output response in the design, and the process signal factor, which indirectly controls the energy transformation occurring in the design.

**Drift** The tendency of a process output or set point to move off-target.

**Dummy Level** A technique for using a two-level control factor in a three- (or more) level array.

**Dynamic Signal-to-Noise Case** A method of studying the linearity, sensitivity, and variability of a design function by relating the performance of a signal factor, numerous control factors, and noise factors to the quality characteristic. These cases relate the power of the proportionality between the signal factor and the quality characteristic to the power of the variance due to noise.

## E

**Economic Coefficient** The economic coefficient is used in the Quality Loss Function. It represents the proportionality constant in the loss function of the average dollars lost ($A_0$) due to a customer reaction and the square of the deviation from the target response ($\Delta_0^2$). This is typically, but not exclusively, calculated when approximately 50% of the customers are motivated to take some course of economic action as a result of poor performance. This is often referred to as the *LD-50 point* in the literature.

**Energy Flow Map** A representation of an engineering system that shows the paths of energy divided into productive and nonproductive work. Analogous to a free body diagram.

**Energy Transformation** The physical process a design or product uses to convert some form of input energy into various other forms of energy that ultimately produce a measurable response. The measurable response may itself be a form of energy, or it may be the consequence of energy transformations that have taken place within the design.

**Engineering Process** A set of disciplined,

planned, and interrelated activities that are employed by engineers to conceive, develop, design, and manufacture a product or process.

**Environmental Noise Factors** Sources of variability that are due to effects external to the design or product, also referred to as *outer noise*. They can be sources of variability that one neighboring subsystem imposes on another neighboring subsystem or component. Examples include vibration, heat, contamination, misuse, and overloading.

**Experiment** An evaluation or series of evaluations that explore, define, quantify, and build data to be used for modeling or predicting functional performance in a component, subsystem, or product. Experiments can be used to build fundamental knowledge for scientific research, or they can be used to optimize product or process performance in the engineering context of a specific commercialization process.

**Experimental Efficiency** This is a process-related improvement that is facilitated by intelligent application of engineering knowledge and the proper use of designed experimental techniques. Examples include the use of fractional factorial arrays, control factors that are engineered for additivity, and compounded noise factors.

**Experimental Error** The variability present in experimental data that is caused by meter error and drift, human inconsistency in taking data, random variability taking place in numerous noise factors not included in the noise array, and control factors that have not been included in the inner array. In the Taguchi approach, variability in the data due to interactive effects is often, but not always, included as experimental error.

**Experimental Factors** Independent parameters that are studied in an orthogonal array experiment. Robust Design classifies experimental factors as either signal factors, control factors, or noise factors.

**Experimental Space** The combination of all the control factor, noise factor, and signal factor levels that produce the range of measured response values in an experiment.

## F

***F*-Ratio** The ratio formed in the ANOVA process by dividing the mean square of each experimental factor effect by the error variance. This is the ratio of variation occurring *between* experimental factors, in comparison to the variation occurring *within* all the experimental factors being evaluated in the experiment. It is a form of signal-to-noise ratio in a statistical sense. The noise in this case is experimental error—not variability due to the noise factors in the noise array. The *F*-ratio can be calculated using S/N values (parameter design approach) or values in regular engineering units (tolerance design approach).

**Factor Effect** The numerical measure of the contribution of an experimental factor to the change in the response (quality characteristic). See **Half Effect.**

**Feedback Control System** A method of compensating for the variability in a process or product by sampling output response and sending a feedback signal that changes an adjustment factor to put the response back on its intended target.

**Feedforward Control System** A method of compensating for the variability in a process or product by sampling specific noise factor levels and sending a signal that changes an adjustment factor to keep the response on its intended target.

**Fisher, Sir Ronald** The inventor of much of what is known today as classical design of experiments. His major focus was on optimizing crop yield in the British agricultural industry.

**Fraction Defective** The portion of an experimental or production sample that is found to be outside of the specification limits.

**Fractional Factorial Design** A family of (typically) two- and three-level orthogonal arrays that greatly aids in experimental efficiency. Depending on how one is loaded with control factors, it can be used to study interactions, or it can be manipulated to promote evaluation of additivity in a design or process. Fractional factorial designs are a subset of all the treatment combinations possible in a full factorial array.

**Frequency Distribution Table** A table of the frequency of occurrence of data set values within specific ranges, used to generate a histogram plot.

**Full Factorial Design** Two- and three-level orthogonal arrays that include every possible combination of the experimental factors. Full factorial experimental designs use degrees of freedom to account for the main effects and all interactions between factors included in the experimental array. Most, if not all, of the higher order interactions are likely to be negligible; thus, there is little need to use large arrays to rigorously evaluate them.

**Functional Signal Factor (*M*)** A signal factor used in the double-dynamic case where engineering knowledge has identified an input variable that controls the major energy transformation in the design. This transformation controls a proportional relationship between the functional signal factor and the quality characteristic. A functional signal factor is used to tune the design's performance onto any desired target. Tunable technology is developed from a concept that possesses a functional signal factor.

**Fundamental** The property of a quality characteristic that expresses the basic or elemental physical activity that is ultimately responsible for delivering customer satisfaction. A response is fundamental if it does not mix mechanisms together and is uninfluenced by factors outside of the component, subsystem, design, or process being optimized.

## G

**Gaussian Distribution** A distribution of data that tends to form a symmetrical, bell-shaped curve. See **Normal Distribution.**

**Goal Post Mentality** A philosophy about quality that accepts anything within the tolerance band as equally good and rejects anything that falls outside of the tolerance band as equally bad. See soccer, hockey, lacrosse, and football rulebooks.

**Grand Total Sum of Squares** The value obtained by taking the response (e.g., the S/N ratio) of each experimental run from a matrix experiment, squaring it, and then adding the squared terms together.

## H

**Half Effect** The change in a response value *either* above or below the overall experimental average obtained for an experimental factor level change. It is called the half effect because it represents just half of the factor effect. An example would be the average numerical response, obtained for a control factor that is evaluated exclusively at either its high set point or at its low set point, minus the overall experimental average.

**Histogram** A graphical display of the frequency distribution of a set of data. Histograms display the shape, dispersion, and central tendency of the distribution of a data set.

**Hypothesis Testing** A statistical evaluation that checks the validity of a statement to a specified degree of certainty. These tests are done using well-known and quantified statistical distributions.

## I

**Ideal Function** The desired, customer-focused response that would be found if there were no noise acting on the product or process.

**Independent Effect** The nature of an experimental factor's effect on the measured response when it is acting independently with respect to any other experimental factor. When all control factors are producing independent effects, then the design is said to be exhibiting an additive response.

**Inference** Drawing some form of conclusion about a measurable functional response based on representative experimental data. Uncertainty and the laws of probability play a major role in making inferences.

**Inner Array** An orthogonal matrix that is used for the control factors in a designed experiment and that is crossed with some form of outer noise array.

**Inspection** The process of examining a component, subsystem, or product for defects either during or after manufacturing. The focus is typically on whether or not the item under inspection is within the allowable tolerances. Like all processes, inspection itself is subject to variability, and out-of-spec parts may pass inspection inadvertently.

**Interaction** The dependence of one experimental factor on the level of another experimental factor for its contribution to the measured response. There are two types of interaction: synergistic (mild to moderate in effect) and antisynergistic (strong in effect).

**Interaction Graph** A plot of the relationship two experimental factors have on the response. The ordinate (vertical axis) represents the response being measured, and the abscissa (horizontal axis) represents one of the two factors being evaluated. The average response values are plotted for the various combinations of the two experimental factors. The points representing the identical levels for the second factor are connected by a line.

## K

***k* factor** See **Economic Constant.**

## L

**Larger-the-Better (LTB)** A case where a larger response value represents a higher level of quality and a lower amount of loss.

**Latin Square Design** A type of fractional factorial orthogonal array that Dr. Taguchi modified for use in developing the experimental component of his system of Quality Engineering.

**Level** The set point at which a control factor, signal factor, or noise factor is placed during a designed experiment.

**Level Average Analysis** See **ANOM (Analysis of Means).**

**Life-Cycle Cost** The costs associated with making, supporting, and servicing a product or process.

**Linearity** The relationship between a dependent variable (the response) and an independent variable (e.g., the signal factor) that is graphically expressed as a straight line. Linearity is typically a topic within the dynamic cases of the robustness process.

**Linear Combination** This term has a general mathematical definition, and a specific mathematical definition associated with the dynamic case. In general, a linear combination is the simple summation of terms. In the dynamic case, it is the specific summation of the product of the signal level and its corresponding response ($M_i y_{i,j}$).

**Linear Graph** A graphical aid used to assign experimental factors to specific columns when evaluating or avoiding specific interactions.

**Local Maximum** A point in the experimental space being studied, within the constraints of the matrix experiment, where the numerical response to a given factor is at a maximum value (either in S/N or engineering units).

**Local Minimum** A point in the experimental space being studied, within the constraints of the matrix experiment, where the numerical response to a given factor is at a minimum value (either in S/N or engineering units).

**Loss to Society** The economic loss that society incurs when a product's functional performance deviates from its targeted value. The loss is often due to action taken by the consumer reacting to poor product performance, but can also be due to the effects that spread throughout society when products fail to perform as expected. For example, a new car breaks down in a busy intersection due to a transmission defect, and 14 people are 15 minutes late for work (cascading loss to many points in society).

**Lower Specification Limit** The lowest functional performance set point that a design or component can attain before functional performance is considered unacceptable.

## M

**Main Effect** The contribution an experimental factor makes to the measured response independent of experimental error and interactive effects. The sum of the half effects for a factor is equal to the main effect.

**Matrix** An array of experimental set points that is derived mathematically. The matrix is composed of rows (containing experimental runs) and columns (containing experimental factors).

**Matrix Experiment** A series of evaluations that is conducted under the constraint of a matrix.

**Mean** The average value of a series of data that may be gathered in an experiment.

**Mean Square Deviation (MSD)** A mathematical calculation that quantifies the average variation a response has with respect to a target value.

**Mean Square Error** A mathematical calculation that quantifies the experimental variance within a set of data.

**Measured Response** The quality characteristic that is a direct measure of functional performance.

**Measurement Error** The variability in a data set that is due to poorly calibrated meters and transducers, human error in reading and recording data, and normal random effects that exist in any measurement system used to quantify data.

**Meter** A measurement device, usually connected to some sort of transducer. The meter supplies a numerical value to quantify functional performance.

**Monotonic** A functional relationship where the dependent parameter (a response) changes unidirectionally with changes in the independent parameter (an experimental factor). In Robust Design, this concept is extended to include consistency in the direction of change of the response even in the presence of potential interactions. Monotonicity is a necessary, but not sufficient, property for additivity.

**Multidisciplinary Team** A group of people possessing a wide variety of technical backgrounds and skills working together in the product commercialization process.

**Multifunctional Team** A group of people possessing a wide variety of business and commercialization responsibilities working together in the product commercialization process. They are the basis of concurrent engineering.

## N

**Noise** Any source of variability. Typically noise is external to the product (such as environmental effects). It may be a function of unit-to-unit variability due to manufacturing, or it may be associated with the effects of deterioration.

**Noise Directionality** A distinct upward or downward trend in the response depending on the level at which the noises are set. Noise factor set points can be compounded depending upon their directional effect on the response.

**Noise Factor** Any factor that promotes variability in a product or process.

**Noise Experiment** An experiment designed to evaluate the strength and directionality of individual noise factors on a product or process.

**Nominal-the-Best (NTB)** A case in which a product or process has a specific nominal or targeted value.

**Normal Distribution** The symmetric distribution of data about an average point. The normal distribution takes the form of a bell-shaped curve. It is a graphic illustration of how randomly selected data points from a product or process response will mostly fall close to the average response, with fewer and fewer data points falling farther and farther away from the mean. The normal distribution can also be expressed as a mathematical function.

## O

**Off-Line Quality Control** The processes included in preproduction commercialization activities. The processes of concept design, parameter design, and tolerance design make up the elements of off-line quality control. It is often viewed as the area where quality is designed into the product or process.

**On-Line Quality Control** The processes included in the production phase of commercialization. The processes of statistical process control (loss function based and traditional), inspection and evolutionary operation (EVOP) are examples of on-line quality control.

**One-Factor-at-a-Time Experiment** An experimental technique that examines one factor at a time, determines the best operational set point, locks in on that factor level, and then moves on to repeat the process for the remaining factors. This technique is widely practiced in scientific circles, but lacks the circumspection and discipline provided by full factorial and fractional factorial experimentation. One-factor-at-a-time experiments are used to build knowledge prior to the main parameter design experiment.

**Optimization** Finding and setting control factor levels at the point where their S/N ratios are at the maximum value. Optimized performance means the control factors are set such that the design is least sensitive to the effects of noise.

**Optimum Level** The set point at which the S/N ratio is at a maximum value for a control factor's response.

**Orthogonality** The property of an array or matrix that gives it balance and the capability of producing data that allow for the independent quantification of independent factor effects.

**Orthogonal Array** A balanced matrix that is used to lay out an experimental plan for the purpose of designing quality into a product or process early in the commercialization process.

**Outer Array** The orthogonal array in parameter design that contains the noise factors and signal factors. Each treatment combination of the control factors specified in the inner array is repeated using each of the treatment combinations specified by the outer array.

## P

**P-Diagram** A schematic representation of the relationship among signal factors, control factors, noise factors, and the measured response. The parameter (P) diagram was introduced to the robustness process by Dr. Madhav Phadke in his book, *Quality Engineering Using Robust Design.*

**Parameter** A factor used in the robustness optimization process. Experimental parameters are signal factors, control factors, and noise factors.

**Parameter Design** The process employed to optimize the levels of control factors against the effect of noise factors. Signal factors (dynamic cases) or tuning factors (NTB cases) are used in the two-step optimization process to adjust the performance onto a specific target during parameter design.

**Parameter Optimization Experiment** This is the main experiment in parameter design that is used to find the optimum levels for the control factors. Usually this experiment is done using the crossed array design.

**Pooled Error** Assigning a certain few experimental factors that have significantly low main effect values to represent or estimate the error taking place in the experiment.

**Population Parameter or Statistic** A statistic such as the mean or standard deviation that is calculated with all the possible values that make up the entire population of data in an experiment. See **Sample Statistics.**

**Predictive Model** A mathematical expression generated from the experimental ANOM data of the initial matrix experiment. The optimum control factor level values are used to predict the optimum S/N ratio that is given by the ANOM of the data from the parameter optimization experiment.

**Probability** The likelihood or chance that an event or response will occur out of some number ($n$) of possible opportunities.

**Process Signal Factor ($M^*$)** A signal factor that is a known adjustment factor in the design or process. Typically, it is a factor that modifies or adjusts the energy transformation but does not cause the function itself. The process signal factor is used in the double-dynamic case to modify the linear relationship of the functional signal factor to the response by changing the slope.

**Proximity Noise** The uncontrolled effect of one component or subsystem in a design on a neighboring component or subsystem.

**Pugh Process** A structured concept selection process used by multidisciplinary teams to converge on superior design concepts. The process uses a matrix consisting of criteria based on the Voice of the Customer and its relationship to specific design concepts. The evaluations are made by comparing the new concepts to a benchmark called the datum. The process uses the classification metrics of "same as the datum," "better than the datum," or "worse than the datum." Several iterations are employed, in which ever-increasing superiority is developed by combining the best features of highly ranked concepts until the best concept emerges and becomes the benchmark.

## Q

**Quadratic Loss Function** The parabolic relationship between the dollars lost by a customer because of off-target product performance and the measured deviation of the product from its intended performance.

**Quality** The degree or grade of excellence [*The American Heritage Dictionary,* 2nd College Edition; Houghton Mifflin Co. (1985)]. In a quality engineering context, it is a product with superior features that performs with consistency throughout its intended life. In an economic context, it is the absence or minimization of costs associated with the purchase and use of a product or process.

**Quality Characteristic** The measured response in an experiment that is directly related to physical energy transformations. The quantifiable measure of functional performance that directly affects the customer's satisfaction.

**Quality Engineering** The processes included in the Taguchi approach to off-line quality control (concept, parameter, and tolerance design) and on-line quality control.

**Quality Function Deployment (QFD)** A disciplined process for obtaining, translating, and deploying the Voice of the Customer into the various phases of the commercialization of products or processes.

**Quality Loss Cost** The costs associated with the loss to society when a product or process performs off the targeted response.

**Quality Loss Function** The relationship between the dollars lost by a customer because of off-target product performance and the measured deviation of the product from its intended performance. Usually described by the quadratic loss function.

## R

**Random Error** The nonsystematic variability that is present in experimental data because of random effects occurring outside of the control-factor main effects, the intentional variation being induced by the noise factors and systematic (nonrandom) error due to human or meter error.

**Randomization** The technique employed to remove any systematic or biased order in running experiments. Randomizing is especially important when applying classical DOE in the construction of math models.

**Reference-Point Proportional** A case in the dynamic family of experiments that uses the same approach as the zero-point proportional case, except that the data does not pass through the origin of the coordinate system. This case is built around the equation $y = mx + b$, where $b$ represents the nonzero $y$ intercept.

**Repetition** The taking of data points where the response is measured multiple times without changing any of the experimental set points. Repeat measurements provide an estimate of measurement error only.

**Replicate** The taking of data in which the set points have all been changed since the previous readings were taken. Often a replicate is taken for the first experimental run and at the middle and end of an experiment (for a total of three replicates of the first experimental run). Replicate measurements provide an estimate of experimental error.

**Reproducibility** The ability of a design to perform as targeted throughout the entire development, design, and production phases of commercialization. The verification tests provide the data on reproducibility in light of the noise imposed on the design.

**Residual Error** A statistical term that means the same as the mean square error. This is the measure of random and nonrandom effects that occur in running experiments and is independent of the main effects of the control factors.

**Response** The measured value taken during an experimental run. Also called the quality characteristic.

**Reliability** The measure of robustness over time. The length of time a product or process performs as intended.

**Resolution III Design** A designed experiment that is structured such that some or all of the experimental factor main effects are confounded with the two-way interactions of the experimental factors. A saturated array is a Resolution III design. This type of experiment uses the fewest degrees of freedom to study the main effects of the experimental factors. Using a Resolution III design in parameter optimization experiments requires the heavy use of engineering principles to promote additivity between the control factors.

**Resolution IV Design** A designed experiment that is structured such that all the experimental factor main effects are free from being confounded with the two-way interactions of the experimental factors. In a Resolution IV design, there is confounding between the two-way interactions of the experimental factors. Using a Resolution IV design in parameter optimization experiments requires more degrees of freedom (treatment combinations) than a Resolution III design to study the same number of control factors.

**Resolution V Design** A designed experiment that is structured such that all the experimental factor main effects are free from being confounded with the two-way interactions of the experimental factors. In a Resolution V design, there is also no confounding between all the two-way interactions. There may be confounding between the two-way interactions and the three-way interactions, which is viewed as being inconsequential. Using a Resolution V design in parameter optimization experiments requires many more degrees of freedom than a Resolution IV design to study the same number of control factors.

**Robust Design** A process within the domain of Quality Engineering of making a product or process insensitive to the effects of variability without actually removing the sources of variability. Sometimes taken to be synonymous with parameter design.

**Robustness Test Fixture** A device used in the off-line quality control processes to stress a representation of the production design. It is heavily instrumented to facilitate the direct physical measurement of function performance as noise is imposed during the designed experiments.

## S

**Sample** A selection of data points that are taken out of a greater population of data.

**Sample Size** The measure of how many data points have been taken from a larger population.

**Sample Statistic** A statistic such as the mean or standard deviation that is calculated using a portion of the values that make up the entire population of data in an experiment.

**Saturated Experiment** The complete loading of an orthogonal array with experimental factors. There is definite confounding of main effects with potential interactions between the experimental factors within a saturated experiment. Also referred to as a Resolution III design.

**Scaling Factor** A control factor that is known to have a strong effect on the mean response and a weak effect on the Type I nominal-the-best S/N response in a design. The scaling factor has the additional property of possessing a proportional relationship to both the standard deviation and the mean. It is used in dynamic cases as well as in Type I NTB cases.

**Scaling Penalty** The increase in sensitivity to noise that occurs during the second step of the two-step optimization process (shifting the mean response onto the target). The size of the penalty depends on the nature of the variability present in the particular design being optimized.

**Scaling Ratio** A ratio expressing the value of the target response vs. the value of the mean response used in the derivation of the Type I S/N ratio. It is used to adjust the quality loss to ultimately make the S/N ratio a quality metric that is independent of the mean response.

**Screening Experiment** Typically a small saturated orthogonal array experiment that is used to determine which factors are important to the response of a product or process. Screening experiments are used to build knowledge prior to the main parameter design experiment.

**Sensitivity** The magnitude (steepness) of the slope between the measured response and the signal factor in a dynamic experiment.

**Signal-to-Noise Ratio** A ratio or value formed by transforming the response data using a logarithm to help make the data more additive. Classically, signal-to-noise is an expression relating the useful part of the response to the nonuseful variation in the response.

**Signal Factor** A parameter that is known to be capable of adjusting the average output response of the design in a linear manner. Signal factors are used exclusively in dynamic cases. Ideally, they are the same or similar to the scaling factor sought out in the NTB cases for use in the two-step optimization process.

**Six Sigma** A corporate quality program developed by the Motorola Corporation. It is also a detailed set of quality processes and tools designed to minimize the effects of variability in product commercialization such that the process capability of the product reaches a level of 3.4 defects per million opportunities. A magnitude of the variability associated with the manufacturing process, six standard deviations represents the interval between the target and the upper and lower specification limits (for a total width of 12 sigma).

**Sliding Level Technique** A method for calibrating the levels of an experimental factor when that factor is known to be interacting with another related experimental factor. Sliding levels are engineered by analysis of the physical relationship between the interacting control factors. This approach is used to improve the additivity of the effects of the control factors.

**Slope** See **Beta.** It quantifies the linear relationship between the measured response and the signal factor.

**Smaller-the-Better (STB)** A static case where the smaller the measured response is, the better the quality of the product or process.

**Specification** A quantifiable set point that typically has a nominal or target value and a tolerance of acceptable values associated with it.

**Standard Deviation** A measure of the variability in a set of data. It is calculated by taking the square root of the variance. Standard deviations are not arithmetically additive, whereas the variances are.

**Static Case** One of the two major types of experimental cases to study product or process responses as related to specific parameters. The static case has no signal factor associated with the response. Thus, the response is considered fixed or static. Control factors and noise factors are used to find local optimum set points for robust performance.

**Subsystem** A group of individual components that perform a specific function within the total product system. Systems consist of two or more subsystems.

**Sum of Squares** A calculation technique used in the ANOVA process to help quantify the effects of the experimental factors and in some cases the mean square error (if replicates have been taken).

**Sum of Squares Due to the Mean** The calculation of the sum of squares of the overall mean response.

**Surrogate Noise Factor** Time, position, and location are not actual noise factors, but stand in nicely as surrogate sources of noise in experiments that do not have clear physical noises. These are typically used in process optimization cases.

**Synergistic Interaction** A mild to moderate form of interactivity between control factors. Synergistic interactions display monotonic but nonparallel relationships between two control factors when their levels are changed.

**System** A compilation of subsystems and components that make up a functioning unit that harmonizes the energy transformations of the parts to provide an overall product output.

## T

**Taguchi, Dr. Genichi** The originator of the system of Quality Engineering presented in this book. Dr. Taguchi is an engineer, former university professor, author, and global quality consultant.

**Target** The ideal point of performance which is known to provide the ultimate in customer satisfaction. Often called the nominal set point or the ideal performance specification.

**Tolerance Design** The final phase of off-line quality control. Orthogonal array experimentation, the application of the loss function, and ANOVA data analysis techniques are used to balance the cost and quality of the product or process.

**Total Sum of Squares** The part of the ANOVA process that calculates the sum of squares due to the combined experimental factor effects and the experimental error. It is the total sum of squares that is decomposed into the sum of squares due to individual experimental factor effects and the sum of squares due to experimental error.

**Transducer** A device that measures the direct physical action that produces functional performance in a design or process. Transducers are typically connected to a meter or data acquisition device to provide a readout and/or storage of the measured response.

**Treatment Combination** A single experimental run from an orthogonal array.

**Tuning Factor** The form of an adjustment factor that is most useful in implementing the two-step optimization process. Tuning factors have the unique property that they can adjust the mean without affecting the S/N ratio. The tuning factor for the Type I nominal-the-best S/N ratio is a scaling factor. For other S/N ratios, the tuning factor may possess a *non*proportional relationship between the standard deviation and the mean.

**Two-Step Optimization Process** The process of first finding the optimum control factor set points to minimize sensitivity to noise, and then adjusting the mean response onto the customer-focused target using a tuning or signal factor.

## U

**UMC** Unit manufacturing cost is the cost associated with making a product or process.

**Unit-to-Unit Variability** Variation in a product or process due to noises in the production process.

**Upper Specification Limit** The largest functional performance set point that a design or component is allowed before functional performance is considered unacceptable.

## V

**Variance** The mean squared deviation of the measured response values from their average value.

**Variation** Changes in parameter values due to systematic or random effects. Variation is the root cause of poor quality and the monetary losses associated with it.

**Verification** The process of computing a predictive model and a predictive S/N ratio for use in building a verification experiment to evaluate the validity of the optimized control factors found in the original parameter design experiment. Verification is attained when three to five replicates of the optimized control factor set points with all the original noises applied during the experimental verification runs come out reasonably close to the predicted S/N ratio.

**Voice of the Customer (VOC)** The wants and needs of customers in their own words. The VOC is used throughout the product commercialization process to keep the engineering targets focused on the needs of the customer.

## Y

**Yates, Frank** British mathematician who had a significant effect on the early development of the science and practice of classical DOE.

**Yield** The percentage of the total number of units produced that is acceptable.

## Z

**Zero-Point Proportional** A specific and fundamental case in the dynamic approach to technology and product development. This case focuses on measuring the linearity, sensitivity, and variability associated with a design that contains a signal factor, various control factors, noise factors, and a response that is linearly related to the signal factor. The model that defines this case is $y = \beta x$. The $y$ intercept is zero.

APPENDIX B

# Quick Start Guide for WinRobust Lite

## Special Upgrade Offer for WinRobust



## Introduction

The following is information you need to quickly get up and running with WinRobust Lite. This covers the basic functions needed to set up and analyze a simple experiment.

For further details, use the online help system furnished with WinRobust Lite. This can be activated by choosing "Help" from the main menu and selecting the appropriate topic.

The furnished documentation describes all operations done by menu selections. There is also a button bar, which will shortcut many of the operations.

Play with the buttons! You can't break anything. Also learn some of the fine details of the software by trying different menu operations.

If you need a complete description of all the WinRobust features, you can get a full manual by upgrading to the full-featured version of WinRobust.

This lite version of WinRobust contains all the functions of the full WinRobust Program, but only contains the L4, L12, and the L18 orthogonal arrays. The full-featured version of WinRobust contains all the arrays found in Appendix C.

## WinRobust Lite Installation

### Hardware and Software Requirements

- IBM PC or compatible PC which can run Microsoft Windows 3.1
- 1 MB hard drive space.
- Mouse or other pointing device
- Windows 3.1 or higher, 1 MB RAM

### Auto-installation Procedure

WinRobust uses an installation program to set up the required directories and programs on your hard disk. You will also have the option to set up a program-group window with a program icon for easy launching of WinRobust from Windows.

To install WinRobust, follow these simple steps:

1. Start Windows.
2. Insert **WinRobust Disk** into the **a:** or **b:** floppy drive.
3. Select **Run** from the **File** menu.
4. Type a:install (or b:install depending on floppy drive).
5. Follow instructions during program installation.

For first-time installation, always choose "Full install." "Custom install" is used for upgrades and special situations which should normally not be needed.

## WinRobust Main Window

### Starting WinRobust

In the Program Group where the WinRobust icon resides, choose the WinRobust icon. (Double-click on the icon.) The main program window (Figure B.1) will appear.

**Figure B.1**

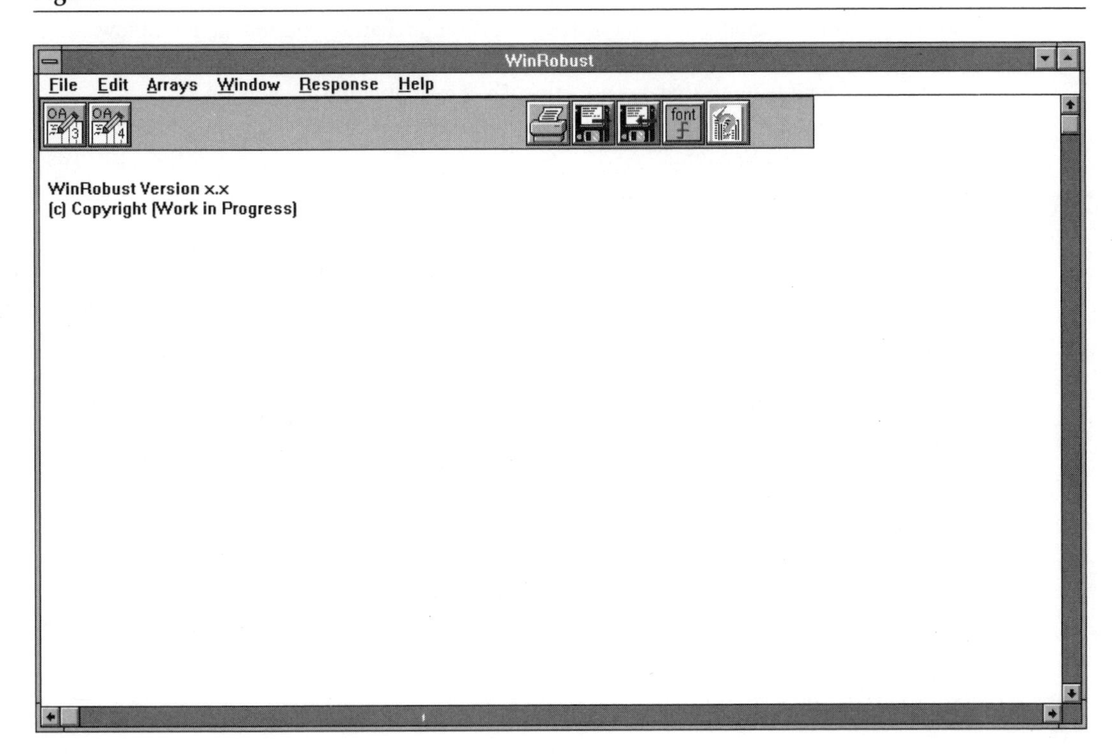


## Main Window Menu

**File**

New Erase current experiment, prepare WinRobust for new experiment.
Open Retrieves previously saved experiment.
Save Saves current experiment.
Save as Saves current experiment into specified file name.
Print Prints current contents of main window.
Exit Quits WinRobust.

**Edit:**

Copy Copies contents of main window to clipboard.
Assign Column Assigns a factor (name, level 1, level 2 . . .) to a column of an experimental array.
Unassign Column Clears a previously assigned column.
Dummy Level Modifies a column for dummy level treatment.
Preferences Sets display and print preferences of items in the main window.
Font Selects display and print font of items in the main window.
Zoom Allows zooming out from main window to display a larger field.

**Arrays:**

Standard L Selects a "standard" Taguchi-L style orthogonal array into the main window.

Modified L Selects a modified Taguchi-L style orthogonal array into the main window. Used for arrays that handle up to four-level factors. (Not available in Lite Version.)

CCD Selects a Central-Composite-Design array into the main window. Useful for modeling. (Not available in Lite Version.)

**Window:**

Results Creates and displays the Result window, which can be used to show tables of factor effects, plots of factor effects, interaction effects, and a probability plot of factor effects.

ANOVA Table Creates an ANOVA window, which displays an Analysis of Variance on the S/N factor effects.

Confirmation Creates and displays a Confirmation window, which is used to predict the response of a set of optimum factor levels and determine if the confirmation run is within acceptable limits.

**Response:**

Add Static Creates a Response window to define a static (no signal) response characteristic and enter data.

Add Dynamic Creates a Response window to define a dynamic response characteristic and enter data.

**Help:**

Accesses hypertext online help.

## Defining an Experiment

### Selecting an Inner (main) Array

To select an inner array, choose **Standard-L** from the Arrays menu. Dialogue boxes will be provided to allow selection of the desired array (Figure B.2).

For the examples that follow, we will select the L8 array from the Standard-L arrays by clicking the L8 button and then clicking on the OK button.

### Assigning Factors to Main Array

The main window should display an experiment main array after selection (Figure B.3). At this point, all the columns are unassigned. The next step is to assign control factors to columns of the main array.

The main array is displayed as shown. The triangle pointer at the top of the array is the *next suggested column* pointer, which suggests the next column to be assigned a factor based on avoiding confounding factors with two-factor interactions.

To select a column for assignment, choose it by double-clicking on it, or by selecting **Assign Column** from the **Edit** menu. This will bring up a dialogue box (Figure B.4) to define the factor name and level values for that factor.

Figure B.2 Standard-L dialogue box

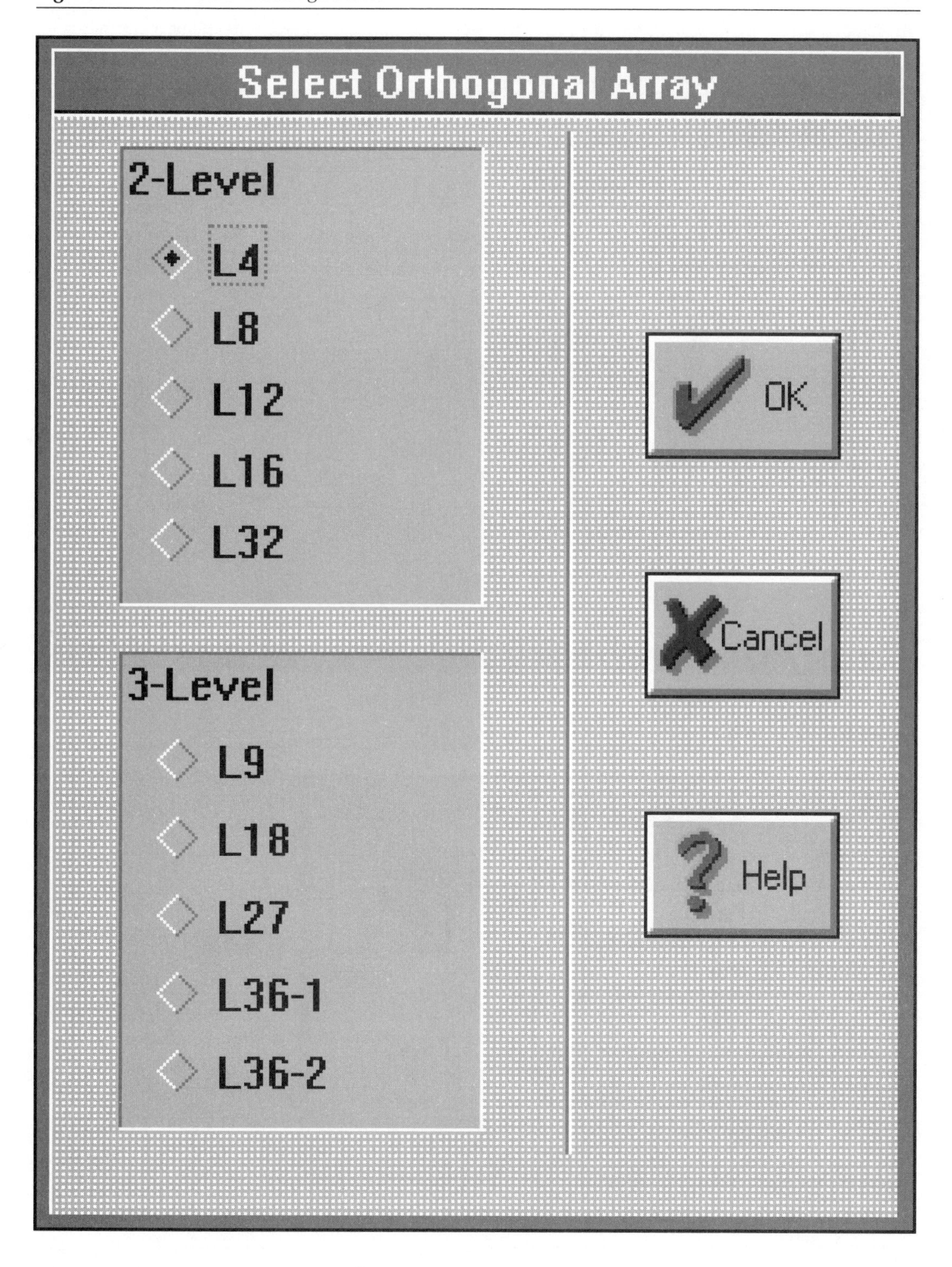

**Figure B.3** Main window showing selected main array

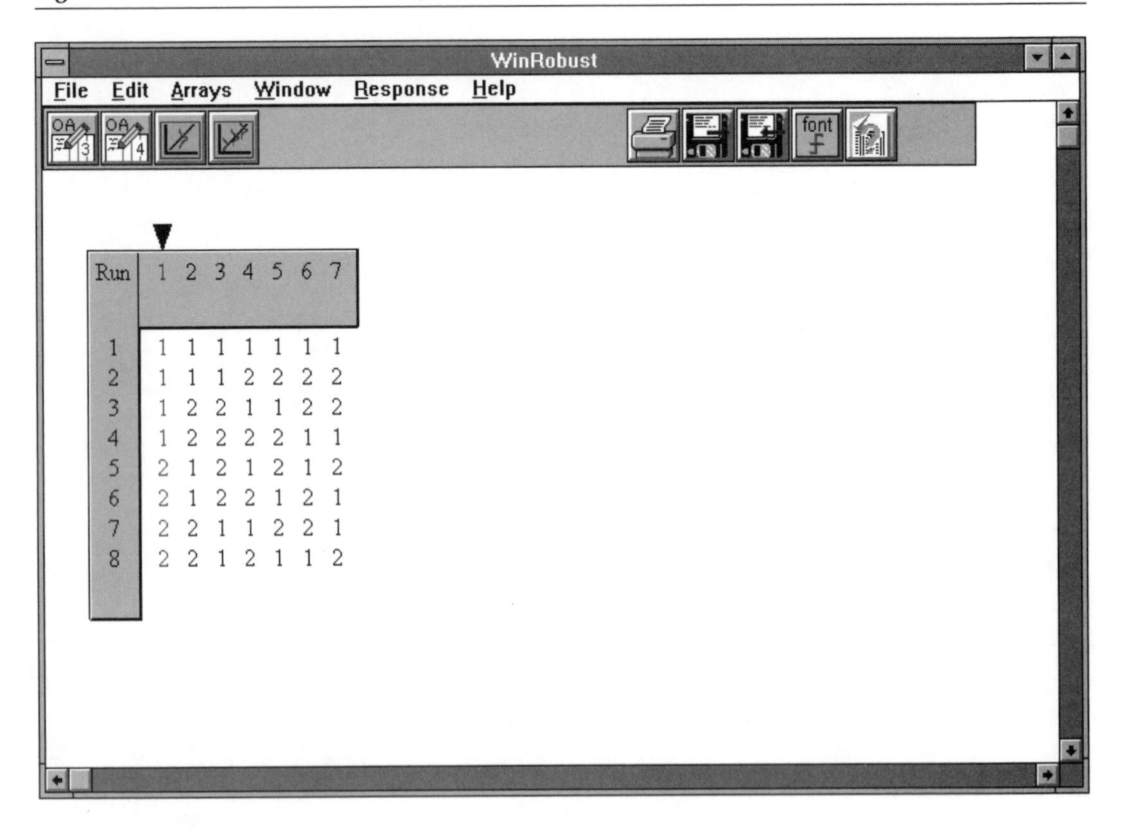


**Figure B.4** Factor dialogue box

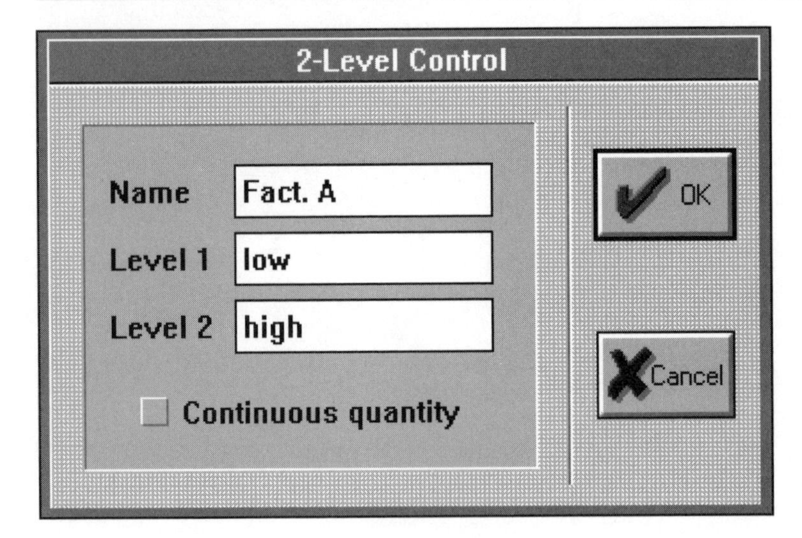

**Figure B.5**

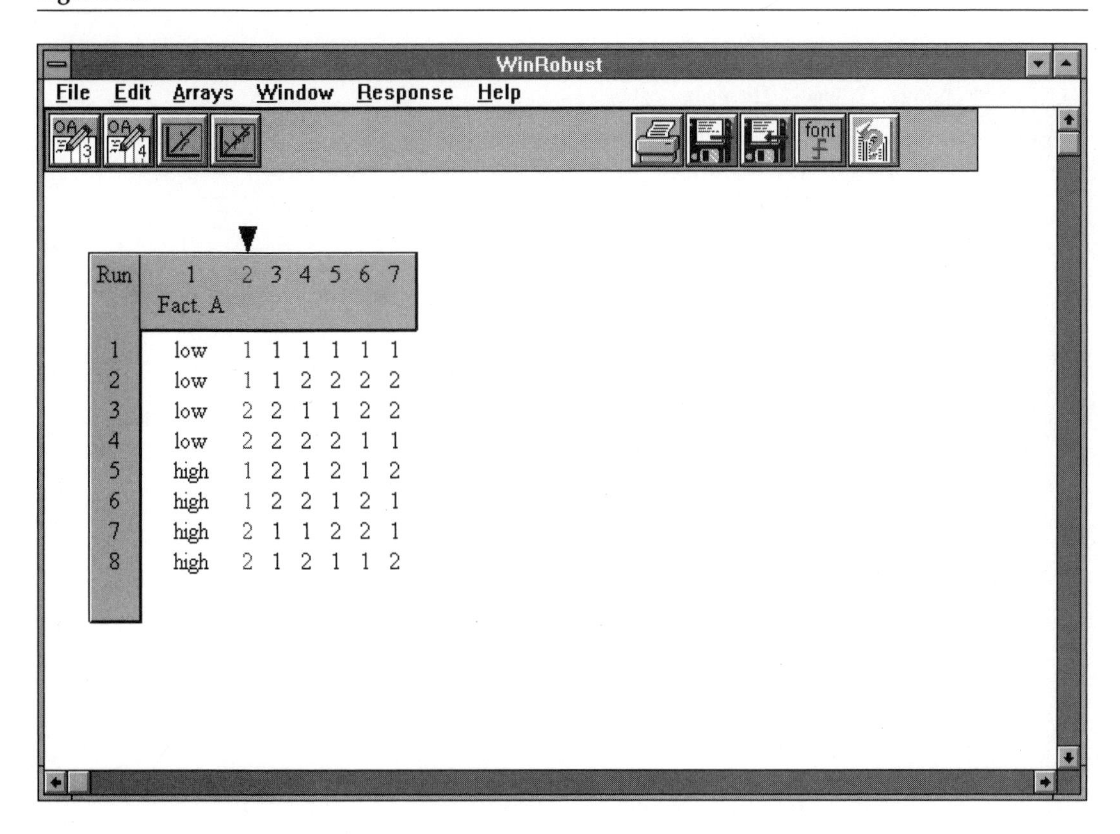


These values are then substituted for column 1 (in this case). Note that the suggested column pointer moves to column 2 (Figure B.5).

Next, another factor is assigned, this time to column 2. The main window after this operation is shown in Figure B.6.

Note that the suggested column pointer skipped column 3 and moved to column 4. Also note that (1×2) appears at the bottom of column 3. The L8 array supports the assignment of interactions, and thus WinRobust informs you that the interaction of column 1 and 2 is aliased with column 3. You may choose to override WinRobust's suggestion and assign the next factor to column 3 if you wish. However, unless you are an expert in assigning experimental arrays, it is a good idea to follow the suggestions.

## Modifying a Column—Dummy Level

Dummy level treatment can be used to assign a two-level factor to a three-level column. This is done by making two of the three levels identical.

To modify a column for dummy level treatment, choose **Dummy Level** from the **Edit** menu. A dialogue box will appear to select the column to be modified (Figure B.7).

Figure B.6

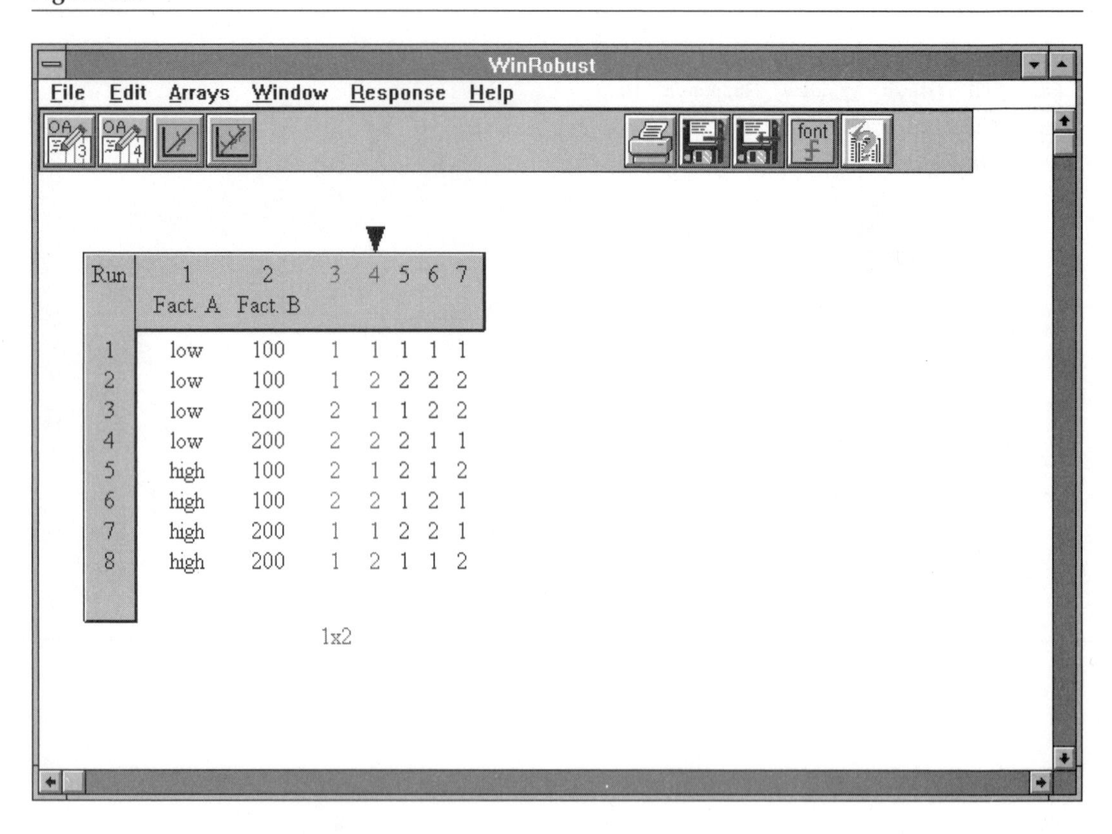


Figure B.7

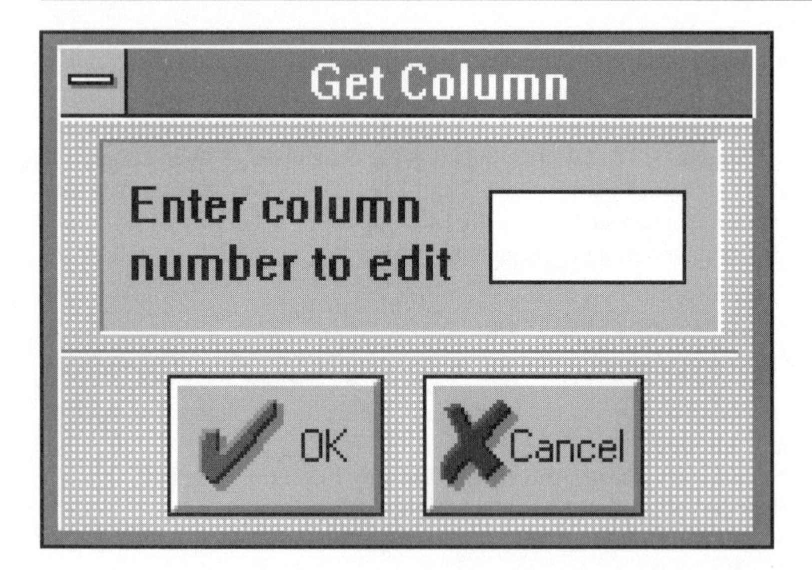

Figure B.8

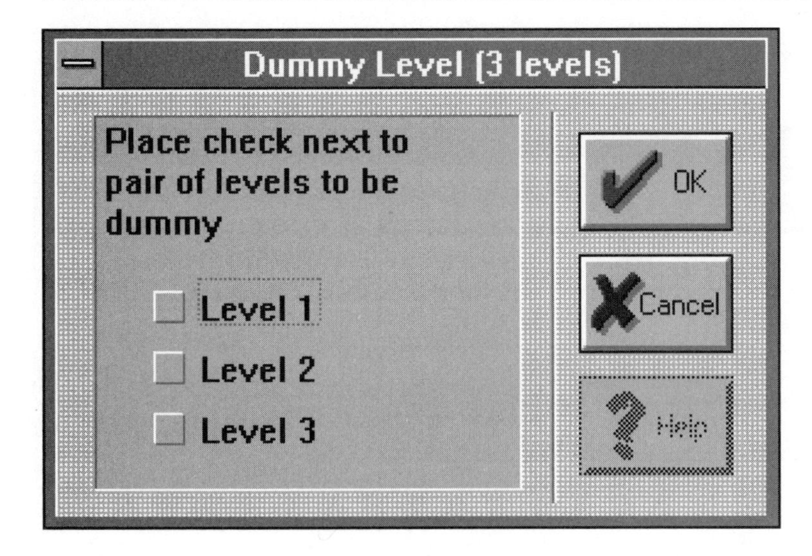


The dummy level dialogue box then appears (Figure B.8). Select the pair of columns to be "dummied" by checking the appropriate pair of boxes.

## Defining a Response

### Adding a Static Response Characteristic

A static response characteristic is a response that has either no signal factor or a fixed signal factor. To add a static response, choose **Add Static** from the **Response Menu** of the main window. A response definition dialogue box will be displayed.

With this dialogue box, the response name, number of different noise conditions, and number of replications per noise condition are defined.

The S/N type for the response characteristic is also defined here. **You must choose one of the S/N types, or an error will occur later in the program.**

Once OK is chosen, the response characteristic will be defined and a response window will be created for the defined response characteristic.

### Response Window

Figure B.9 is an example of a Response window for a static response. It consists of a spreadsheet-like interface for entering the experimental data, a "Crunch" button, and a menu.

**Figure B.9**

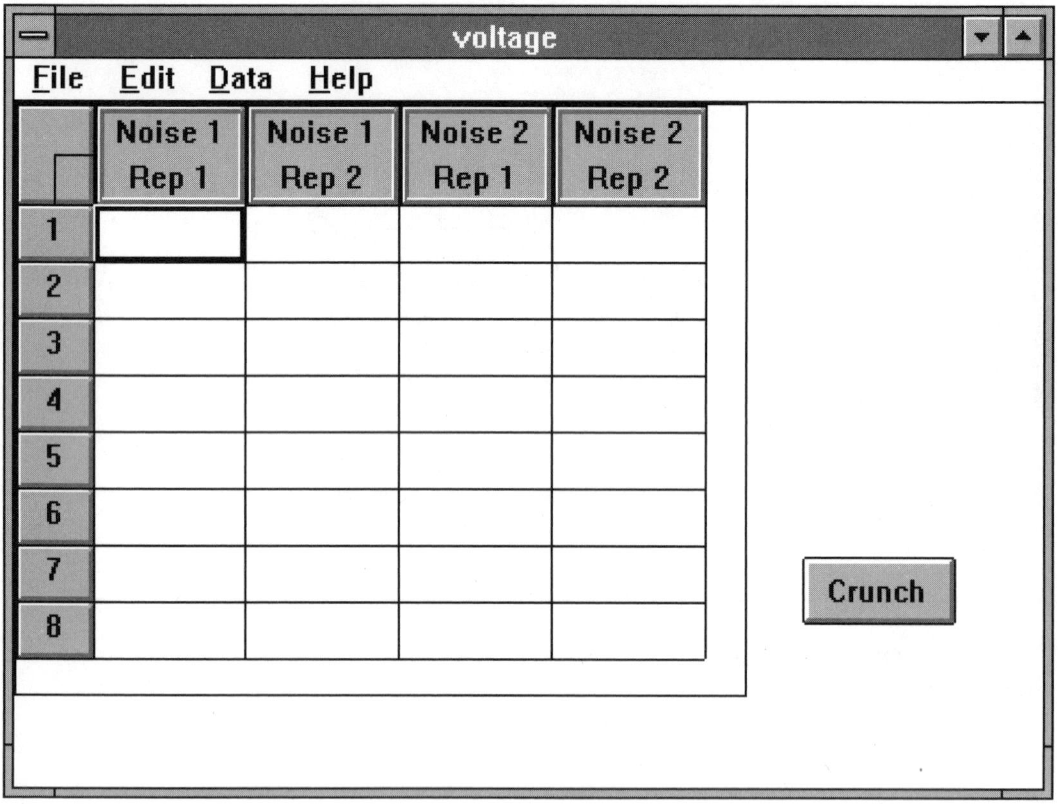

voltage

File Edit Data Help

| | Noise 1 Rep 1 | Noise 1 Rep 2 | Noise 2 Rep 1 | Noise 2 Rep 2 |
|---|---|---|---|---|
| 1 | | | | |
| 2 | | | | |
| 3 | | | | |
| 4 | | | | |
| 5 | | | | |
| 6 | | | | |
| 7 | | | | |
| 8 | | | | |

Crunch

## Entering Experimental Data Directly

To enter experimental data directly from the keyboard, take the following steps:

1. Select the block of the response window spreadsheet to receive the data by (1) selecting the whole spreadsheet by clicking on the upper left-hand corner of the spreadsheet, (2) select a column by clicking on a column heading, (3) select a row by clicking on the row number, or (4) select blocks with the mouse by clicking on a cell and dragging the mouse over the cells to be included in the block.
2. Type the data in the cell (these must be numerical data) and hit *enter* (or *return*). The cursor will automatically move to the next cell. Make certain that you hit *enter* after each data entry, or the data will not be recorded.
3. After all cells have data entered, hit the "Crunch" button to calculate statistics.

## Entering Observation Data from Clipboard

Data from other applications (spreadsheets, databases, etc.) can be entered into a Response window spreadsheet via the Clipboard. The source application must put the data into the clipboard in tab-delimited text format. Most applications do this.

1. The data block copied from the source application must be the same size as the spreadsheet in the Response window (same number of rows, same number of columns).
2. Choose **Paste** from the **Edit** menu of the Response window. Data will be transferred from the clipboard to the cells of the Response window spreadsheet.
3. Hit the "Crunch" button to calculate statistics.

## Transferring Observation Data to the Clipboard

Data can be transferred to other applications through the clipboard. Choose **Copy** from the **Edit** menu, and the Response window spreadsheet contents will be copied to the clipboard.

## Clearing Data

The Response window spreadsheet is cleared by choosing **Clear** from the **Edit** menu.

## Hiding the Response Window

The Response window is removed from view by clicking on the close box in the upper left-hand corner of the window. It is reshown by selecting the response name from the **Response** item in the main window menu.

## Adding a Dynamic Response Characteristic

A dynamic response characteristic is a response where there exists a desired relationship between a signal factor and the output. WinRobust can calculate dynamic S/N's of linear dynamic characteristics, where the output is supposed to be proportional to the signal factor. Nonlinear characteristics have to be handled as a custom S/N.

To add a dynamic response characteristic, choose **Add dynamic** from the **Response** menu of the main window. A response definition dialogue box will be displayed.

The response name, the number of signal factor levels, and the number of noise conditions per signal factor are defined here. If there are replicates per noise factor, treat the replicates as additional noise factors. **There must be a minimum of two signal factor levels and two noise factor levels per signal factor to calculate a dynamic S/N.** Currently, there are two choices for S/N calculation.

Zero-point proportional is used for dynamic characteristics where the ideal function should pass through the origin (signal = 0, output = 0). The reference-point proportional does not impose this constraint.

## Entering Signal Factor Values

Signal factors must be entered in order for the dynamic S/N ratio calculation to be done correctly. For each signal, overwrite the contents in the signal box with the numerical value of the signal factor. After all the experimental data and signal factor data are entered, hit the "Crunch" button to calculate the response statistics.

## Displaying Results and Analysis

WinRobust allows many options for presenting results of the experiment. The following are the options that are available:

- Display factor effects (response averages) in the form of a response table.
- Display factor effects in the form of response plots.
- Plot effects of two-way interactions (if experimental design allows).
- Plot factor effects on a normal probability axis for visual determination of the significance of the factor effects.

To get into the results mode of WinRobust, select the **Results** item from the **Window** menu item of the main window. The window is initially blank, because no display options have been selected yet.

### Display Response Table

To display a table of effects factor by factor, choose **Table** from the **Display** item in the Result window menu.

### Display Response Plots

To display plots of factor effects by factor, choose **Plots** from the **Display** item in the Result Window menu.

WinRobust will autosize the plots so that all the factors are displayed (printed) within the display device width.

# APPENDIX C

# Orthogonal Arrays

## The Two-Level Standard Arrays

L4*

| *Run* | *1* | *2* | *3* |
|---|---|---|---|
| 1 | 1 | 1 | 1 |
| 2 | 1 | 2 | 2 |
| 3 | 2 | 1 | 2 |
| 4 | 2 | 2 | 1 |
| | 2×3 | 1×3 | 1×2 |

L8

| *Run* | *1* | *2* | *3* | *4* | *5* | *6* | *7* |
|---|---|---|---|---|---|---|---|
| 1 | 1 | 1 | 1 | 1 | 1 | 1 | 1 |
| 2 | 1 | 1 | 1 | 2 | 2 | 2 | 2 |
| 3 | 1 | 2 | 2 | 1 | 1 | 2 | 2 |
| 4 | 1 | 2 | 2 | 2 | 2 | 1 | 1 |
| 5 | 2 | 1 | 2 | 1 | 2 | 1 | 2 |
| 6 | 2 | 1 | 2 | 2 | 1 | 2 | 1 |
| 7 | 2 | 2 | 1 | 1 | 2 | 2 | 1 |
| 8 | 2 | 2 | 1 | 2 | 1 | 1 | 2 |
| | 2×3<br>4×5<br>6×7 | 1×3<br>4×6<br>5×7 | 1×2<br>4×7<br>5×6 | 1×5<br>2×6<br>3×7 | 1×4<br>2×7<br>3×6 | 1×7<br>2×4<br>3×5 | 1×6<br>2×5<br>3×4 |

* The L4, L12, and L18 arrays are included in the WinRobust Lite Software that is included in this book. The additional arrays displayed in Appendix C are available in the extended version of WinRobust available through the address given in Appendix B.

## L12*

| *Run* | *1* | *2* | *3* | *4* | *5* | *6* | *7* | *8* | *9* | *10* | *11* |
|---|---|---|---|---|---|---|---|---|---|---|---|
| 1 | 1 | 1 | 1 | 1 | 1 | 1 | 1 | 1 | 1 | 1 | 1 |
| 2 | 1 | 1 | 1 | 1 | 1 | 2 | 2 | 2 | 2 | 2 | 2 |
| 3 | 1 | 1 | 2 | 2 | 2 | 1 | 1 | 1 | 2 | 2 | 2 |
| 4 | 1 | 2 | 1 | 2 | 2 | 1 | 2 | 2 | 1 | 1 | 2 |
| 5 | 1 | 2 | 2 | 1 | 2 | 2 | 1 | 2 | 1 | 2 | 1 |
| 6 | 1 | 2 | 2 | 2 | 1 | 2 | 2 | 1 | 2 | 1 | 1 |
| 7 | 2 | 1 | 2 | 2 | 1 | 1 | 2 | 2 | 1 | 2 | 1 |
| 8 | 2 | 1 | 2 | 1 | 2 | 2 | 2 | 1 | 1 | 1 | 2 |
| 9 | 2 | 1 | 1 | 2 | 2 | 2 | 1 | 2 | 2 | 1 | 1 |
| 10 | 2 | 2 | 2 | 1 | 1 | 1 | 1 | 2 | 2 | 1 | 2 |
| 11 | 2 | 2 | 1 | 2 | 1 | 2 | 1 | 1 | 1 | 2 | 2 |
| 12 | 2 | 2 | 1 | 1 | 2 | 1 | 2 | 1 | 2 | 2 | 1 |

The L12 is a special array in which all columns have their interactions more or less evenly distributed between them. This array cannot be used to study specific interactions. It is an excellent array to use in noise factor evaluations (see Chapter 9).

* The L4, L12, and L18 arrays are included in the WinRobust Lite Software that is included in this book. The additional arrays displayed in Appendix C are available in the extended version of WinRobust available through the address given in Appendix B.

L16

| Run | 1 | 2 | 3 | 4 | 5 | 6 | 7 | 8 | 9 | 10 | 11 | 12 | 13 | 14 | 15 |
|---|---|---|---|---|---|---|---|---|---|---|---|---|---|---|---|
| 1 | 1 | 1 | 1 | 1 | 1 | 1 | 1 | 1 | 1 | 1 | 1 | 1 | 1 | 1 | 1 |
| 2 | 1 | 1 | 1 | 1 | 1 | 1 | 1 | 2 | 2 | 2 | 2 | 2 | 2 | 2 | 2 |
| 3 | 1 | 1 | 1 | 2 | 2 | 2 | 2 | 1 | 1 | 1 | 1 | 2 | 2 | 2 | 2 |
| 4 | 1 | 1 | 1 | 2 | 2 | 2 | 2 | 2 | 2 | 2 | 2 | 1 | 1 | 1 | 1 |
| 5 | 1 | 2 | 2 | 1 | 1 | 2 | 2 | 1 | 1 | 2 | 2 | 1 | 1 | 2 | 2 |
| 6 | 1 | 2 | 2 | 1 | 1 | 2 | 2 | 2 | 2 | 1 | 1 | 2 | 2 | 1 | 1 |
| 7 | 1 | 2 | 2 | 2 | 2 | 1 | 1 | 1 | 1 | 2 | 2 | 2 | 2 | 1 | 1 |
| 8 | 1 | 2 | 2 | 2 | 2 | 1 | 1 | 2 | 2 | 1 | 1 | 1 | 1 | 2 | 2 |
| 9 | 2 | 1 | 2 | 1 | 2 | 1 | 2 | 1 | 2 | 1 | 2 | 1 | 2 | 1 | 2 |
| 10 | 2 | 1 | 2 | 1 | 2 | 1 | 2 | 2 | 1 | 2 | 1 | 2 | 1 | 2 | 1 |
| 11 | 2 | 1 | 2 | 2 | 1 | 2 | 1 | 1 | 2 | 1 | 2 | 2 | 1 | 2 | 1 |
| 12 | 2 | 1 | 2 | 2 | 1 | 2 | 1 | 2 | 1 | 2 | 1 | 1 | 2 | 1 | 2 |
| 13 | 2 | 2 | 1 | 1 | 2 | 2 | 1 | 1 | 2 | 2 | 1 | 1 | 2 | 2 | 1 |
| 14 | 2 | 2 | 1 | 1 | 2 | 2 | 1 | 2 | 1 | 1 | 2 | 2 | 1 | 1 | 2 |
| 15 | 2 | 2 | 1 | 2 | 1 | 1 | 2 | 1 | 2 | 2 | 1 | 2 | 1 | 1 | 2 |
| 16 | 2 | 2 | 1 | 2 | 1 | 1 | 2 | 2 | 1 | 1 | 2 | 1 | 2 | 2 | 1 |
|  | 2×3 | 1×3 | 1×2 | 1×5 | 1×4 | 1×7 | 1×6 | 1×9 | 1×8 | 1×11 | 1×10 | 1×13 | 1×12 | 1×15 | 1×14 |
|  | 4×5 | 4×6 | 4×7 | 2×6 | 2×7 | 2×4 | 2×5 | 2×10 | 2×11 | 2×8 | 2×9 | 2×14 | 2×15 | 2×12 | 2×13 |
|  | 6×7 | 5×7 | 5×6 | 3×7 | 3×6 | 3×5 | 3×4 | 3×11 | 3×10 | 3×9 | 3×8 | 3×15 | 3×14 | 3×13 | 3×12 |
|  | 8×9 | 8×10 | 8×11 | 8×12 | 8×13 | 8×14 | 8×15 | 4×12 | 4×13 | 4×14 | 4×15 | 4×8 | 4×9 | 4×10 | 4×11 |
|  | 10×11 | 9×11 | 9×10 | 9×13 | 9×12 | 9×15 | 9×14 | 5×13 | 5×12 | 5×15 | 5×14 | 5×9 | 5×8 | 5×11 | 5×10 |
|  | 12×13 | 12×14 | 12×15 | 10×14 | 10×15 | 10×12 | 10×13 | 6×14 | 6×15 | 6×12 | 6×13 | 6×10 | 6×11 | 6×8 | 6×9 |
|  | 14×15 | 13×15 | 13×14 | 11×15 | 11×14 | 11×13 | 11×12 | 7×15 | 7×14 | 7×13 | 7×12 | 7×11 | 7×10 | 7×9 | 7×8 |

L32

| Run | 1 | 2 | 3 | 4 | 5 | 6 | 7 | 8 | 9 | 10 | 11 | 12 | 13 | 14 | 15 | 16 | 17 | 18 | 19 | 20 | 21 | 22 | 23 | 24 | 25 | 26 | 27 | 28 | 29 | 30 | 31 |
|---|---|---|---|---|---|---|---|---|---|---|---|---|---|---|---|---|---|---|---|---|---|---|---|---|---|---|---|---|---|---|---|
| 1 | 1 | 1 | 1 | 1 | 1 | 1 | 1 | 1 | 1 | 1 | 1 | 1 | 1 | 1 | 1 | 1 | 1 | 1 | 1 | 1 | 1 | 1 | 1 | 1 | 1 | 1 | 1 | 1 | 1 | 1 | 1 |
| 2 | 1 | 1 | 1 | 1 | 1 | 1 | 1 | 1 | 1 | 1 | 1 | 1 | 1 | 1 | 1 | 2 | 2 | 2 | 2 | 2 | 2 | 2 | 2 | 2 | 2 | 2 | 2 | 2 | 2 | 2 | 2 |
| 3 | 1 | 1 | 1 | 1 | 1 | 1 | 1 | 2 | 2 | 2 | 2 | 2 | 2 | 2 | 2 | 1 | 1 | 1 | 1 | 1 | 1 | 1 | 1 | 2 | 2 | 2 | 2 | 2 | 2 | 2 | 2 |
| 4 | 1 | 1 | 1 | 1 | 1 | 1 | 1 | 2 | 2 | 2 | 2 | 2 | 2 | 2 | 2 | 2 | 2 | 2 | 2 | 2 | 2 | 2 | 2 | 1 | 1 | 1 | 1 | 1 | 1 | 1 | 1 |
| 5 | 1 | 1 | 1 | 2 | 2 | 2 | 2 | 1 | 1 | 1 | 1 | 2 | 2 | 2 | 2 | 1 | 1 | 1 | 1 | 2 | 2 | 2 | 2 | 1 | 1 | 1 | 1 | 2 | 2 | 2 | 2 |
| 6 | 1 | 1 | 1 | 2 | 2 | 2 | 2 | 1 | 1 | 1 | 1 | 2 | 2 | 2 | 2 | 2 | 2 | 2 | 2 | 1 | 1 | 1 | 1 | 2 | 2 | 2 | 2 | 1 | 1 | 1 | 1 |
| 7 | 1 | 1 | 1 | 2 | 2 | 2 | 2 | 2 | 2 | 2 | 2 | 1 | 1 | 1 | 1 | 1 | 1 | 1 | 1 | 2 | 2 | 2 | 2 | 2 | 2 | 2 | 2 | 1 | 1 | 1 | 1 |
| 8 | 1 | 1 | 1 | 2 | 2 | 2 | 2 | 2 | 2 | 2 | 2 | 1 | 1 | 1 | 1 | 2 | 2 | 2 | 2 | 1 | 1 | 1 | 1 | 1 | 1 | 1 | 1 | 2 | 2 | 2 | 2 |
| 9 | 1 | 2 | 2 | 1 | 1 | 2 | 2 | 1 | 1 | 2 | 2 | 1 | 1 | 2 | 2 | 1 | 1 | 2 | 2 | 1 | 1 | 2 | 2 | 1 | 1 | 2 | 2 | 1 | 1 | 2 | 2 |
| 10 | 1 | 2 | 2 | 1 | 1 | 2 | 2 | 1 | 1 | 2 | 2 | 1 | 1 | 2 | 2 | 2 | 2 | 1 | 1 | 2 | 2 | 1 | 1 | 2 | 2 | 1 | 1 | 2 | 2 | 1 | 1 |
| 11 | 1 | 2 | 2 | 1 | 1 | 2 | 2 | 2 | 2 | 1 | 1 | 2 | 2 | 1 | 1 | 1 | 1 | 2 | 2 | 1 | 1 | 2 | 2 | 2 | 2 | 1 | 1 | 2 | 2 | 1 | 1 |
| 12 | 1 | 2 | 2 | 1 | 1 | 2 | 2 | 2 | 2 | 1 | 1 | 2 | 2 | 1 | 1 | 2 | 2 | 1 | 1 | 2 | 2 | 1 | 1 | 1 | 1 | 2 | 2 | 1 | 1 | 2 | 2 |
| 13 | 1 | 2 | 2 | 2 | 2 | 1 | 1 | 1 | 1 | 2 | 2 | 2 | 2 | 1 | 1 | 1 | 1 | 2 | 2 | 2 | 2 | 1 | 1 | 1 | 1 | 2 | 2 | 2 | 2 | 1 | 1 |
| 14 | 1 | 2 | 2 | 2 | 2 | 1 | 1 | 1 | 1 | 2 | 2 | 2 | 2 | 1 | 1 | 2 | 2 | 1 | 1 | 1 | 1 | 2 | 2 | 2 | 2 | 1 | 1 | 1 | 1 | 2 | 2 |
| 15 | 1 | 2 | 2 | 2 | 2 | 1 | 1 | 2 | 2 | 1 | 1 | 1 | 1 | 2 | 2 | 1 | 1 | 2 | 2 | 2 | 2 | 1 | 1 | 2 | 2 | 1 | 1 | 1 | 1 | 2 | 2 |
| 16 | 1 | 2 | 2 | 2 | 2 | 1 | 1 | 2 | 2 | 1 | 1 | 1 | 1 | 2 | 2 | 2 | 2 | 1 | 1 | 1 | 1 | 2 | 2 | 1 | 1 | 2 | 2 | 2 | 2 | 1 | 1 |

| | | | | | | | | | | | | | | | | | | | | | | | | | | | | | | | |
|---|---|---|---|---|---|---|---|---|---|---|---|---|---|---|---|---|---|---|---|---|---|---|---|---|---|---|---|---|---|---|---|
| 17 | 2 | 1 | 2 | 1 | 2 | 1 | 2 | 1 | 2 | 1 | 2 | 1 | 2 | 1 | 2 | 1 | 2 | 1 | 2 | 1 | 2 | 1 | 2 | 1 | 2 | 1 | 2 | 1 | 2 | 1 | 2 |
| 18 | 2 | 1 | 2 | 1 | 2 | 1 | 2 | 1 | 2 | 1 | 2 | 1 | 2 | 1 | 2 | 2 | 1 | 2 | 1 | 2 | 1 | 2 | 1 | 2 | 1 | 2 | 1 | 2 | 1 | 2 | 1 |
| 19 | 2 | 1 | 2 | 1 | 2 | 1 | 2 | 2 | 1 | 2 | 1 | 2 | 1 | 2 | 1 | 1 | 2 | 1 | 2 | 1 | 2 | 1 | 2 | 2 | 1 | 2 | 1 | 2 | 1 | 2 | 1 |
| 20 | 2 | 1 | 2 | 1 | 2 | 1 | 2 | 2 | 1 | 2 | 1 | 2 | 1 | 2 | 1 | 2 | 1 | 2 | 1 | 2 | 1 | 2 | 1 | 1 | 2 | 1 | 2 | 1 | 2 | 1 | 2 |
| 21 | 2 | 1 | 2 | 2 | 1 | 2 | 1 | 1 | 2 | 1 | 2 | 2 | 1 | 2 | 1 | 1 | 2 | 1 | 2 | 2 | 1 | 2 | 1 | 1 | 2 | 1 | 2 | 2 | 1 | 2 | 1 |
| 22 | 2 | 1 | 2 | 2 | 1 | 2 | 1 | 1 | 2 | 1 | 2 | 2 | 1 | 2 | 1 | 2 | 1 | 2 | 1 | 1 | 2 | 1 | 2 | 2 | 1 | 2 | 1 | 1 | 2 | 1 | 2 |
| 23 | 2 | 1 | 2 | 2 | 1 | 2 | 1 | 2 | 1 | 2 | 1 | 1 | 2 | 1 | 2 | 1 | 2 | 1 | 2 | 2 | 1 | 2 | 1 | 2 | 1 | 2 | 1 | 1 | 2 | 1 | 2 |
| 24 | 2 | 1 | 2 | 2 | 1 | 2 | 1 | 2 | 1 | 2 | 1 | 1 | 2 | 1 | 2 | 2 | 1 | 2 | 1 | 1 | 2 | 1 | 2 | 1 | 2 | 1 | 2 | 2 | 1 | 2 | 1 |
| 25 | 2 | 2 | 1 | 1 | 2 | 2 | 1 | 1 | 2 | 2 | 1 | 1 | 2 | 2 | 1 | 1 | 2 | 2 | 1 | 1 | 2 | 2 | 1 | 1 | 2 | 2 | 1 | 1 | 2 | 2 | 1 |
| 26 | 2 | 2 | 1 | 1 | 2 | 2 | 1 | 1 | 2 | 2 | 1 | 1 | 2 | 2 | 1 | 2 | 1 | 1 | 2 | 2 | 1 | 1 | 2 | 2 | 1 | 1 | 2 | 2 | 1 | 1 | 2 |
| 27 | 2 | 2 | 1 | 1 | 2 | 2 | 1 | 2 | 1 | 1 | 2 | 2 | 1 | 1 | 2 | 1 | 2 | 2 | 1 | 1 | 2 | 2 | 1 | 2 | 1 | 1 | 2 | 2 | 1 | 1 | 2 |
| 28 | 2 | 2 | 1 | 1 | 2 | 2 | 1 | 2 | 1 | 1 | 2 | 2 | 1 | 1 | 2 | 2 | 1 | 1 | 2 | 2 | 1 | 1 | 2 | 1 | 2 | 2 | 1 | 1 | 2 | 2 | 1 |
| 29 | 2 | 2 | 1 | 2 | 1 | 1 | 2 | 1 | 2 | 2 | 1 | 2 | 1 | 1 | 2 | 1 | 2 | 2 | 1 | 2 | 1 | 1 | 2 | 1 | 2 | 2 | 1 | 2 | 1 | 1 | 2 |
| 30 | 2 | 2 | 1 | 2 | 1 | 1 | 2 | 1 | 2 | 2 | 1 | 2 | 1 | 1 | 2 | 2 | 1 | 1 | 2 | 1 | 2 | 2 | 1 | 2 | 1 | 1 | 2 | 1 | 2 | 2 | 1 |
| 31 | 2 | 2 | 1 | 2 | 1 | 1 | 2 | 2 | 1 | 1 | 2 | 1 | 2 | 2 | 1 | 1 | 2 | 2 | 1 | 2 | 1 | 1 | 2 | 2 | 1 | 1 | 2 | 1 | 2 | 2 | 1 |
| 32 | 2 | 2 | 1 | 2 | 1 | 1 | 2 | 2 | 1 | 1 | 2 | 1 | 2 | 2 | 1 | 2 | 1 | 1 | 2 | 1 | 2 | 2 | 1 | 1 | 2 | 2 | 1 | 2 | 1 | 1 | 2 |

## Interactions for Columns of the L32 Array

| 1 | 2 | 3 | 4 | 5 | 6 | 7 | 8 | 9 | 10 | 11 | 12 | 13 | 14 | 15 |
|---|---|---|---|---|---|---|---|---|---|---|---|---|---|---|
| 2×3 | 1×3 | 1×2 | 1×5 | 1×4 | 1×7 | 1×6 | 1×9 | 1×8 | 1×11 | 1×10 | 1×13 | 1×12 | 1×15 | 1×14 |
| 4×5 | 4×6 | 4×7 | 2×6 | 2×7 | 2×4 | 2×5 | 2×10 | 2×11 | 2×8 | 2×9 | 2×14 | 2×15 | 2×12 | 2×13 |
| 6×7 | 5×7 | 5×6 | 3×7 | 3×6 | 3×5 | 3×4 | 3×11 | 3×10 | 3×9 | 3×8 | 3×15 | 3×14 | 3×13 | 3×12 |
| 8×9 | 8×10 | 8×11 | 8×12 | 8×13 | 8×14 | 8×15 | 4×12 | 4×13 | 4×14 | 4×15 | 4×8 | 4×9 | 4×10 | 4×11 |
| 10×11 | 9×11 | 9×10 | 9×13 | 9×12 | 9×15 | 9×14 | 5×13 | 5×12 | 5×15 | 5×14 | 5×9 | 5×8 | 5×11 | 5×10 |
| 12×13 | 12×14 | 12×15 | 10×14 | 10×15 | 10×12 | 10×13 | 6×14 | 6×15 | 6×12 | 6×13 | 6×10 | 6×11 | 6×8 | 6×9 |
| 14×15 | 13×15 | 13×14 | 11×15 | 11×14 | 11×13 | 11×12 | 7×15 | 7×14 | 7×13 | 7×12 | 7×11 | 7×10 | 7×9 | 7×8 |
| 16×17 | 16×18 | 16×19 | 16×20 | 16×21 | 16×22 | 16×23 | 16×24 | 16×25 | 16×26 | 16×27 | 16×28 | 16×29 | 16×30 | 16×31 |
| 18×19 | 17×19 | 17×18 | 17×21 | 17×20 | 17×23 | 17×22 | 17×25 | 17×24 | 17×27 | 17×26 | 17×29 | 17×28 | 17×31 | 17×30 |
| 20×21 | 20×22 | 20×23 | 18×22 | 18×23 | 18×20 | 18×21 | 18×26 | 18×27 | 18×24 | 18×25 | 18×30 | 18×31 | 18×28 | 18×29 |
| 22×23 | 21×23 | 21×22 | 19×23 | 19×22 | 19×21 | 19×20 | 19×27 | 19×26 | 19×25 | 19×24 | 19×31 | 19×30 | 19×29 | 19×28 |
| 24×25 | 24×26 | 24×27 | 24×28 | 24×29 | 24×30 | 24×31 | 20×28 | 20×29 | 20×30 | 20×31 | 20×24 | 20×25 | 20×26 | 20×27 |
| 26×27 | 25×27 | 25×26 | 25×29 | 25×28 | 25×31 | 25×30 | 21×29 | 21×28 | 21×31 | 21×30 | 21×25 | 21×24 | 21×27 | 21×26 |
| 28×29 | 28×30 | 28×31 | 26×30 | 26×31 | 26×28 | 26×29 | 22×30 | 22×31 | 22×28 | 22×29 | 22×26 | 22×27 | 22×24 | 22×25 |
| 30×31 | 29×31 | 29×30 | 27×31 | 27×30 | 27×29 | 27×28 | 23×31 | 23×30 | 23×29 | 23×28 | 23×27 | 23×26 | 23×25 | 23×24 |

| *16* | *17* | *18* | *19* | *20* | *21* | *22* | *23* | *24* | *25* | *26* | *27* | *28* | *29* | *30* | *31* |
|---|---|---|---|---|---|---|---|---|---|---|---|---|---|---|---|
| 1×17 | 1×16 | 1×19 | 1×18 | 1×21 | 1×20 | 1×23 | 1×22 | 1×25 | 1×24 | 1×27 | 1×26 | 1×29 | 1×28 | 1×31 | 1×30 |
| 2×18 | 2×19 | 2×16 | 2×17 | 2×22 | 2×23 | 2×20 | 2×21 | 2×26 | 2×27 | 2×24 | 2×25 | 2×30 | 2×31 | 2×28 | 2×29 |
| 3×19 | 3×18 | 3×17 | 3×16 | 3×23 | 3×22 | 3×21 | 3×20 | 3×27 | 3×26 | 3×25 | 3×24 | 3×31 | 3×30 | 3×29 | 3×28 |
| 4×20 | 4×21 | 4×22 | 4×23 | 4×16 | 4×17 | 4×18 | 4×19 | 4×28 | 4×29 | 4×30 | 4×31 | 4×24 | 4×25 | 4×26 | 4×27 |
| 5×21 | 5×20 | 5×23 | 5×22 | 5×17 | 5×16 | 5×19 | 5×18 | 5×29 | 5×28 | 5×31 | 5×30 | 5×25 | 5×24 | 5×27 | 5×26 |
| 6×22 | 6×23 | 6×20 | 6×21 | 6×18 | 6×19 | 6×16 | 6×17 | 6×30 | 6×31 | 6×28 | 6×29 | 6×26 | 6×27 | 6×24 | 6×25 |
| 7×23 | 7×22 | 7×21 | 7×20 | 7×19 | 7×18 | 7×17 | 7×16 | 7×31 | 7×30 | 7×29 | 7×28 | 7×27 | 7×26 | 7×25 | 7×24 |
| 8×24 | 8×25 | 8×26 | 8×27 | 8×28 | 8×29 | 8×30 | 8×31 | 8×16 | 8×17 | 8×18 | 8×19 | 8×20 | 8×21 | 8×22 | 8×23 |
| 9×25 | 9×24 | 9×27 | 9×26 | 9×29 | 9×28 | 9×31 | 9×30 | 9×17 | 9×16 | 9×19 | 9×18 | 9×21 | 9×20 | 9×23 | 9×22 |
| 10×26 | 10×27 | 10×24 | 10×25 | 10×30 | 10×31 | 10×28 | 10×29 | 10×18 | 10×19 | 10×16 | 10×17 | 10×22 | 10×23 | 10×20 | 10×21 |
| 11×27 | 11×26 | 11×25 | 11×24 | 11×31 | 11×30 | 11×29 | 11×28 | 11×19 | 11×18 | 11×17 | 11×16 | 11×23 | 11×22 | 11×21 | 11×20 |
| 12×28 | 12×29 | 12×30 | 12×31 | 12×24 | 12×25 | 12×26 | 12×27 | 12×20 | 12×21 | 12×22 | 12×23 | 12×16 | 12×17 | 12×18 | 12×19 |
| 13×29 | 13×28 | 13×31 | 13×30 | 13×25 | 13×24 | 13×27 | 13×26 | 13×21 | 13×20 | 13×23 | 13×22 | 13×17 | 13×16 | 13×19 | 13×18 |
| 14×30 | 14×31 | 14×28 | 14×29 | 14×26 | 14×27 | 14×24 | 14×25 | 14×22 | 14×23 | 14×20 | 14×21 | 14×18 | 14×19 | 14×16 | 14×17 |
| 15×31 | 15×30 | 15×29 | 15×28 | 15×27 | 15×26 | 15×25 | 15×24 | 15×23 | 15×22 | 15×21 | 15×20 | 15×19 | 15×18 | 15×17 | 15×16 |

## The Three-Level Arrays

L9

| Run | 1 | 2 | 3 | 4 |
|---|---|---|---|---|
| 1 | 1 | 1 | 1 | 1 |
| 2 | 1 | 2 | 2 | 2 |
| 3 | 1 | 3 | 3 | 3 |
| 4 | 2 | 1 | 2 | 3 |
| 5 | 2 | 2 | 3 | 1 |
| 6 | 2 | 3 | 1 | 2 |
| 7 | 3 | 1 | 3 | 2 |
| 8 | 3 | 2 | 1 | 3 |
| 9 | 3 | 3 | 2 | 1 |
| | 2×3<br>2×4<br>3×4 | 1×3<br>1×4<br>3×4 | 1×2<br>1×4<br>2×4 | 1×2<br>1×3<br>2×3 |

Notice the need to consume two columns (3 and 4) to estimate the interaction between columns 1 and 2.

## L18*

| *Run* | *1* | *2* | *3* | *4* | *5* | *6* | *7* | *8* |
|---|---|---|---|---|---|---|---|---|
| 1 | 1 | 1 | 1 | 1 | 1 | 1 | 1 | 1 |
| 2 | 1 | 1 | 2 | 2 | 2 | 2 | 2 | 2 |
| 3 | 1 | 1 | 3 | 3 | 3 | 3 | 3 | 3 |
| 4 | 1 | 2 | 1 | 1 | 2 | 2 | 3 | 3 |
| 5 | 1 | 2 | 2 | 2 | 3 | 3 | 1 | 1 |
| 6 | 1 | 2 | 3 | 3 | 1 | 1 | 2 | 2 |
| 7 | 1 | 3 | 1 | 2 | 1 | 3 | 2 | 3 |
| 8 | 1 | 3 | 2 | 3 | 2 | 1 | 3 | 1 |
| 9 | 1 | 3 | 3 | 1 | 3 | 2 | 1 | 2 |
| 10 | 2 | 1 | 1 | 3 | 3 | 2 | 2 | 1 |
| 11 | 2 | 1 | 2 | 1 | 1 | 3 | 3 | 2 |
| 12 | 2 | 1 | 3 | 2 | 2 | 1 | 1 | 3 |
| 13 | 2 | 2 | 1 | 2 | 3 | 1 | 3 | 2 |
| 14 | 2 | 2 | 2 | 3 | 1 | 2 | 1 | 3 |
| 15 | 2 | 2 | 3 | 1 | 2 | 3 | 2 | 1 |
| 16 | 2 | 3 | 1 | 3 | 2 | 3 | 1 | 2 |
| 17 | 2 | 3 | 2 | 1 | 3 | 1 | 2 | 3 |
| 18 | 2 | 3 | 3 | 2 | 1 | 2 | 3 | 1 |

The only interaction that can be studied with the L18 array is between columns 1 and 2. There is no need to consume any other columns to do this evaluation. The rest of the columns are uniformly confounded in a fashion similar to the L12. Interactions between columns 3–8 cannot be evaluated.

* The L4, L12, and L18 arrays are included in the WinRobust Lite Software that is included in this book. The additional arrays displayed in Appendix C are available in the extended version of WinRobust available through the address given in Appendix B.

## L27

| *Run* | *1* | *2* | *3* | *4* | *5* | *6* | *7* | *8* | *9* | *10* | *11* | *12* | *13* |
|---|---|---|---|---|---|---|---|---|---|---|---|---|---|
| 1 | 1 | 1 | 1 | 1 | 1 | 1 | 1 | 1 | 1 | 1 | 1 | 1 | 1 |
| 2 | 1 | 1 | 1 | 1 | 2 | 2 | 2 | 2 | 2 | 2 | 2 | 2 | 2 |
| 3 | 1 | 1 | 1 | 1 | 3 | 3 | 3 | 3 | 3 | 3 | 3 | 3 | 3 |
| 4 | 1 | 2 | 2 | 2 | 1 | 1 | 1 | 2 | 2 | 2 | 3 | 3 | 3 |
| 5 | 1 | 2 | 2 | 2 | 2 | 2 | 2 | 3 | 3 | 3 | 1 | 1 | 1 |
| 6 | 1 | 2 | 2 | 2 | 3 | 3 | 3 | 1 | 1 | 1 | 2 | 2 | 2 |
| 7 | 1 | 3 | 3 | 3 | 1 | 1 | 1 | 3 | 3 | 3 | 2 | 2 | 2 |
| 8 | 1 | 3 | 3 | 3 | 2 | 2 | 2 | 1 | 1 | 1 | 3 | 3 | 3 |
| 9 | 1 | 3 | 3 | 3 | 3 | 3 | 3 | 2 | 2 | 2 | 1 | 1 | 1 |
| 10 | 2 | 1 | 2 | 3 | 1 | 2 | 3 | 1 | 2 | 3 | 1 | 2 | 3 |
| 11 | 2 | 1 | 2 | 3 | 2 | 3 | 1 | 2 | 3 | 1 | 2 | 3 | 1 |
| 12 | 2 | 1 | 2 | 3 | 3 | 1 | 2 | 3 | 1 | 2 | 3 | 1 | 2 |
| 13 | 2 | 2 | 3 | 1 | 1 | 2 | 3 | 2 | 3 | 1 | 3 | 1 | 2 |
| 14 | 2 | 2 | 3 | 1 | 2 | 3 | 1 | 3 | 1 | 2 | 1 | 2 | 3 |
| 15 | 2 | 2 | 3 | 1 | 3 | 1 | 2 | 1 | 2 | 3 | 2 | 3 | 1 |
| 16 | 2 | 3 | 1 | 2 | 1 | 2 | 3 | 3 | 1 | 2 | 2 | 3 | 1 |
| 17 | 2 | 3 | 1 | 2 | 2 | 3 | 1 | 1 | 2 | 3 | 3 | 1 | 2 |
| 18 | 2 | 3 | 1 | 2 | 3 | 1 | 2 | 2 | 3 | 1 | 1 | 2 | 3 |
| 19 | 3 | 1 | 3 | 2 | 1 | 3 | 2 | 1 | 3 | 2 | 1 | 3 | 2 |
| 20 | 3 | 1 | 3 | 2 | 2 | 1 | 3 | 2 | 1 | 3 | 2 | 1 | 3 |
| 21 | 3 | 1 | 3 | 2 | 3 | 2 | 1 | 3 | 2 | 1 | 3 | 2 | 1 |
| 22 | 3 | 2 | 1 | 3 | 1 | 3 | 2 | 2 | 1 | 3 | 3 | 2 | 1 |
| 23 | 3 | 2 | 1 | 3 | 2 | 1 | 3 | 3 | 2 | 1 | 1 | 3 | 2 |
| 24 | 3 | 2 | 1 | 3 | 3 | 2 | 1 | 1 | 3 | 2 | 2 | 1 | 3 |
| 25 | 3 | 3 | 2 | 1 | 1 | 3 | 2 | 3 | 2 | 1 | 2 | 1 | 3 |
| 26 | 3 | 3 | 2 | 1 | 2 | 1 | 3 | 1 | 3 | 2 | 3 | 2 | 1 |
| 27 | 3 | 3 | 2 | 1 | 3 | 2 | 1 | 2 | 1 | 3 | 1 | 3 | 2 |

## Interactions for Columns of the L27 array

| 1 | 2 | 3 | 4 | 5 | 6 | 7 | 8 | 9 | 10 | 11 | 12 | 13 |
|---|---|---|---|---|---|---|---|---|---|---|---|---|
| 2×3 | 1×3 | 1×2 | 1×2 | 1×6 | 1×5 | 1×5 | 1×9 | 1×8 | 1×8 | 1×12 | 1×11 | 1×11 |
| 2×4 | 1×4 | 1×4 | 1×3 | 1×7 | 1×7 | 1×6 | 1×10 | 1×10 | 1×9 | 1×13 | 1×13 | 1×12 |
| 3×4 | 3×4 | 2×4 | 2×3 | 2×8 | 2×9 | 2×10 | 2×5 | 2×6 | 2×7 | 2×5 | 2×6 | 2×7 |
| 5×6 | 5×8 | 5×9 | 5×10 | 2×11 | 2×12 | 2×13 | 2×11 | 2×12 | 2×13 | 2×8 | 2×9 | 2×10 |
| 5×7 | 5×11 | 5×13 | 5×12 | 3×9 | 3×10 | 3×8 | 3×7 | 3×5 | 3×6 | 3×6 | 3×7 | 3×5 |
| 6×7 | 6×9 | 6×10 | 6×8 | 3×13 | 3×11 | 3×12 | 3×12 | 3×13 | 3×11 | 3×10 | 3×8 | 3×9 |
| 8×9 | 6×12 | 6×11 | 6×13 | 4×10 | 4×8 | 4×9 | 4×6 | 4×10 | 4×5 | 4×7 | 4×5 | 4×6 |
| 8×10 | 7×10 | 7×8 | 7×9 | 4×12 | 4×13 | 4×11 | 4×13 | 4×12 | 4×12 | 4×9 | 4×10 | 4×8 |
| 9×10 | 7×13 | 7×12 | 7×11 | 6×7 | 5×7 | 5×6 | 5×11 | 5×13 | 5×12 | 5×8 | 5×10 | 5×9 |
| 11×12 | 8×11 | 8×12 | 8×13 | 8×11 | 8×13 | 8×12 | 6×13 | 6×12 | 6×11 | 6×10 | 6×9 | 6×8 |
| 11×13 | 9×12 | 9×13 | 9×11 | 9×13 | 9×12 | 9×11 | 7×12 | 7×11 | 7×13 | 7×9 | 7×8 | 7×10 |
| 12×13 | 10×13 | 10×11 | 10×12 | 10×12 | 10×11 | 10×13 | 9×10 | 8×10 | 8×9 | 12×13 | 11×13 | 11×12 |

## Mixed Two- and Three-Level Arrays

### L36-1 (11 Two-Level Columns and 12 Three-Level Columns)

| *Run* | *1* | *2* | *3* | *4* | *5* | *6* | *7* | *8* | *9* | *10* | *11* | *12* | *13* | *14* | *15* | *16* | *17* | *18* | *19* | *20* | *21* | *22* | *23* |
|---|---|---|---|---|---|---|---|---|---|---|---|---|---|---|---|---|---|---|---|---|---|---|---|
| 1 | 1 | 1 | 1 | 1 | 1 | 1 | 1 | 1 | 1 | 1 | 1 | 1 | 1 | 1 | 1 | 1 | 1 | 1 | 1 | 1 | 1 | 1 | 1 |
| 2 | 1 | 1 | 1 | 1 | 1 | 1 | 1 | 1 | 1 | 1 | 1 | 2 | 2 | 2 | 2 | 2 | 2 | 2 | 2 | 2 | 2 | 2 | 2 |
| 3 | 1 | 1 | 1 | 1 | 1 | 1 | 1 | 1 | 1 | 1 | 1 | 3 | 3 | 3 | 3 | 3 | 3 | 3 | 3 | 3 | 3 | 3 | 3 |
| 4 | 1 | 1 | 1 | 1 | 1 | 2 | 2 | 2 | 2 | 2 | 2 | 1 | 1 | 1 | 1 | 2 | 2 | 2 | 2 | 3 | 3 | 3 | 3 |
| 5 | 1 | 1 | 1 | 1 | 1 | 2 | 2 | 2 | 2 | 2 | 2 | 2 | 2 | 2 | 2 | 3 | 3 | 3 | 3 | 1 | 1 | 1 | 1 |
| 6 | 1 | 1 | 1 | 1 | 1 | 2 | 2 | 2 | 2 | 2 | 2 | 3 | 3 | 3 | 3 | 1 | 1 | 1 | 1 | 2 | 2 | 2 | 2 |
| 7 | 1 | 1 | 2 | 2 | 2 | 1 | 1 | 1 | 2 | 2 | 2 | 1 | 1 | 2 | 3 | 1 | 2 | 3 | 3 | 1 | 2 | 2 | 3 |
| 8 | 1 | 1 | 2 | 2 | 2 | 1 | 1 | 1 | 2 | 2 | 2 | 2 | 2 | 3 | 1 | 2 | 3 | 1 | 1 | 2 | 3 | 3 | 1 |
| 9 | 1 | 1 | 2 | 2 | 2 | 1 | 1 | 1 | 2 | 2 | 2 | 3 | 3 | 1 | 2 | 3 | 1 | 2 | 2 | 3 | 1 | 1 | 2 |
| 10 | 1 | 2 | 1 | 2 | 2 | 1 | 2 | 2 | 1 | 1 | 2 | 1 | 1 | 3 | 2 | 1 | 3 | 2 | 3 | 2 | 1 | 3 | 2 |
| 11 | 1 | 2 | 1 | 2 | 2 | 1 | 2 | 2 | 1 | 1 | 2 | 2 | 2 | 1 | 3 | 2 | 1 | 3 | 1 | 3 | 2 | 1 | 3 |
| 12 | 1 | 2 | 1 | 2 | 2 | 1 | 2 | 2 | 1 | 1 | 2 | 3 | 3 | 2 | 1 | 3 | 2 | 1 | 2 | 1 | 3 | 2 | 1 |
| 13 | 1 | 2 | 2 | 1 | 2 | 2 | 1 | 2 | 1 | 2 | 1 | 1 | 2 | 3 | 1 | 3 | 2 | 1 | 3 | 3 | 2 | 1 | 2 |
| 14 | 1 | 2 | 2 | 1 | 2 | 2 | 1 | 2 | 1 | 2 | 1 | 2 | 3 | 1 | 2 | 1 | 3 | 2 | 1 | 1 | 3 | 2 | 3 |
| 15 | 1 | 2 | 2 | 1 | 2 | 2 | 1 | 2 | 1 | 2 | 1 | 3 | 1 | 2 | 3 | 2 | 1 | 3 | 2 | 2 | 1 | 3 | 1 |
| 16 | 1 | 2 | 2 | 2 | 1 | 2 | 2 | 1 | 2 | 1 | 1 | 1 | 2 | 3 | 2 | 1 | 1 | 3 | 2 | 3 | 3 | 2 | 1 |
| 17 | 1 | 2 | 2 | 2 | 1 | 2 | 2 | 1 | 2 | 1 | 1 | 2 | 3 | 1 | 3 | 2 | 2 | 1 | 3 | 1 | 1 | 3 | 2 |
| 18 | 1 | 2 | 2 | 2 | 1 | 2 | 2 | 1 | 2 | 1 | 1 | 3 | 1 | 2 | 1 | 3 | 3 | 2 | 1 | 2 | 2 | 1 | 3 |
| 19 | 2 | 1 | 2 | 2 | 1 | 1 | 2 | 2 | 1 | 2 | 1 | 1 | 2 | 1 | 3 | 3 | 3 | 1 | 2 | 2 | 1 | 2 | 3 |
| 20 | 2 | 1 | 2 | 2 | 1 | 1 | 2 | 2 | 1 | 2 | 1 | 2 | 3 | 2 | 1 | 1 | 1 | 2 | 3 | 3 | 2 | 3 | 1 |
| 21 | 2 | 1 | 2 | 2 | 1 | 1 | 2 | 2 | 1 | 2 | 1 | 3 | 1 | 3 | 2 | 2 | 2 | 3 | 1 | 1 | 3 | 1 | 2 |
| 22 | 2 | 1 | 2 | 1 | 2 | 2 | 2 | 1 | 1 | 1 | 2 | 1 | 2 | 2 | 3 | 3 | 1 | 2 | 1 | 1 | 3 | 3 | 2 |
| 23 | 2 | 1 | 2 | 1 | 2 | 2 | 2 | 1 | 1 | 1 | 2 | 2 | 3 | 3 | 1 | 1 | 2 | 3 | 2 | 2 | 1 | 1 | 3 |
| 24 | 2 | 1 | 2 | 1 | 2 | 2 | 2 | 1 | 1 | 1 | 2 | 3 | 1 | 1 | 2 | 2 | 3 | 1 | 3 | 3 | 2 | 2 | 1 |
| 25 | 2 | 1 | 1 | 2 | 2 | 2 | 1 | 2 | 2 | 1 | 1 | 1 | 3 | 2 | 1 | 2 | 3 | 3 | 1 | 3 | 1 | 2 | 2 |
| 26 | 2 | 1 | 1 | 2 | 2 | 2 | 1 | 2 | 2 | 1 | 1 | 2 | 1 | 3 | 2 | 3 | 1 | 1 | 2 | 1 | 2 | 3 | 3 |
| 27 | 2 | 1 | 1 | 2 | 2 | 2 | 1 | 2 | 2 | 1 | 1 | 3 | 2 | 1 | 3 | 1 | 2 | 2 | 3 | 2 | 3 | 1 | 1 |
| 28 | 2 | 2 | 2 | 1 | 1 | 1 | 1 | 2 | 2 | 1 | 2 | 1 | 3 | 2 | 2 | 2 | 1 | 1 | 3 | 2 | 3 | 1 | 3 |
| 29 | 2 | 2 | 2 | 1 | 1 | 1 | 1 | 2 | 2 | 1 | 2 | 2 | 1 | 3 | 3 | 3 | 2 | 2 | 1 | 3 | 1 | 2 | 1 |
| 30 | 2 | 2 | 2 | 1 | 1 | 1 | 1 | 2 | 2 | 1 | 2 | 3 | 2 | 1 | 1 | 1 | 3 | 3 | 2 | 1 | 2 | 3 | 2 |
| 31 | 2 | 2 | 1 | 2 | 1 | 2 | 1 | 1 | 1 | 2 | 2 | 1 | 3 | 3 | 3 | 2 | 3 | 2 | 2 | 1 | 2 | 1 | 1 |
| 32 | 2 | 2 | 1 | 2 | 1 | 2 | 1 | 1 | 1 | 2 | 2 | 2 | 1 | 1 | 1 | 3 | 1 | 3 | 3 | 2 | 3 | 2 | 2 |
| 33 | 2 | 2 | 1 | 2 | 1 | 2 | 1 | 1 | 1 | 2 | 2 | 3 | 2 | 2 | 2 | 1 | 2 | 1 | 1 | 3 | 1 | 3 | 3 |
| 34 | 2 | 2 | 1 | 1 | 2 | 1 | 2 | 1 | 2 | 2 | 1 | 1 | 3 | 1 | 2 | 3 | 2 | 3 | 1 | 2 | 2 | 3 | 1 |
| 35 | 2 | 2 | 1 | 1 | 2 | 1 | 2 | 1 | 2 | 2 | 1 | 2 | 1 | 2 | 3 | 1 | 3 | 1 | 2 | 3 | 3 | 1 | 2 |
| 36 | 2 | 2 | 1 | 1 | 2 | 1 | 2 | 1 | 2 | 2 | 1 | 3 | 2 | 3 | 1 | 2 | 1 | 2 | 3 | 1 | 1 | 2 | 3 |

## L36-2 (Three Two-Level Columns and 13 Three-Level Columns)

| *Run* | *1* | *2* | *3* | *4* | *5* | *6* | *7* | *8* | *9* | *10* | *11* | *12* | *13* | *14* | *15* | *16* |
|---|---|---|---|---|---|---|---|---|---|---|---|---|---|---|---|---|
| 1 | 1 | 1 | 1 | 1 | 1 | 1 | 1 | 1 | 1 | 1 | 1 | 1 | 1 | 1 | 1 | 1 |
| 2 | 1 | 1 | 1 | 1 | 2 | 2 | 2 | 2 | 2 | 2 | 2 | 2 | 2 | 2 | 2 | 2 |
| 3 | 1 | 1 | 1 | 1 | 3 | 3 | 3 | 3 | 3 | 3 | 3 | 3 | 3 | 3 | 3 | 3 |
| 4 | 1 | 2 | 2 | 1 | 1 | 1 | 1 | 1 | 2 | 2 | 2 | 2 | 3 | 3 | 3 | 3 |
| 5 | 1 | 2 | 2 | 1 | 2 | 2 | 2 | 2 | 3 | 3 | 3 | 3 | 1 | 1 | 1 | 1 |
| 6 | 1 | 2 | 2 | 1 | 3 | 3 | 3 | 3 | 1 | 1 | 1 | 1 | 2 | 2 | 2 | 2 |
| 7 | 2 | 1 | 2 | 1 | 1 | 1 | 2 | 3 | 1 | 2 | 3 | 3 | 1 | 2 | 2 | 3 |
| 8 | 2 | 1 | 2 | 1 | 2 | 2 | 3 | 1 | 2 | 3 | 1 | 1 | 2 | 3 | 3 | 1 |
| 9 | 2 | 1 | 2 | 1 | 3 | 3 | 1 | 2 | 3 | 1 | 2 | 2 | 3 | 1 | 1 | 2 |
| 10 | 2 | 2 | 1 | 1 | 1 | 1 | 3 | 2 | 1 | 3 | 2 | 3 | 2 | 1 | 3 | 2 |
| 11 | 2 | 2 | 1 | 1 | 2 | 2 | 1 | 3 | 2 | 1 | 3 | 1 | 3 | 2 | 1 | 3 |
| 12 | 2 | 2 | 1 | 1 | 3 | 3 | 2 | 1 | 3 | 2 | 1 | 2 | 1 | 3 | 2 | 1 |
| 13 | 1 | 1 | 1 | 2 | 1 | 2 | 3 | 1 | 3 | 2 | 1 | 3 | 3 | 2 | 1 | 2 |
| 14 | 1 | 1 | 1 | 2 | 2 | 3 | 1 | 2 | 1 | 3 | 2 | 1 | 1 | 3 | 2 | 3 |
| 15 | 1 | 1 | 1 | 2 | 3 | 1 | 2 | 3 | 2 | 1 | 3 | 2 | 2 | 1 | 3 | 1 |
| 16 | 1 | 2 | 2 | 2 | 1 | 2 | 3 | 2 | 1 | 1 | 3 | 2 | 3 | 3 | 2 | 1 |
| 17 | 1 | 2 | 2 | 2 | 2 | 3 | 1 | 3 | 2 | 2 | 1 | 3 | 1 | 1 | 3 | 2 |
| 18 | 1 | 2 | 2 | 2 | 3 | 1 | 2 | 1 | 3 | 3 | 2 | 1 | 2 | 2 | 1 | 3 |
| 19 | 2 | 1 | 2 | 2 | 1 | 2 | 1 | 3 | 3 | 3 | 1 | 2 | 2 | 1 | 2 | 3 |
| 20 | 2 | 1 | 2 | 2 | 2 | 3 | 2 | 1 | 1 | 1 | 2 | 3 | 3 | 2 | 3 | 1 |
| 21 | 2 | 1 | 2 | 2 | 3 | 1 | 3 | 2 | 2 | 2 | 3 | 1 | 1 | 3 | 1 | 2 |
| 22 | 2 | 2 | 1 | 2 | 1 | 2 | 2 | 3 | 3 | 1 | 2 | 1 | 1 | 3 | 3 | 2 |
| 23 | 2 | 2 | 1 | 2 | 2 | 3 | 3 | 1 | 1 | 2 | 3 | 2 | 2 | 1 | 1 | 3 |
| 24 | 2 | 2 | 1 | 2 | 3 | 1 | 1 | 2 | 2 | 3 | 1 | 3 | 3 | 2 | 2 | 1 |
| 25 | 1 | 1 | 1 | 3 | 1 | 3 | 2 | 1 | 2 | 3 | 3 | 1 | 3 | 1 | 2 | 2 |
| 26 | 1 | 1 | 1 | 3 | 2 | 1 | 3 | 2 | 3 | 1 | 1 | 2 | 1 | 2 | 3 | 3 |
| 27 | 1 | 1 | 1 | 3 | 3 | 2 | 1 | 3 | 1 | 2 | 2 | 3 | 2 | 3 | 1 | 1 |
| 28 | 1 | 2 | 2 | 3 | 1 | 3 | 2 | 2 | 2 | 1 | 1 | 3 | 2 | 3 | 1 | 3 |
| 29 | 1 | 2 | 2 | 3 | 2 | 1 | 3 | 3 | 3 | 2 | 2 | 1 | 3 | 1 | 2 | 1 |
| 30 | 1 | 2 | 2 | 3 | 3 | 2 | 1 | 1 | 1 | 3 | 3 | 2 | 1 | 2 | 3 | 2 |
| 31 | 2 | 1 | 2 | 3 | 1 | 3 | 3 | 3 | 2 | 3 | 2 | 2 | 1 | 2 | 1 | 1 |
| 32 | 2 | 1 | 2 | 3 | 2 | 1 | 1 | 1 | 3 | 1 | 3 | 3 | 2 | 3 | 2 | 2 |
| 33 | 2 | 1 | 2 | 3 | 3 | 2 | 2 | 2 | 1 | 2 | 1 | 1 | 3 | 1 | 3 | 3 |
| 34 | 2 | 2 | 1 | 3 | 1 | 3 | 1 | 2 | 3 | 2 | 3 | 1 | 2 | 2 | 3 | 1 |
| 35 | 2 | 2 | 1 | 3 | 2 | 1 | 2 | 3 | 1 | 3 | 1 | 2 | 3 | 3 | 1 | 2 |
| 36 | 2 | 2 | 1 | 3 | 3 | 2 | 3 | 1 | 2 | 1 | 2 | 3 | 1 | 1 | 2 | 3 |

## Custom Arrays

### Modified L8 (One Factor at Four Levels, Four Factors at Two Levels)

| *Run* | *1* | *2* | *3* | *4* | *5* |
|---|---|---|---|---|---|
| 1 | 1 | 1 | 1 | 1 | 1 |
| 2 | 1 | 2 | 2 | 2 | 2 |
| 3 | 2 | 1 | 1 | 2 | 2 |
| 4 | 2 | 2 | 2 | 1 | 1 |
| 5 | 3 | 1 | 2 | 1 | 2 |
| 6 | 3 | 2 | 1 | 2 | 1 |
| 7 | 4 | 1 | 2 | 2 | 1 |
| 8 | 4 | 2 | 1 | 1 | 2 |

### Modified L16 (One Factor at Four Levels, 12 Factors at Two Levels)

| *Run* | *1* | *2* | *3* | *4* | *5* | *6* | *7* | *8* | *9* | *10* | *11* | *12* | *13* |
|---|---|---|---|---|---|---|---|---|---|---|---|---|---|
| 1 | 1 | 1 | 1 | 1 | 1 | 1 | 1 | 1 | 1 | 1 | 1 | 1 | 1 |
| 2 | 1 | 1 | 1 | 1 | 1 | 2 | 2 | 2 | 2 | 2 | 2 | 2 | 2 |
| 3 | 1 | 2 | 2 | 2 | 2 | 1 | 1 | 1 | 1 | 2 | 2 | 2 | 2 |
| 4 | 1 | 2 | 2 | 2 | 2 | 2 | 2 | 2 | 2 | 1 | 1 | 1 | 1 |
| 5 | 2 | 1 | 1 | 2 | 2 | 1 | 1 | 2 | 2 | 1 | 1 | 2 | 2 |
| 6 | 2 | 1 | 1 | 2 | 2 | 2 | 2 | 1 | 1 | 2 | 2 | 1 | 1 |
| 7 | 2 | 2 | 2 | 1 | 1 | 1 | 1 | 2 | 2 | 2 | 2 | 1 | 1 |
| 8 | 2 | 2 | 2 | 1 | 1 | 2 | 2 | 1 | 1 | 1 | 1 | 2 | 2 |
| 9 | 3 | 1 | 2 | 1 | 2 | 1 | 2 | 1 | 2 | 1 | 2 | 1 | 2 |
| 10 | 3 | 1 | 2 | 1 | 2 | 2 | 1 | 2 | 1 | 2 | 1 | 2 | 1 |
| 11 | 3 | 2 | 1 | 2 | 1 | 1 | 2 | 1 | 2 | 2 | 1 | 2 | 1 |
| 12 | 3 | 2 | 1 | 2 | 1 | 2 | 1 | 2 | 1 | 1 | 2 | 1 | 2 |
| 13 | 4 | 1 | 2 | 2 | 1 | 1 | 2 | 2 | 1 | 1 | 2 | 2 | 1 |
| 14 | 4 | 1 | 2 | 2 | 1 | 2 | 1 | 1 | 2 | 2 | 1 | 1 | 2 |
| 15 | 4 | 2 | 1 | 1 | 2 | 1 | 2 | 2 | 1 | 2 | 1 | 1 | 2 |
| 16 | 4 | 2 | 1 | 1 | 2 | 2 | 1 | 1 | 2 | 1 | 2 | 2 | 1 |

## Modified L16 (Two Factors at Four Levels, Nine Factors at Two Levels)

| *Run* | *1* | *2* | *3* | *4* | *5* | *6* | *7* | *8* | *9* | *10* | *11* |
|---|---|---|---|---|---|---|---|---|---|---|---|
| 1 | 1 | 1 | 1 | 1 | 1 | 1 | 1 | 1 | 1 | 1 | 1 |
| 2 | 1 | 2 | 1 | 1 | 1 | 2 | 2 | 2 | 2 | 2 | 2 |
| 3 | 1 | 3 | 2 | 2 | 2 | 1 | 1 | 1 | 2 | 2 | 2 |
| 4 | 1 | 4 | 2 | 2 | 2 | 2 | 2 | 2 | 1 | 1 | 1 |
| 5 | 2 | 1 | 1 | 2 | 2 | 1 | 2 | 2 | 1 | 2 | 2 |
| 6 | 2 | 2 | 1 | 2 | 2 | 2 | 1 | 1 | 2 | 1 | 1 |
| 7 | 2 | 3 | 2 | 1 | 1 | 1 | 2 | 2 | 2 | 1 | 1 |
| 8 | 2 | 4 | 2 | 1 | 1 | 2 | 1 | 1 | 1 | 2 | 2 |
| 9 | 3 | 1 | 2 | 1 | 2 | 2 | 1 | 2 | 2 | 1 | 2 |
| 10 | 3 | 2 | 2 | 1 | 2 | 1 | 2 | 1 | 1 | 2 | 1 |
| 11 | 3 | 3 | 1 | 2 | 1 | 2 | 1 | 2 | 1 | 2 | 1 |
| 12 | 3 | 4 | 1 | 2 | 1 | 1 | 2 | 1 | 2 | 1 | 2 |
| 13 | 4 | 1 | 2 | 2 | 1 | 2 | 2 | 1 | 2 | 2 | 1 |
| 14 | 4 | 2 | 2 | 2 | 1 | 1 | 1 | 2 | 1 | 1 | 2 |
| 15 | 4 | 3 | 1 | 1 | 2 | 2 | 2 | 1 | 1 | 1 | 2 |
| 16 | 4 | 4 | 1 | 1 | 2 | 1 | 1 | 2 | 2 | 2 | 1 |

## Modified L16 (Three Factors at Four Levels, Six Factors at Two Levels)

| *Run* | *1* | *2* | *3* | *4* | *5* | *6* | *7* | *8* | *9* |
|---|---|---|---|---|---|---|---|---|---|
| 1 | 1 | 1 | 1 | 1 | 1 | 1 | 1 | 1 | 1 |
| 2 | 1 | 2 | 2 | 1 | 1 | 2 | 2 | 2 | 2 |
| 3 | 1 | 3 | 3 | 2 | 2 | 1 | 1 | 2 | 2 |
| 4 | 1 | 4 | 4 | 2 | 2 | 2 | 2 | 1 | 1 |
| 5 | 2 | 1 | 2 | 2 | 2 | 1 | 2 | 1 | 2 |
| 6 | 2 | 2 | 1 | 2 | 2 | 2 | 1 | 2 | 1 |
| 7 | 2 | 3 | 4 | 1 | 1 | 1 | 2 | 2 | 1 |
| 8 | 2 | 4 | 3 | 1 | 1 | 2 | 1 | 1 | 2 |
| 9 | 3 | 1 | 3 | 1 | 2 | 2 | 2 | 2 | 1 |
| 10 | 3 | 2 | 4 | 1 | 2 | 1 | 1 | 1 | 2 |
| 11 | 3 | 3 | 1 | 2 | 1 | 2 | 2 | 1 | 2 |
| 12 | 3 | 4 | 2 | 2 | 1 | 1 | 1 | 2 | 1 |
| 13 | 4 | 1 | 4 | 2 | 1 | 2 | 1 | 2 | 2 |
| 14 | 4 | 2 | 3 | 2 | 1 | 1 | 2 | 1 | 1 |
| 15 | 4 | 3 | 2 | 1 | 2 | 2 | 1 | 1 | 1 |
| 16 | 4 | 4 | 1 | 1 | 2 | 1 | 2 | 2 | 2 |

## Modified L16 (Four Factors at Four Levels, Three Factors at Two Levels)

| *Run* | *1* | *2* | *3* | *4* | *5* | *6* | *7* |
|---|---|---|---|---|---|---|---|
| 1 | 1 | 1 | 1 | 1 | 1 | 1 | 1 |
| 2 | 1 | 2 | 2 | 1 | 2 | 2 | 2 |
| 3 | 1 | 3 | 3 | 2 | 3 | 1 | 2 |
| 4 | 1 | 4 | 4 | 2 | 4 | 2 | 1 |
| 5 | 2 | 1 | 2 | 2 | 1 | 2 | 1 |
| 6 | 2 | 2 | 1 | 2 | 2 | 1 | 2 |
| 7 | 2 | 3 | 4 | 1 | 4 | 2 | 2 |
| 8 | 2 | 4 | 3 | 1 | 3 | 1 | 1 |
| 9 | 3 | 1 | 3 | 1 | 4 | 2 | 2 |
| 10 | 3 | 2 | 4 | 1 | 3 | 1 | 1 |
| 11 | 3 | 3 | 1 | 2 | 2 | 2 | 1 |
| 12 | 3 | 4 | 2 | 2 | 1 | 1 | 2 |
| 13 | 4 | 1 | 4 | 2 | 2 | 1 | 2 |
| 14 | 4 | 2 | 3 | 2 | 1 | 2 | 1 |
| 15 | 4 | 3 | 2 | 1 | 4 | 1 | 1 |
| 16 | 4 | 4 | 1 | 1 | 3 | 2 | 2 |

## Modified L16 (Five Factors at Four Levels)

| *Run* | *1* | *2* | *3* | *4* | *5* |
|---|---|---|---|---|---|
| 1 | 1 | 1 | 1 | 1 | 1 |
| 2 | 1 | 2 | 2 | 2 | 2 |
| 3 | 1 | 3 | 3 | 3 | 3 |
| 4 | 1 | 4 | 4 | 4 | 4 |
| 5 | 2 | 1 | 2 | 3 | 4 |
| 6 | 2 | 2 | 1 | 4 | 3 |
| 7 | 2 | 3 | 4 | 1 | 2 |
| 8 | 2 | 4 | 3 | 2 | 1 |
| 9 | 3 | 1 | 3 | 4 | 2 |
| 10 | 3 | 2 | 4 | 3 | 1 |
| 11 | 3 | 3 | 1 | 2 | 4 |
| 12 | 3 | 4 | 2 | 1 | 3 |
| 13 | 4 | 1 | 4 | 2 | 3 |
| 14 | 4 | 2 | 3 | 1 | 4 |
| 15 | 4 | 3 | 2 | 4 | 1 |
| 16 | 4 | 4 | 1 | 3 | 2 |

# APPENDIX D

# Bibliography

## A


(A1) Abraham, B., and MacKay, J. "Designed Experiments for Reduction of Variation." *IIQP Research Report,* RR-91-09, University of Waterloo, Oct. 1991.

(A2) Akao, Y. A. *QFD: Integrating Customer Requirements into Product Design.* Productivity Press, 1990.

(A3) Altov (Altshuller), G. *The Art of Inventing.* Technical Innovation Center, 1992.


## B


(B1) Banks, D. "Is Industrial Statistics Out of Control?" *Statistical Science* **8** (4), 356–409 (1993).

(B2) Banks, J. *Principles of Quality Control.* John Wiley & Sons, 1989.

(B3) Barker, T. *Quality by Experimental Design,* 2nd Ed. Marcel Dekker, 1994.

(B4) Barker, T. *Engineering Quality by Design.* Marcel Dekker, 1990.

(B5) Beard, J., and Sutherland, J. "Robust Suspension System Design." *ASME* **65.1,** *Advances in Design Automation—Vol. 1,* pp. 387–395.

(B6) Bendell, A., Disney, J., and Pridmore, W. A. *Taguchi Methods: Applications in World Industry.* IFS Publications, 1989.

(B7) Bhote, K.R. *World Class Quality: Using Design of Experiments to Make It Happen.* AMACOM, 1991.

(B8) Bires, J. T. "Accuracy Improvement of a Disposable Blood Oxygen Sensor Used for Open Heart Surgery." *Proceedings from the 11th Annual Taguchi Symposium.* ASI Press, 1993.

(B9) Box, G. E. P. "Signal-to-Noise Ratios, Performance Criteria, and Transformations." *Technometrics* **30** (1), 1–36 (1988).

(B10) Box, G. E. P., Hunter, W. G., and Hunter, J. S. *Statistics for Experimenters.* JohnWiley & Sons, 1978.

(B11) Box, G., and Hicks, W. "A Question of Quality." *Harvard Business Review,* March–April, pp. 225–229 (1990).
(B12) Boza, L., *et al.* "Achieving Customer Satisfaction Through Robust Design." *AT&T Technical Journal* (1), 48–58 (1994).
(B13) Bueche, F. J. *Introduction to Physics for Scientists and Engineers.* McGraw-Hill, 1969, pp. 44–49.

## C

(C1) Carter, D. E., and Baker, B. S. *Concurrent Engineering: The Product Development Environment for the 1990's.* Addison-Wesley, 1992.
(C2) Cohen, L. *Quality Function Deployment.* Addison-Wesley, 1995.
(C3) Colunga, J. S., Lau, D., Otterman., J., Prodin, B., Turner, K., and King, J. "Fuel Delivery System Robustness." *Proceedings from the 11th Annual Taguchi Symposium.* ASI Press, 1993.
(C4) Condra, L. *Reliability Improvement with Design of Experiments.* Marcel Dekker, 1993.
(C5) Corbett, J., Dooner, M., Meleka, J., and Pym, C. *Design for Manufacture: Strategies, Principles and Techniques.* Addison-Wesley, 1991.
(C6) Cross, N. *Engineering Design Methods,* John Wiley & Sons, 1989.
(C7) Custer, L., McCarville, D. R., Harry, M., and Prins, J. *Analysis of Variance.* Addison-Wesley, 1993.

## D

(D1) Dandy, G. C., and Warner, R. F. *Planning and Design of Engineering Systems.* Unwin Hyman, 1989.
(D2) Dehnad, K. *Quality Control, Robust Design and the Taguchi Method.* Wadsworth and Brooks/Cole, 1989.

## E

(E1) Ealey, L. *Quality by Design: Taguchi Methods and the U.S. Industry.* ASI Press, 1988.
(E2) Enrick, N. L. *Quality, Reliability and Process Improvement.* Industrial Press Inc., 1985.
(E3) Evans, M., Hastings, N., and Peacock, B. *Statistical Distributions,* 2nd Ed. John Wiley & Sons, 1993.

## F

(F1) Fathi, Y. "Producer–Consumer Tolerances." *Journal of Quality Technology* **22** (2), 138–145 (1990).
(F2) Fisher, R. A. *The Design of Experiments.* Oliver and Boyd, 1935.
(F3) Fowlkes, W. Y., and Rushing, A. "Utilization of the Quality Loss Function on an Electrophotographic Printer." *Proceedings from the 11th Annual Taguchi Symposium.* ASI Press, 1993.
(F4) Freeny, A., and Nair, V. "Robust Parameter Design with Uncontrolled Noise Variables," *Statistica Sinica* **2,** 313–334 (1992).
(F5) Freund. *Modern Elementary Statistics.* Prentice Hall, 1988.

## G

(G1) Goh, T. N. "Taguchi Methods: Some Technical, Cultural and Pedagogical Perspectives." *Quality and Reliability International* **9,** 185–202 (1993).

(G2) Grove, D. M., and Davis, T. P. "Taguchi's Idle Column Method." *Technometrics* **33** (3), 349–353 (1991).

(G3) Grove, D. M., and Davis, T. P. *Engineering Quality and Experimental Design.* Longman Scientific and Technical, 1992.

## H

(H1) Hamada, M., and Wu, C. "The Treatment of Related Experimental Factors by Sliding Levels." *Journal of Quality Technology* **27,** 45–55 (1995).

(H2) Harry, M. J., and Lawson, J. R. *Six Sigma Producibility Analysis and Process Characterization.* Addison-Wesley, 1992.

(H3) Hartley, J. R. *Concurrent Engineering.* Productivity Press, 1990.

(H4) Hauser, J. R. "How Puritan-Bennett Used the House of Quality." *Sloan Management Review,* Spring 1993, pp. 61–70.

(H5) Hinckley, C. M., and Barken, P. "Selecting the Best Defect Reduction Methodology," Sandia Report SAND94-8621-VC-406, 1994.

## I

(I1) Ishikawa, K. *Guide to Quality Control,* Asian Productivity Organization, 1982.

## J

(J1) Juran, J. M., and Gryna, F. M. *Juran's Quality Control Handbook,* 4th Ed. McGraw-Hill, 1988.

## K

(K1) Kackar, R. N., Lagergren, E. S., and Filliben, J. J. "Taguchi's Fixed-Element Arrays Are Fractional Factorials." *Journal of Quality Technology* **23** (2 ), 107–116 (1991).

(K2) Kacker, R. N. "Off-Line Quality Control, Parameter Design, and the Taguchi Method." *Journal of Quality Technology* **17** (4), 176–209 (1985).

## L

(L1) Leon, R. V., Shoemaker, A. C., and Kacker, R. N. "Performance Measures Independent of Adjustment." *Technometrics* **29** (3), 253–285 (1987).

(L2) Leon, R. V., Shoemaker, A. C., and Tsui, K. *Technometrics* **35,** 21 (1993).

## M

(M1) Maghsoodloo, S. "The Exact Relationship of Taguchi's Signal-to-Noise Ratio to His Quality Loss Function." *Journal of Quality Technology* **22** (1), 57–67 (1991).

(M2) Mason, R. L., Gunst, R. F., and Hess, J. L. *Statistical Design & Analysis of Experiments.* John Wiley & Sons, 1989.
(M3) Meon, R. D., Nolan, T. W., and Provost, L. P. *Improving Quality through Planned Experimentation.* McGraw-Hill, 1991.
(M4) Montgomery, D. *Statistical Quality Control,* 2nd Ed. John Wiley & Sons, 1991.
(M5) Montgomery, D. *Design and Analysis of Experiments,* 3rd Ed. John Wiley & Sons, 1991.
(M6) Mori, T. *Taguchi Techniques for Image and Pattern Developing Technology.* Prentice Hall, 1995.
(M7) Mori, T. *The New Experimental Design: Taguchi's Approach to Quality Engineering.* ASI Press, 1990.
(M8) Moskowitz, H., and Tang, K. "Bayesian Variables Acceptance-Sampling Plans: Quadratic Loss Function and Step-Loss Function." *Technometrics* **34** (3), 340–347 (1992).

## N

(N1) Nair, V. N., and Pregibon, D. "A Data Analysis Strategy for Quality Engineering Experiments." *AT&T Technical Journal* **65** (3), 73–84 (1986).
(N2) Nair, V. N. "Taguchi's Parameter Design: A Panel Discussion." *Technometrics* **34** (2), 127–161 (1992).
(N3) North, S., and Fowlkes, W. Y. "Tolerance Certification of a Film Feeding Device." *Proceedings of the 12th Annual Taguchi Symposium.* ASI Press, 1994.

## O

(O1) O'Conner, P. D. T. *Practical Reliability Engineering,* 3rd Ed. John Wiley & Sons, 1991.

## P

(P1) Peace, G. S. *Taguchi Methods: A Hands-On Approach to Quality Engineering.* Addison-Wesley, 1993.
(P2) Phadke, M. *Quality Engineering Using Robust Design.* Prentice Hall, 1989.
(P3) Pugh, S. *Total Design.* Addison-Wesley, 1990.

## R

(R1) Robinson, G. K. "Improving Taguchi's Packaging of Fractional Factorial Designs." *Journal of Quality Technology* **25** (1), 1–11 (1993).
(R2) Ross, P. *Taguchi Techniques for Quality Engineering.* McGraw-Hill; 1988.
(R3) Roy, R. *A Primer on the Taguchi Method.* Van Nostrand Reinhold, 1990.

## S

(S1) Schmidt, S., and Launsby, R. *Understanding Industrial Designed Experiments,* 3rd Ed. Air Academy Press, 1991.
(S2) Shoemaker, A. C., Tsui, K. L., and Wu, C. F. "Economical Experimentation Methods for Robust Design." *Technometrics* **33** (4), 415–427 (1991).
(S3) Shoemaker, A. C., and Kacker, R. N. "A Methodology for Planning Experiments in Robust

Product and Process Design." *Quality and Reliability Engineering International* **4,** 95–103 (1988).
(S4) Swift, K. G., and Allen, A. J. "Product Variability Risks and Robust Design." *Proceedings of the Institution of Mechanical Engineers* **208,** 9–19.

## T

(T1) Taguchi, G., Yokohama, Y., and Wu, Y. *Quality Engineering Series,* Vol. 2, *Taguchi Methods, On-Line Production.* ASI Press, 1994.
(T2) Taguchi, G., Yokohama, Y., and Wu, Y. *Quality Engineering Series,* Vol. 4, *Taguchi Methods, Design of Experiments.* ASI Press, 1993.
(T3) Taguchi, G. *Taguchi on Robust Technology Development.* ASME Press, 1993.
(T4) Taguchi, G., and Konishi, S. *Quality Engineering Series,* Vol. 1, *Taguchi Methods, Signal-to-Noise Ratio for Quality Evaluation.* ASI Press, 1991.
(T5) Taguchi, G., and Konishi, S. *Orthogonal Arrays and Linear Graphs: Tools for Quality Engineering.* ASI Press, 1987.
(T6) Taguchi, G. *System of Experimental Design,* Vols. 1 and 2. ASI & Quality Resources, 1987.
(T7) Taguchi, G., and Clausing, D. "Robust Design." *Harvard Business Review,* p. 229 (1990).
(T8) Taguchi, G., Elsayed, E., and Hsiang, T. *Quality Engineering in Production Systems.* McGraw-Hill, 1989.
(T9) Taguchi, G. *Introduction to Quality Engineering.* Asian Productivity Organization, 1989.
(T10) Taguchi, G., and Konishi, S. *Quality Engineering Series,* Vol. 3, *Taguchi Methods, Research and Development.* ASI Press, 1992.
(T11) Taguchi, G., and Wu, Y. "Quality Engineering for Chemical and Biological Industries." *Proceedings of the12th Annual Taguchi Symposium.* ASI Press, 1994.
(T12) Taylor, W. A. *Optimization & Variation Reduction in Quality.* McGraw-Hill, 1991.

## U

(U1) Ullman, R. *The Mechanical Design Process.* McGraw-Hill, 1992.

## W

(W1) Wadsworth, H. M. *Handbook of Statistical Methods for Scientists and Engineers.* McGraw-Hill, 1990.
(W2) Wu, C. F., and Chen, Y. "A Graph-Aided Method for Planning Two-Level Experiments When Certain Interactions Are Important." *Technometrics* **34** (2), 162–175 (1992).
(W3) Wu, Y. *Proceedings from the 12th Annual Taguchi Symposium.* ASI Press, 1994.

# INDEX

## -D-

## -E-



## -F-



## -G-



## -H-

## -Q-

## -R-



## -S-

## -T-

Addison-Wesley warrants the enclosed disk to be free of defects in materials and faulty workmanship under normal use for a period of ninety days after purchase. If a defect is discovered in the disk during this warranty period, a replacement disk can be obtained at no charge by sending the defective disk, postage prepaid, with proof of purchase to:

Addison-Wesley Publishing Company
Editorial Department
Corporate & Professional Publishing Group
One Jacob Way
Reading, Massachusetts 01867

After the 90-day period, a replacement will be sent upon receipt of the defective disk and a check or money order for $10.00, payable to Addison-Wesley Publishing Company.

Addison-Wesley makes no warranty or representation, either express or implied, with respect to this software, its quality, performance, merchantability, or fitness for a particular purpose. In no event will Addison-Wesley, its distributors, or dealers be liable for direct, indirect, special, incidental, or consequential damages arising out of the use or inability to use the software. The exclusion of implied warranties is not permitted in some states. Therefore, the above exclusion may not apply to you. This warranty provides you with specific legal rights. There may be other rights that you may have that vary from state to state.

Note to Apple Macintosh users:

WinRobust Lite can be run under 3.1 emulation on PowerMacs (PowerPC based) with SoftWindows software.

For non-PowerPC based Macintosh systems, such as those powered by 68030 (IIci, IIcx, IIfx, and some Power Books) and 68040 (Quadra, Centras, LC, Performa, and some Power Books) processors, 23MB hard disk space and 16MB RAM (12 for System 7.5) are required. A 68040 processor is recommended for best performance.

SoftWindows is published by Insignia Solutions, Inc.

Questions, comments, or technical support regarding WinRobust Lite contact:

Abacus Digital
phone/fax (716) 223-1368
e-mail AbD95@aol.com